Dualities and Representations of Lie Superalgebras

Dualities and Representations of Lie Superalgebras

Shun-Jen Cheng
Weiqiang Wang

Graduate Studies in Mathematics

Volume 144

American Mathematical Society
Providence, Rhode Island

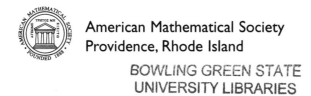

2010 *Mathematics Subject Classification.* Primary 17B10, 17B20.

For additional information and updates on this book, visit
www.ams.org/bookpages/gsm-144

Library of Congress Cataloging-in-Publication Data

Cheng, Shun-Jen, 1963–
Dualities and representations of Lie superalgebras / Shun-Jen Cheng, Weiqiang Wang.
pages cm. — (Graduate studies in mathematics ; volume 144)
Includes bibliographical references and index.
ISBN 978-0-8218-9118-6 (alk. paper)
1. Lie superalgebras. 2. Duality theory (Mathematics) I. Wang, Weiqiang, 1970– II. Title.

QA252.3.C44 2013
512′.482—dc23

2012031989

∞ The paper used in this book is acid-free and falls within the guidelines
established to ensure permanence and durability.
Visit the AMS home page at `http://www.ams.org/`

10 9 8 7 6 5 4 3 2 1 17 16 15 14 13 12

To Mei-Hui, Xiaohui, Isabelle, and our parents

Contents

Preface

Lie algebras, Lie groups, and their representation theories are parts of the mathematical language describing symmetries, and they have played a central role in modern mathematics. An early motivation of studying Lie superalgebras as a generalization of Lie algebras came from supersymmetry in mathematical physics. Ever since a Cartan-Killing type classification of finite-dimensional complex Lie superalgebras was obtained by Kac [**60**] in 1977, the theory of Lie superalgebras has established itself as a prominent subject in modern mathematics. An independent classification of the finite-dimensional complex simple Lie superalgebras whose even subalgebras are reductive (called simple Lie superalgebras of classical type) was given by Scheunert, Nahm, and Rittenberg in [**106**].

The goal of this book is a systematic account of the structure and representation theory of finite-dimensional complex Lie superalgebras of classical type. The book intends to serve as a rigorous introduction to representation theory of Lie superalgebras on one hand, and, on the other hand, it covers a new approach developed in the past few years toward understanding the Bernstein-Gelfand-Gelfand (BGG) category for classical Lie superalgebras. In spite of much interest in representations of Lie superalgebras stimulated by mathematical physics, these basic topics have not been treated in depth in book form before. The reason seems to be that the representation theory of Lie superalgebras is dramatically different from that of complex semisimple Lie algebras, and a systematic, yet accessible, approach toward the basic problem of finding irreducible characters for Lie superalgebras was not available in a great generality until very recently.

We are aware that there is an enormous literature with numerous partial results for Lie superalgebras, and it is not our intention to make this book an encyclopedia. Rather, we treat in depth the representation theory of the three most important classes of Lie superalgebras, namely, the general linear Lie superalgebras $\mathfrak{gl}(m|n)$, the ortho-symplectic Lie superalgebras $\mathfrak{osp}(m|2n)$, and the queer Lie superalgebras $\mathfrak{q}(n)$. To a large extent, representations of $\mathfrak{sl}(m|n)$ can be understood via $\mathfrak{gl}(m|n)$. The lecture notes [**32**] by the authors can be considered as a prototype for this book. The presentation in this book is organized around three dualities with a unifying theme of determining irreducible characters:

Schur duality, Howe duality, and super duality.

The new book of Musson [**90**] treats in detail the ring theoretical aspects of the universal enveloping algebras of Lie superalgebras as well as the basic structures of simple Lie superalgebras.

There are two superalgebra generalizations of Schur duality. The first one, due to Sergeev [**110**] and independently Berele-Regev [**7**], is an interplay between the general linear Lie superalgebra $\mathfrak{gl}(m|n)$ and the symmetric group, which incorporates the trivial and sign modules in a unified framework. The irreducible polynomial characters of $\mathfrak{gl}(m|n)$ arising this way are given by the super Schur polynomials. The second one, called Sergeev duality, is an interplay between the queer Lie superalgebra $\mathfrak{q}(n)$ and a twisted hyperoctahedral group algebra. The Schur Q-functions and related combinatorics of shifted tableaux appear naturally in the description of the irreducible polynomial characters of $\mathfrak{q}(n)$.

It has been observed that much of the classical invariant theory for the polynomial algebra has a parallel theory for the exterior algebra as well. The First Fundamental Theorem (FFT) for both polynomial invariants and skew invariants for classical groups admits natural reformulation and extension in the theory of Howe's reductive dual pairs [**51, 52**]. Lie superalgebras allow an elegant and uniform treatment of Howe duality on the polynomial and exterior algebras (cf. Cheng-Wang [**29**]). For the general linear Lie groups, Schur duality, Howe duality, and FFT are equivalent. Unlike Schur duality, Howe duality treats classical Lie groups and (super)algebras beyond type A equally well. The Howe dualities allow us to determine the character formulas for the irreducible modules appearing in the dualities.

The third duality, super duality, has a completely different flavor. It views the representation theories of Lie superalgebras and Lie algebras as two sides of the same coin, and it is an unexpected and rather powerful approach developed in the past few years by the authors and their collaborators, culminating in Cheng-Lam-Wang [**24**]. The super duality approach allows one to overcome in a conceptual way various major obstacles in super representation theory via an equivalence of module categories of Lie algebras and Lie superalgebras.

Schur, Howe, and super dualities provide approaches to the irreducible character problem in increasing generality and sophistication. Schur and Howe dualities only offer a solution to the irreducible character problem for modules in some semisimple subcategories. On the other hand, super duality provides a conceptual solution to the long-standing irreducible character problem in fairly general BGG categories (including all finite-dimensional modules) over classical Lie superalgebras in terms of the usual Kazhdan-Lusztig polynomials of classical Lie algebras. Totally different and independent approaches to the irreducible character problem of *finite-dimensional* $\mathfrak{gl}(m|n)$-modules have been developed by Serganova [**107**] and Brundan [**11**]. Also Brundan's conjecture on irreducible characters of $\mathfrak{gl}(m|n)$ in the full BGG category $\mathcal{O}$ has recently been proved in [**26**]. Super duality again plays a crucial role in the proof. However, this latest approach to the full BGG category $\mathcal{O}$ is beyond the scope of this book.

The book is largely self-contained and should be accessible to graduate students and non-experts as well. Besides assuming basic knowledge of entry-level graduate algebra (and some familiarity with basic homological algebra in the final Chapter 6), the other prerequisite is a one-semester course in the theory of finite-dimensional semisimple Lie algebras. For example, either the book by Humphreys or the first half of the book by Carter on semisimple Lie algebras is sufficient. Some familiarity with symmetric functions and representations of symmetric groups can be sometimes useful, and Appendix A provides a quick summary for our purpose. It is possible that super experts may also benefit from the book, as several "folklore" results are rigorously proved and occasionally corrected in great detail here, sometimes with new proofs. The proofs of some of these results can be at times rather difficult to trace or read in the literature (and not merely because they might be in a different language).

Here is a broad outline of the book chapter by chapter. Each chapter ends with exercises and historical notes. Though we have tried to attribute the main results accurately and fairly, we apologize beforehand for any unintended omissions and mistakes.

Chapter 1 starts by defining various classes of Lie superalgebras. For the basic Lie superalgebras, we introduce the invariant bilinear forms, root systems, fundamental systems, and Weyl groups. Positive systems and fundamental systems for basic Lie superalgebras are not conjugate under the Weyl group, and the notion of odd reflections is introduced to relate non-conjugate positive systems. The PBW theorem for the universal enveloping algebra of a Lie superalgebra is formulated, and highest weight theory for basic Lie superalgebras and $\mathfrak{q}(n)$ is developed.

In Chapter 2, we focus on Lie superalgebras of types $\mathfrak{gl}$, $\mathfrak{osp}$ and $\mathfrak{q}$. We classify their finite-dimensional simple modules using odd reflection techniques. We then formulate and establish precisely the images of the respective Harish-Chandra homomorphisms and linkage principles. We end with a Young diagrammatic description of the extremal weights in the simple polynomial $\mathfrak{gl}(m|n)$-modules and finite-dimensional simple $\mathfrak{osp}(m|2n)$-modules. It takes considerably more effort to formulate and prove these results for Lie superalgebras than for semisimple Lie algebras because of the existence of non-conjugate Borel subalgebras and the limited role of Weyl groups for Lie superalgebras.

Schur duality for Lie superalgebras is developed in Chapter 3. We start with some results on the structure of associative superalgebras including the super variants of the Wedderburn theorem, Schur's lemma, and the double centralizer property. The Schur-Sergeev duality for $\mathfrak{gl}(m|n)$ is proved, and it provides a classification of irreducible polynomial $\mathfrak{gl}(m|n)$-modules. As a consequence, the characters of the simple polynomial $\mathfrak{gl}(m|n)$-modules are given by the super Schur polynomials. On the algebraic combinatorial level, there is a natural super generalization of the notion of semistandard tableau, which is a hybrid of the traditional version and its conjugate counterpart. The Schur-Sergeev duality for $\mathfrak{q}(n)$ requires understanding the representation theory of a twisted hyperoctahedral group algebra, which we develop from scratch. The characters of the simple polynomial $\mathfrak{q}(n)$-modules are given by the Schur Q-polynomials up to some 2-powers.

In Chapter 4, we give a quick introduction to classical invariant theory, which serves as a preparation for Howe duality in the next chapter. We describe several versions of the FFT for the classical groups, i.e., a tensor algebra version, a polynomial algebra version, and a supersymmetric algebra version.

Howe duality is the main topic of Chapter 5. Like Schur duality, Howe duality involves commuting actions of a classical Lie group G and a classical superalgebra $\mathfrak{g}'$ on a supersymmetric algebra. The precise relation between the classical Lie superalgebras and Weyl-Clifford algebras $\mathfrak{WC}$ is established. According to the FFT for classical invariant theory in Chapter 4 when applied to the G-action on the associated graded algebra $\mathrm{gr}\,\mathfrak{WC}$, the basic invariants generating $(\mathrm{gr}\,\mathfrak{WC})^G$ turn out to form the associated graded space for a Lie superalgebra $\mathfrak{g}'$. From this it follows that the algebra of G-invariants $\mathfrak{WC}^G$ is generated by $\mathfrak{g}'$. Multiplicity-free decompositions for various $(G,\mathfrak{g}')$-Howe dualities are obtained explicitly. Character formulas for the irreducible $\mathfrak{g}'$-modules appearing in $(G,\mathfrak{g}')$-Howe duality are then obtained via a comparison with Howe duality involving classical groups G and infinite-dimensional Lie algebras, which we develop in detail.

Finally in Chapter 6, we develop a super duality approach to obtain a complete and conceptual solution of the irreducible character problem in certain parabolic Bernstein-Gelfand-Gelfand categories for general linear and ortho-symplectic Lie superalgebras. This chapter is technically more sophisticated than the earlier chapters. Super duality is an equivalence of categories between parabolic categories for Lie superalgebras and their Lie algebra counterparts at an infinite-rank limit, and it matches the corresponding parabolic Verma modules, irreducible modules, Kostant $\mathfrak{u}$-homology groups, and Kazhdan-Lusztig-Vogan polynomials. Truncation functors are introduced to relate the BGG categories for infinite-rank and finite-rank Lie superalgebras. In this way, we obtain a solution à la Kazhdan-Lusztig of the irreducible character problem in the corresponding parabolic BGG categories for finite-dimensional basic Lie superalgebras.

There is an appendix in the book. In Appendix A, we have included a fairly self-contained treatment of some elementary aspects of symmetric function theory, including Schur functions, supersymmetric functions and Schur Q-functions. The celebrated boson-fermion correspondence serves as a prominent example relating superalgebras to mathematical physics and algebraic combinatorics. The Fock space therein is used in setting up the Howe duality for infinite-dimensional Lie algebras in Chapter 5.

For a one-semester introductory course on Lie superalgebras, we recommend two plausible ways of using this book. A first approach uses Chapters 1, 2, 3, with possible supplements from Chapter 5 and Appendix A. A second approach uses Chapters 1, 3, 5 with possible supplements from Chapter 4 and Appendix A. It is also possible to use this book for a course on the interaction between representations of Lie superalgebras and algebraic combinatorics. The more advanced Chapter 6 can be used in a research seminar.

Acknowledgment. The book project started with the lecture notes [**32**] of the authors, which were an expanded written account of a series of lectures delivered by the second-named author in the summer school at East China Normal University, Shanghai, in July 2009. In a graduate course at the University of Virginia in Spring 2010, the second-named author lectured on what became a large portion of Chapters 3, 4, and 5 of the book. The materials in Chapter 3 on Schur duality have been used by the second-named author in the winter school in Taipei in December 2010. Part of the materials in the first three chapters have also been used by the first-named author in a lecture series in Shanghai in March 2011, and then in a lecture series by both authors in a workshop in Tehran in May 2011. We thank the participants in all these occasions for their helpful suggestions and feedback, and we especially thank Constance Baltera, Jae-Hoon Kwon, Li Luo, Jinkui Wan, and Youjie Wang for their corrections. We are grateful to Ngau Lam for his collaboration which has changed our way of thinking about the subject of Lie superalgebras.

The first-named author gratefully acknowledges the support from the National Science Council, Taiwan, and the second-named author gratefully acknowledges the continuing support of the National Science Foundation, USA.

Shun-Jen Cheng

Weiqiang Wang

Chapter 1

Lie superalgebra ABC

We start by introducing the basic notions and definitions in the theory of Lie superalgebras, such as basic and queer Lie superalgebras, Cartan and Borel subalgebras, root systems, positive and fundamental systems. We formulate the main structure results for the basic Lie superalgebras and the queer Lie superalgebras. We describe in detail the structures of Lie superalgebras of type $\mathfrak{gl}$, $\mathfrak{osp}$ and $\mathfrak{q}$. A distinguishing feature for Lie superalgebras is that Borel subalgebras, positive systems, or fundamental systems of a simple finite-dimensional Lie superalgebra may not be conjugate under the action of the corresponding Weyl group; rather, they are shown to be related to each other by real and odd reflections. A highest weight theory is developed for Lie superalgebras. We describe how fundamental systems are related and how highest weights are transformed by an odd reflection.

1.1. Lie superalgebras: Definitions and examples

Throughout this book we will work over the field $\mathbb{C}$ of complex numbers. Let

$$\mathbb{Z}_2 = \{\bar{0}, \bar{1}\}$$

denote the group of two elements and let $\mathfrak{S}_n$ denote the symmetric group in n letters.

In this section, we introduce many examples of Lie superalgebras. The examples, most relevant to this book, of the general linear and ortho-symplectic Lie superalgebras are introduced first. Other series of finite-dimensional simple Lie superalgebras of classical type, namely, the queer and the periplectic Lie superalgebras, along with the three exceptional ones, are then described. The finite-dimensional Cartan type Lie superalgebras are then realized explicitly as subalgebras of the Lie superalgebra of polynomial vector fields on a purely odd dimensional superspace. The section ends with Kac's classification theorem of the finite-dimensional simple Lie superalgebras over $\mathbb{C}$, which we state without proof.

1.1.1. Basic definitions. A **vector superspace** V is a vector space endowed with a $\mathbb{Z}_2$-gradation: $V = V_{\bar{0}} \oplus V_{\bar{1}}$. The **dimension** of the vector superspace V is the tuple $\dim V = (\dim V_{\bar{0}} | \dim V_{\bar{1}})$ or sometimes $\dim V = \dim V_{\bar{0}} + \dim V_{\bar{1}}$ (which should be clear from the context). The **superdimension** of V is defined to be $\mathrm{sdim} V := \dim V_{\bar{0}} - \dim V_{\bar{1}}$. We denote the superspace with even subspace $\mathbb{C}^m$ and odd subspace $\mathbb{C}^n$ by $\mathbb{C}^{m|n}$. It has dimension $(m|n)$. The **parity** of a homogeneous element $a \in V_i$ is denoted by $|a| = i$, $i \in \mathbb{Z}_2$. An element in $V_{\bar{0}}$ is called **even**, while an element in $V_{\bar{1}}$ is called **odd**. A **subspace** of a vector superspace $V = V_{\bar{0}} \oplus V_{\bar{1}}$ is a vector superspace $W = W_{\bar{0}} \oplus W_{\bar{1}} \subseteq V$ with compatible $\mathbb{Z}_2$-gradation, i.e., $W_i \subseteq V_i$, for $i \in \mathbb{Z}_2$.

Let V be a superspace. Throughout the book, when we write $|v|$ for an element $v \in V$, we will always implicitly assume that v is a homogeneous element and automatically extend the relevant formulas by linearity (whenever applicable). Also, note that if V and W are superspaces, then the space of linear transformations from V to W is naturally a vector superspace. In particular, the space of endomorphisms of V, denoted by $\mathrm{End}(V)$, is a vector superspace. When $V = \mathbb{C}^{m|n}$ we write $I = I_{m|n} = I_V$ for the **identity matrix** on V.

There is a **parity reversing functor** Π on the category of vector superspaces. For a vector superspace $V = V_{\bar{0}} \oplus V_{\bar{1}}$, we let

$$\Pi(V) = \Pi(V)_{\bar{0}} \oplus \Pi(V)_{\bar{1}}, \quad \Pi(V)_i = V_{i+\bar{1}}, \forall i \in \mathbb{Z}_2.$$

Clearly, $\Pi^2 = \mathrm{I}$.

Definition 1.1. A **superalgebra** A, sometimes also called a $\mathbb{Z}_2$-graded algebra, is a vector superspace $A = A_{\bar{0}} \oplus A_{\bar{1}}$ equipped with a bilinear multiplication satisfying $A_i A_j \subseteq A_{i+j}$, for $i, j \in \mathbb{Z}_2$. A **module** M over a superalgebra A is always understood in the $\mathbb{Z}_2$-graded sense, that is $M = M_{\bar{0}} \oplus M_{\bar{1}}$ such that $A_i M_j \subseteq M_{i+j}$, for $i, j \in \mathbb{Z}_2$. **Subalgebras** and **ideals** of superalgebras are also understood in the $\mathbb{Z}_2$-graded sense. A superalgebra that has no nontrivial ideal is called **simple**. A **homomorphism** between A-modules M and N is a linear map $f : M \to N$ satisfying that $f(am) = af(m)$, for all $a \in A$, $m \in M$. A homomorphism $f : M \to N$ is **of degree** $|\mathbf{f}| \in \mathbb{Z}_2$ if $f(M_i) \subseteq M_{i+|f|}$ for $i \in \mathbb{Z}_2$.

A homomorphism between modules M and N of a superalgebra A is sometimes understood in the literature as a linear map $f : M \to N$ of parity $|f| \in \mathbb{Z}_2$ which satisfies $(\star)$ $f(am) = (-1)^{|a| \cdot |f|} af(m)$, for homogeneous $a \in A$, $m \in M$. Let us call such a map a $\star$-homomorphism. These two definitions can be converted to each other as follows. Given a homomorphism $f: M \to N$ of degree $|f|$ in the sense of Definition 1.1, we define $f^\dagger: M \to N$ by the formula

$$f^\dagger(x) := (-1)^{|f|\cdot|x|} f(x). \tag{1.1}$$

Then $f^\dagger$ is a $\star$-homomorphism. Conversely, (1.1) also converts a $\star$-homomorphism into a homomorphism as in Definition 1.1.

Now we come to the definition of the main object of our study.

Definition 1.2. A **Lie superalgebra** is a superalgebra $\mathfrak{g} = \mathfrak{g}_{\bar{0}} \oplus \mathfrak{g}_{\bar{1}}$ with bilinear multiplication $[\cdot,\cdot]$ satisfying the following two axioms: for homogeneous elements $a,b,c \in \mathfrak{g}$,

(1) Skew-supersymmetry: $[a,b] = -(-1)^{|a|\cdot|b|}[b,a]$.
(2) Super Jacobi identity: $[a,[b,c]] = [[a,b],c] + (-1)^{|a|\cdot|b|}[b,[a,c]]$.

A bilinear form $(\cdot,\cdot): \mathfrak{g} \times \mathfrak{g} \to \mathbb{C}$ on a Lie superalgebra $\mathfrak{g}$ is called **invariant** if $([a,b],c) = (a,[b,c])$, for all $a,b,c \in \mathfrak{g}$.

For a Lie superalgebra $\mathfrak{g} = \mathfrak{g}_{\bar{0}} \oplus \mathfrak{g}_{\bar{1}}$, the even part $\mathfrak{g}_{\bar{0}}$ is a Lie algebra. Hence, if $\mathfrak{g}_{\bar{1}} = 0$, then $\mathfrak{g}$ is just a usual Lie algebra. A Lie superalgebra $\mathfrak{g}$ with purely odd part, i.e., $\mathfrak{g}_{\bar{0}} = 0$, has to be **abelian**, i.e., $[\mathfrak{g},\mathfrak{g}] = 0$.

Definition 1.3. Let $\mathfrak{g}$ and $\mathfrak{g}'$ be Lie superalgebras. A **homomorphism of Lie superalgebras** is an even linear map $f: \mathfrak{g} \to \mathfrak{g}'$ satisfying

$$f([a,b]) = [f(a), f(b)], \quad a,b \in \mathfrak{g}.$$

Example 1.4. (1) Let $A = A_{\bar{0}} \oplus A_{\bar{1}}$ be an associative superalgebra. We can make A into a Lie superalgebra by letting

$$[a,b] := ab - (-1)^{|a|\cdot|b|} ba,$$

for homogeneous $a,b \in A$ and extending $[\cdot,\cdot]$ by bilinearity.

(2) Let $\mathfrak{g}$ be a Lie superalgebra. Then $\mathrm{End}(\mathfrak{g})$ is an associative superalgebra, and hence it carries a structure of a Lie superalgebra by (1). We define the **adjoint map** $\mathrm{ad}: \mathfrak{g} \to \mathrm{End}(\mathfrak{g})$ by

$$\mathrm{ad}(a)(b) := [a,b], \quad a,b \in \mathfrak{g}.$$

Then ad is a homomorphism of Lie superalgebras due to the super Jacobi identity. The resulting action of $\mathfrak{g}$ on itself is called the **adjoint action**.

(3) Let $A = A_{\bar{0}} \oplus A_{\bar{1}}$ be a superalgebra. An endomorphism $D \in \mathrm{End}(A)_s$, for $s \in \mathbb{Z}_2$, is called a **derivation** of degree s if it satisfies that

$$D(ab) = D(a)b + (-1)^{s|a|} aD(b), \quad a,b \in A.$$

Denote by $\mathrm{Der}(A)_s$ the space of derivations on A of degree s. One verifies that the superspace of derivations of A, $\mathrm{Der}(A) = \mathrm{Der}(A)_{\bar{0}} \oplus \mathrm{Der}(A)_{\bar{1}}$, is a subalgebra of the Lie superalgebra $(\mathrm{End}(A), [\cdot,\cdot])$.

In the case when $\mathfrak{g}$ is a Lie superalgebra we have $\mathrm{ad}g \in \mathrm{Der}(\mathfrak{g})$, for all $g \in \mathfrak{g}$, by the super Jacobi identity. Indeed, such derivations are called **inner derivations**. The inner derivations form an ideal in $\mathrm{Der}(\mathfrak{g})$.

(4) Let $\mathfrak{g} = \mathfrak{g}_{\bar{0}} \oplus \mathfrak{g}_{\bar{1}}$ be a superspace such that $\mathfrak{g}_{\bar{0}} = \mathbb{C}z$ is one-dimensional. Suppose that we have a symmetric bilinear form $B(\cdot,\cdot)$ on $\mathfrak{g}_{\bar{1}}$. We can make $\mathfrak{g}$ into a Lie superalgebra by letting z commute with $\mathfrak{g}$ and declaring

$$[v,w] := B(v,w)z, \quad v,w \in \mathfrak{g}_{\bar{1}}.$$

The special cases when $B(\cdot,\cdot)$ is zero and when $B(\cdot,\cdot)$ is non-degenerate, respectively, are of particular interest. Indeed, their corresponding universal enveloping superalgebras (see Section 1.5.1) are isomorphic to the exterior and Clifford superalgebras of Section 1.1.6 and Definition 3.33, respectively.

For a Lie superalgebra $\mathfrak{g} = \mathfrak{g}_{\bar{0}} \oplus \mathfrak{g}_{\bar{1}}$, the restriction of the adjoint homomorphism $\mathrm{ad}|_{\mathfrak{g}_{\bar{0}}} : \mathfrak{g}_{\bar{0}} \to \mathrm{End}(\mathfrak{g}_{\bar{1}})$ is a homomorphism of Lie algebras. That is, $\mathfrak{g}_{\bar{1}}$ is a $\mathfrak{g}_{\bar{0}}$-module under the adjoint action.

Remark 1.5. To a Lie superalgebra $\mathfrak{g} = \mathfrak{g}_{\bar{0}} \oplus \mathfrak{g}_{\bar{1}}$ we associate the following data:

(1) A Lie algebra $\mathfrak{g}_{\bar{0}}$.

(2) A $\mathfrak{g}_{\bar{0}}$-module $\mathfrak{g}_{\bar{1}}$ induced by the adjoint action.

(3) A $\mathfrak{g}_{\bar{0}}$-homomorphism $S^2(\mathfrak{g}_{\bar{1}}) \to \mathfrak{g}_{\bar{0}}$ induced by the Lie bracket.

(4) The condition coming from Definition 1.2(2) with $a,b,c \in \mathfrak{g}_{\bar{1}}$.

Conversely, the above data determine a Lie superalgebra structure on $\mathfrak{g}_{\bar{0}} \oplus \mathfrak{g}_{\bar{1}}$.

1.1.2. The general and special linear Lie superalgebras. Let $V = V_{\bar{0}} \oplus V_{\bar{1}}$ be a vector superspace so that $\mathrm{End}(V)$ is an associative superalgebra. As in Example 1.4(1), $\mathrm{End}(V)$, equipped with the supercommutator, forms a Lie superalgebra called the **general linear Lie superalgebra** and denoted by $\mathfrak{gl}(V)$. When $V = \mathbb{C}^{m|n}$ we also write $\mathfrak{gl}(m|n)$ for $\mathfrak{gl}(V)$.

Choose ordered bases for $V_{\bar{0}}$ and $V_{\bar{1}}$ that combine to a homogeneous ordered basis for V. We will make it a convention to parameterize such a basis by the set

$$I(m|n) = \{\bar{1},\ldots,\bar{m};1,\ldots,n\} \tag{1.2}$$

with total order

$$\bar{1} < \ldots < \bar{m} < 0 < 1 < \ldots < n. \tag{1.3}$$

Here 0 is inserted for notational convenience later on. The elementary matrices are accordingly denoted by E_{ij}, with $i,j \in I(m|n)$. With respect to such an ordered basis, End(V) and $\mathfrak{gl}(V)$ can be realized as $(m+n)\times(m+n)$ complex matrices of the block form

$$g = \begin{pmatrix} a & b \\ c & d \end{pmatrix}, \tag{1.4}$$

where a, b, c, and d are respectively $m\times m$, $m\times n$, $n\times m$, and $n\times n$ matrices. The even subalgebra $\mathfrak{gl}(V)_{\bar 0}$ consists of matrices of the form (1.4) with $b=c=0$, while the odd subspace $\mathfrak{gl}(V)_{\bar 1}$ consists of those with $a=d=0$. In particular, $\mathfrak{gl}(V)_{\bar 0}\cong \mathfrak{gl}(m)\oplus\mathfrak{gl}(n)$, and as a $\mathfrak{gl}(V)_{\bar 0}$-module, $\mathfrak{gl}(V)_{\bar 1}$ is self-dual and is isomorphic to $(\mathbb{C}^m\otimes\mathbb{C}^{n*})\oplus(\mathbb{C}^{m*}\otimes\mathbb{C}^n)$. Here and below, $\mathbb{C}^{n*}$ denotes the dual space of $\mathbb{C}^n$.

Remark 1.6. Let Π be the parity reversing functor defined in Section 1.1.1. We have an isomorphism of Lie superalgebras from $\mathfrak{gl}(V)$ to $\mathfrak{gl}(\Pi V)$ by sending T to $\Pi T\Pi^{-1}$. When $\dim V=(m|n)$, we obtain an isomorphism of Lie superalgebras $\mathfrak{gl}(m|n)\cong\mathfrak{gl}(n|m)$.

For each element $g\in\mathfrak{gl}(m|n)$ of the form (1.4) we define the **supertrace** as

$$\mathrm{str}(g) := \mathrm{tr}(a)-\mathrm{tr}(d),$$

where $\mathrm{tr}(x)$ denotes the trace of the square matrix x. One checks that

$$\mathrm{str}([g,g'])=0, \qquad \text{for } g,g'\in\mathfrak{gl}(m|n).$$

Thus, the subspace

$$\mathfrak{sl}(m|n) := \{g\in\mathfrak{gl}(m|n)\mid \mathrm{str}(g)=0\}$$

is a subalgebra of $\mathfrak{gl}(m|n)$, and it is called the **special linear Lie superalgebra**. One verifies directly that $[\mathfrak{gl}(m|n),\mathfrak{gl}(m|n)]=\mathfrak{sl}(m|n)$. Furthermore, $\mathfrak{sl}(m|n)\cong\mathfrak{sl}(n|m)$, and when $m\neq n$ and $m+n\geq 2$, $\mathfrak{sl}(m|n)$ is simple. When $m=n$, $\mathfrak{sl}(m|m)$ contains a nontrivial center generated by the identity matrix $I_{m|m}$. For $m\geq 2$, $\mathfrak{sl}(m|m)/\mathbb{C}I_{m|m}$ is simple.

Example 1.7. Let $\mathfrak{g}=\mathfrak{gl}(1|1)$ and consider the following basis for $\mathfrak{g}$:

$$e=\begin{pmatrix}0&1\\0&0\end{pmatrix},\qquad f=\begin{pmatrix}0&0\\1&0\end{pmatrix},\qquad E_{\bar 1,\bar 1}=\begin{pmatrix}1&0\\0&0\end{pmatrix},\qquad E_{11}=\begin{pmatrix}0&0\\0&1\end{pmatrix}.$$

Set $h:=E_{\bar 1,\bar 1}+E_{11}=I_{1|1}$. Then h is central, $[e,f]=h$, and $\mathfrak{sl}(1|1)$ has a basis $\{e,h,f\}$.

Let $\mathbb{I}=\mathbb{I}_{\bar 0}\sqcup\mathbb{I}_{\bar 1}$ be a parametrization of a homogeneous basis of the superspace $V=V_{\bar 0}\oplus V_{\bar 1}$, where $\mathbb{I}_{\bar 0}$ and $\mathbb{I}_{\bar 1}$ parameterize the corresponding bases of $V_{\bar 0}$ and $V_{\bar 1}$, respectively. For an element i in $\mathbb{I}$ we define

$$|i| := \begin{cases} 0, & \text{if } i\in\mathbb{I}_{\bar 0},\\ 1, & \text{if } i\in\mathbb{I}_{\bar 1}.\end{cases}$$

For example, for the parametrization $I(m|n)$ with ordering (1.3), we have $|i|=0$ for $i<0$, and $|i|=1$ for $i>0$. Choosing a total ordering of the homogeneous basis we may identify $\mathfrak{gl}(V)$ with the space of $|\mathbb{I}| \times |\mathbb{I}|$ matrices. For such a matrix $A = \sum_{i,j\in\mathbb{I}} a_{ij}E_{ij}$, $a_{ij} \in \mathbb{C}$, we define the **supertranspose** of A to be

$$A^{\mathrm{st}} := \sum_{i,j\in\mathbb{I}} (-1)^{|j|(|i|+|j|)} a_{ij}E_{ji}. \tag{1.5}$$

We define the **Chevalley automorphism** $\tau : \mathfrak{gl}(V) \to \mathfrak{gl}(V)$ by the formula

$$\tau(A) := -A^{\mathrm{st}}. \tag{1.6}$$

It is straightforward to check that τ is an automorphism of Lie superalgebras. We note that τ restricts to an automorphism of $\mathfrak{sl}(V)$. Also, for m,n both nonzero, τ has order 4 and hence is, in general, not an involution.

1.1.3. The ortho-symplectic Lie superalgebras.

Definition 1.8. Let $V = V_{\bar{0}} \oplus V_{\bar{1}}$ be a vector superspace. A bilinear form

$$B(\cdot,\cdot) : V \times V \longrightarrow V$$

is called **even** (respectively, **odd**), if $B(V_i,V_j) = 0$ unless $i+j = \bar{0}$ (respectively, $i+j = \bar{1}$). An even bilinear form B is said to be **supersymmetric** if $B|_{V_{\bar{0}}\times V_{\bar{0}}}$ is symmetric and $B|_{V_{\bar{1}}\times V_{\bar{1}}}$ is skew-symmetric, and it is called **skew-supersymmetric** if $B|_{V_{\bar{0}}\times V_{\bar{0}}}$ is skew-symmetric and $B|_{V_{\bar{1}}\times V_{\bar{1}}}$ is symmetric.

Let B be a non-degenerate even supersymmetric bilinear form on a vector superspace $V = V_{\bar{0}} \oplus V_{\bar{1}}$. It follows that $\dim V_{\bar{1}}$ is necessarily even. For $s \in \mathbb{Z}_2$, let

$$\begin{aligned}\mathfrak{osp}(V)_s &:= \{g \in \mathfrak{gl}(V)_s \mid B(g(x),y) = -(-1)^{s\cdot|x|}B(x,g(y)), \forall x,y \in V\},\\ \mathfrak{osp}(V) &:= \mathfrak{osp}(V)_{\bar{0}} \oplus \mathfrak{osp}(V)_{\bar{1}}.\end{aligned}$$

One checks that $\mathfrak{osp}(V)$ is a Lie superalgebra, called the **ortho-symplectic Lie superalgebra**; that is, $\mathfrak{osp}(V)$ is the subalgebra of $\mathfrak{gl}(V)$ that preserves a non-degenerate supersymmetric bilinear form. Its even subalgebra is isomorphic to $\mathfrak{so}(V_{\bar{0}}) \oplus \mathfrak{sp}(V_{\bar{1}})$, a direct sum of the orthogonal Lie algebra on $V_{\bar{0}}$ and the symplectic Lie algebra on $V_{\bar{1}}$. When $V = \mathbb{C}^{\ell|2m}$, we write $\mathfrak{osp}(\ell|2m)$ for $\mathfrak{osp}(V)$. Note that when ℓ (respectively, m) is zero, the ortho-symplectic Lie superalgebra reduces to the classical Lie algebra $\mathfrak{sp}(2m)$ (respectively, $\mathfrak{so}(\ell)$).

Similarly, we define the Lie superalgebra $\mathfrak{spo}(V)$ as the subalgebra of $\mathfrak{gl}(V)$ that preserves a non-degenerate skew-supersymmetric bilinear form on V (here $\dim V_{\bar{0}}$ has to be even). When $V = \mathbb{C}^{2m|\ell}$, we write $\mathfrak{spo}(2m|\ell)$ for $\mathfrak{spo}(V)$.

Remark 1.9. A non-degenerate supersymmetric bilinear form B on V induces a non-degenerate skew-supersymmetric bilinear form B^{Π} on $\Pi(V)$, defined by

$$B^{\Pi}(\Pi(v),\Pi(v')) := (-1)^{|v|}B(v,v'), \qquad \text{for } v,v' \in V.$$

The restriction of the isomorphism $\mathfrak{gl}(V) \cong \mathfrak{gl}(\Pi V)$ in Remark 1.6 gives rise to a Lie superalgebra isomorphism between $\mathfrak{osp}(V)$ (with respect to B) and $\mathfrak{spo}(\Pi V)$ (with respect to B^{Π}); see Exercise 1.3. It follows that $\mathfrak{osp}(\ell|2m) \cong \mathfrak{spo}(2m|\ell)$.

We now give an explicit matrix realization of the ortho-symplectic Lie superalgebra. To this end, we first observe that the supertranspose (1.5) of a matrix in the block form (1.4) is equal to

$$\begin{pmatrix} a & b \\ c & d \end{pmatrix}^{\mathrm{st}} = \begin{pmatrix} a^t & c^t \\ -b^t & d^t \end{pmatrix},$$

where x^t denotes the usual transpose of a matrix x.

Define the $(2m+2n+1)\times(2m+2n+1)$ matrix in the $(m|m|n|n|1)$-block form

$$\mathfrak{J}_{2m|2n+1} := \begin{pmatrix} 0 & I_m & 0 & 0 & 0 \\ -I_m & 0 & 0 & 0 & 0 \\ 0 & 0 & 0 & I_n & 0 \\ 0 & 0 & I_n & 0 & 0 \\ 0 & 0 & 0 & 0 & 1 \end{pmatrix}. \tag{1.7}$$

Let $\mathfrak{J}_{2m|2n}$ denote the $(2m+2n)\times(2m+2n)$ matrix obtained from $\mathfrak{J}_{2m|2n+1}$ by deleting the last row and column. For $\ell = 2n$ or $2n+1$, by definition $\mathfrak{spo}(2m|\ell)$ is the subalgebra of $\mathfrak{gl}(2m|\ell)$ that preserves the bilinear form on $\mathbb{C}^{2m|\ell}$ with matrix $\mathfrak{J}_{2m|\ell}$ relative to the standard basis of $\mathbb{C}^{2m|\ell}$, and hence

$$\mathfrak{spo}(2m|\ell) = \{g \in \mathfrak{gl}(2m|\ell) \mid g^{st}\mathfrak{J}_{2m|\ell} + \mathfrak{J}_{2m|\ell}\, g = 0\}.$$

By a direct computation, $\mathfrak{spo}(2m|2n+1)$ consists of the $(2m+2n+1)\times(2m+2n+1)$ matrices of the following $(m|m|n|n|1)$-block form

$$\begin{pmatrix} d & e & y_1^t & x_1^t & z_1^t \\ f & -d^t & -y^t & -x^t & -z^t \\ x & x_1 & a & b & -v^t \\ y & y_1 & c & -a^t & -u^t \\ z & z_1 & u & v & 0 \end{pmatrix}, \quad b, c \text{ skew-symmetric, } e, f \text{ symmetric.} \tag{1.8}$$

Note that $\mathfrak{spo}(2m|2n+1)_{\bar{1}} \cong \mathbb{C}^{2m} \otimes \mathbb{C}^{2n+1}$ (which is self-dual) as a module over $\mathfrak{spo}(2m|2n+1)_{\bar{0}} \cong \mathfrak{sp}(2m) \oplus \mathfrak{so}(2n+1)$.

The Lie superalgebra $\mathfrak{spo}(2m|2n)$ consists of matrices (1.8) with the last row and column removed. Note that $\mathfrak{spo}(2m|2n)_{\bar{1}} \cong \mathbb{C}^{2m} \otimes \mathbb{C}^{2n}$ (which is self-dual) as a module over $\mathfrak{spo}(2m|2n)_{\bar{0}} \cong \mathfrak{sp}(2m) \oplus \mathfrak{so}(2n)$.

Here and below, the rows and columns of the matrices $\mathfrak{J}_{2m|\ell}$ and (1.8) (or its modification) are indexed by the finite set $I(2m|\ell)$.

Proposition 1.10. *The automorphism* τ *in* (1.6) *restricts to an automorphism of* $\mathfrak{spo}(2m|\ell)$.

Proof. Take an element $g \in \mathfrak{spo}(2m|\ell)$. Thus, $\mathfrak{J}_{2m|\ell}\, g + g^{\mathrm{st}}\mathfrak{J}_{2m|\ell} = 0$, and hence

$$g^{\mathrm{st}}\mathfrak{J}_{2m|\ell}^{\mathrm{st}} + \mathfrak{J}_{2m|\ell}^{\mathrm{st}}(g^{\mathrm{st}})^{\mathrm{st}} = 0. \tag{1.9}$$

Observe that $\mathfrak{J}_{2m|\ell}^{\mathrm{st}} = (\mathfrak{J}_{2m|\ell})^{-1}$. So if we multiply (1.9) on the left and on the right by $\mathfrak{J}_{2m|\ell}$, we obtain the identity

$$\mathfrak{J}_{2m|\ell}g^{\mathrm{st}} + (g^{\mathrm{st}})^{\mathrm{st}}\mathfrak{J}_{2m|\ell} = 0.$$

This implies that $\mathfrak{J}_{2m|\ell}\tau(g) + \tau(g)^{\mathrm{st}}\mathfrak{J}_{2m|\ell} = 0$, and so $\tau(g) \in \mathfrak{spo}(2m|\ell)$. □

1.1.4. The queer Lie superalgebras. Let $V = V_{\bar{0}} \oplus V_{\bar{1}}$ be a vector superspace with $\dim V_{\bar{0}} = \dim V_{\bar{1}}$. Choose $P \in \mathrm{End}(V)_{\bar{1}}$ such that $P^2 = I_{n|n}$. The subspace

$$\mathfrak{q}(V) = \{T \in \mathrm{End}(V) \mid [T,P] = 0\}$$

is a subalgebra of $\mathfrak{gl}(V)$ called the **queer Lie superalgebra**. Different choices of P give rise to isomorphic queer Lie superalgebras. If $V = \mathbb{C}^{n|n}$, then $\mathfrak{q}(V)$ is also denoted by $\mathfrak{q}(n)$.

To give an explicit matrix realization of $\mathfrak{q}(n)$, let us take P to be the $2n \times 2n$ matrix

$$P := \sqrt{-1}\begin{pmatrix} 0 & I_n \\ -I_n & 0 \end{pmatrix}. \tag{1.10}$$

Then, for $g \in \mathfrak{gl}(n|n)$ of the form (1.4), we have $g \in \mathfrak{q}(n)$ if and only if $gP - (-1)^{|g|}Pg = 0$, and in turn, if and only if

$$g = \begin{pmatrix} a & b \\ b & a \end{pmatrix}, \tag{1.11}$$

where a, b are arbitrary complex $n \times n$ matrices. Thus we have $\mathfrak{q}(n)_{\bar{0}} \cong \mathfrak{gl}(n)$ as Lie algebras, and $\mathfrak{q}(n)_{\bar{1}} \cong \mathfrak{gl}(n)$ as the adjoint $\mathfrak{q}(n)_{\bar{0}}$-module. A linear basis for $\mathfrak{q}(n)$ consists of the following elements:

$$\widetilde{E}_{ij} := E_{\bar{i}\bar{j}} + E_{ij}, \ \overline{E}_{ij} := E_{i\bar{j}} + E_{\bar{i}j}, \quad 1 \le i, j \le n. \tag{1.12}$$

The derived superalgebra $[\mathfrak{q}(n), \mathfrak{q}(n)]$ consists of matrices of the form (1.11), with $a \in \mathfrak{gl}(n)$ and $b \in \mathfrak{sl}(n)$, and so it contains a one-dimensional center generated by the identity matrix $I_{n|n}$. The quotient superalgebra $[\mathfrak{q}(n), \mathfrak{q}(n)]/\mathbb{C}I_{n|n}$ has even part isomorphic to $\mathfrak{sl}(n)$ and odd part isomorphic to the adjoint module, and one can show that it is simple for $n \ge 3$. For $n = 2$, the odd part of the quotient Lie superalgebra is an abelian ideal, since the adjoint module of $\mathfrak{sl}(2)$ does not appear in the symmetric square of the adjoint module.

Remark 1.11. Consider the subspace $\widetilde{\mathfrak{q}}(n)$ of $\mathfrak{gl}(n|n)$ consisting of elements that commute with P. That is,

$$\widetilde{\mathfrak{q}}(n) := \{g \in \mathfrak{gl}(n|n) | gP - Pg = 0\}.$$

In matrix form, $\widetilde{\mathfrak{q}}(n)$ consists of the following $n|n$-block matrices:

$$\begin{pmatrix} a & b \\ -b & a \end{pmatrix}, \tag{1.13}$$

where a and b are arbitrary $n \times n$ matrices. One checks that $\widetilde{\mathfrak{q}}(n)$ is closed under the Lie bracket and hence is a subalgebra of $\mathfrak{gl}(n|n)$. Indeed $\widetilde{\mathfrak{q}}(n)$ is isomorphic to $\mathfrak{q}(n)$, since the map τ in (1.6) sends $\mathfrak{q}(n)$ to $\widetilde{\mathfrak{q}}(n)$, and vice versa. Thus, (1.13) gives another realization of the queer Lie superalgebra.

1.1.5. The periplectic and exceptional Lie superalgebras. Let us describe more examples of Lie superalgebras.

The periplectic Lie superalgebras. Let $V = V_{\bar{0}} \oplus V_{\bar{1}}$ be a vector superspace with $\dim V_{\bar{0}} = \dim V_{\bar{1}}$. Let $C(\cdot,\cdot)$ be a non-degenerate odd symmetric bilinear form on V. One checks that the subspace of $\mathfrak{gl}(V)$ preserving C is closed under the Lie bracket and hence is a Lie subalgebra of $\mathfrak{gl}(V)$. This superalgebra is called the **periplectic Lie superalgebra** and will be denoted by $\mathfrak{p}(V)$. Different choices of C give rise to isomorphic periplectic Lie superalgebras. In the case $V = \mathbb{C}^{n|n}$, $\mathfrak{p}(V)$ is also denoted by $\mathfrak{p}(n)$.

To write down an explicit matrix realization of $\mathfrak{p}(n)$ as a subalgebra of $\mathfrak{gl}(n|n)$, let us take the $2n \times 2n$ matrix

$$\mathfrak{P} := \begin{pmatrix} 0 & I_n \\ I_n & 0 \end{pmatrix}, \tag{1.14}$$

which determines an odd symmetric bilinear form C on $\mathbb{C}^{n|n}$. Then, $g \in \mathfrak{p}(n)$ if and only if $g^{st}\mathfrak{P} + \mathfrak{P}g = 0$. It follows that

$$\mathfrak{p}(n) = \left\{ \begin{pmatrix} a & b \\ c & -a^t \end{pmatrix}, \text{ where } b \text{ is symmetric and } c \text{ is skew-symmetric} \right\}. \tag{1.15}$$

We have

$$\mathfrak{p}(n)_{\bar{0}} \cong \mathfrak{gl}(n), \quad \mathfrak{p}(n)_{\bar{1}} \cong S^2(\mathbb{C}^n) \oplus \wedge^2(\mathbb{C}^{n*}). \tag{1.16}$$

For $n \geq 3$, the derived superalgebra $[\mathfrak{p}(n), \mathfrak{p}(n)]$ is simple, and it consists of matrices of the form $\begin{pmatrix} a & b \\ c & -a^t \end{pmatrix}$ with $\mathrm{tr}(a) = 0$, $b^t = b$, and $c^t = -c$.

Remark 1.12. One checks that the Lie subalgebra $\widetilde{\mathfrak{p}}(n)$ of $\mathfrak{gl}(n|n)$ preserving the non-degenerate odd *skew-symmetric* bilinear form corresponding to P in (1.10) consists of matrices of the form $\begin{pmatrix} a & b \\ c & -a^t \end{pmatrix}$ with $b^t = -b$ and $c^t = c$. Similar to Remark 1.11, the map τ in (1.6) restricts to an isomorphism between $\mathfrak{p}(n)$ and $\widetilde{\mathfrak{p}}(n)$.

The exceptional Lie superalgebra $D(2|1,\alpha)$.

We take three copies of the Lie algebra $\mathfrak{sl}(2)$ denoted by $\mathfrak{g}_i$ ($i = 1,2,3$), and we associate to each $\mathfrak{g}_i$ a copy of the standard $\mathfrak{sl}(2)$-module V_i.

Clearly, as $\mathfrak{g}_i$-modules, we have an isomorphism $S^2(V_i) \cong \mathfrak{g}_i$. By Schur's Lemma we may associate a nonzero scalar $\alpha_i \in \mathbb{C}$ to any such isomorphism. Now consider $\mathfrak{g} = \mathfrak{g}_{\bar{0}} \oplus \mathfrak{g}_{\bar{1}}$, where $\mathfrak{g}_{\bar{0}} \cong \mathfrak{g}_1 \oplus \mathfrak{g}_2 \oplus \mathfrak{g}_3$, and $\mathfrak{g}_{\bar{1}}$ is the irreducible $\mathfrak{g}_{\bar{0}}$-module

$V_1 \otimes V_2 \otimes V_3$. We can associate three nonzero complex numbers α_i, $i = 1,2,3$, to any surjective $\mathfrak{g}_{\bar{0}}$-homomorphism from $S^2(\mathfrak{g}_{\bar{1}})$ to $\mathfrak{g}_{\bar{0}}$. Thus, our vector superspace $\mathfrak{g}$ satisfies Conditions (1)–(3) of Remark 1.5. It is easy to see that Condition (4) of Remark 1.5 is equivalent to $\sum_{i=1}^3 \alpha_i = 0$. Thus, we obtain a Lie superalgebra $\mathfrak{g}(\alpha_1, \alpha_2, \alpha_3)$ depending on three nonzero parameters α_i, $i = 1,2,3$. For $\sigma \in \mathfrak{S}_3$, we have $\mathfrak{g}(\alpha_1, \alpha_2, \alpha_3) \cong \mathfrak{g}(\alpha_{\sigma(1)}, \alpha_{\sigma(2)}, \alpha_{\sigma(3)})$. Also, $\mathfrak{g}(\alpha_1, \alpha_2, \alpha_3) \cong \mathfrak{g}(\lambda\alpha_1, \lambda\alpha_2, \lambda\alpha_3)$, for any nonzero $\lambda \in \mathbb{C}$. Thus, we have a one-parameter family of Lie superalgebras $D(2|1,\alpha) := \mathfrak{g}(\alpha, 1, -1-\alpha)$ that are simple for $\alpha \neq 0, -1$.

We have $\mathfrak{g}(\alpha, 1, -1-\alpha) \cong \mathfrak{g}(1, \alpha, -1-\alpha) \cong \mathfrak{g}(1, \alpha^{-1}, -\alpha^{-1} - 1)$, and also $\mathfrak{g}(\alpha, 1, -1-\alpha) \cong \mathfrak{g}(-1-\alpha, 1, \alpha)$, which imply

$$D(2|1,\alpha) \cong D(2|1,\alpha^{-1}) \cong D(2|1,-1-\alpha).$$

The maps $\alpha \mapsto \alpha^{-1}$ and $\alpha \mapsto (-1-\alpha)$ generate an action of $\mathfrak{S}_3$ on $\mathbb{C} \setminus \{0,-1\}$. We have $D(2|1,\alpha) \cong D(2|1,\beta)$ if and only if $\beta \in \mathfrak{S}_3 \cdot \alpha$. This gives additional isomorphisms

$$D(2|1,\alpha) \cong D(2|1,-(1+\alpha)^{-1}\alpha) \cong D(2|1,-1-\alpha^{-1}) \cong D(2|1,-(1+\alpha)^{-1}).$$

Thus, any orbit of $(\mathbb{C} \setminus \{0,-1\})/\mathfrak{S}_3$ consists of six points, except for the orbit corresponding to the three points $\alpha = 1, -2, -\frac{1}{2}$, and the orbit corresponding to the two points $\alpha = -\frac{1}{2} \pm \sqrt{-\frac{3}{4}}$. Finally, note that $D(2|1,1) \cong \mathfrak{osp}(4|2)$.

The exceptional Lie superalgebra $F(3|1)$. There is a simple Lie superalgebra $F(3|1)$ with $F(3|1)_{\bar{0}} \cong \mathfrak{sl}(2) \oplus \mathfrak{so}(7)$. The odd part, as an $F(3|1)_{\bar{0}}$-module, is isomorphic to the tensor product of the standard $\mathfrak{sl}(2)$-module and the simple $\mathfrak{so}(7)$-spin module. Hence, $\dim F(3|1) = (24|16)$. In the literature, $F(3|1)$ is often denoted by $F(4)$, which could be confused with the simple Lie algebra of type F_4.

The exceptional Lie superalgebra $G(3)$. There is a simple Lie superalgebra $G(3)$ with $G(3)_{\bar{0}} \cong \mathfrak{sl}(2) \oplus G_2$. The odd part as a $G(3)_{\bar{0}}$-module is isomorphic to the tensor product of the standard $\mathfrak{sl}(2)$-module and the fundamental 7-dimensional G_2-module. Hence, $\dim G(3) = (17|14)$.

1.1.6. The Cartan series. In this subsection, we describe the Cartan series of finite-dimensional simple Lie superalgebras without proof. This part is included for the sake of presenting a complete classification of finite-dimensional simple Lie superalgebras in Theorem 1.13 and will not be used elsewhere in the book.

Lie superalgebra $W(n)$. Let $\wedge(n)$ be the **exterior algebra** in n indeterminates $\xi_1, \xi_2, \ldots, \xi_n$. We have $\xi_i\xi_j = -\xi_j\xi_i$, for all i, j, and in particular, $\xi_i^2 = 0$, for all i. Setting $|\xi_i| = \bar{1}$, for all i, the algebra $\wedge(n)$ becomes a superalgebra which we also refer to as an **exterior superalgebra**.

By general construction in Example 1.4, we have a finite-dimensional Lie superalgebra of derivations on $\wedge(n)$, which will be denoted by $W(n)$.

For $i = 1, \dots, n$, the derivation $\frac{\partial}{\partial \xi_i} : \wedge(n) \to \wedge(n)$ of degree $\bar{1}$ is uniquely determined by

$$\frac{\partial}{\partial \xi_i}(\xi_j) = \delta_{ij}, \quad j = 1, \dots, n.$$

Given an element $f = (f_1, f_2, \dots, f_n) \in \wedge(n)^n$, with $|f_i| = |f_j|$, for all i, j, the linear map $D_f : \wedge(n) \to \wedge(n)$ of the form

$$D_f = \sum_{i=1}^{n} f_i \frac{\partial}{\partial \xi_i}$$

is a derivation of $\wedge(n)$. Furthermore, all homogeneous derivations of $\wedge(n)$ are of this form, since a derivation is determined by its values at ξ_i for all i. Therefore, sending $f \mapsto D_f$ defines a linear isomorphism from $\wedge(n)^n$ to $W(n)$, and so $W(n)$ has dimension $2^n n$.

Setting $\deg \xi_i = 1$ and $\deg \frac{\partial}{\partial \xi_i} = -1$, for all i, gives rise to a $\mathbb{Z}$-gradation on $W(n)$, called the **principal gradation**. We have

$$W(n) = \bigoplus_{j=-1}^{n-1} W(n)_j.$$

The $\mathbb{Z}$-gradation is compatible with the super structure on $W(n)$; that is, $W(n)_s = \oplus_{j \equiv s \mod 2} W(n)_j$, for $s \in \mathbb{Z}_2$. The 0th degree component $W(n)_0$ is a Lie algebra isomorphic to $\mathfrak{gl}(n)$, and each $W(n)_j$ is a $\mathfrak{gl}(n)$-module isomorphic to $\wedge^{j+1}(\mathbb{C}^n) \otimes \mathbb{C}^{n*}$. In particular, when $n = 2$, we have $W(2)_0 \cong \mathfrak{gl}(2)$, $W(2)_{-1} \cong \mathbb{C}^{2*}$ and $W(2)_1 \cong \mathbb{C}^2$ as $\mathfrak{gl}(2)$-modules. Indeed, we have isomorphisms of Lie superalgebras $W(2) \cong \mathfrak{osp}(2|2) \cong \mathfrak{sl}(2|1)$ (see Exercises 1.4 and 1.5).

The Lie superalgebra $W(n)$ is simple, for $n \geq 2$. Moreover, $W(n)$ contains the following three series of simple Lie superalgebras as subalgebras that we shall describe.

Lie superalgebra $S(n)$**.** The first series is the super-analogue of the Lie algebra of divergence-free vector fields given by

$$S(n) := \Big\{ \sum_{j=1}^{n} f_j \frac{\partial}{\partial \xi_j} \in W(n) \mid \sum_{j=1}^{n} \frac{\partial}{\partial \xi_j}(f_j) = 0 \Big\}.$$

The Lie superalgebra $S(n)$ is a $\mathbb{Z}$-graded subalgebra of $W(n)$ and we have $S(n) = \bigoplus_{j=-1}^{n-2} S(n)_j$. The Lie algebra $S(n)_0$ is isomorphic to $\mathfrak{sl}(n)$, and the jth degree component $S(n)_j$ is isomorphic to the top irreducible summand of the $\mathfrak{sl}(n)$-module $\wedge^{j+1}(\mathbb{C}^n) \otimes \mathbb{C}^{n*}$. The Lie superalgebra $S(n)$ is simple, for $n \geq 3$.

Lie superalgebra $\widetilde{S}(n)$**.** Let n be even so that $\omega = 1 + \xi_1 \xi_2 \cdots \xi_n$ is a $\mathbb{Z}_2$-homogeneous invertible element in $\wedge(n)$. Consider the subspace of $W(n)$ given by

$$\widetilde{S}(n) := \Big\{ \sum_{j=1}^{n} f_j \frac{\partial}{\partial \xi_j} \in W(n) \mid \sum_{j=1}^{n} \frac{\partial}{\partial \xi_j}(\omega f_j) = 0 \Big\}.$$

It can be shown that $\widetilde{S}(n)$ is a subalgebra of the Lie superalgebra $W(n)$. The Lie superalgebra $\widetilde{S}(n)$ is no longer $\mathbb{Z}$-graded, as the defining condition is not homogeneous. However, $\widetilde{S}(n)$ inherits a natural filtration from the filtration on $W(n)$ induced by the principal gradation. The associated graded Lie superalgebra of $\widetilde{S}(n)$ is isomorphic to $S(n)$. Explicitly, $\widetilde{S}(n)$ is the following direct sum of vector spaces inside $W(n)$:

$$\widetilde{S}(n) = \bigoplus_{j=-1}^{n-2} \widetilde{S}(n)_j, \tag{1.17}$$

where $\widetilde{S}(n)_{-1}$ is spanned by $\{(1 - \xi_1\xi_2 \cdots \xi_n)\frac{\partial}{\partial \xi_i} | i = 1, \ldots, n\}$, and $\widetilde{S}(n)_j = S(n)_j$, for $j = 0, \ldots, n-2$. For $n \geq 2$ and n even, $\widetilde{S}(n)$ is simple. Note that $\widetilde{S}(2) \cong \mathfrak{spo}(2|1)$.

Lie superalgebra $H(n)$. As in the classical setting, $W(n)$ contains a subalgebra $H(n)$ as defined below, which is a super-analogue of the Lie algebra of Hamiltonian vector fields. For $f, g \in \wedge(n)$, we define the Poisson bracket by

$$\{f, g\} := (-1)^{|f|} \sum_{j=1}^{n} \frac{\partial f}{\partial \xi_j} \frac{\partial g}{\partial \xi_j}.$$

The Poisson bracket makes $\wedge(n)$ into a Lie superalgebra, which we will denote by $\widetilde{H}(n)$. Now putting $\deg f := k - 2$, for $f \in \wedge(n)_k$, $\widetilde{H}(n)$ becomes a $\mathbb{Z}$-graded Lie superalgebra. The superalgebra $\widetilde{H}(n)$ is not simple, as it has center $\mathbb{C}1$. However, the derived superalgebra of $\widetilde{H}(n)/\mathbb{C}1$, which we denote by $H(n)$, is simple, for $n \geq 4$. Moreover, $H(n) = \bigoplus_{j=-1}^{n-3} H(n)_j$ is a graded Lie superalgebra. The 0th degree component is a Lie algebra isomorphic to $\mathfrak{so}(n)$. As an $\mathfrak{so}(n)$-module we have $H(n)_j \cong \wedge^{j+2}(\mathbb{C}^n)$, for $-1 \leq j \leq n-3$. Finally, $H(n)$ can be viewed as a subalgebra of $W(n)$, since the assignment $f \mapsto (-1)^{|f|} \sum_{j=1}^{n} \frac{\partial f}{\partial \xi_j} \frac{\partial}{\partial \xi_j}$, for $f \in \wedge(n)$, gives rise to an embedding of Lie superalgebras from $\widetilde{H}(n)/\mathbb{C}1$ into $W(n)$.

1.1.7. The classification theorem. The following theorem of Kac [**60**] gives a classification of finite-dimensional complex simple Lie superalgebras. Note the following isomorphisms of Lie superalgebras (see Exercises 1.5, 1.4, and 1.7):

$$\mathfrak{osp}(2|2) \cong \mathfrak{sl}(2|1) \cong W(2), \ \mathfrak{sl}(2|2)/\mathbb{C}I_{2|2} \cong H(4), \ [\mathfrak{p}(3), \mathfrak{p}(3)] \cong S(3).$$

Theorem 1.13. *The following is a complete list of pairwise non-isomorphic finite-dimensional simple Lie superalgebras over $\mathbb{C}$.*

(1) *A finite-dimensional simple Lie algebra in the Killing-Cartan list.*

(2) $\mathfrak{sl}(m|n)$, *for* $m > n \geq 1$ *(excluding* $(m,n) = (2,1)$*);* $\mathfrak{sl}(m|m)/\mathbb{C}I_{m|m}$, *for* $m \geq 3$*;* $\mathfrak{spo}(2m|n)$, *for* $m,n \geq 1$.

(3) $D(2|1,\alpha)$, *for* $\alpha \in \left(\mathbb{C}\setminus\{0,\pm 1,-2,-\frac{1}{2}\}\right)/\mathfrak{S}_3$*;* $F(3|1)$*; and* $G(3)$.

(4) $[\mathfrak{p}(n),\mathfrak{p}(n)]$ *and* $[\mathfrak{q}(n),\mathfrak{q}(n)]/\mathbb{C}I_{n|n}$, *for* $n \geq 3$.

(5) $W(n)$, *for* $n \geq 3$*;* $S(n)$, *for* $n \geq 4$*;* $\widetilde{S}(2n)$, *for* $n \geq 2$*; and* $H(n)$, *for* $n \geq 4$.

A Lie superalgebra $\mathfrak{g} = \mathfrak{g}_{\bar{0}} \oplus \mathfrak{g}_{\bar{1}}$ in Theorem 1.13(1)-(4) has the property that $\mathfrak{g}_{\bar{0}}$ is a reductive Lie algebra and the adjoint $\mathfrak{g}_{\bar{0}}$-module $\mathfrak{g}_{\bar{1}}$ is semisimple. To distinguish between such a Lie superalgebra from one in the Cartan series of Theorem 1.13(5), we introduce the following terminology.

Definition 1.14. A Lie superalgebra $\mathfrak{g} = \mathfrak{g}_{\bar{0}} \oplus \mathfrak{g}_{\bar{1}}$ in Theorem 1.13(2)-(4) is called **classical**. A classical Lie superalgebra in Theorem 1.13(2)-(3) is called **basic**. The Lie superalgebra $\mathfrak{gl}(m|n)$ for $m,n \geq 1$ is also declared to be **basic**.

Remark 1.15. The basic Lie superalgebras admit non-degenerate even supersymmetric bilinear forms, and this property characterizes the simple basic Lie superalgebras among all simple Lie superalgebras in the list of Theorem 1.13.

Remark 1.16. Let us comment on the simplicity of the Lie superalgebras in Theorem 1.13. It is not difficult to check directly the simplicity of the $\mathfrak{sl}$ series, $\mathfrak{spo}(2m|2)$, and those in (4). A Lie superalgebra $\mathfrak{g}$ in the remaining cases in (2) and (3), with the exception of $\mathfrak{spo}(2m|2)$, satisfies the properties that the adjoint $\mathfrak{g}_{\bar{0}}$-module $\mathfrak{g}_{\bar{1}}$ is irreducible and faithful, and $[\mathfrak{g}_{\bar{1}},\mathfrak{g}_{\bar{1}}] = \mathfrak{g}_{\bar{0}}$. A Lie superalgebra $\mathfrak{g}$ with such properties can be easily shown to be simple (see Exercise 1.9).

For the $\mathbb{Z}$-graded Cartan type Lie superalgebras $\mathfrak{g} = \bigoplus_{j\geq -1}\mathfrak{g}_j$ in (5), we first note that they are all *transitive* (i.e., $[\mathfrak{g}_{-1},x] = 0$ implies that $x = 0$, for $x \in \mathfrak{g}_j$ and $j \geq 0$) and *irreducible* (i.e., the $\mathfrak{g}_0$-module $\mathfrak{g}_{-1}$ is irreducible). Furthermore, we have $[\mathfrak{g}_{-1},\mathfrak{g}_1] = \mathfrak{g}_0$, $[\mathfrak{g}_0,\mathfrak{g}_1] = \mathfrak{g}_1$, and $\mathfrak{g}_j$ is *generated* by $\mathfrak{g}_1$ (i.e., $\mathfrak{g}_j = [\mathfrak{g}_{j-1},\mathfrak{g}_1]$), for $j \geq 2$. A Lie superalgebra $\mathfrak{g}$ with these properties is simple (see Exercise 1.10). Finally, the simplicity of $\widetilde{S}(n)$ can be verified with a bit of extra work using the explicit realization given in (1.17).

1.2. Structures of classical Lie superalgebras

In this section, Cartan subalgebras, root systems, Weyl groups, and invariant bilinear forms for basic Lie superalgebras are introduced and described in detail for type $\mathfrak{gl}$, $\mathfrak{osp}$ and $\mathfrak{q}$. A structure theorem is formulated for the basic and type $\mathfrak{q}$ Lie superalgebras.

1.2.1. A basic structure theorem. In this subsection, we shall assume that $\mathfrak{g} = \mathfrak{g}_{\bar{0}} \oplus \mathfrak{g}_{\bar{1}}$ is a basic Lie superalgebra (see Definition 1.14).

A **Cartan subalgebra** $\mathfrak{h}$ of $\mathfrak{g}$ is defined to be a Cartan subalgebra of the even subalgebra $\mathfrak{g}_{\bar{0}}$. Since every inner automorphism of $\mathfrak{g}_{\bar{0}}$ extends to one of Lie superalgebra $\mathfrak{g}$ and Cartan subalgebras of $\mathfrak{g}_{\bar{0}}$ are conjugate under inner automorphisms, we conclude that the Cartan subalgebras of $\mathfrak{g}$ are conjugate under inner automorphisms.

Let $\mathfrak{h}$ be a Cartan subalgebra of $\mathfrak{g}$. For $\alpha \in \mathfrak{h}^*$, let

$$\mathfrak{g}_\alpha = \{g \in \mathfrak{g} \mid [h, g] = \alpha(h)g, \forall h \in \mathfrak{h}\}.$$

The **root system** for $\mathfrak{g}$ is defined to be

$$\Phi = \{\alpha \in \mathfrak{h}^* \mid \mathfrak{g}_\alpha \neq 0, \alpha \neq 0\}.$$

Define the sets of **even** and **odd roots**, respectively, to be

$$\Phi_{\bar{0}} := \{\alpha \in \Phi | \mathfrak{g}_\alpha \cap \mathfrak{g}_{\bar{0}} \neq 0\}, \quad \Phi_{\bar{1}} := \{\alpha \in \Phi | \mathfrak{g}_\alpha \cap \mathfrak{g}_{\bar{1}} \neq 0\}.$$

Definition 1.17. For a basic or queer Lie superalgebra $\mathfrak{g} = \mathfrak{g}_{\bar{0}} \oplus \mathfrak{g}_{\bar{1}}$, the **Weyl group** W of $\mathfrak{g}$ is defined to be the Weyl group of the reductive Lie algebra $\mathfrak{g}_{\bar{0}}$.

As we shall see, the Weyl groups play a less vital though still important role in determining central characters and the linkage principle for Lie superalgebras that is somewhat different from the theory of semisimple Lie algebras.

The following theorem shows that the structures of the basic Lie superalgebras are similar to those of semisimple Lie algebras.

Theorem 1.18. *Let $\mathfrak{g}$ be a basic Lie superalgebra with a Cartan subalgebra $\mathfrak{h}$.*

(1) *We have a root space decomposition of $\mathfrak{g}$ with respect to $\mathfrak{h}$:*

$$\mathfrak{g} = \mathfrak{h} \oplus \bigoplus_{\alpha \in \Phi} \mathfrak{g}_\alpha, \qquad \text{and } \mathfrak{g}_0 = \mathfrak{h}.$$

(2) $\dim \mathfrak{g}_\alpha = 1$, *for $\alpha \in \Phi$. (Now fix some nonzero $e_\alpha \in \mathfrak{g}_\alpha$.)*

(3) $[\mathfrak{g}_\alpha, \mathfrak{g}_\beta] \subseteq \mathfrak{g}_{\alpha+\beta}$, *for $\alpha, \beta, \alpha + \beta \in \Phi$.*

(4) *Φ, $\Phi_{\bar{0}}$ and $\Phi_{\bar{1}}$ are invariant under the action of the Weyl group W on $\mathfrak{h}^*$.*

(5) *There exists a non-degenerate even invariant supersymmetric bilinear form $(\cdot,\cdot)$ on $\mathfrak{g}$.*

(6) $(\mathfrak{g}_\alpha, \mathfrak{g}_\beta) = 0$ *unless $\alpha = -\beta \in \Phi$.*

(7) *The restriction of the bilinear form $(\cdot,\cdot)$ on $\mathfrak{h} \times \mathfrak{h}$ is non-degenerate and W-invariant.*

(8) $[e_\alpha, e_{-\alpha}] = (e_\alpha, e_{-\alpha})h_\alpha$, *where h_α is the **coroot** determined by $(h_\alpha, h) = \alpha(h)$ for $h \in \mathfrak{h}$.*

(9) $\Phi = -\Phi$, $\Phi_{\bar{0}} = -\Phi_{\bar{0}}$, *and* $\Phi_{\bar{1}} = -\Phi_{\bar{1}}$.

(10) *Let $\alpha \in \Phi$. Then, $k\alpha \in \Phi$ for some integer $k \neq \pm 1$ if and only if α is an odd root such that $(\alpha, \alpha) \neq 0$; in this case, we must have $k = \pm 2$.*

Proof. For $\mathfrak{g} = \mathfrak{gl}(m|n)$ or $\mathfrak{osp}(2m|\ell)$, we will describe explicitly the root systems and trace forms below in this section, and so the theorem in these cases follows by inspection. The theorem for the other infinite series basic Lie superalgebras $\mathfrak{sl}(m|n)$ follows by some easy modification of the $\mathfrak{gl}(m|n)$ case.

Part (4) follows by the invariance, under the action of the Weyl group W, of the sets of weights for the adjoint $\mathfrak{g}_{\bar{0}}$-modules $\mathfrak{g}_{\bar{0}}$ and $\mathfrak{g}_{\bar{1}}$, respectively.

Since the remaining three exceptional Lie superalgebras $D(2|1,\alpha)$, $G(3)$ and $F(3|1)$ will not be studied in detail in the book, we will be sketchy. Most parts of the theorem, with the exception of (5), again follow by inspection from the constructions of these superalgebras and standard arguments as for simple Lie algebras (with the help of (5)). As the exceptional Lie superalgebras are simple, any nonzero invariant supersymmetric bilinear form must be non-degenerate. For $G(3)$ and $F(3|1)$, the Killing form is such a nonzero even form. A direct and ad hoc construction of a nonzero invariant bilinear form on $D(2|1,\alpha)$ is possible. □

It follows by Theorem 1.18(1) that $\mathfrak{h}$ is self-normalizing in $\mathfrak{g}$ (and $\mathfrak{h}$ is abelian), justifying the terminology of a Cartan subalgebra. Since $\mathfrak{h} \subseteq \mathfrak{g}_{\bar{0}}$ and $\dim \mathfrak{g}_\alpha = 1$ for each $\alpha \in \Phi$ by Theorem 1.18(2), there exists $i \in \mathbb{Z}_2$ such that $\mathfrak{g}_\alpha \subseteq \mathfrak{g}_i$. Hence Φ is a disjoint union of $\Phi_{\bar{0}}$ and $\Phi_{\bar{1}}$, and we have

$$\Phi_i = \{\alpha \in \Phi \mid \mathfrak{g}_\alpha \subseteq \mathfrak{g}_i\}, \quad i \in \mathbb{Z}_2.$$

Remark 1.19. One uniform approach to establish Theorem 1.18 is as follows (see [**60**, Proposition 2.5.3]). One follows the standard construction of contragredient (i.e., Kac-Moody) (super)algebras (see [**18**, Chapter 14]) to show that any Lie superalgebra in Theorem 1.18 is a Kac-Moody superalgebra (or rather the quotient by its possibly nontrivial center) associated to some generalized Cartan matrix with $\mathbb{Z}_2$-grading. These (quotients of) Kac-Moody superalgebras carry non-degenerate even invariant supersymmetric bilinear forms by a standard Kac-Moody type argument (see [**18**, Chapter 16]).

A root $\alpha \in \Phi$ is called **isotropic** if $(\alpha, \alpha) = 0$. An isotropic root is necessarily an odd root. Denote the set of isotropic odd roots by

$$\begin{aligned} \bar{\Phi}_{\bar{1}} &:= \{\alpha \in \Phi_{\bar{1}} \mid (\alpha,\alpha) = 0\} \\ &= \{\alpha \in \Phi_{\bar{1}} \mid 2\alpha \notin \Phi\}. \end{aligned} \tag{1.18}$$

The second equation above follows by Theorem 1.18(10). Moreover, we have

$$e_\alpha^2 = \frac{1}{2}[e_\alpha, e_\alpha] = 0, \qquad \text{for } \alpha \in \bar{\Phi}_{\bar{1}}.$$

We also introduce the following set of roots

$$\bar{\Phi}_{\bar{0}} = \{\alpha \in \Phi_{\bar{0}} \mid \alpha/2 \notin \Phi\}. \tag{1.19}$$

1.2.2. Invariant bilinear forms for $\mathfrak{gl}$ and $\mathfrak{osp}$. In contrast to the semisimple Lie algebras, the Killing form for a basic Lie superalgebra may be zero, and even when it is nonzero, it may not be positive definite on the real vector space spanned by Φ. In this subsection, we give a down-to-earth description of an invariant non-degenerate even supersymmetric bilinear form for Lie superalgebras of type $\mathfrak{gl}$ and $\mathfrak{osp}$.

The supertrace str on the general linear Lie superalgebra gives rise to a non-degenerate supersymmetric bilinear form

$$(\cdot,\cdot):\mathfrak{gl}(m|n)\times\mathfrak{gl}(m|n)\to\mathbb{C},\quad (a,b)=\mathrm{str}(ab),$$

where ab denotes the matrix multiplication. It is straightforward to check that this form is invariant. Restricting to the Cartan subalgebra $\mathfrak{h}$ of diagonal matrices, we obtain a non-degenerate symmetric bilinear form on $\mathfrak{h}$:

$$(E_{ii},E_{jj})=\begin{cases}1 & \text{if } \bar{1}\le i=j\le \bar{m},\\ -1 & \text{if } 1\le i=j\le n,\\ 0 & \text{if } i\ne j,\end{cases}$$

where $i,j\in I(m|n)$. We recall here that $I(m|n)$ is defined in (1.2). Denote by $\{\delta_i,\varepsilon_j\}_{i,j}$ the basis of $\mathfrak{h}^*$ dual to $\{E_{\bar{i}\bar{i}},E_{jj}\}_{i,j}$, where $1\le i\le m$ and $1\le j\le n$. Using the bilinear form $(\cdot,\cdot)$ we can identify δ_i with $(E_{\bar{i},\bar{i}},\cdot)$ and ε_j with $-(E_{jj},\cdot)$. When it is convenient we also use the notation

$$\varepsilon_{\bar{i}}:=\delta_i,\quad \text{for } 1\le i\le m. \tag{1.20}$$

The form $(\cdot,\cdot)$ on $\mathfrak{h}$ induces a non-degenerate bilinear form on $\mathfrak{h}^*$, which will be denoted by $(\cdot,\cdot)$ as well. Then, for $i,j\in I(m|n)$, we have

$$(\varepsilon_i,\varepsilon_j)=\begin{cases}1 & \text{if } \bar{1}\le i=j\le \bar{m},\\ -1 & \text{if } 1\le i=j\le n,\\ 0 & \text{if } i\ne j.\end{cases} \tag{1.21}$$

Such a bilinear form on $\mathfrak{gl}(2n|\ell)$ restricts to a non-degenerate invariant supersymmetric bilinear form on the subalgebra $\mathfrak{spo}(2n|\ell)$, which will also be denoted by $(\cdot,\cdot)$. The further restriction to a Cartan subalgebra of $\mathfrak{spo}(2n|\ell)$ remains non-degenerate. This allows us to identify a Cartan subalgebra $\mathfrak{h}$ with its dual $\mathfrak{h}^*$, and one also obtains a bilinear form on $\mathfrak{h}^*$.

1.2.3. Root system and Weyl group for $\mathfrak{gl}(m|n)$. Let $\mathfrak{g}=\mathfrak{gl}(m|n)$ and $\mathfrak{h}$ be the Cartan subalgebra of diagonal matrices. Its root system $\Phi=\Phi_{\bar{0}}\cup\Phi_{\bar{1}}$ is given by

$$\Phi_{\bar{0}}=\{\varepsilon_i-\varepsilon_j\mid i\ne j\in I(m|n),\ i,j>0 \text{ or } i,j<0\},$$
$$\Phi_{\bar{1}}=\{\pm(\varepsilon_i-\varepsilon_j)\mid i,j\in I(m|n),\ i<0<j\}.$$

Observe that E_{ij} is a root vector corresponding to the root $\varepsilon_i-\varepsilon_j$, for $i\ne j\in I(m|n)$.

The Weyl group of $\mathfrak{gl}(m|n)$, which is by definition the Weyl group of the even subalgebra $\mathfrak{g}_{\bar{0}} = \mathfrak{gl}(m) \oplus \mathfrak{gl}(n)$, is isomorphic to $\mathfrak{S}_m \times \mathfrak{S}_n$, where we recall that $\mathfrak{S}_n$ denotes the symmetric group of n letters.

1.2.4. Root system and Weyl group for $\mathfrak{spo}(2m|2n+1)$. Now we describe the root system for the ortho-symplectic Lie superalgebra $\mathfrak{spo}(2m|2n+1)$, which is defined in matrix form (1.8) in Section 1.1.3. Recall that the rows and columns of the matrices are indexed by $I(2m|2n+1)$. The subalgebra $\mathfrak{h}$ of $\mathfrak{g}$ of diagonal matrices has a basis given by

$$H_{\bar{i}} := E_{\bar{i},\bar{i}} - E_{\overline{m+i},\overline{m+i}}, \quad 1 \le i \le m,$$
$$H_j := E_{jj} - E_{n+j,n+j}, \quad 1 \le j \le n,$$

and it is the **standard Cartan subalgebra** for $\mathfrak{spo}(2m|2n+1)$. Let $\{\delta_i, \varepsilon_j \mid 1 \le i \le m, 1 \le j \le n\}$ be the corresponding dual basis in $\mathfrak{h}^*$. With respect to $\mathfrak{h}$, the root system $\Phi = \Phi_{\bar{0}} \cup \Phi_{\bar{1}}$ for $\mathfrak{spo}(2m|2n+1)$ is

$$\{\pm\delta_i \pm \delta_j, \pm 2\delta_p, \pm\varepsilon_k \pm \varepsilon_l, \pm\varepsilon_q\} \cup \{\pm\delta_p \pm \varepsilon_q, \pm\delta_p\},$$

where $1 \le i < j \le m$, $1 \le k < l \le n$, $1 \le p \le m$, $1 \le q \le n$.

A root vector is a nonzero vector in $\mathfrak{g}_\alpha$ for $\alpha \in \Phi$ and will be denoted by e_α. The root vectors for $\mathfrak{spo}(2m|2n+1)$ can be chosen explicitly as follows in (1.22)-(1.29) ($1 \le i \ne j \le m, 1 \le k \ne l \le n$):

$$e_{\varepsilon_k} = E_{2n+1,k+n} - E_{k,2n+1}, \qquad e_{-\varepsilon_k} = E_{2n+1,k} - E_{k+n,2n+1}, \tag{1.22}$$
$$e_{2\delta_i} = E_{\bar{i},\overline{i+m}}, \qquad e_{-2\delta_i} = E_{\overline{i+m},\bar{i}}, \tag{1.23}$$
$$e_{\delta_i+\delta_j} = E_{\bar{i},\overline{j+m}} + E_{\bar{j},\overline{i+m}}, \qquad e_{-\delta_i-\delta_j} = E_{\overline{j+m},\bar{i}} + E_{\overline{i+m},\bar{j}}, \tag{1.24}$$
$$e_{\delta_i-\delta_j} = E_{\bar{i},\bar{j}} - E_{\overline{j+m},\overline{i+m}}, \qquad e_{\varepsilon_k-\varepsilon_l} = E_{kl} - E_{l+n,k+n}, \tag{1.25}$$
$$e_{\varepsilon_k+\varepsilon_l} = E_{k,l+n} - E_{l,k+n}, \qquad e_{-\varepsilon_k-\varepsilon_l} = E_{k+n,l} - E_{l+n,k}, \tag{1.26}$$

$$e_{\delta_i+\varepsilon_k} = E_{k,\overline{i+m}} + E_{\bar{i},k+n}, \qquad e_{-\delta_i-\varepsilon_k} = E_{k+n,\bar{i}} - E_{\overline{i+m},k}, \tag{1.27}$$
$$e_{\delta_i-\varepsilon_k} = E_{k+n,\overline{i+m}} + E_{\bar{i},k}, \qquad e_{-\delta_i+\varepsilon_k} = E_{k,\bar{i}} - E_{\overline{i+m},k+n}, \tag{1.28}$$
$$e_{\delta_i} = E_{2n+1,\overline{i+m}} + E_{\bar{i},2n+1}, \qquad e_{-\delta_i} = E_{2n+1,\bar{i}} - E_{\overline{i+m},2n+1}. \tag{1.29}$$

The Weyl group of $\mathfrak{spo}(2m|2n+1)$, which is by definition the Weyl group of $\mathfrak{g}_{\bar{0}} = \mathfrak{sp}(2m) \oplus \mathfrak{so}(2n+1)$, is isomorphic to $(\mathbb{Z}_2^m \rtimes \mathfrak{S}_m) \times (\mathbb{Z}_2^n \rtimes \mathfrak{S}_n)$.

1.2.5. Root system and Weyl group for $\mathfrak{spo}(2m|2n)$. Let $\mathfrak{g} = \mathfrak{spo}(2m|2n)$. The abelian Lie subalgebra $\mathfrak{h}$ spanned by $\{H_{\bar{i}}, H_j | 1 \le i \le m, 1 \le j \le n\}$ is a Cartan subalgebra for $\mathfrak{spo}(2m|2n)$. Again, when it is convenient, we will use the notation $\varepsilon_{\bar{i}} = \delta_i$, for $1 \le i \le m$. With respect to $\mathfrak{h}$, the root system $\Phi = \Phi_{\bar{0}} \cup \Phi_{\bar{1}}$ is given by

$$\{\pm\delta_i \pm \delta_j, \pm 2\delta_p, \pm\varepsilon_k \pm \varepsilon_l\} \cup \{\pm\delta_p \pm \varepsilon_q\},$$

where $1 \le i < j \le m$, $1 \le k < l \le n$, $1 \le p \le m$, $1 \le q \le n$.

The root vectors for $\mathfrak{spo}(2m|2n)$ are given by (1.23)–(1.28).

The Weyl group of $\mathfrak{spo}(2m|2n)$, which is by definition the Weyl group of the even subalgebra $\mathfrak{g}_{\bar{0}} = \mathfrak{sp}(2m) \oplus \mathfrak{so}(2n)$, is an index 2 subgroup of $(\mathbb{Z}_2^m \rtimes \mathfrak{S}_m) \times (\mathbb{Z}_2^n \rtimes \mathfrak{S}_n)$, with only an even number of signs in $\mathbb{Z}_2^n$ permitted.

1.2.6. Root system and odd invariant form for $\mathfrak{q}(n)$. In this subsection let $\mathfrak{g}$ be the queer Lie superalgebra $\mathfrak{q}(n)$ (see Section 1.1.4). Since the case of $\mathfrak{q}(n)$ is different from the basic Lie superalgebra case, we will describe altogether its Cartan subalgebras, root systems, positive systems, Borel subalgebras, and invariant bilinear form.

Recall that $\mathfrak{q}(n)$ can be realized as matrices in the $n|n$ block form $\begin{pmatrix} a & b \\ b & a \end{pmatrix}$ indexed by $I(n|n)$. The subalgebra consisting of matrices with a, b being diagonal (which we refer to as "block diagonal matrices") will be called the **standard Cartan subalgebra**. We define a Cartan subalgebra $\mathfrak{h}$ to be any subalgebra conjugate to the standard Cartan subalgebra by the adjoint action of some element in the group $GL(n)$ associated to $\mathfrak{q}(n)_{\bar{0}}$. Note that $\mathfrak{h}$ is self-normalizing in $\mathfrak{g}$ and $\mathfrak{h}$ is nilpotent, justifying the terminology of a Cartain subalgebra. However, $\mathfrak{h} = \mathfrak{h}_{\bar{0}} \oplus \mathfrak{h}_{\bar{1}}$ is not abelian, since $[\mathfrak{h}_{\bar{0}}, \mathfrak{h}] = 0$ and $[\mathfrak{h}_{\bar{1}}, \mathfrak{h}_{\bar{1}}] = \mathfrak{h}_{\bar{0}}$.

Now fix $\mathfrak{h} = \mathfrak{h}_{\bar{0}} \oplus \mathfrak{h}_{\bar{1}}$ to be the standard Cartan subalgebra. The vectors

$$H_i := E_{\bar{i}\bar{i}} + E_{ii}, \quad i = 1, \ldots, n \tag{1.30}$$

form a basis for $\mathfrak{h}_{\bar{0}}$, while the vectors

$$\overline{H}_i := E_{\bar{i}i} + E_{i\bar{i}}, \quad i = 1, \ldots, n \tag{1.31}$$

form a basis for $\mathfrak{h}_{\bar{1}}$. We let $\{\varepsilon_i \mid i = 1, \ldots, n\}$ denote the basis in $\mathfrak{h}_{\bar{0}}^*$ dual to $\{H_i \mid i = 1, \ldots, n\}$. With respect to $\mathfrak{h}_{\bar{0}}$, we have the root space decomposition $\mathfrak{g} = \mathfrak{h} \oplus \bigoplus_{\alpha \in \Phi} \mathfrak{g}_\alpha$ with root system $\Phi = \Phi_{\bar{0}} \cup \Phi_{\bar{1}}$, where $\Phi_{\bar{0}}$ and $\Phi_{\bar{1}}$ are understood as *distinct* isomorphic copies of the root system $\{\varepsilon_i - \varepsilon_j \mid 1 \le i \ne j \le n\}$ for $\mathfrak{gl}(n)$. We have $\dim_{\mathbb{C}} \mathfrak{g}_\alpha = 1$, for each $\alpha \in \Phi$, and $\mathfrak{g}_\alpha \subseteq \mathfrak{g}_i$, for $\alpha \in \Phi_i$ and $i \in \mathbb{Z}_2$. The Weyl group of $\mathfrak{q}(n)$ is identified with the symmetric group $\mathfrak{S}_n$.

The matrices $\begin{pmatrix} a & b \\ b & a \end{pmatrix}$ with a, b being upper triangular (which we refer to as "block upper triangular matrices") form a solvable subalgebra $\mathfrak{b}$, which will be called the **standard Borel subalgebra** of $\mathfrak{g}$. The positive system corresponding to the Borel subalgebra $\mathfrak{b}$ is $\Phi^+ = \Phi_{\bar{0}}^+ \cup \Phi_{\bar{1}}^+$, where $\Phi_{\bar{0}}^+ = \Phi_{\bar{1}}^+ = \{\varepsilon_i - \varepsilon_j \mid 1 \le i < j \le n\}$. Let $\Phi^- = -\Phi^+$ and so $\Phi = \Phi^+ \cup \Phi^-$. Let

$$\mathfrak{n}^+ = \bigoplus_{\alpha \in \Phi^+} \mathfrak{g}_\alpha, \quad \mathfrak{n}^- = \bigoplus_{\alpha \in \Phi^-} \mathfrak{g}_\alpha.$$

Then, we have $\mathfrak{b} = \mathfrak{h} \oplus \mathfrak{n}^+$, and we have a triangular decomposition

$$\mathfrak{g} = \mathfrak{n}^- \oplus \mathfrak{h} \oplus \mathfrak{n}^+.$$

To be compatible with the definition of Weyl vector ρ in (1.35) for basic Lie superalgebras later on, it makes sense to set $\rho = 0$ for $\mathfrak{q}(n)$.

Any subalgebra of $\mathfrak{g}$ that is conjugate to $\mathfrak{b}$ by $GL(n)$ will be referred to as a **Borel subalgebra** of $\mathfrak{g}$.

For $g = \begin{pmatrix} a & b \\ b & a \end{pmatrix}$ in $\mathfrak{q}(n)$, the **odd trace** is defined to be

$$\operatorname{otr}(g) := \operatorname{tr}(b). \tag{1.32}$$

Using this, we obtain an odd non-degenerate invariant symmetric bilinear form $(\cdot,\cdot)$ on $\mathfrak{q}(n)$ defined by

$$(g, g') = \operatorname{otr}(gg'), \quad g, g' \in \mathfrak{q}(n). \tag{1.33}$$

Here "odd" is understood in the sense of Definition 1.8.

1.3. Non-conjugate positive systems and odd reflections

In this section, positive systems, fundamental systems, and Dynkin diagrams for basic Lie superalgebras are defined and classified, along with Borel subalgebras. In contrast to semisimple Lie algebras, the fundamental systems for a Lie superalgebra may not be conjugate under the Weyl group action.

1.3.1. Positive systems and fundamental systems. Let Φ be a root system for a basic Lie superalgebra $\mathfrak{g}$ with a given Cartan subalgebra $\mathfrak{h}$, and let E be the real vector space spanned by Φ. We have $E \otimes_{\mathbb{R}} \mathbb{C} = \mathfrak{h}^*$, for $\mathfrak{g} \neq \mathfrak{gl}(m|n)$. For $\mathfrak{g} = \mathfrak{gl}(m|n)$ the space $E \otimes_{\mathbb{R}} \mathbb{C}$ is a subspace of $\mathfrak{h}^*$ of codimension one.

A total ordering $\geq$ on E below is always assumed to be compatible with the real vector space structure; that is, $v \geq w$ and $v' \geq w'$ imply that $v + v' \geq w + w'$, $-w \geq -v$, and $cv \geq cw$ for $c \in \mathbb{R}$ and $c > 0$.

A **positive system** Φ^+ is a subset of Φ consisting precisely of all those roots $\alpha \in \Phi$ satisfying $\alpha > 0$ for some total ordering of E. Given a positive system Φ^+, we define the **fundamental system** $\Pi \subset \Phi^+$ to be the set of $\alpha \in \Phi^+$ which cannot be written as a sum of two roots in Φ^+. We refer to elements in Φ^+ as **positive roots** and elements in Π as **simple roots**. Similarly, we denote by Φ^- the corresponding set of negative roots. Set $\Phi_i^+ = \Phi^+ \cap \Phi_i$ and $\Phi_i^- = \Phi^- \cap \Phi_i$, for $i \in \mathbb{Z}_2$. By Theorem 1.18(9), we have $\Phi^- = -\Phi^+$ and $\Phi_i^- = -\Phi_i^+$, for $i \in \mathbb{Z}_2$. Then, we have

$$\Phi^+ = \Phi_{\bar{0}}^+ \cup \Phi_{\bar{1}}^+.$$

Recall $\bar{\Phi}_{\bar{1}}$ from (1.18). Associated to a positive system Φ^+, we let

$$\bar{\Phi}_{\bar{1}}^+ := \bar{\Phi}_{\bar{1}} \cap \Phi^+. \tag{1.34}$$

Proposition 1.20. *Let $\mathfrak{g}$ be a basic Lie superalgebra with a Cartan subalgebra $\mathfrak{h}$. There is a one-to-one correspondence between the set of positive systems for $(\mathfrak{g},\mathfrak{h})$ and the set of fundamental systems for $(\mathfrak{g},\mathfrak{h})$. The Weyl group of $\mathfrak{g}$ acts naturally on the set of the positive systems (respectively, fundamental systems).*

Proof. It follows by definition that the fundamental system exists and is unique for a given positive system. A positive root, if not simple, can be written as a sum of two positive roots. Continuing this way, any positive root is a $\mathbb{Z}_+$-linear combination of simple roots, and hence a positive system is determined by its fundamental system.

By Theorem 1.18, $\Phi = -\Phi$ and Φ is W-invariant. Then W acts naturally on the set of positive systems, and then on the set of fundamental systems by the above correspondence. □

We define

$$\mathfrak{n}^+ = \bigoplus_{\alpha\in\Phi^+} \mathfrak{g}_\alpha, \qquad \mathfrak{n}^- = \bigoplus_{\alpha\in\Phi^-} \mathfrak{g}_\alpha.$$

Then $\mathfrak{n}^\pm$ are $\mathrm{ad}\,\mathfrak{h}$-stable nilpotent subalgebras of $\mathfrak{g}$ and we obtain a triangular decomposition

$$\mathfrak{g} = \mathfrak{n}^- \oplus \mathfrak{h} \oplus \mathfrak{n}^+.$$

The solvable subalgebra $\mathfrak{b} = \mathfrak{h} \oplus \mathfrak{n}^+$ is called a **Borel subalgebra** of $\mathfrak{g}$ (corresponding to Φ^+). We have $\mathfrak{b} = \mathfrak{b}_{\bar{0}} \oplus \mathfrak{b}_{\bar{1}}$, where $\mathfrak{b}_i = \mathfrak{b} \cap \mathfrak{g}_i$ for $i \in \mathbb{Z}_2$.

Remark 1.21. The rank one subalgebra corresponding to an isotropic simple root is isomorphic to $\mathfrak{sl}(1|1)$, which is solvable. Therefore, if we enlarge a Borel subalgebra by adding the root space corresponding to a negative isotropic simple root, then the resulting subalgebra is still solvable. Thus, a Borel subalgebra is not a maximal solvable subalgebra for Lie superalgebras in general.

Given a positive system $\Phi^+ = \Phi^+_{\bar{0}} \cup \Phi^+_{\bar{1}}$, the **Weyl vector** ρ is defined by

$$\rho = \rho_{\bar{0}} - \rho_{\bar{1}}, \tag{1.35}$$

where

$$\rho_{\bar{0}} = \frac{1}{2}\sum_{\alpha\in\Phi^+_{\bar{0}}} \alpha, \quad \rho_{\bar{1}} = \frac{1}{2}\sum_{\beta\in\Phi^+_{\bar{1}}} \beta.$$

Denote

$$\mathbf{1}_{m|n} = (\delta_1 + \ldots + \delta_m) - (\varepsilon_1 + \ldots + \varepsilon_n). \tag{1.36}$$

The following lemma is proved by a direct computation.

Lemma 1.22. *We have the following formulas for the Weyl vector ρ for the standard positive system Φ^+:*

(1) $\rho = \sum_{i=1}^m (m-i+1)\delta_i - \sum_{j=1}^n j\varepsilon_j - \frac{1}{2}(m+n+1)\mathbf{1}_{m|n}$, *for* $\mathfrak{g} = \mathfrak{gl}(m|n)$.

(2) $\rho = \sum_{i=1}^{m}(m-n-i+\frac{1}{2})\delta_i + \sum_{j=1}^{n}(n-j+\frac{1}{2})\varepsilon_j$, for $\mathfrak{g} = \mathfrak{spo}(2m|2n+1)$.

(3) $\rho = \sum_{i=1}^{m}(m-n-i+1)\delta_i + \sum_{j=1}^{n}(n-j)\varepsilon_j$, for $\mathfrak{g} = \mathfrak{spo}(2m|2n)$.

We shall next describe completely the positive and fundamental systems for Lie superalgebras of type $\mathfrak{gl}$ and $\mathfrak{osp}$ case-by-case.

1.3.2. Positive and fundamental systems for $\mathfrak{gl}(m|n)$. Recall the root system Φ for $\mathfrak{gl}(m|n)$ and the standard Cartan subalgebra $\mathfrak{h}$ described in Section 1.2.3. The subalgebra of upper triangular matrices is the **standard Borel subalgebra** of $\mathfrak{g}$ that contains $\mathfrak{h}$, and the corresponding **standard positive system** of Φ is given by $\{\varepsilon_i - \varepsilon_j \mid i,j \in I(m|n), i < j\}$. Bearing in mind $\varepsilon_{\bar{i}} = \delta_i$, the standard fundamental system for $\mathfrak{gl}(m|n)$ is

$$\{\delta_i - \delta_{i+1}, \varepsilon_j - \varepsilon_{j+1}, \delta_m - \varepsilon_1 \mid 1 \leq i \leq m-1, 1 \leq j \leq n-1\},$$

with the corresponding simple root vectors $e_i := E_{i,i+1}$, for $i \in I(m-1|n-1)$, and $e_{\bar{m}} := E_{\bar{m},1}$. The simple coroots are $h_j := E_{jj} - E_{j+1,j+1}$, for $j \in I(m-1|n-1)$, and $h_{\bar{m}} := E_{\bar{m},\bar{m}} + E_{11}$. Denote $f_i := E_{i+1,i}$, for $i \in I(m-1|n-1)$, and $f_{\bar{m}} := E_{1,\bar{m}}$. (Here we have slightly abused notation to let $i+1$ mean $\bar{\iota+1}$, for $i = \bar{\iota}$ with $1 \leq \iota \leq m-1$.) Then $\{e_i, h_i, f_i \mid i \in I(m|n-1)\}$ is a set of **Chevalley generators** for $\mathfrak{sl}(m|n)$.

Note that

$$(\delta_i - \delta_{i+1}, \delta_i - \delta_{i+1}) = 2, \quad 1 \leq i \leq m-1,$$
$$(\delta_m - \varepsilon_1, \delta_m - \varepsilon_1) = 0,$$
$$(\varepsilon_j - \varepsilon_{j+1}, \varepsilon_j - \varepsilon_{j+1}) = -2, \quad 1 \leq j \leq n-1.$$

Thus $\delta_m - \varepsilon_1$ is an isotropic simple root. Following the usual convention for Lie algebras, we draw the corresponding **standard Dynkin diagram** with its fundamental system attached:

(1.37) $\underset{\delta_1-\delta_2}{\bigcirc}\!-\!\underset{\delta_2-\delta_3}{\bigcirc}\!-\cdots-\!\underset{\delta_m-\varepsilon_1}{\otimes}\!-\!\underset{\varepsilon_1-\varepsilon_2}{\bigcirc}\!-\cdots-\!\underset{\varepsilon_{n-2}-\varepsilon_{n-1}}{\bigcirc}\!-\!\underset{\varepsilon_{n-1}-\varepsilon_n}{\bigcirc}$

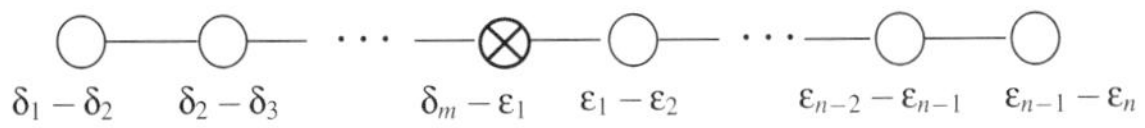

Here, as usual, we denote by $\bigcirc$ an even simple root α such that $\frac{1}{2}\alpha$ is not a root. Following Kac's notation, $\otimes$ denotes an odd isotropic simple root.

Let us classify all possible positive systems for $\mathfrak{gl}(m|n)$, keeping in mind $\varepsilon_{\bar{i}} = \delta_i$, for $1 \leq i \leq m$. If we ignore the parity of roots for the moment, the root system of $\mathfrak{gl}(m|n)$ is the same as the root system for $\mathfrak{gl}(m+n)$. Hence, by definition, their positive systems (respectively, fundamental systems) are exactly described in the same way, and so there are $(m+n)!$ of them in total. It follows from the well-known classification for $\mathfrak{gl}(m+n)$ that a fundamental system for $\mathfrak{gl}(m|n)$ consists of $(m+n-1)$ roots $\varepsilon_{i_1} - \varepsilon_{i_2}, \varepsilon_{i_2} - \varepsilon_{i_3}, \ldots, \varepsilon_{i_{m+n-1}} - \varepsilon_{i_{m+n}}$, where $\{i_1, i_2, \ldots, i_{m+n}\} = I(m|n)$. Then we restore the parity of the simple roots in a fundamental system for $\mathfrak{gl}(m|n)$. The corresponding Dynkin diagram is of the form

(1.38)
$$\bigcirc\!\!-\!\!\bigcirc\!\!-\cdots-\!\!\bigcirc\!\!-\!\!\bigcirc\!-\cdots-\!\!\bigcirc\!\!-\!\!\bigcirc$$
$$\varepsilon_{i_1}-\varepsilon_{i_2}\quad \varepsilon_{i_2}-\varepsilon_{i_3}\qquad \varepsilon_{i_k}-\varepsilon_{i_{k+1}}\qquad\qquad \varepsilon_{i_{m+n-1}}-\varepsilon_{i_{m+n}}$$

where $\odot$ is either $\bigcirc$ or $\otimes$, depending on whether the corresponding simple root is even or odd.

Example 1.23. For $n = m$, there exists a fundamental system consisting of only odd roots $\{\delta_1 - \varepsilon_1, \varepsilon_1 - \delta_2, \delta_2 - \varepsilon_2, \ldots, \delta_m - \varepsilon_m\}$, whose corresponding Dynkin diagram is

$$\otimes\!\!-\!\!\otimes\!\!-\cdots-\!\!\otimes\!\!-\!\!\otimes\!-\cdots-\!\!\otimes\!\!-\!\!\otimes$$
$$\delta_1-\varepsilon_1\quad \varepsilon_1-\delta_2\qquad \delta_k-\varepsilon_k\quad \varepsilon_k-\delta_{k+1}\qquad \varepsilon_{m-1}-\delta_m\quad \delta_m-\varepsilon_m$$

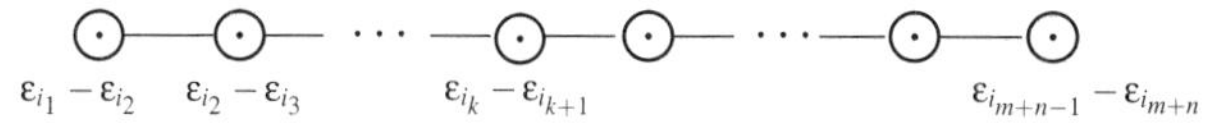

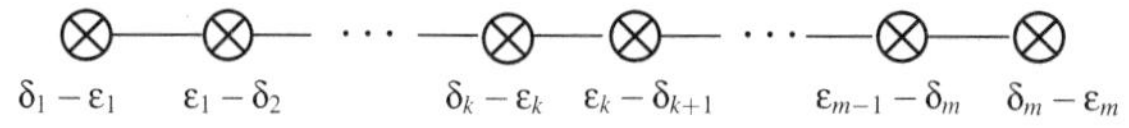

The **$\varepsilon\delta$-sequence** for a fundamental system Π as in (1.38) is obtained by switching the ordered sequence $\varepsilon_{i_1}\varepsilon_{i_2}\ldots\varepsilon_{i_{m+n}}$ for Π to the $\varepsilon\delta$-notation via the identification $\varepsilon_{\bar{i}} = \delta_i$ and then dropping the indices. Clearly, an $\varepsilon\delta$-sequence has m δ's and n ε's. In general, there exist positive systems for Φ that are not conjugate to each other under the action of the Weyl group, in contrast to the semisimple Lie algebra case.

Example 1.24. (1) The standard Borel subalgebra of $\mathfrak{gl}(m|n)$ corresponds to the sequence $\underbrace{\delta\cdots\delta}_{m}\underbrace{\varepsilon\cdots\varepsilon}_{n}$ while the Borel opposite to the standard one corresponds to $\underbrace{\varepsilon\cdots\varepsilon}_{n}\underbrace{\delta\cdots\delta}_{m}$.

(2) The three W-conjugacy classes of fundamental systems for $\mathfrak{gl}(1|2)$ correspond to the three sequences $\delta\varepsilon\varepsilon$, $\varepsilon\delta\varepsilon$, $\varepsilon\varepsilon\delta$, respectively.

1.3.3. Positive and fundamental systems for $\mathfrak{spo}(2m|2n+1)$. Now we describe the positive/fundamental systems and Dynkin diagrams for $\mathfrak{spo}(2m|2n+1)$ whose root system is described in Section 1.2.4. The standard positive system $\Phi^+ = \Phi_{\bar{0}}^+ \cup \Phi_{\bar{1}}^+$ corresponding to the **standard Borel subalgebra** for $\mathfrak{spo}(2m|2n+1)$ is

$$\{\delta_i \pm \delta_j, 2\delta_p, \varepsilon_k \pm \varepsilon_l, \varepsilon_q\} \cup \{\delta_p \pm \varepsilon_q, \delta_p\},$$

where $1 \leq i < j \leq m$, $1 \leq k < l \leq n$, $1 \leq p \leq m$, $1 \leq q \leq n$. The fundamental system Π of Φ^+ contains one odd simple root $\delta_m - \varepsilon_1$, and it is given by

$$\Pi = \{\delta_i - \delta_{i+1}, \delta_m - \varepsilon_1, \varepsilon_k - \varepsilon_{k+1}, \varepsilon_n \mid 1 \leq i \leq m-1, 1 \leq k \leq n-1\}.$$

The corresponding **standard Dynkin diagram** for $\mathfrak{spo}(2m|2n+1)$ is

(1.39)
$$\bigcirc\!\!-\!\!\bigcirc\!\!-\cdots-\!\!\otimes\!\!-\!\!\bigcirc\!-\cdots-\!\!\bigcirc\!\!\Rightarrow\!\!\bigcirc$$
$$\delta_1-\delta_2\quad \delta_2-\delta_3\qquad \delta_m-\varepsilon_1\quad \varepsilon_1-\varepsilon_2\qquad \varepsilon_{n-1}-\varepsilon_n\quad \varepsilon_n$$

Another often used positive system in Φ for $\mathfrak{spo}(2m|2n+1)$ is given by

$$\{\delta_i \pm \delta_j, 2\delta_p, \varepsilon_k \pm \varepsilon_l, \varepsilon_q\} \cup \{\varepsilon_q \pm \delta_p, \delta_p\},$$

where $1 \leq i < j \leq m$, $1 \leq k < l \leq n$, $1 \leq p \leq m$, $1 \leq q \leq n$, with the following fundamental system

$$\{\varepsilon_k - \varepsilon_{k+1}, \varepsilon_n - \delta_1, \delta_i - \delta_{i+1}, \delta_m | 1 \leq i \leq m-1, 1 \leq k \leq n-1\}.$$

The corresponding Dynkin diagram is

(1.40) 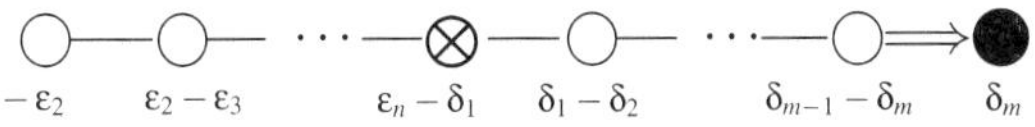

where we follow Kac's convention and use ● to denote a non-isotropic odd simple root, as $(\delta_m, \delta_m) = 1$.

Now let us classify all the possible fundamental systems for $\mathfrak{spo}(2m|2n+1)$ with given Cartan subalgebra $\mathfrak{h}$, keeping in mind $\varepsilon_{\bar{i}} = \delta_i$. Note that $2\varepsilon_{\bar{p}} \in \Phi^+$ if and only if $\varepsilon_{\bar{p}} \in \Phi^+$, and that $\pm 2\varepsilon_{\bar{p}}$ are never in any fundamental system by definition. Hence, for the sake of classification of positive systems and fundamental systems in Φ, it suffices to consider the subset $\tilde{\Phi} := \Phi \setminus \{\pm 2\varepsilon_{\bar{p}} \mid 1 \leq p \leq m\}$. Ignoring the parity of the roots, $\tilde{\Phi}$ may be identified with the root system of the classical Lie algebra $\mathfrak{so}(2m+2n+1)$, whose fundamental systems are completely known and acted upon simply transitively by the Weyl group of $\mathfrak{so}(2m+2n+1)$ (which is $\cong \mathbb{Z}_2^{m+n} \rtimes \mathfrak{S}_{m+n}$). The number of W-conjugacy classes of fundamental systems for $\mathfrak{spo}(2m|2n+1)$ is $|W(\mathfrak{so}(2m+2n+1))|/|W| = \binom{m+n}{m}$.

The **$\varepsilon\delta$-sequence** associated to a fundamental system (or a Dynkin diagram) for $\mathfrak{spo}(2m|2n+1)$ is defined as for $\mathfrak{gl}(m|n)$, starting from the type A end of the Dynkin diagram (to fix the ambiguity). For example, the $\varepsilon\delta$-sequence associated to the standard Dynkin diagram above is m δ's followed by n ε's.

Example 1.25. Let $\mathfrak{g} = \mathfrak{osp}(1|2)$. Then its even subalgebra $\mathfrak{g}_{\bar{0}}$ is isomorphic to $\mathfrak{sl}(2) = \mathbb{C}\langle e, h, f\rangle$, and as a $\mathfrak{g}_{\bar{0}}$-module $\mathfrak{g}_{\bar{1}}$ is isomorphic to the 2-dimensional natural $\mathfrak{sl}(2)$-module $\mathbb{C}E + \mathbb{C}F$. The simple root consists of a (unique) odd non-isotropic root δ_1 so that $2\delta_1$ is an even root. The Dynkin diagram is ●. The root vectors E and F associated to the odd roots δ_1 and $-\delta_1$ can be chosen such that $[E,E] = 2e, [F,F] = -2f, [E,F] = h$.

1.3.4. Positive and fundamental systems for $\mathfrak{spo}(2m|2n)$.

Now we consider $\mathfrak{g} = \mathfrak{spo}(2m|2n)$, whose root system is described in Section 1.2.5. The **standard positive system** $\Phi^+ = \Phi^+_{\bar{0}} \cup \Phi^+_{\bar{1}}$ in Φ corresponding to the standard Borel subalgebra is

$$\{\delta_i \pm \delta_j, 2\delta_p, \varepsilon_k \pm \varepsilon_l\} \cup \{\delta_p \pm \varepsilon_q\},$$

where $1 \le i < j \le m$, $1 \le k < l \le n$, $1 \le p \le m$, $1 \le q \le n$, with its fundamental system being

$$\Pi = \{\delta_i - \delta_{i+1}, \delta_m - \varepsilon_1, \varepsilon_k - \varepsilon_{k+1}, \varepsilon_{n-1} + \varepsilon_n \mid 1 \le i \le m-1,\ 1 \le k \le n-1\}.$$

The corresponding **standard Dynkin diagram** is

(1.41) 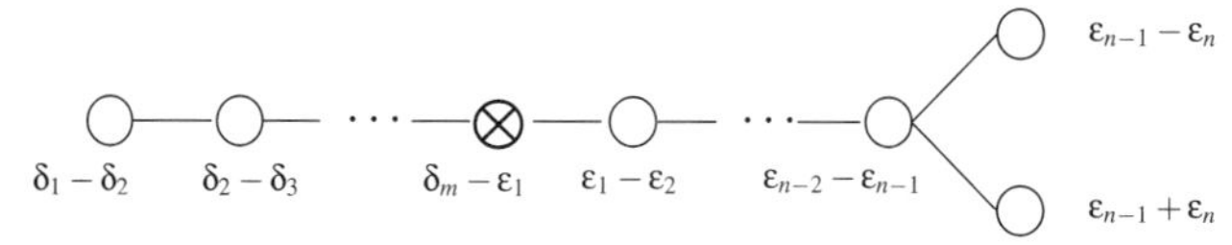

Another often used positive system in Φ is given by

$$\{\delta_i \pm \delta_j, 2\delta_p, \varepsilon_k \pm \varepsilon_l\} \cup \{\varepsilon_q \pm \delta_p\},$$

where $1 \le i < j \le m$, $1 \le k < l \le n$, $1 \le p \le m$, $1 \le q \le n$, with its fundamental system being

$$\{\varepsilon_k - \varepsilon_{k+1}, \varepsilon_n - \delta_1, \delta_i - \delta_{i+1}, 2\delta_m \mid 1 \le i \le m-1,\ 1 \le k \le n-1\}.$$

The corresponding Dynkin diagram is

(1.42) 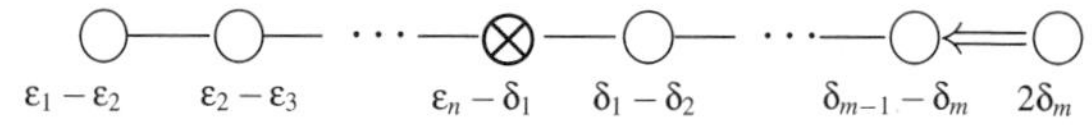

As we have already observed above, there are (at least) two Dynkin diagrams for $\mathfrak{spo}(2m|2n)$ of different shapes. The classification of all fundamental systems for the root system Φ of $\mathfrak{spo}(2m|2n)$ is divided into 2 cases below. As usual, we keep in mind $\varepsilon_{\bar{i}} = \delta_i$.

(1) First, we classify the fundamental systems Π in Φ that do not contain any long root (i.e., a root of the form $\pm 2\varepsilon_{\bar{p}}$). With parity ignored, the subset $\tilde{\Phi} := \Phi \setminus \{\pm 2\varepsilon_{\bar{p}} \mid 1 \le p \le m\}$ may be identified with the root system of the classical Lie algebra $\mathfrak{so}(2m+2n)$, whose fundamental systems are completely known, and they are acted upon simply transitively by the Weyl group of $\mathfrak{so}(2m+2n)$ (which is an index 2 subgroup of $\mathbb{Z}_2^{m+n} \rtimes \mathfrak{S}_{m+n}$). Observe that the positive system Φ^+ for Φ corresponding to Π is completely determined by the positive system $\tilde{\Phi} \cap \Phi^+$ for $\tilde{\Phi}$ (and vice versa). We conclude that the fundamental systems for Φ that do not contain any long root are exactly the fundamental systems for $\tilde{\Phi}$. However, not every fundamental system of $\tilde{\Phi}$ gives rise to a fundamental system of Φ. To be precise, a fundamental system of Φ cannot contain a pair of roots of the form $\{\pm\varepsilon_i \pm \varepsilon_{\bar{p}}, \pm\varepsilon_i \mp \varepsilon_{\bar{p}}\}$, $i \neq \bar{p}$ and $1 \le p \le m$. It follows that there are $\frac{n}{m+n} \cdot |W(\mathfrak{so}(2m+2n))|$ such fundamental systems of Φ, and hence the number of W-conjugacy classes for $\mathfrak{g}$ in this case is $\frac{n}{m+n} \cdot |W(\mathfrak{so}(2m+2n))|/|W| = \binom{m+n-1}{m}$.

(2) Next, we classify the fundamental systems Π that contain some long root. In this case, we consider $\widehat{\Phi} := \Phi \cup \{\pm 2\varepsilon_j, 1 \le j \le n\}$. With the parity ignored, $\widehat{\Phi}$ may be identified with the root system of $\mathfrak{sp}(2m+2n)$, whose fundamental systems

are completely described. Observe that the positive system Φ^+ for Φ that corresponds to Π is completely determined by the positive system $\widehat{\Phi}^+$ of $\widehat{\Phi}$ that contains Φ^+. We conclude that the fundamental systems for Φ that contain some long root are exactly the fundamental systems for $\widehat{\Phi}$ whose long root is of the form $\pm 2\varepsilon_{\overline{p}}$. It follows that the number of W-conjugacy classes of such fundamental systems for $\mathfrak{g}$ is $\frac{m}{m+n} \cdot |W(\mathfrak{sp}(2m+2n))|/|W| = 2\binom{m+n-1}{n}$.

The **$\varepsilon\delta$-sequence for** $\mathfrak{spo}(2m|2n)$ associated to a positive system is now defined as follows. We can first obtain an ordered sequence of ε's and δ's just as for $\mathfrak{spo}(2m|2n+1)$. If this sequence has an ε as its last member, then it is the $\varepsilon\delta$-sequence. If the last member in this sequence is a δ, then the $\varepsilon\delta$-sequence is obtained from this sequence by attaching a sign to the last ε.

Example 1.26. For $\mathfrak{spo}(4|4)$, the $\varepsilon\delta$-sequences $\varepsilon\delta\delta\varepsilon$, $\varepsilon\delta\varepsilon\delta$, and $\varepsilon\delta(-\varepsilon)\delta$ are distinct.

1.3.5. Conjugacy classes of fundamental systems. We now describe the classification of the W-conjugacy classes of fundamental systems for these Lie superalgebras via the $\varepsilon\delta$-sequences.

Proposition 1.27. *Let Φ be the root system and let W be the Weyl group for a Lie superalgebra $\mathfrak{g}$ of type $\mathfrak{gl}$ or $\mathfrak{osp}$. Then the W-conjugacy classes of fundamental systems in Φ are in one-to-one correspondence with the associated $\varepsilon\delta$-sequences. In particular, there are $\binom{m+n}{m}$ W-conjugacy classes of fundamental systems for $\mathfrak{gl}(m|n)$ and $\mathfrak{spo}(2m|2n+1)$, while there are $\binom{m+n}{m}+\binom{m+n-1}{n}$ W-conjugacy classes of fundamental systems for $\mathfrak{spo}(2m|2n)$.*

Proof. By the case-by-case classification of fundamental systems in Φ and definition of $\varepsilon\delta$-sequences, we clearly have a well-defined map

$$\Theta : \{\text{fundamental systems in } \Phi\}/W \longrightarrow \{\varepsilon\delta\text{-sequences for } \mathfrak{g}\}.$$

As we can easily construct a fundamental system Π for a given $\varepsilon\delta$-sequence, Θ is surjective. To show Θ is a bijection, it remains to show that the two finite sets have the same cardinalities.

The number of W-conjugacy classes of fundamental systems equals $|W'|/|W|$, where W' denotes the Weyl group of $\mathfrak{gl}(m+n)$ when $\mathfrak{g} = \mathfrak{gl}(m|n)$ and W' denotes the Weyl group of $\mathfrak{so}(2m+2n+1)$ when $\mathfrak{g} = \mathfrak{spo}(2m|2n+1)$. In either case, $|W'|/|W| = \binom{m+n}{m}$. On the other hand, in either case, an $\varepsilon\delta$-sequence is simply an ordered arrangement of m δ's and n ε's. Thus the total number is also $\binom{m+n}{m}$.

In the case of $\mathfrak{spo}(2m|2n)$, when the last slot of an $\varepsilon\delta$-sequence is an ε, the previous $(m+n-1)$ slots can be filled with m δ's and $(n-1)$ ε's. Thus we obtain $\binom{m+n-1}{m}$ different $\varepsilon\delta$-sequences this way. When the last slot in an $\varepsilon\delta$-sequence is a δ, then the previous $(m+n-1)$ slots are filled with $(m-1)$ δ's and n ε's, with the last ε having either a positive or negative sign. This way we obtain additional

$2\binom{m+n-1}{n}$ distinct $\varepsilon\delta$-sequences. Now $\binom{m+n-1}{m}+2\binom{m+n-1}{n}$ is exactly the number of W-conjugacy classes of fundamental systems for $\mathfrak{spo}(2m|2n)$ that we have computed earlier. □

The fundamental and positive systems for the exceptional Lie superalgebras $G(3), F(3|1), D(2|1,\alpha)$ are also listed completely by Kac [**60**] (with one missing for $F(3|1)$; see [**130**, 5.1]).

From the classification of the fundamental and positive systems we immediately obtain the following proposition and lemma.

Proposition 1.28. *Let $\mathfrak{g}$ be a basic Lie superalgebra, excluding $\mathfrak{sl}(n|n)/\mathbb{C}I_{n|n}$ and $\mathfrak{gl}(m|n)$. Let $\mathfrak{h}$ be a Cartan subalgebra of $\mathfrak{g}$. Then any fundamental system for the root system of $(\mathfrak{g},\mathfrak{h})$ forms a basis in $\mathfrak{h}^*$. (For $\mathfrak{gl}(m|n)$, any fundamental system is linearly independent in $\mathfrak{h}^*$.)*

Lemma 1.29. *Let $\mathfrak{g}$ be a basic Lie superalgebra. Let Π be the fundamental system in a positive system Φ^+, and let α be an isotropic odd root in Φ^+. Then there exists $w \in W$ such that $w(\alpha) \in \Pi$.*

1.4. Odd and real reflections

In this section, beside real reflections associated to even roots as for semisimple Lie algebras, we introduce odd reflections associated to isotropic odd simple roots. Both real and odd reflections permute the fundamental systems of a root system.

1.4.1. A fundamental lemma. We have seen that the fundamental systems of a root system Φ are not always W-conjugate due to the existence of odd roots. Recall a root $\alpha \in \Phi$ is isotropic if $(\alpha,\alpha)=0$, and an isotropic root must be odd. The following lemma plays a fundamental role in the representation theory of Lie superalgebras.

Lemma 1.30. *Let $\mathfrak{g}$ be a basic Lie superalgebra and let Π be a fundamental system of a positive system Φ^+. Let α be an odd isotropic simple root. Then,*

$$\Phi^+_\alpha := \{-\alpha\} \cup \Phi^+ \backslash \{\alpha\} \tag{1.43}$$

is a new positive system whose corresponding fundamental system Π_α is given by

$$\Pi_\alpha = \{\beta \in \Pi \mid (\beta,\alpha)=0, \beta \neq \alpha\} \cup \{\beta+\alpha \mid \beta \in \Pi, (\beta,\alpha) \neq 0\} \cup \{-\alpha\}. \tag{1.44}$$

Proof. For $\mathfrak{g} = \mathfrak{gl}(m|n)$, $\mathfrak{spo}(2m|2n)$, or $\mathfrak{spo}(2m|2n+1)$, which we are mostly concerned about in this book, we have already obtained complete descriptions of all fundamental systems. For most Π's, the set $\{\beta \in \Pi \mid (\beta,\alpha) \neq 0\}$ has cardinality at most 2 and consists of roots of the form $\pm\varepsilon_i \pm \varepsilon_j$ for $i \neq j \in I(m|n)$. By inspection we see that Π_α is a fundamental system. There are a few extra cases when the root α corresponds to a node in the corresponding Dynkin diagram which is either connected to a long simple root, or connected to a short non-isotropic root, or is a branching node, or is one of the (short) end nodes of a branching node, as follows:

Here the vertical dashed lines indicate an edge when it connects two $\otimes$'s and no edge otherwise, $\oplus$ means either $\bigcirc$ or $\bullet$, and $\odot$ means either $\otimes$ or $\bigcirc$. It can be checked directly in these cases that Π_α is also a fundamental system.

Take $\beta \in \Phi_\alpha^+ \cap \Phi^+$. Since Π is a fundamental system for Φ^+, by Proposition 1.28 we have a unique expression $\beta = \sum_{\gamma\in\Pi\setminus\{\alpha\}} m_\gamma\gamma + m_\alpha\alpha$ for $m_\gamma \in \mathbb{Z}_+$ and $m_\alpha \in \mathbb{Z}_+$. It follows by definition of Π_α that β can be expressed as a linear combination of Π_α of the form $\beta = \sum_{\kappa\in\Pi_\alpha\setminus\{-\alpha\}} m'_\kappa\kappa + m'_\alpha(-\alpha)$ for some suitable integer m'_α. By choice of β we have $m'_\kappa > 0$ for some κ, and hence $m'_\alpha \in \mathbb{Z}_+$, since Π_α is a fundamental system. This shows that Φ_α^+ is the positive system corresponding to Π_α.

For an exceptional Lie superalgebra $\mathfrak{g} = D(2|1,\alpha)$, $G(3)$, or $F(3|1)$, which will not be studied in any detail in this book, our proof shall be rather sketchy. A conceptual approach (see [**66**]) would be to follow Remark 1.19 to regard $\mathfrak{g}$ as a Kac-Moody superalgebra with Chevalley generators e_i, f_i, h_i for $i \in I$ associated to a Cartan matrix A (and a fundamental system $\Pi = \{\alpha_i \mid i \in I\}$), where I is $\mathbb{Z}_2$-graded. For an isotropic odd root $\alpha \in \Pi$, one can construct a new set of Chevalley generators e'_i, f'_i, h'_i associated to Π_α, which gives rise to a new Cartan matrix A'. By standard machinery of the Kac-Moody theory, the Kac-Moody superalgebra associated to A and A' coincide with $\mathfrak{g}$. From this it follows that Π_α is a fundamental system for $\mathfrak{g}$. The same argument as above shows that Φ_α is the positive system associated to the fundamental system Π_α. This uniform approach is applicable to all basic Lie superalgebras. □

1.4.2. Odd reflections. Let $\mathfrak{b} = \mathfrak{h} \oplus \mathfrak{n}^+$, where $\mathfrak{n}^+$ corresponds to the positive system Φ^+. Then the new Borel subalgebra corresponding to Φ_α^+ in (1.44) for an isotropic odd simple root α is given by

$$\mathfrak{b}^\alpha := \mathfrak{h} \oplus \bigoplus_{\beta\in\Phi_\alpha^+} \mathfrak{g}_\beta. \tag{1.45}$$

Observe that $\mathfrak{b}_{\bar{0}} = \mathfrak{b}_{\bar{0}}^\alpha$ by Lemma 1.30. The process of obtaining Π_α (respectively, Φ_α^+ or $\mathfrak{b}^\alpha$) from Π (respectively, Φ^+ or $\mathfrak{b}$) will be referred to as **odd reflection** (with respect to α) and will be denoted by r_α. We shall write

$$r_\alpha(\Pi) = \Pi_\alpha, \quad r_\alpha(\Phi^+) = \Phi_\alpha^+, \quad r_\alpha(\mathfrak{b}) = \mathfrak{b}^\alpha. \tag{1.46}$$

Note that $r_{-\alpha}r_\alpha = 1$.

Remark 1.31. In contrast to a real reflection associated to an even root (see below), an odd reflection r_α with respect to an isotropic odd root α may not be extended to a linear transformation on $\mathfrak{h}^*$ which sends a simple root in Π to a simple root in Π_α. For example, for $\mathfrak{spo}(2m|2n)$ with a fundamental system Π corresponding to the first Dynkin diagram below, take $\alpha = \delta_m - \varepsilon_n$. The odd reflection transforms Π to Π_α corresponding to the second Dynkin diagram below (with the $\cdots$ portion unchanged). A plausible transformation would have to fix those δ_i, ε_j (for $i < m, j < n$) and interchange ε_n and δ_m, but this transformation would not send $\delta_m + \varepsilon_n$ to $2\delta_m$.

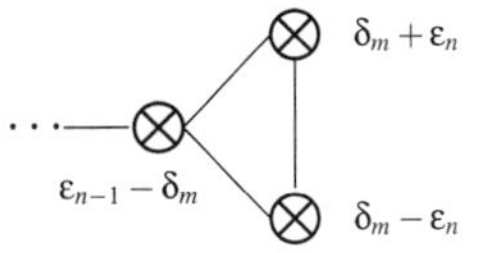

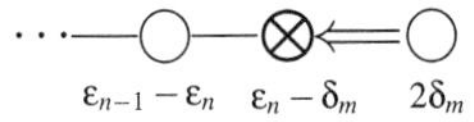

1.4.3. Real reflections. By Theorem 1.18, a basic Lie superalgebra $\mathfrak{g}$ admits an even non-degenerate supersymmetric bilinear form $(\cdot,\cdot)$, which restricts to a non-degenerate form $(\cdot,\cdot)$ on $\mathfrak{h}$ and on $\mathfrak{h}^*$. For an even root α (which is automatically non-isotropic), we define the **real reflection** r_α as a linear map on $\mathfrak{h}^*$ given by

$$r_\alpha(x) = x - 2\frac{(x,\alpha)}{(\alpha,\alpha)}\alpha, \qquad \text{for } x \in \mathfrak{h}^*.$$

In particular, r_α preserves Φ, $\Phi_{\bar{0}}$, and $\Phi_{\bar{1}}$, respectively. The group generated by real reflections r_α, for $\alpha \in \Phi_{\bar{0}}$, is precisely the Weyl group W of $\mathfrak{g}_{\bar{0}}$ (and hence the Weyl group of $\mathfrak{g}$ by Definition 1.17).

For an even *simple* root α, we must have $\alpha/2 \notin \Phi$, that is, $\alpha \in \bar{\Phi}_{\bar{0}}$ as defined in (1.19). In this case, (1.46) can be understood with Φ^+_α as in (1.43), $\mathfrak{b}^\alpha$ as in (1.45), and Π_α as the image of Π under r_α.

For an *odd* root α such that $2\alpha \in \Phi$, we define the reflection r_α as the real reflection $r_{2\alpha}$ associated to 2α. In this case, it is understood that $\Pi_\alpha = r_\alpha(\Pi)$, and

$$\Phi^+_\alpha = \{-\alpha, -2\alpha\} \cup \Phi^+ \backslash \{\alpha, 2\alpha\} \tag{1.47}$$

is the new positive system associated to Π_α.

1.4.4. Reflections and fundamental systems.

Proposition 1.32. *For two fundamental systems Π and $'\Pi$ of a basic Lie superalgebra $\mathfrak{g}$, there exists a sequence consisting of real and odd reflections $r_1, r_2, \ldots, r_k$ such that $r_k \ldots r_2 r_1(\Pi) = {}'\Pi$.*

Proof. Denote by Φ^+ and $'\Phi^+$ the positive systems associated to Π and $'\Pi$ respectively. We prove the corollary by induction on $|\Phi^+ \cap {}'\Phi^-|$. If $|\Phi^+ \cap {}'\Phi^-| = 0$, then $\Pi = {}'\Pi$. Assume now $|\Phi^+ \cap {}'\Phi^-| > 0$. Then $\Pi \neq {}'\Pi$, and we pick $\alpha \in \Pi \cap {}'\Phi^- \neq \emptyset$

and apply the real or odd simple reflection r_α. Observe that Φ_α^+ is the positive system with fundamental system $\Pi_\alpha = r_\alpha(\Pi)$, regardless of the parity of the simple root α. Note that

$$|\Phi_\alpha^+ \cap {}'\Phi^-| < |\Phi^+ \cap {}'\Phi^-|.$$

By the inductive assumption, there exists a sequence of real and odd reflections $r_2, \ldots, r_k$ such that $r_k \ldots r_2(\Pi_\alpha) = {}'\Pi$. Hence $r_k \ldots r_2 r_\alpha(\Pi) = {}'\Pi$. □

Proposition 1.33. *Let $\mathfrak{g}$ be a basic Lie superalgebra. Let Φ^+ be a positive system with Π as its fundamental system and ρ as its associated Weyl vector. Then, $(\rho,\beta) = \frac{1}{2}(\beta,\beta)$ for every simple root $\beta \in \Pi$.*

Proof. Let $\alpha \in \Pi$. Let Π_α and Φ_α^+ be respectively the fundamental and positive systems obtained by a real or odd reflection r_α defined above. Let ρ_α denote the Weyl vector for the positive system Φ_α^+. Recall Φ_α^+ is given in (1.43) for α odd or for $\alpha \in \bar{\Phi}_{\bar{0}}$, and in (1.47) otherwise. We compute that

$$\rho_\alpha = \begin{cases} \rho - \alpha, & \text{for } \alpha \in \bar{\Phi}_{\bar{0}} \text{ or non-isotropic odd,} \\ \rho + \alpha, & \text{for } \alpha \text{ isotropic odd.} \end{cases}$$

Claim. Assuming that the proposition holds for a fundamental system Π, it holds for the new fundamental system Π_α for $\alpha \in \Pi$.

Let us prove the claim. First assume $\alpha \in \Pi$ is even or non-isotropic odd. In this case, $\rho_\alpha = r_\alpha(\rho)$, $\Pi_\alpha = r_\alpha(\Pi)$, and so we have for each $\beta \in \Pi$ that

$$(\rho_\alpha, r_\alpha(\beta)) = (\rho,\beta) = \frac{1}{2}(\beta,\beta) = \frac{1}{2}(r_\alpha(\beta), r_\alpha(\beta)).$$

Now assume $\alpha \in \Pi$ is isotropic odd. We will check case-by-case, recalling the definition of Π_α from (1.44). If $(\beta,\alpha) = 0$ for $\beta \in \Pi$, then $\beta \in \Pi_\alpha$. By the assumption of the claim, we have

$$(\rho_\alpha,\beta) = (\rho+\alpha,\beta) = (\rho,\beta) = \frac{1}{2}(\beta,\beta).$$

If $(\beta,\alpha) \neq 0$ for $\beta \in \Pi$, then $\beta+\alpha \in \Pi_\alpha$. By the assumption of the claim, $(\rho,\alpha) = \frac{1}{2}(\alpha,\alpha) = 0$, and hence

$$(\rho_\alpha, \beta+\alpha) = (\rho+\alpha,\beta+\alpha) = (\rho+\alpha,\beta) = \frac{1}{2}(\beta,\beta) + (\alpha,\beta) = \frac{1}{2}(\beta+\alpha,\beta+\alpha).$$

This completes the proof of the claim.

By the claim and Proposition 1.32, it suffices to prove the proposition for just one fundamental system, e.g., the standard fundamental system for type $\mathfrak{gl}$ and $\mathfrak{osp}$. Assume that β is an even simple root or a non-isotropic odd simple root. In either case, one checks that $r_\beta(\rho) = \rho - \beta$. Since $(\cdot,\cdot)$ is W-invariant, we have

$$(\rho,\beta) = \big(r_\beta(\rho), r_\beta(\beta)\big) = (\rho-\beta,-\beta) = -(\rho,\beta) + (\beta,\beta).$$

It follows that $(\rho,\beta) = \frac{1}{2}(\beta,\beta)$ just as for semisimple Lie algebras.

For isotropic odd simple root β, we do a case-by-case inspection. For type $\mathfrak{gl}$ or $\mathfrak{osp}$, the equation $(\rho,\beta) = \frac{1}{2}(\beta,\beta)$ follows directly from Lemma 1.22. For the three exceptional cases, it can be checked directly by using a fundamental system with exactly one isotropic simple root and the detailed description of root systems (cf. Kac [**60**]). We skip the details. □

1.4.5. Examples. Below we illustrate the notion of odd reflections by examples.

Example 1.34. Associated with $\mathfrak{gl}(1|2)$, we have $\Phi_{\bar{0}} = \{\pm(\varepsilon_1 - \varepsilon_2)\}$ and $\Phi_{\bar{1}} = \{\pm(\delta_1 - \varepsilon_1), \pm(\delta_1 - \varepsilon_2)\}$. There are 6 fundamental systems that are related by real and odd reflections as follows. There are three conjugacy classes of Borel subalgebras corresponding to the three columns below, and each vertical pair corresponds to such a conjugacy class.

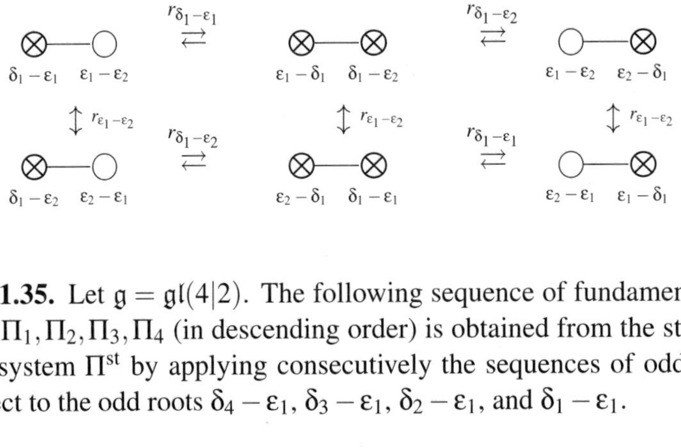

Example 1.35. Let $\mathfrak{g} = \mathfrak{gl}(4|2)$. The following sequence of fundamental systems $\Pi_0 = \Pi^{\mathrm{st}}, \Pi_1, \Pi_2, \Pi_3, \Pi_4$ (in descending order) is obtained from the standard fundamental system Π^{st} by applying consecutively the sequences of odd reflections with respect to the odd roots $\delta_4 - \varepsilon_1$, $\delta_3 - \varepsilon_1$, $\delta_2 - \varepsilon_1$, and $\delta_1 - \varepsilon_1$.

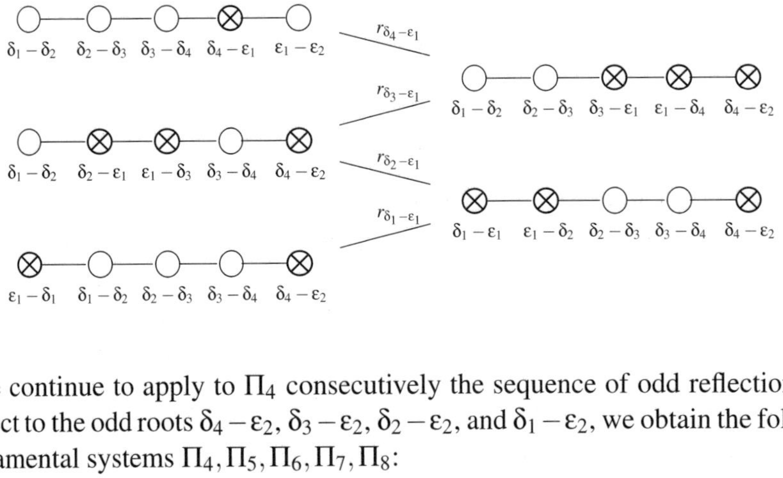

If we continue to apply to Π_4 consecutively the sequence of odd reflections with respect to the odd roots $\delta_4 - \varepsilon_2$, $\delta_3 - \varepsilon_2$, $\delta_2 - \varepsilon_2$, and $\delta_1 - \varepsilon_2$, we obtain the following fundamental systems $\Pi_4, \Pi_5, \Pi_6, \Pi_7, \Pi_8$:

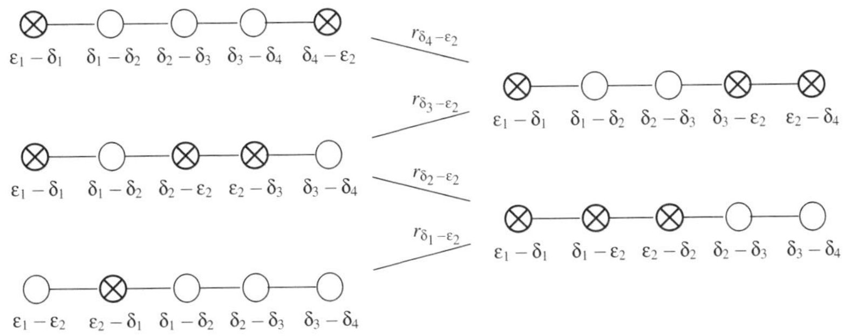

1.5. Highest weight theory

In this section, we formulate the Poincaré-Birkhoff-Witt Theorem for Lie superalgebras. Finite-dimensional irreducible modules over certain solvable Lie superalgebras, including all Borel subalgebras, are classified. This is then used to develop a highest weight theory of the basic and queer Lie superalgebras.

1.5.1. The Poincaré-Birkhoff-Witt (PBW) Theorem. Let $\mathfrak{g} = \mathfrak{g}_{\bar{0}} \oplus \mathfrak{g}_{\bar{1}}$ be a Lie superalgebra. A **universal enveloping algebra** of $\mathfrak{g}$ is an associative superalgebra with unity $U(\mathfrak{g})$ together with a homomorphism of Lie superalgebras $\iota : \mathfrak{g} \to U(\mathfrak{g})$ characterized by the following universal property. Given an associative superalgebra A and a homomorphism of Lie superalgebras $\varphi : \mathfrak{g} \to A$, there exists a unique homomorphism of associative superalgebras $\psi : U(\mathfrak{g}) \to A$ such that the following diagram commutes:

$$\begin{array}{ccc} \mathfrak{g} & \xrightarrow{\iota} & U(\mathfrak{g}) \\ & \searrow^{\varphi} & \vdots\ \exists\psi \\ & & A \end{array}$$

In particular, this implies that representations of $\mathfrak{g}$ are representations of $U(\mathfrak{g})$ and vice versa. It follows by a standard tensor algebra construction that a universal enveloping algebra exists, and it is unique up to isomorphism by the defining universal property.

Theorem 1.36 (Poincaré-Birkhoff-Witt Theorem). *Let $\{x_1, x_2, \dots, x_p\}$ be a basis for $\mathfrak{g}_{\bar{0}}$ and let $\{y_1, y_2, \dots, y_q\}$ be a basis for $\mathfrak{g}_{\bar{1}}$. Then the set*

$$\{x_1^{r_1} x_2^{r_2} \dots x_p^{r_p} y_1^{s_1} y_2^{s_2} \dots y_q^{s_q} \mid r_1, r_2, \dots, r_p \in \mathbb{Z}_+, s_1, s_2, \dots, s_q \in \{0,1\}\}$$

is a basis for $U(\mathfrak{g})$.

The proof for Theorem 1.36 is a straightforward super generalization of the Lie algebra case and will be omitted (see [**102**, Theorem 2.1] for detail).

The superalgebra $U(\mathfrak{g})$ carries a filtered algebra structure by letting

$$U_k(\mathfrak{g}) = \operatorname{span}\{x_1^{r_1}x_2^{r_2}\dots x_p^{r_p}y_1^{s_1}y_2^{s_2}\dots y_q^{s_q} \mid \sum_i r_i + \sum_j s_j \leq k\}, \quad k \in \mathbb{Z}_+.$$

Its associated graded algebra is isomorphic to $S(\mathfrak{g}_{\bar{0}}) \otimes \wedge(\mathfrak{g}_{\bar{1}})$.

Given a Lie subalgebra $\mathfrak{l}$ of a finite-dimensional Lie superalgebra $\mathfrak{g}$ and an $\mathfrak{l}$-module V, we define the induced module as

$$\mathrm{Ind}_{\mathfrak{l}}^{\mathfrak{g}} V = U(\mathfrak{g}) \otimes_{U(\mathfrak{l})} V.$$

By the PBW Theorem, if V is finite-dimensional, then so is $\mathrm{Ind}_{\mathfrak{g}_{\bar{0}}}^{\mathfrak{g}} V$.

1.5.2. Representations of solvable Lie superalgebras. Just as for Lie algebras, a finite-dimensional Lie superalgebra $\mathfrak{g} = \mathfrak{g}_{\bar{0}} \oplus \mathfrak{g}_{\bar{1}}$ is called **solvable** if $\mathfrak{g}^{(n)} = 0$ for some $n \geq 1$, where we define inductively $\mathfrak{g}^{(n)} = [\mathfrak{g}^{(n-1)}, \mathfrak{g}^{(n-1)}]$ and $\mathfrak{g}^{(0)} = \mathfrak{g}$.

Let $\mathfrak{g} = \mathfrak{g}_{\bar{0}} \oplus \mathfrak{g}_{\bar{1}}$ be a finite-dimensional solvable Lie superalgebra such that $[\mathfrak{g}_{\bar{1}}, \mathfrak{g}_{\bar{1}}] \subseteq [\mathfrak{g}_{\bar{0}}, \mathfrak{g}_{\bar{0}}]$. Given $\lambda \in \mathfrak{g}_{\bar{0}}^*$ with $\lambda([\mathfrak{g}_{\bar{0}}, \mathfrak{g}_{\bar{0}}]) = 0$, we define a one-dimensional $\mathfrak{g}$-module $\mathbb{C}_\lambda = \mathbb{C}v_\lambda$ by

$$xv_\lambda = \lambda(x)v_\lambda, \quad \text{for } x \in \mathfrak{g}_{\bar{0}},$$
$$yv_\lambda = 0, \quad \text{for } y \in \mathfrak{g}_{\bar{1}}.$$

There is a canonical linear isomorphism $(\mathfrak{g}_{\bar{0}}/[\mathfrak{g}_{\bar{0}}, \mathfrak{g}_{\bar{0}}])^* \cong \{\lambda \in \mathfrak{g}_{\bar{0}}^* \mid \lambda([\mathfrak{g}_{\bar{0}}, \mathfrak{g}_{\bar{0}}]) = 0\}$.

Lemma 1.37. *Let $\mathfrak{g} = \mathfrak{g}_{\bar{0}} \oplus \mathfrak{g}_{\bar{1}}$ be a finite-dimensional solvable Lie superalgebra such that $[\mathfrak{g}_{\bar{1}}, \mathfrak{g}_{\bar{1}}] \subseteq [\mathfrak{g}_{\bar{0}}, \mathfrak{g}_{\bar{0}}]$. Then every finite-dimensional irreducible $\mathfrak{g}$-module is one-dimensional. A complete list of finite-dimensional irreducible $\mathfrak{g}$-modules is given by $\mathbb{C}_\lambda$, for $\lambda \in (\mathfrak{g}_{\bar{0}}/[\mathfrak{g}_{\bar{0}}, \mathfrak{g}_{\bar{0}}])^*$.*

Proof. We have already shown that any such $\mathbb{C}_\lambda$ is a $\mathfrak{g}$-module. Clearly every one-dimensional $\mathfrak{g}$-module is isomorphic to $\mathbb{C}_\lambda$, for some $\lambda \in (\mathfrak{g}_{\bar{0}}/[\mathfrak{g}_{\bar{0}}, \mathfrak{g}_{\bar{0}}])^*$. So it suffices to show that every finite-dimensional irreducible $\mathfrak{g}$-module V is one dimensional.

The even subalgebra $\mathfrak{g}_{\bar{0}}$ is solvable since $\mathfrak{g}$ is solvable and $\mathfrak{g}_{\bar{0}}^{(n)} \subseteq \mathfrak{g}^{(n)}$ for every n. By applying Lie's theorem to $\mathfrak{g}_{\bar{0}}$, V contains a nonzero $\mathfrak{g}_{\bar{0}}$-invariant vector v_λ, where $\lambda \in \mathfrak{g}_{\bar{0}}^*$ with $\lambda([\mathfrak{g}_{\bar{0}}, \mathfrak{g}_{\bar{0}}]) = 0$. Here, by a $\mathfrak{g}_{\bar{0}}$-invariant vector we mean that $xv_\lambda = \lambda(x)v_\lambda, \forall x \in \mathfrak{g}_{\bar{0}}$. Now Frobenius reciprocity says that $\mathrm{Hom}_{\mathfrak{g}}(\mathrm{Ind}_{\mathfrak{g}_{\bar{0}}}^{\mathfrak{g}} \mathbb{C}v_\lambda, V) \cong \mathrm{Hom}_{\mathfrak{g}_{\bar{0}}}(\mathbb{C}v_\lambda, V)$, and thus, by the irreducibility of V, there exists a $\mathfrak{g}$-epimorphism from $\mathrm{Ind}_{\mathfrak{g}_{\bar{0}}}^{\mathfrak{g}} \mathbb{C}v_\lambda \cong \wedge(\mathfrak{g}_{\bar{1}}) \otimes \mathbb{C}v_\lambda$ onto V. Thus, to prove that V is one dimensional, it suffices to prove that $\mathrm{Ind}_{\mathfrak{g}_{\bar{0}}}^{\mathfrak{g}} \mathbb{C}v_\lambda$ has a composition series with one-dimensional composition factors. We shall construct such a composition series explicitly.

Set $\dim \mathfrak{g}_{\bar{1}} = n$. By Lie's theorem, the $\mathfrak{g}_{\bar{0}}$-module $\mathfrak{g}_{\bar{1}}$ has an ordered basis $\{y_j | j = 1, \ldots, n\}$ such that for all $1 \leq i \leq n$ we have

$$\operatorname{ad} \mathfrak{g}_{\bar{0}}(y_i) \subseteq \sum_{j=i}^{n} \mathbb{C} y_j, \quad \operatorname{ad}[\mathfrak{g}_{\bar{0}}, \mathfrak{g}_{\bar{0}}](y_i) \subseteq \sum_{j=i+1}^{n} \mathbb{C} y_j. \tag{1.48}$$

Let $\mathfrak{B}_1 = \{1, y_1\}$, and define inductively $\mathfrak{B}_k = \{\mathfrak{B}_{k-1}, \mathfrak{B}_{k-1} y_k\}$, for $2 \leq k \leq n$, where the set $\mathfrak{B}_{k-1} y_k$ denotes the ordered set of elements obtained by multiplying all elements in $\mathfrak{B}_{k-1}$ on the right by y_k. Then $\mathfrak{B}_n v_\lambda$ is an ordered basis for $\operatorname{Ind}_{\mathfrak{g}_{\bar{0}}}^{\mathfrak{g}} \mathbb{C} v_\lambda$ of cardinality 2^n. Denote this ordered basis by $\{v_1, \cdots, v_{2^n}\}$. Set $V_i := \oplus_{j=i}^{2^n} \mathbb{C} v_j$. Thanks to (1.48), we have a filtration of $\mathfrak{g}_{\bar{0}}$-modules

$$V = V_1 \supseteq V_2 \supseteq V_3 \supseteq \cdots \supseteq \cdots V_{2^n} \supseteq 0.$$

Clearly we have, for $1 \leq k, \ell \leq n$ and $i_1 < \ldots < i_\ell \leq n$,

$$y_k y_{i_1} y_{i_2} \ldots y_{i_\ell} v_\lambda = \begin{cases} [y_k, y_{i_1}] y_{i_2} \ldots y_{i_\ell} v_\lambda - y_{i_1} y_k y_{i_2} \ldots y_{i_\ell} v_\lambda, & \text{if } k \neq i_1, \\ \frac{1}{2} [y_k, y_{i_1}] y_{i_2} \ldots y_{i_\ell} v_\lambda, & \text{if } k = i_1. \end{cases} \tag{1.49}$$

Using the assumption that $[\mathfrak{g}_{\bar{1}}, \mathfrak{g}_{\bar{1}}] \subseteq [\mathfrak{g}_{\bar{0}}, \mathfrak{g}_{\bar{0}}]$, it follows from (1.48) and (1.49) that every y_k leaves V_i invariant, and hence each V_i is a $\mathfrak{g}$-module. Thus, the above filtration is a composition series with one-dimensional composition factors. □

Example 1.38. The Lie superalgebra $\mathfrak{g} = \mathbb{C} z \oplus \mathfrak{g}_{\bar{1}}$ in Example 1.4(4) associated to a non-degenerated symmetric bilinear form B on a nonzero space $\mathfrak{g}_{\bar{1}}$ is solvable. But an irreducible module of $\mathfrak{g}$ with the central element z acting as a nonzero scalar has dimension more than one. So the condition $[\mathfrak{g}_{\bar{1}}, \mathfrak{g}_{\bar{1}}] \subseteq [\mathfrak{g}_{\bar{0}}, \mathfrak{g}_{\bar{0}}]$ in Lemma 1.37 cannot be dropped.

1.5.3. Highest weight theory for basic Lie superalgebras. Let $\mathfrak{g}$ be a basic Lie superalgebra. Let $\mathfrak{h}$ be the standard Cartan subalgebra and let Φ be the root system. Let $\mathfrak{b} = \mathfrak{h} \oplus \mathfrak{n}^+$ be a Borel subalgebra of $\mathfrak{g} = \mathfrak{b} \oplus \mathfrak{n}^-$ and let Φ^+ be the associated positive system. The condition of Lemma 1.37 is satisfied for the solvable Lie superalgebra $\mathfrak{b}$, since we have $\mathfrak{b}_{\bar{1}} = \mathfrak{n}_{\bar{1}}^+$, and

$$[\mathfrak{b}_{\bar{1}}, \mathfrak{b}_{\bar{1}}] = [\mathfrak{n}_{\bar{1}}^+, \mathfrak{n}_{\bar{1}}^+] \subseteq \mathfrak{n}_{\bar{0}}^+ = [\mathfrak{h}, \mathfrak{n}_{\bar{0}}^+] \subseteq [\mathfrak{b}_{\bar{0}}, \mathfrak{b}_{\bar{0}}].$$

Let V be a finite-dimensional irreducible representation of $\mathfrak{g}$. Then by Lemma 1.37 V contains a one-dimensional $\mathfrak{b}$-module that is of the form $\mathbb{C}_\lambda = \mathbb{C} v_\lambda$, for $\lambda \in \mathfrak{h}^* \cong (\mathfrak{b}/[\mathfrak{b}, \mathfrak{b}])^*$. That is,

$$h v_\lambda = \lambda(h) v_\lambda \ (h \in \mathfrak{h}), \qquad x v_\lambda = 0 \ (x \in \mathfrak{n}^+).$$

By the PBW theorem and the irreducibility of V, we obtain $V = U(\mathfrak{n}^-) v_\lambda$ and thus a weight space decomposition

$$V = \bigoplus_{\mu \in \mathfrak{h}^*} V_\mu, \tag{1.50}$$

where the μ-weight space V_μ is given by

$$V_\mu := \{v \in V \mid hv = \mu(h)v, \forall h \in \mathfrak{h}\}.$$

By (1.50), $V_\mu = 0$ unless $\lambda - \mu$ is a $\mathbb{Z}_+$-linear combination of positive roots. The weight λ is called the $\mathfrak{b}$-**highest weight** (and sometime called an **extremal weight**) of V, the space $\mathbb{C}v_\lambda$ is called the $\mathfrak{b}$-**highest weight space**, and the vector v_λ is called a $\mathfrak{b}$-**highest weight vector** for V. When no confusion arises, we will simply say highest weight by dropping $\mathfrak{b}$. Hence, we have established the following.

Proposition 1.39. *Let $\mathfrak{g}$ be a basic Lie superalgebra with a Borel subalgebra $\mathfrak{b}$. Then every finite-dimensional irreducible $\mathfrak{g}$-module is a $\mathfrak{b}$-highest weight module.*

We shall denote the highest weight irreducible module of highest weight λ by $L(\lambda)$, $L(\mathfrak{g},\lambda)$, or $L(\mathfrak{g},\mathfrak{b},\lambda)$, depending on whether $\mathfrak{b}$ and $\mathfrak{g}$ are clear from the context.

Recall from Section 1.4 the notations Π_α and $\mathfrak{b}^\alpha$ associated to an isotropic odd simple root α. Denote by $\langle \cdot,\cdot \rangle : \mathfrak{h}^* \times \mathfrak{h} \to \mathbb{C}$ the standard bilinear pairing. Denote by h_α the corresponding coroot for α, and denote by e_α and f_α the root vectors of roots α and $-\alpha$ so that $[e_\alpha, f_\alpha] = h_\alpha$.

Lemma 1.40. *Let L be a (not necessarily finite-dimensional) simple $\mathfrak{g}$-module and let v be a $\mathfrak{b}$-highest weight vector of L of $\mathfrak{b}$-highest weight λ. Let α be an isotropic odd simple root.*

(1) *If $\langle \lambda, h_\alpha \rangle = 0$, then L is a $\mathfrak{g}$-module of $\mathfrak{b}^\alpha$-highest weight λ and v is a $\mathfrak{b}^\alpha$-highest weight vector.*

(2) *If $\langle \lambda, h_\alpha \rangle \neq 0$, then L is a $\mathfrak{g}$-module of $\mathfrak{b}^\alpha$-highest weight $(\lambda - \alpha)$ and $f_\alpha v$ is a $\mathfrak{b}^\alpha$-highest weight vector.*

Proof. We first observe three simple identities:

(i) $e_\alpha f_\alpha v = [e_\alpha, f_\alpha]v = h_\alpha v = \langle \lambda, h_\alpha \rangle v$.

(ii) $e_\beta f_\alpha v = [e_\beta, f_\alpha]v = 0$ for any $\beta \in \Phi^+ \cap \Phi^+_\alpha$, since either $\beta - \alpha$ is not a root or it belongs to $\Phi^+ \cap \Phi^+_\alpha$.

(iii) $f_\alpha^2 v = 0$, since α is an isotropic odd root and so $f_\alpha^2 = 0$.

Now, we consider two cases separately.

(1) Assume that $\langle \lambda, h_\alpha \rangle = 0$. Then we must have $f_\alpha v = 0$, for otherwise $f_\alpha v$ would be a $\mathfrak{b}$-singular vector in the simple $\mathfrak{g}$-module L by (i) and (ii). Also $e_\beta v = 0$, for $\beta \in \Phi^+ \cap \Phi^+_\alpha$. Thus Lemma 1.30 implies that v is a $\mathfrak{b}^\alpha$-highest weight vector of weight λ in the $\mathfrak{g}$-module L.

(2) Assume that $\langle \lambda, h_\alpha \rangle \neq 0$. Then (i) above implies that $f_\alpha v$ is nonzero. Now it follows by (ii), (iii), and Lemma 1.30 that $f_\alpha v$ is a $\mathfrak{b}^\alpha$-highest weight vector of weight $\lambda - \alpha$ in L. □

Example 1.41. Let $\mathfrak{g} = \mathfrak{gl}(4|2)$. Denote the sequence of Borel subalgebras corresponding to the fundamental systems Π_i in Example 1.35 as $\mathfrak{b}_i$, for $0 \le i \le 8$, with $\mathfrak{b}^{\mathrm{st}} = \mathfrak{b}_0$. Consider the finite-dimensional irreducible $\mathfrak{gl}(4|2)$-module $L(\lambda)$, where $\lambda = a_1\delta_1 + a_2\delta_2 + a_3\delta_3 + a_4\delta_4 + b_1\varepsilon_1 + b_2\varepsilon_2$ with $a_1 \ge a_2 \ge a_3 \ge a_4 \ge 2$ and $b_1 \ge b_2 \ge 0$. We identify $\lambda = (a_1, a_2, a_3, a_4 | b_1, b_2)$. The $\mathfrak{b}_i$-extremal weights, denoted by λ^i for $0 \le i \le 4$, are computed as follows:

$$\begin{aligned}
\lambda^0 &= \lambda = (a_1, a_2, a_3, a_4 | b_1, b_2),\\
\lambda^1 &= (a_1, a_2, a_3, a_4 - 1 | b_1 + 1, b_2),\\
\lambda^2 &= (a_1, a_2, a_3 - 1, a_4 - 1 | b_1 + 2, b_2),\\
\lambda^3 &= (a_1, a_2 - 1, a_3 - 1, a_4 - 1 | b_1 + 3, b_2),\\
\lambda^4 &= (a_1 - 1, a_2 - 1, a_3 - 1, a_4 - 1 | b_1 + 4, b_2).
\end{aligned}$$

If we continue to consecutively apply to λ^4 the sequence of odd reflections with respect to the odd roots $\delta_4 - \varepsilon_2$, $\delta_3 - \varepsilon_2$, $\delta_2 - \varepsilon_2$, and $\delta_1 - \varepsilon_2$, we obtain the following $\mathfrak{b}_i$-extremal weights λ^i, for $5 \le i \le 8$:

$$\begin{aligned}
\lambda^5 &= (a_1 - 1, a_2 - 1, a_3 - 1, a_4 - 2 | b_1 + 4, b_2 + 1),\\
\lambda^6 &= (a_1 - 1, a_2 - 1, a_3 - 2, a_4 - 2 | b_1 + 4, b_2 + 2),\\
\lambda^7 &= (a_1 - 1, a_2 - 2, a_3 - 2, a_4 - 2 | b_1 + 4, b_2 + 3),\\
\lambda^8 &= (a_1 - 2, a_2 - 2, a_3 - 2, a_4 - 2 | b_1 + 4, b_2 + 4).
\end{aligned}$$

We shall see in Section 2.4 that all these weights λ^i afford very simple visualization in terms of Young diagrams.

1.5.4. Highest weight theory for $\mathfrak{q}(n)$. Let $\mathfrak{g} = \mathfrak{q}(n)$ be the queer Lie superalgebra. We recall several subalgebras of $\mathfrak{g}$ from Section 1.2.6. Let $\mathfrak{h}$ be the standard Cartan subalgebra consisting of block diagonal matrices and let $\mathfrak{b}$ be the standard Borel subalgebra of block upper triangular matrices of $\mathfrak{g}$. We have $\mathfrak{b} = \mathfrak{h} \oplus \mathfrak{n}^+$ and $\mathfrak{g} = \mathfrak{b} \oplus \mathfrak{n}^-$. The Cartan subalgebra $\mathfrak{h} = \mathfrak{h}_{\bar 0} \oplus \mathfrak{h}_{\bar 1}$ is a solvable but nonabelian Lie superalgebra, since $[\mathfrak{h}_{\bar 1}, \mathfrak{h}_{\bar 1}] = \mathfrak{h}_{\bar 0}$ and $[\mathfrak{h}_{\bar 0}, \mathfrak{h}_{\bar 0} + \mathfrak{h}_{\bar 1}] = 0$.

For $\lambda \in \mathfrak{h}_{\bar 0}^*$, define a symmetric bilinear form $\langle \cdot, \cdot \rangle_\lambda$ on $\mathfrak{h}_{\bar 1}$ by

$$\langle v, w \rangle_\lambda := \lambda([v, w]), \qquad \text{for } v, w \in \mathfrak{h}_{\bar 1}.$$

Denote by $\mathrm{Rad}\langle \cdot, \cdot \rangle_\lambda$ the radical of the form $\langle \cdot, \cdot \rangle_\lambda$. Then $\langle \cdot, \cdot \rangle_\lambda$ descends to a nondegenerate symmetric bilinear form on $\mathfrak{h}_{\bar 1}/\mathrm{Rad}\langle \cdot, \cdot \rangle_\lambda$, and it gives rise to a superalgebra $\mathcal{C}_\lambda$ as follows. Choosing an orthonormal basis $\{e_i\}$ of $\in \mathfrak{h}_{\bar 1}/\mathrm{Rad}\langle \cdot, \cdot \rangle_\lambda$ with respect to the form $\langle \cdot, \cdot \rangle_\lambda$, the associative superalgebra $\mathcal{C}_\lambda$ is generated by $\{e_i\}$ subject to the relations (3.23), and hence is isomorphic to the Clifford superalgebra $\mathcal{C}_k$, where $k = \dim \mathfrak{h}_{\bar 1}/\mathrm{Rad}\langle \cdot, \cdot \rangle_\lambda$ (see Definition 3.33). By definition we have an isomorphism of associative superalgebras

$$\mathcal{C}_\lambda \cong U(\mathfrak{h})/I_\lambda, \tag{1.51}$$

where I_λ denotes the ideal of $U(\mathfrak{h})$ generated by $\mathrm{Rad}\langle\cdot,\cdot\rangle_\lambda$ and $a-\lambda(a)$ for $a\in\mathfrak{h}_{\bar 0}$.

Let $\mathfrak{h}'_{\bar 1}\subseteq\mathfrak{h}_{\bar 1}$ be a maximal isotropic subspace with respect to $\langle\cdot,\cdot\rangle_\lambda$, and define the Lie subalgebra $\mathfrak{h}':=\mathfrak{h}_{\bar 0}\oplus\mathfrak{h}'_{\bar 1}$. The one-dimensional $\mathfrak{h}_{\bar 0}$-module $\mathbb{C}v_\lambda$, defined by $hv_\lambda=\lambda(h)v_\lambda$, extends to an $\mathfrak{h}'$-module by letting $\mathfrak{h}'_{\bar 1}.v_\lambda=0$. Define the induced $\mathfrak{h}$-module

$$W_\lambda:=\mathrm{Ind}_{\mathfrak{h}'}^{\mathfrak{h}}\mathbb{C}v_\lambda.$$

Recall the well-known fact that a Clifford superalgebra admits a unique irreducible ($\mathbb{Z}_2$-graded) module $\widetilde{W}_\lambda$ (see, for example, Exercise 3.11 in Chapter 3).

Lemma 1.42. *For $\lambda\in\mathfrak{h}_{\bar 0}^*$, the $\mathfrak{h}$-module W_λ is isomorphic to $\widetilde{W}_\lambda$ (viewed as an $\mathfrak{h}$-module via the pullback through* (1.51)*) and is irreducible. Furthermore, every finite-dimensional irreducible $\mathfrak{h}$-module is isomorphic to W_λ, for some $\lambda\in\mathfrak{h}_{\bar 0}^*$.*

Lemma 1.42 shows that the $\mathfrak{h}$-module W_λ is independent of a choice of a maximal isotropic subspace $\mathfrak{h}'_{\bar 1}$.

Proof. The action of $U(\mathfrak{h})$ on W_λ descends to an action of $U(\mathfrak{h})/I_\lambda$, and via (1.51) we identify W_λ with the unique irreducible module $\widetilde{W}_\lambda$ of the Clifford superalgebra $\mathcal{C}_\lambda$. Hence, the $\mathfrak{h}$-module W_λ is irreducible.

Suppose we are given an irreducible $\mathfrak{h}$-module U. Then it contains an $\mathfrak{h}_{\bar 0}$-weight vector v'_λ of weight $\lambda\in\mathfrak{h}_{\bar 0}^*$. Recall that $\mathfrak{h}'=\mathfrak{h}_{\bar 0}\oplus\mathfrak{h}'_{\bar 1}$, and consider the $\mathfrak{h}'$-submodule $\big(\pi,U(\mathfrak{h}')v'_\lambda\big)$ of U such that $\pi([\mathfrak{h}'_{\bar 1},\mathfrak{h}'_{\bar 1}])=\pi([\mathfrak{h}'_{\bar 0},\mathfrak{h}'_{\bar 0}])=0$. By applying Lemma 1.37 to Lie superalgebra $\pi(\mathfrak{h}')$, there exists a one-dimensional $\mathfrak{h}'$-submodule $\mathbb{C}v_\lambda$ of $U(\mathfrak{h}')v'_\lambda\subseteq U$. By Frobenius reciprocity we have a surjective $\mathfrak{h}$-homomorphism from W_λ onto U. Since W_λ is irreducible, we have $U\cong W_\lambda$. □

Let V be a finite-dimensional irreducible $\mathfrak{g}$-module. Pick an irreducible $\mathfrak{h}$-module W_λ in V, where $\lambda\in\mathfrak{h}_{\bar 0}^*$ can be taken to be maximal in the partial order induced by the positive system Φ^+ by the finite dimensionality of V. By definition, W_λ is $\mathfrak{h}_{\bar 0}$-semisimple of weight λ. For any $\alpha\in\Phi^+$ with associated even root vector e_α and odd root vector $\overline{e}_\alpha$ in $\mathfrak{n}^+$, the space $\mathbb{C}e_\alpha W_\lambda+\mathbb{C}\overline{e}_\alpha W_\lambda$ is an $\mathfrak{h}$-module that is $\mathfrak{h}_{\bar 0}$-semisimple of weight $\lambda+\alpha$. If $\mathbb{C}e_\alpha W_\lambda+\mathbb{C}\overline{e}_\alpha W_\lambda\neq 0$ for some $\alpha\in\Phi^+$, then it contains an isomorphic copy of $W_{\lambda+\alpha}$ as an $\mathfrak{h}$-submodule, contradicting the maximal weight assumption of λ. Hence, we have $\mathfrak{n}^+W_\lambda=0$. By irreducibility of V we must have $U(\mathfrak{n}^-)W_\lambda=V$, which gives rise to a weight space decomposition of $V=\bigoplus_{\mu\in\mathfrak{h}_{\bar 0}^*}V_\mu$. The space $W_\lambda=V_\lambda$ is the **highest weight space** of V, and it completely determines the irreducible module V. We denote V by $L(\mathfrak{g},\lambda)$, or $L(\lambda)$, if $\mathfrak{g}$ is evident from the context. Summarizing, we have proved the following.

Proposition 1.43. *Let $\mathfrak{g}=\mathfrak{q}(n)$. Any finite-dimensional irreducible $\mathfrak{g}$-module is a highest weight module.*

Let $\ell(\lambda)$ be the number of nonzero parts in a composition λ (generalizing the notation for length of a partition λ). We set

$$\delta(\lambda) = \begin{cases} 0, & \text{if } \ell(\lambda) \text{ is even,} \\ 1, & \text{if } \ell(\lambda) \text{ is odd.} \end{cases} \tag{1.52}$$

Now for $\lambda \in \mathfrak{h}^*$, recalling the notation H_i from (1.30), we identify λ with the composition $\lambda = (\lambda(H_1), \ldots, \lambda(H_n))$, and hence, $\ell(\lambda)$ is equal to the dimension of the space $\mathfrak{h}_{\bar{1}}/\mathrm{Rad}\langle\cdot|\cdot\rangle_\lambda$. We remark that the highest weight space W_λ of $L(\lambda)$ has dimension $2^{(\ell(\lambda)+\delta(\lambda))/2}$. Note that the Clifford superalgebra $\mathcal{C}_\lambda$ admits an odd automorphism if and only if $\ell(\lambda)$ is odd. Hence, the $\mathfrak{h}$-module W_λ, or equivalently the irreducible $\mathcal{C}_\lambda$-module W_λ, has an odd automorphism if and only if $\ell(\lambda)$ is an odd integer. An automorphism of the irreducible $\mathfrak{g}$-module $L(\lambda)$ clearly induces an $\mathfrak{h}$-module automorphism of its highest weight space. Conversely, any $\mathfrak{h}$-module automorphism on W_λ induces an automorphism of the $\mathfrak{g}$-module $\mathrm{Ind}_{\mathfrak{b}}^{\mathfrak{g}} W_\lambda$. Since an automorphism preserves the maximal submodule, it induces an automorphism of the unique irreducible quotient $\mathfrak{g}$-module. We have proved the following.

Lemma 1.44. *Let $\mathfrak{g} = \mathfrak{q}(n)$ be the queer Lie superalgebra with a Cartan subalgebra $\mathfrak{h}$. Let $L(\lambda)$ be the irreducible $\mathfrak{g}$-module of highest weight $\lambda \in \mathfrak{h}_{\bar{0}}^*$. Then,*

$$\dim End_{\mathfrak{g}}(L(\lambda)) = 2^{\delta(\lambda)}.$$

This suggests that Schur's Lemma requires modification for superalgebras. This will be discussed in depth in Chapter 3, Lemma 3.4.

1.6. Exercises

Exercises 1.13, 1.14, 1.15, 1.16, and 1.24 below indicate that various classical theorems in the theory of Lie algebras fail for Lie superalgebras.

Exercise 1.1. Let φ be an automorphism of Lie superalgebra $\mathfrak{gl}(m|n)$, and let $J \in \mathfrak{gl}(m|n)_{\bar{0}}$. Define

$$\mathfrak{g}(\varphi, J) := \{g \in \mathfrak{gl}(m|n) \mid Jg - \varphi(g)J = 0\}.$$

Prove that $\mathfrak{g}(\varphi, J)$ is a subalgebra of $\mathfrak{gl}(m|n)$.

Exercise 1.2. Let $\mathfrak{J}_{2m|\ell}$ be as defined in 1.1.3, and let $\mathfrak{J}'_{2m|\ell}$ be the matrix obtained from $\mathfrak{J}_{2m|\ell}$ by substituting I_m with $-I_m$. Prove:

(1) There exists an automorphism φ of Lie superalgebra $\mathfrak{gl}(2m|\ell)$ given by $\varphi(g) = -g^{\mathrm{st}}$.

(2) $\mathfrak{g}(\varphi, \mathfrak{J}_{2m|\ell}) = \mathfrak{spo}(2m|\ell)$.

(3) $\mathfrak{g}(\varphi^3, \mathfrak{J}_{2m|\ell}) = \mathfrak{g}(\varphi, \mathfrak{J}'_{2m|\ell})$.

Exercise 1.3. Prove that $\Pi g^{\mathrm{st}} \Pi^{-1} = (\Pi g \Pi^{-1})^{\mathrm{st}^3}$, for $g \in \mathfrak{gl}(m|n)$; conclude that $\mathfrak{osp}(m|n) \cong \mathfrak{spo}(n|m)$, for n even (as stated in Remark 1.9).

Exercise 1.4. The space $\wedge(2)$ is naturally a $W(2)$-module with submodule $\mathbb{C}1$. Show that the action of $W(2)$ on $\wedge(2)/\mathbb{C}1$ is faithful, and thus induces, by identifying $\wedge(2)/\mathbb{C}1$ with $\mathbb{C}^{1|2}$, an isomorphism of Lie superalgebras $W(2) \cong \mathfrak{sl}(1|2)$.

Exercise 1.5. Prove that the following linear map $\psi : \mathfrak{spo}(2|2) \to \mathfrak{sl}(1|2)$ is an isomorphism of Lie superalgebras:

$$\begin{pmatrix} d & e & y_1 & x_1 \\ f & -d & -y & x \\ -x & x_1 & -a & 0 \\ y & y_1 & 0 & a \end{pmatrix} \stackrel{\psi}{\longmapsto} \begin{pmatrix} 2a & \sqrt{2}y_1 & \sqrt{2}y \\ \sqrt{2}x_1 & d+a & e \\ \sqrt{2}x & f & -d+a \end{pmatrix}.$$

Exercise 1.6. Suppose that $\mathfrak{g} = \mathfrak{g}_{-1} \oplus \mathfrak{g}_0 \oplus \mathfrak{g}_1$ is a $\mathbb{Z}$-graded Lie superalgebra such that $\mathfrak{g}_{\bar{0}} = \mathfrak{g}_0$, $\mathfrak{g}_{\bar{1}} = \mathfrak{g}_1 \oplus \mathfrak{g}_{-1}$, and $[\mathfrak{g}_1, \mathfrak{g}_{-1}] = \mathfrak{g}_0$. Assume further that $\mathfrak{g}_0$ is a simple Lie algebra and $\mathfrak{g}_{\pm 1}$ are irreducible $\mathfrak{g}_0$-modules such that $\mathrm{Hom}_{\mathfrak{g}_0}(\mathfrak{g}_0, \mathfrak{g}_1 \otimes \mathfrak{g}_{-1}) \cong \mathbb{C}$. Prove that these data determine the Lie superalgebra structure on $\mathfrak{g}$ uniquely.

Exercise 1.7. Prove:

(1) The simple Lie superalgebras $S(3)$ and $[\mathfrak{p}(3), \mathfrak{p}(3)]$ are isomorphic. (Hint: use Exercise 1.6.)

(2) The Lie superalgebra $H(4)$ and $\mathfrak{sl}(2|2)/\mathbb{C}I_{2|2}$ are isomorphic.

Exercise 1.8. Let $\mathfrak{g} = \mathfrak{g}_{\bar{0}} \oplus \mathfrak{g}_{\bar{1}}$ be a finite-dimensional simple Lie superalgebra with $\mathfrak{g}_{\bar{1}} \neq 0$. Prove:

(1) $[\mathfrak{g}_{\bar{0}}, \mathfrak{g}_{\bar{1}}] = \mathfrak{g}_{\bar{1}}$.

(2) $[\mathfrak{g}_{\bar{1}}, \mathfrak{g}_{\bar{1}}] = \mathfrak{g}_{\bar{0}}$.

(3) The $\mathfrak{g}_{\bar{0}}$-module $\mathfrak{g}_{\bar{1}}$ is faithful.

Exercise 1.9. Let $\mathfrak{g} = \mathfrak{g}_{\bar{0}} \oplus \mathfrak{g}_{\bar{1}}$ be a Lie superalgebra such that $[\mathfrak{g}_{\bar{1}}, \mathfrak{g}_{\bar{1}}] = \mathfrak{g}_{\bar{0}}$ and the adjoint $\mathfrak{g}_{\bar{0}}$-module $\mathfrak{g}_{\bar{1}}$ is faithful and irreducible. Prove that $\mathfrak{g}$ is simple.

Exercise 1.10. Let $\mathfrak{g} = \bigoplus_{j=-1}^{n} \mathfrak{g}_j$ be a $\mathbb{Z}$-graded Lie superalgebra satisfying

(1) $[\mathfrak{g}_{-1}, x] = 0$ implies that $x = 0$, for all $x \in \mathfrak{g}_j$ with $j \geq 0$;

(2) The adjoint $\mathfrak{g}_0$-module $\mathfrak{g}_{-1}$ is irreducible;

(3) $[\mathfrak{g}_{-1}, \mathfrak{g}_1] = \mathfrak{g}_0$, $[\mathfrak{g}_0, \mathfrak{g}_1] = \mathfrak{g}_1$, and $\mathfrak{g}_{j+1} = [\mathfrak{g}_j, \mathfrak{g}_1]$ for $j \geq 1$.

Prove that $\mathfrak{g}$ is simple.

Exercise 1.11. Let $\mathfrak{g} = \mathfrak{sl}(2|2)/\mathbb{C}I_{2|2}$. Prove that $\dim \mathfrak{g}_\alpha = 2$, for $\alpha \in \Phi_{\bar{1}}$.

Exercise 1.12. Let $\mathfrak{g}$ be a Lie superalgebra. Prove that $\mathfrak{g} \otimes \wedge(n)$ is a Lie superalgebra with Lie bracket defined by

$$[a \otimes \lambda, b \otimes \mu] = (-1)^{|\lambda||b|}[a,b] \otimes \lambda\mu, \quad a,b \in \mathfrak{g}; \lambda, \mu \in \wedge(n).$$

Exercise 1.13. Let $\mathfrak{g}$ be a finite-dimensional simple Lie superalgebra. The Lie superalgebra $W(n)$ acts naturally on $\mathfrak{g} \otimes \wedge(n)$ so that the semidirect product $\mathcal{G} = (\mathfrak{g} \otimes \wedge(n)) \rtimes W(n)$ is a Lie superalgebra. Prove that $\mathcal{G}$ is **semisimple** (which by definition means that $\mathcal{G}$ has no nontrivial solvable ideal.)

Exercise 1.14. Find a filtration of subalgebras for $\mathcal{G}$ in Exercise 1.13 of the form

$$0 = \mathcal{G}_0 \subsetneq \mathcal{G}_1 \subsetneq \mathcal{G}_2 \subsetneq \cdots \subsetneq \mathcal{G}_{k-1} \subsetneq \mathcal{G}_k = \mathcal{G},$$

for some k, such that $\mathcal{G}_{i-1}$ is an ideal in $\mathcal{G}_i$ wtih $\mathcal{G}_i/\mathcal{G}_{i-1}$ simple, for all $i = 1, \ldots, k$. Prove that $\mathcal{G}$ is not a direct sum of simple Lie superalgebras.

Exercise 1.15. Let $\mathcal{G}$ be constructed as in Exercise 1.13 from a not necessarily simple Lie superalgebra $\mathfrak{g}$. Suppose that V is a faithful irreducible representation of $\mathfrak{g}$. Prove that, as a representation of $\mathcal{G}$, $V \otimes \wedge(n)$ is faithful and irreducible.

Exercise 1.16. Continuing Exercise 1.15, assume that $\mathfrak{g}$ is the one-dimensional abelian Lie algebra and V is a nontrivial irreducible representation of $\mathfrak{g}$. Prove that $\mathcal{G}$ is not reductive, i.e., $\mathcal{G}$ is not a direct sum of a semisimple Lie superalgebra and a one-dimensional even subalgebra. (It is known that a finite-dimensional Lie algebra possessing a finite-dimensional faithful representation is reductive.)

Exercise 1.17. Let $\mathfrak{g}$ be a simple Lie superalgebra whose Killing form is non-degenerate. Prove that every derivation of $\mathfrak{g}$ is inner.

Exercise 1.18. Prove that the Killing forms on the Lie superalgebras $\mathfrak{sl}(k|\ell)$ with $k \neq \ell$, $k + \ell \geq 2$, and $\mathfrak{osp}(m|2n)$ with $m - 2n \neq 2$, $m + 2n \geq 2$ are non-degenerate. Conclude that the derivations of these Lie superalgebras are all inner.

Exercise 1.19. Let $\mathfrak{g} = \mathfrak{g}_{\bar{0}} \oplus \mathfrak{g}_{\bar{1}}$ be a finite-dimensional simple Lie superalgebra. Suppose that $\mathfrak{g}_{\bar{0}}$ is a semisimple Lie algebra and let $\mathrm{Der}(\mathfrak{g}) = \mathrm{ad}(\mathfrak{g}) \oplus W$ be a decomposition of $\mathrm{ad}(\mathfrak{g}_{\bar{0}})$-modules. Prove:

(1) W is a trivial $\mathrm{ad}(\mathfrak{g}_{\bar{0}})$-module.

(2) For $D \in W$, the map $D : \mathfrak{g} \to \mathfrak{g}$ is an $\mathrm{ad}(\mathfrak{g}_{\bar{0}})$-homomorphism vanishing on $\mathfrak{g}_{\bar{0}}$.

Exercise 1.20. Use Exercise 1.19(2) to prove the following:

(1) Every derivation of $\mathfrak{osp}(m|2n)$ is inner.

(2) The space $\mathrm{Der}(\mathfrak{g})/\mathrm{ad}(\mathfrak{g})$ for $\mathfrak{sl}(m+1|m+1)/\mathbb{C}I_{m+1|m+1}$, $[\mathfrak{p}(m), \mathfrak{p}(m)]$, and $[\mathfrak{q}(m), \mathfrak{q}(m)]/\mathbb{C}I_{m|m}$, for $m \geq 2$, are all one-dimensional.

(3) The space $\mathrm{Der}(\mathfrak{g})/\mathrm{ad}(\mathfrak{g})$ for $\mathfrak{sl}(2|2)/\mathbb{C}I_{2|2}$ is three-dimensional.

Exercise 1.21. Let $\mathfrak{g} = \bigoplus_{j \geq -1} \mathfrak{g}_j$ be the simple $\mathbb{Z}$-graded Lie superalgebra $W(m)$, $S(m)$, or $H(m+1)$, for $m \geq 3$, with principal gradation. Let $\mathfrak{b}_0$ be a Borel subalgebra of $\mathfrak{g}_0$. Set $\mathfrak{g}_{\geq 0} = \bigoplus_{j \geq 0} \mathfrak{g}_j$, $\mathfrak{g}_{>0} = \bigoplus_{j > 0} \mathfrak{g}_j$, and $\mathfrak{b} = \mathfrak{b}_0 \oplus \mathfrak{g}_{>0}$. Prove:

(1) $\mathfrak{b}$ is a solvable Lie superalgebra satisfying $[\mathfrak{b}_{\bar{1}}, \mathfrak{b}_{\bar{1}}] \subseteq [\mathfrak{b}_{\bar{0}}, \mathfrak{b}_{\bar{0}}]$.

(2) A finite-dimensional irreducible representation of $\mathfrak{g}_{\geq 0}$ is an irreducible representation of $\mathfrak{g}_0$ on which the subalgebra $\mathfrak{g}_{>0}$ acts trivially.

(3) Any finite-dimensional irreducible module of $\mathfrak{g}$ is a quotient of $\mathrm{Ind}_{\mathfrak{g}_{\geq 0}}^{\mathfrak{g}} V$, where V is a finite-dimensional irreducible representation of $\mathfrak{g}_{\geq 0}$.

Conclude that, up to isomorphism, finite-dimensional irreducible $\mathfrak{g}_0$-modules are in one-to-one correspondence with finite-dimensional irreducible $\mathfrak{g}$-modules.

Exercise 1.22. Let $\mathfrak{g} = [\mathfrak{p}(2), \mathfrak{p}(2)]$. Prove:

(1) $[\mathfrak{g}, \mathfrak{g}] \subsetneq \mathfrak{g}$, and hence $\mathfrak{g}$ is not simple.

(2) $\mathfrak{g}$ is semisimple.

(3) Any nontrivial finite-dimensional irreducible $\mathfrak{g}$-module is a direct sum of two copies of the same irreducible $\mathfrak{g}_{\bar{0}}$-module.

Exercise 1.23. It is known that every finite-dimensional Lie algebra has a finite-dimensional faithful representation. Prove that every finite-dimensional Lie superalgebra has a finite-dimensional faithful representation.

Exercise 1.24. Prove that the exact sequence of Lie superalgebras

$$0 \longrightarrow \mathbb{C}I_{m|m} \to \mathfrak{sl}(m|m) \longrightarrow \mathfrak{sl}(m|m)/\mathbb{C}I_{m|m} \longrightarrow 0$$

is non-split, though $\mathfrak{sl}(m|m)/\mathbb{C}I_{m|m}$ is simple, for $m \geq 2$. (Hence Levi's theorem fails for Lie superalgebras.)

Exercise 1.25. Prove that the bilinear form (1.33) induced from the odd trace form (1.32) is a symmetric non-degenerate invariant form on $\mathfrak{q}(n)$.

Notes

Section 1.1. The classification of finite-dimensional simple complex Lie superalgebras was first announced by Kac in [**59**]. The detailed proof of the classification, along with many other fundamental results on Lie superalgebras, appeared in Kac [**60**] two years later. An independent proof of the classification of the finite-dimensional complex simple Lie superalgebras whose even subalgebras are reductive was given in the two papers by Scheunert, Nahm, and Rittenberg in [**106**] around the same time.

Section 1.2. The structure theory of root systems, root space decompositions, and invariant bilinear forms of basic Lie superalgebras presented here is fairly standard. More details can be found in the standard references (see Kac [**60, 62**] and Scheunert [**105**]).

Section 1.3. The concept of odd reflections was introduced by Leites, Saveliev, and Serganova [**78**] to relate non-conjugate Borel subalgebras, fundamental and positive systems. The list of conjugacy classes of fundamental systems under the Weyl group action for the basic Lie algebras was given by Kac [**60**]. We introduce a notion of $\varepsilon\delta$-sequences (which appeared in Cheng-Wang [**32**]) to facilitate the parametrization of the conjugacy classes of fundamental systems for Lie superalgebras of type $\mathfrak{gl}$ and $\mathfrak{osp}$. For definitions of Borel subalgebras for general Lie superalgebras, see [**90, 95**].

Section 1.4. Lemma 1.30 on odd reflections appeared for more general Kac-Moody Lie superalgebras in Kac-Wakimoto [**66**, Lemma 1.2], and it is sometimes attributed to Serganova's 1988 thesis (cf., e.g., [**108**]).

Section 1.5. A detailed proof of the PBW Theorem for Lie superalgebras, Theorem 1.36, can be found in Milnor-Moore [**85**, Theorem 6.20] and Ross [**102**, Theorem 2.1]. As in the classical theory, the first step to develop a highest weight theory for the basic and queer Lie superalgebras is to study representations of Borel subalgebras. Lemma 1.37 on solvable Lie superalgebras appeared in Kac [**60**, Proposition 5.2.4]. The proof given here is different. Lemma 1.40 on the change of the extremal weights of an irreducible module under an odd reflection (which appeared in Penkov-Serganova [**94**, Lemma 1]) plays a fundamental role in representation theory of Lie superalgebras developed in the book.

Exercises 1.8, 1.9, and 1.17–1.20 are taken from [**60**] and [**105**], while Exercise 1.21 was implicit in [**60**]. Exercise 1.16 was inspired by similar examples for Lie algebras in prime characteristic, which we learned from A. Premet.

Chapter 2

Finite-dimensional modules

In this chapter, we mainly work on Lie superalgebras of type $\mathfrak{gl}, \mathfrak{osp}$, and $\mathfrak{q}$. The finite-dimensional irreducible modules for these Lie superalgebras are classified in terms of highest weights. In contrast to semisimple Lie algebras, there is more work to do to achieve such classifications for Lie superalgebras. We describe the images of the Harish-Chandra homomorphisms in terms of (super)symmetric functions and formulate the linkage principle on composition factors of a Verma module. As an application, we present a Weyl-type character formula for the so-called typical finite-dimensional irreducible modules. The chapter concludes with a study of extremal weights, i.e., highest weights with respect to (not necessarily conjugate) Borel subalgebras, of various finite-dimensional irreducible modules.

2.1. Classification of finite-dimensional simple modules

In this section, the highest weights of finite-dimensional irreducible modules over Lie superalgebras of type $\mathfrak{gl}$, $\mathfrak{osp}$, and $\mathfrak{q}$ are determined.

2.1.1. Finite-dimensional simple modules of $\mathfrak{gl}(m|n)$. Let $\mathfrak{g}$ be the Lie superalgebra $\mathfrak{gl}(m|n)$ and let $\mathfrak{h}$ be the Cartan subalgebra of diagonal matrices spanned by the basis elements $\{E_{ii} | i \in I(m|n)\}$. Let $\mathfrak{n}^+$ (respectively, $\mathfrak{n}^-$) be the subalgebra of strictly upper (respectively, strictly lower) triangular matrices of $\mathfrak{gl}(m|n)$ so that Φ^+ corresponds to (1.37). Then we have the standard triangular decomposition

$$\mathfrak{g} = \mathfrak{n}^- \oplus \mathfrak{h} \oplus \mathfrak{n}^+.$$

The even subalgebra admits a compatible triangular decomposition

$$\mathfrak{g}_{\bar{0}} = \mathfrak{n}_{\bar{0}}^- \oplus \mathfrak{h} \oplus \mathfrak{n}_{\bar{0}}^+,$$

where $\mathfrak{n}_{\bar{0}}^{\pm} = \mathfrak{g}_{\bar{0}} \cap \mathfrak{n}^{\pm}$. Let $\mathfrak{b} = \mathfrak{h} \oplus \mathfrak{n}^+$ and $\mathfrak{b}_{\bar{0}} = \mathfrak{h} \oplus \mathfrak{n}_{\bar{0}}^+$.

Moreover, the Lie superalgebra $\mathfrak{g}$ admits a $\mathbb{Z}$-gradation

$$\mathfrak{g} = \mathfrak{g}_{-1} \oplus \mathfrak{g}_0 \oplus \mathfrak{g}_1, \tag{2.1}$$

where $\mathfrak{g}_{-1}$ (respectively, $\mathfrak{g}_1$) is spanned by all E_{ij} with $i, j \in I(m|n)$ such that $i > 0 > j$ (respectively, $i < 0 < j$). Note that $\mathfrak{g}_{-1}$ and $\mathfrak{g}_1$ are abelian Lie superalgebras, and that the $\mathbb{Z}$-degree zero subspace $\mathfrak{g}_0$ coincides with the $\mathbb{Z}_2$-degree zero subspace $\mathfrak{g}_{\bar{0}}$.

For $\lambda \in \mathfrak{h}^*$, let $L^0(\lambda)$ be the simple $\mathfrak{g}_{\bar{0}}$-module of highest weight $\lambda \in \mathfrak{h}^*$ (relative to the Borel $\mathfrak{b}_{\bar{0}}$). Then $L^0(\lambda)$ may be extended trivially to a $\mathfrak{g}_{\bar{0}} \oplus \mathfrak{g}_1$-module due to (2.1). Define the **Kac module** over $\mathfrak{g}$ by

$$K(\lambda) = \mathrm{Ind}_{\mathfrak{g}_0 \oplus \mathfrak{g}_1}^{\mathfrak{g}} L^0(\lambda),$$

which, as a vector space, can be identified by the PBW Theorem 1.36 with

$$K(\lambda) = \wedge(\mathfrak{g}_{-1}) \otimes L^0(\lambda). \tag{2.2}$$

Proposition 2.1. *Let $\mathfrak{g} = \mathfrak{gl}(m|n)$. There exists a surjective $\mathfrak{g}$-module homomorphism (unique up to a scalar multiple) $K(\lambda) \twoheadrightarrow L(\lambda)$. Moreover, the following are equivalent:*

(1) *$L(\lambda)$ is finite dimensional.*
(2) *$L^0(\lambda)$ is finite dimensional.*
(3) *$K(\lambda)$ is finite dimensional.*

Proof. The existence of a surjective homomorphism $K(\lambda) \twoheadrightarrow L(\lambda)$ follows by the embedding of $\mathfrak{g}_{\bar{0}}$-modules $L^0(\lambda) \hookrightarrow L(\lambda)$ and Frobenius reciprocity.

$(1) \Rightarrow (2)$. Note that $L^0(\lambda)$ is an irreducible direct summand of $L(\lambda)$ regarded as a $\mathfrak{g}_{\bar{0}}$-module.

$(2) \Rightarrow (3)$. Follows from (2.2).

$(3) \Rightarrow (1)$. Follows from the surjectivity of the map $K(\lambda) \twoheadrightarrow L(\lambda)$. □

According to Proposition 1.39, every finite-dimensional simple $\mathfrak{g}$-module is a highest weight module $L(\lambda)$, for some $\lambda \in \mathfrak{h}^*$. Moreover, $L(\lambda) \not\cong L(\mu)$, if $\lambda \neq \mu$. By Proposition 2.1, the classification of finite-dimensional simple $\mathfrak{g}$-modules is the same as the classification of finite-dimensional simple modules of $\mathfrak{g}_{\bar{0}} = \mathfrak{gl}(m) \oplus \mathfrak{gl}(n)$, which is well known. Hence, we have established the following.

Proposition 2.2. *A complete list of pairwise non-isomorphic finite-dimensional simple $\mathfrak{gl}(m|n)$-modules are $L(\lambda)$, for $\lambda = \sum_{i=1}^m \lambda_i \delta_i + \sum_{j=1}^n \nu_j \varepsilon_j \in \mathfrak{h}^*$ satisfying $\lambda_i - \lambda_{i+1} \in \mathbb{Z}_+$ and $\nu_j - \nu_{j+1} \in \mathbb{Z}_+$ for all possible i, j.*

Remark 2.3. Note that $\mathfrak{g}_{\bar{0}}$ is a Levi subalgebra of $\mathfrak{g} = \mathfrak{gl}(m|n)$ (corresponding to the removal of the odd simple root from the standard Dynkin diagram of $\mathfrak{g}$), and hence the Kac module $K(\lambda)$ is a distinguished parabolic Verma module relative to the standard Borel subalgebra. Since $K(\lambda)$ is a highest weight module, it is indecomposable. However, for $m \geq 1$ and $n \geq 1$, $K(\lambda)$ is reducible for suitable λ, e.g., when $\lambda(h_\alpha) = 0$ for the odd simple root α by Lemma 1.40. Hence, the category of finite-dimensional $\mathfrak{gl}(m|n)$-modules is not semisimple when $m \geq 1$ and $n \geq 1$.

2.1.2. Finite-dimensional simple modules of $\mathfrak{spo}(2m|2)$. The classification of finite-dimensional simple modules for $\mathfrak{g} = \mathfrak{spo}(2m|2)$ is carried out similarly to the $\mathfrak{gl}(m|n)$ case. Let Φ^+ be the positive system of the root system Φ of $\mathfrak{g}$ corresponding to the Dynkin diagram (1.42). Associated to Φ^+ we have the triangular decomposition $\mathfrak{g} = \mathfrak{n}^- \oplus \mathfrak{h} \oplus \mathfrak{n}^+$, compatible with that of its even subalgebra $\mathfrak{g}_{\bar{0}} = \mathfrak{n}^-_{\bar{0}} \oplus \mathfrak{h} \oplus \mathfrak{n}^+_{\bar{0}}$, where $\mathfrak{n}^\pm_{\bar{0}} = \mathfrak{g}_{\bar{0}} \cap \mathfrak{n}^\pm$. In addition, $\mathfrak{g} = \mathfrak{spo}(2m|2)$ admits the following two favorable properties that are shared by $\mathfrak{gl}(m|n)$ but not by other $\mathfrak{spo}$ superalgebras: first, $\mathfrak{g}$ admits a $\mathbb{Z}$-graded Lie superalgebra structure of the form $\mathfrak{g} = \mathfrak{g}_{-1} \oplus \mathfrak{g}_0 \oplus \mathfrak{g}_1$, where $\mathfrak{g}_0 = \mathfrak{g}_{\bar{0}}$, $\mathfrak{g}_{\bar{1}} = \mathfrak{g}_{-1} \oplus \mathfrak{g}_1$, and the set of roots for $\mathfrak{g}_1$ is given by $\Phi^+_{\bar{1}} = \{\varepsilon_1 \pm \delta_i, 1 \leq i \leq m\}$. Secondly, $\mathfrak{g}_{\bar{0}}$ is a Levi subalgebra of $\mathfrak{g}$ corresponding to the removal of the odd simple root from the Dynkin diagram (1.42).

Hence we define the Kac module $K(\lambda)$ of $\mathfrak{g}$ as before: $K(\lambda) = \mathrm{Ind}^{\mathfrak{g}}_{\mathfrak{g}_0 \oplus \mathfrak{g}_1} L^0(\lambda)$, where $L^0(\lambda)$ denotes the simple $\mathfrak{g}_{\bar{0}}$-module of highest weight λ relative to $\Phi^+_{\bar{0}}$. Proposition 2.1 remains valid in the current setting, and it implies that the classification of finite-dimensional simple $\mathfrak{g}$-modules is the same as the classification of finite-dimensional simple modules of $\mathfrak{g}_{\bar{0}} \cong \mathfrak{sp}(2m) \oplus \mathbb{C}$.

Proposition 2.4. *A complete list of pairwise non-isomorphic finite-dimensional simple $\mathfrak{spo}(2m|2)$-modules are $L(\lambda)$, for $\lambda = \sum_{i=1}^m \lambda_i \delta_i + \upsilon_1 \varepsilon_1 \in \mathfrak{h}^*$, such that $\upsilon_1 \in \mathbb{C}$ and $(\lambda_1, \ldots, \lambda_m)$ is a partition.*

2.1.3. A virtual character formula. Let $\mathfrak{g}$ be a Lie superalgebra of type $\mathfrak{gl}$ or $\mathfrak{osp}$. With respect to a positive system $\Phi^+ = \Phi^+_{\bar{0}} \cup \Phi^+_{\bar{1}}$ (see Sections 1.3.2, 1.3.3, and 1.3.4), we have the triangular decomposition $\mathfrak{g} = \mathfrak{n}^- \oplus \mathfrak{h} \oplus \mathfrak{n}^+$. Letting $\mathfrak{b} = \mathfrak{n}^+ \oplus \mathfrak{h}$, we define the Verma module

$$\Delta(\lambda) = \mathrm{Ind}^{\mathfrak{g}}_{\mathfrak{b}} \mathbb{C}v_\lambda$$

associated to a weight $\lambda \in \mathfrak{h}^*$. Here, as usual, $\mathbb{C}v_\lambda$ stands for the one-dimensional $\mathfrak{b}$-module on which $\mathfrak{h}$ acts by the character λ, and $\mathfrak{n}^+$ acts trivially. Then $\Delta(\lambda)$ admits a unique simple quotient $\mathfrak{g}$-module, which is the irreducible $\mathfrak{g}$-module $L(\lambda)$ of highest weight λ.

A weight $\lambda \in \mathfrak{h}^*$ is called **dominant integral with respect to $\Phi^+_{\bar{0}}$** or simply **$\Phi^+_{\bar{0}}$-dominant integral** if $\langle \lambda, h_\alpha \rangle \in \mathbb{Z}_+$, for all $\alpha \in \Phi^+_{\bar{0}}$. A difference of two characters of finite-dimensional $\mathfrak{g}$-modules is called a **virtual character** of finite-dimensional $\mathfrak{g}$-modules. Also recall the Weyl vector $\rho = \rho_{\bar{0}} - \rho_{\bar{1}}$ from (1.35). We shall need the following proposition (see Santos [**104**, Proposition 5.7]) to determine if certain weights could be highest weights of finite-dimensional $\mathfrak{g}$-modules.

Proposition 2.5. *Let $\mathfrak{g}$ be of type $\mathfrak{gl}$ or $\mathfrak{osp}$. For any $\Phi_{\bar{0}}^+$-dominant integral weight $\lambda \in \mathfrak{h}^*$, the expression*

$$\frac{\prod_{\beta \in \Phi_{\bar{1}}^+}(e^{\beta/2} + e^{-\beta/2})}{\prod_{\alpha \in \Phi_{\bar{0}}^+}(e^{\alpha/2} - e^{-\alpha/2})} \sum_{w \in W} (-1)^{\ell(w)} e^{w(\lambda+\rho)} \tag{2.3}$$

is a virtual character of finite-dimensional $\mathfrak{g}$-modules.

Proof. Following the proof of [**104**, Proposition 5.7], we shall make use of the so-called Bernstein functor $\mathcal{L}_0$ [**104**, (12)] from the category of $\mathfrak{h}$-semisimple $\mathfrak{g}$-modules to the category of $\mathfrak{g}_{\bar{0}}$-semisimple $\mathfrak{g}$-modules. Actually, when restricting to the category of $\mathfrak{h}$-semisimple $\mathfrak{g}_{\bar{0}}$-modules, the functor $\mathcal{L}_0$ is just the classical Bernstein functor, which is a special case of the functor P defined in Knapp and Vogan [**72**, (2.8)]. As we shall only need some standard properties of the Bernstein functor, which can be found in [**72, 104**], we will not recall here the precise definition, which is rather involved.

An immediate consequence of the definition of the functor $\mathcal{L}_0$ is that every $\mathfrak{g}_{\bar{0}}$-composition factor of $\mathcal{L}_0(V)$ is also a $\mathfrak{g}_{\bar{0}}$-composition factor of V. Hence $\mathcal{L}_0$ takes a finitely generated $\mathfrak{h}$-semisimple $\mathfrak{g}_{\bar{0}}$-module in the BGG category to a finite-dimensional $\mathfrak{g}_{\bar{0}}$-module. Denote by $\mathcal{L}_i$ the ith derived functor of $\mathcal{L}_0$. Then, each $\mathcal{L}_i(V)$ is a finite-dimensional $\mathfrak{g}$-module for any Verma $\mathfrak{g}$-module V (which lies in the BGG category of $\mathfrak{g}_{\bar{0}}$-modules). Now it is a classical result (cf. [**72**, Corollary 4.160]) that the resulting Euler characteristic when applying $\mathcal{L}_0$ to a $\mathfrak{g}_{\bar{0}}$-Verma module is given by the Weyl character formula, i.e.,

$$\sum_{i=0}^{\infty}(-1)^i \mathrm{ch}\,\mathcal{L}_i(\mathrm{Ind}_{\mathfrak{b}_{\bar{0}}}^{\mathfrak{g}_{\bar{0}}}\mathbb{C}_\lambda) = \frac{\sum_{\mathrm{w}\in \mathrm{W}}(-1)^{\ell(\mathrm{w})}\mathrm{e}^{\mathrm{w}(\lambda+\rho_{\bar{0}})}}{\prod_{\alpha\in\Phi_{\bar{0}}^+}(\mathrm{e}^{\alpha/2}-\mathrm{e}^{-\alpha/2})},$$

and $\mathcal{L}_i(\mathrm{Ind}_{\mathfrak{b}_{\bar{0}}}^{\mathfrak{g}_{\bar{0}}}\mathbb{C}_\lambda) = 0$ for $i \gg 0$. This can be viewed as an algebraic version of the Borel-Weil-Bott theorem.

Given a short exact sequence of finite-dimensional $\mathfrak{h}$-semisimple $\mathfrak{g}_{\bar{0}}$-modules

$$0 \longrightarrow M \longrightarrow E \longrightarrow N \longrightarrow 0,$$

it follows by the Euler-Poincaré principle that

$$\sum_{i=0}^{\infty}(-1)^i \mathrm{ch}\,\mathcal{L}_i(M) + \sum_{i=0}^{\infty}(-1)^i \mathrm{ch}\,\mathcal{L}_i(N) = \sum_{i=0}^{\infty}(-1)^i \mathrm{ch}\,\mathcal{L}_i(E).$$

As a $\mathfrak{g}_{\bar{0}}$-module, $\Delta(\lambda)$ and $\mathrm{Ind}_{\mathfrak{b}_{\bar{0}}}^{\mathfrak{g}_{\bar{0}}}\mathbb{C}_\lambda \otimes \Lambda(\mathfrak{n}_{\bar{1}}^-)$ have finite $\mathfrak{g}_{\bar{0}}$-Verma flags, where a Verma flag is a filtration whose sections are Verma modules. Since these two $\mathfrak{g}_{\bar{0}}$-modules also have identical characters, their $\mathfrak{g}_{\bar{0}}$-Verma flags have the same Verma module multiplicities. Now the character of the $\mathfrak{b}_{\bar{0}}$-module $\mathbb{C}_\lambda \otimes \Lambda(\mathfrak{n}_{\bar{1}}^-)$ equals $e^\lambda \prod_{\beta\in\Phi_{\bar{1}}^+}(1+e^{-\beta})$. Noting the W-invariance of $\prod_{\beta\in\Phi_{\bar{1}}^+}(e^{\beta/2}+e^{-\beta/2})$, we have

$$
\begin{aligned}
\sum_{i=0}^{\infty}(-1)^i \mathrm{ch}\,\mathcal{L}_i(\Delta(\lambda)) &= \sum_{i=0}^{\infty}(-1)^i \mathrm{ch}\,\mathcal{L}_i\big(\mathrm{Ind}_{\mathfrak{b}_{\bar{0}}}^{\mathfrak{g}_{\bar{0}}}\mathbb{C}_\lambda \otimes \Lambda(\mathfrak{n}_{\bar{1}}^-)\big)\\
&= \frac{\sum_{w\in W}(-1)^{\ell(w)} w\big(e^{\lambda+\rho_{\bar{0}}}\prod_{\beta\in\Phi_{\bar{1}}^+}(1+e^{-\beta})\big)}{\prod_{\alpha\in\Phi_{\bar{0}}^+}(e^{\alpha/2}-e^{-\alpha/2})}\\
&= \frac{\sum_{w\in W}(-1)^{\ell(w)} w\big(e^{\lambda+\rho_{\bar{0}}-\rho_{\bar{1}}}\prod_{\beta\in\Phi_{\bar{1}}^+}(e^{\beta/2}+e^{-\beta/2})\big)}{\prod_{\alpha\in\Phi_{\bar{0}}^+}(e^{\alpha/2}-e^{-\alpha/2})}\\
&= \frac{\prod_{\beta\in\Phi_{\bar{1}}^+}(e^{\beta/2}+e^{-\beta/2})}{\prod_{\alpha\in\Phi_{\bar{0}}^+}(e^{\alpha/2}-e^{-\alpha/2})}\sum_{w\in W}(-1)^{\ell(w)} e^{w(\lambda+\rho)}.
\end{aligned}
$$

Since the left-hand side is a virtual character of finite-dimensional $\mathfrak{g}$-modules, so is the right-hand side. This proves the proposition. □

Remark 2.6. Upon changing the sign $+$ in the numerator of (2.3) to $-$, the resulting expression

$$
\frac{\prod_{\beta\in\Phi_{\bar{1}}^+}(e^{\beta/2}-e^{-\beta/2})}{\prod_{\alpha\in\Phi_{\bar{0}}^+}(e^{\alpha/2}-e^{-\alpha/2})}\sum_{w\in W}(-1)^{\ell(w)} e^{w(\lambda+\rho)}
$$

is a virtual supercharacter (see Section 2.2.4) of finite-dimensional $\mathfrak{g}$-modules.

2.1.4. Finite-dimensional simple modules of $\mathfrak{spo}(2m|2n+1)$. We start with a general remark for a Lie superalgebra $\mathfrak{g}$ of type $\mathfrak{osp}$. By Proposition 1.39, every finite-dimensional irreducible $\mathfrak{g}$-module is necessarily of the form $L(\lambda)$, for some $\lambda \in \mathfrak{h}^*$. Moreover, we have $L(\lambda) \cong L(\mu)$ if and only if $\lambda = \mu$. However, the classification of finite-dimensional simple $\mathfrak{g}$-modules is nontrivial, partly because the even subalgebra $\mathfrak{g}_{\bar{0}}$ of $\mathfrak{g}$ is not a Levi subalgebra.

Lemma 2.7. *Let $\mathfrak{g}$ be a basic Lie superalgebra. A necessary condition for the finite dimensionality of the $\mathfrak{g}$-module $L(\lambda)$ is that λ is $\Phi_{\bar{0}}^+$-dominant integral.*

Proof. The proof follows by noting that λ is a highest weight with respect to $\Phi_{\bar{0}}^+$ for $L(\lambda)$, regarded as a $\mathfrak{g}_{\bar{0}}$-module. □

In this subsection we consider $\mathfrak{g} = \mathfrak{spo}(2m|2n+1)$ and let Φ^+ be the standard positive system (1.39) for $\mathfrak{g}$. The next lemma is a reformulation of the well-known dominance integral (DI) conditions for Lie algebras $\mathfrak{sp}(2m)$ and $\mathfrak{so}(2n+1)$.

Lemma 2.8. *Let $\mathfrak{g} = \mathfrak{spo}(2m|2n+1)$ with $\mathfrak{g}_{\bar{0}} = \mathfrak{sp}(2m) \oplus \mathfrak{so}(2n+1)$. A weight $\lambda = \sum_{i=1}^m \lambda_i \delta_i + \sum_{j=1}^n \nu_j \varepsilon_j$ is $\Phi_{\bar{0}}^+$-dominant integral if and only if the following two conditions are satisfied:*

(DI.b-1) $\lambda_1 \geq \ldots \geq \lambda_m$ *with all* $\lambda_i \in \mathbb{Z}_+$.

(DI.b-2) $\nu_1 \geq \ldots \geq \nu_n$, *with either* (i) *all* $\nu_j \in \mathbb{Z}_+$ *or* (ii) *all* $\nu_j \in \frac{1}{2} + \mathbb{Z}_+$.

For $\mathfrak{g} = \mathfrak{spo}(2m|2n+1)$, a $\Phi_{\bar{0}}^+$-dominant integral weight is called an **integer weight** if it satisfies (DI.b-1), and it is called a **half-integer weight** if it satisfies (DI.b-2). The $L(\lambda)$'s for integer and half-integer weights λ turn out to be quite different, and they are analyzed separately.

Proposition 2.9. *Let $\mathfrak{g} = \mathfrak{spo}(2m|2n+1)$ and let $\lambda = \sum_{i=1}^m \lambda_i \delta_i + \sum_{j=1}^n \nu_j \varepsilon_j \in \mathfrak{h}^*$ be a half-integer $\Phi_{\bar{0}}^+$-dominant integral weight. Then the highest weight $\mathfrak{g}$-module $L(\lambda)$ is finite dimensional if and only if $\lambda_m \geq n$.*

Proof. Take a half-integer $\Phi_{\bar{0}}^+$-dominant integral weight λ such that $\lambda_m \geq n$. We will look for the highest weight appearing in the virtual character (2.3) of finite-dimensional $\mathfrak{g}$-modules in Proposition 2.5. Recalling the Weyl vector $\rho = \rho_{\bar{0}} - \rho_{\bar{1}}$ from (1.35), we compute that

$$\rho_{\bar{1}} = \frac{2n+1}{2} \sum_{i=1}^m \delta_i,$$

$$\rho_{\bar{0}} = m\delta_1 + (m-1)\delta_2 + \ldots + \delta_m + (n-1/2)\varepsilon_1 + (n-3/2)\varepsilon_2 + \ldots + (1/2)\varepsilon_n.$$

Since the Weyl group of $\mathfrak{sp}(2m)$ is $\cong \mathbb{Z}_2^m \rtimes \mathfrak{S}_m$ and it consists of the signed permutations among $\delta_1, \ldots, \delta_m$, we conclude by the assumption $\lambda_m \geq n$ that the weight $\lambda + \rho$ remains $\Phi_{\bar{0}}^+$-dominant and regular. Thus the highest weight occurring in the virtual character of finite-dimensional $\mathfrak{g}$-modules (2.3) is $(\lambda + \rho) + \rho_{\bar{1}} - \rho_{\bar{0}} = \lambda$, and so $L(\lambda)$ must be finite dimensional by Proposition 2.5.

Conversely, suppose that the module $L(\lambda)$ is finite dimensional for a half-integer $\Phi_{\bar{0}}^+$-dominant integral weight λ. We apply the following sequence of mn odd reflections to the standard fundamental and positive systems of $\mathfrak{g}$ in (1.39). We first apply the n odd reflections with respect to $\delta_m - \varepsilon_1, \delta_m - \varepsilon_2, \ldots, \delta_m - \varepsilon_n$, then the n odd reflections with respect to $\delta_{m-1} - \varepsilon_1, \ldots, \delta_{m-1} - \varepsilon_n$ etc., until we finally apply the n odd reflections with respect to $\delta_1 - \varepsilon_1, \delta_1 - \varepsilon_2, \ldots, \delta_1 - \varepsilon_n$. After a computation similar to Example 1.35, the resulting fundamental system is that of the diagram (1.40). By the half-integer weight assumption on λ, the value of λ at any isotropic odd coroot belongs to $\frac{1}{2} + \mathbb{Z}$ and hence is nonzero. By Lemma 1.40, after an odd reflection with respect to an isotropic odd root α, the new highest weight is $\lambda - \alpha$, which again takes value in $\frac{1}{2} + \mathbb{Z}$ at any odd isotropic coroot. Continuing this way and using a computation very similar to Example 1.41, we see that the new highest weight corresponding to the above Dynkin diagram is equal to

$\lambda - n\sum_{i=1}^m \delta_i + m\sum_{j=1}^n \varepsilon_j$. Now this weight must take non-negative integer value at $h_{2\delta_m}$ by the finite dimensionality of $L(\lambda)$, from which we conclude that $\lambda_m \geq n$. □

We next determine for which integer $\Phi_{\bar{0}}^+$-dominant integral weight λ the $\mathfrak{g}$-module $L(\lambda)$ is finite dimensional. Recall μ' denotes the conjugate partition of a partition μ. The following definition plays a fundamental role in the book.

Definition 2.10. A partition $\mu = (\mu_1, \mu_2, \ldots)$ is called an $(m|n)$-**hook partition** (or simply a **hook partition** when m,n are implicitly understood) if $\mu_{m+1} \leq n$. Equivalently, μ is an $(m|n)$-hook partition if $\mu'_{n+1} \leq m$.

Diagrammatically, the hook condition means that the $(m+1, n+1)$-box in the Young diagram is missing; it can be visualized as follows:

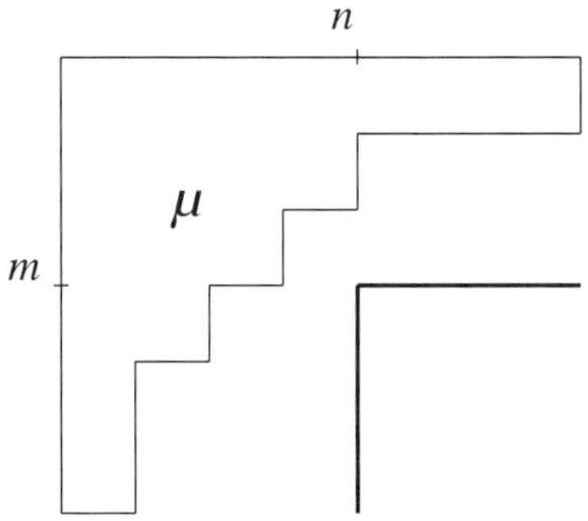

Given an $(m|n)$-hook partition μ, we denote by $\mu^+ = (\mu_{m+1}, \mu_{m+2}, \ldots)$ and write its conjugate, which is necessarily of length $\leq n$, as $\nu = (\mu^+)' = (\nu_1, \ldots, \nu_n)$. We define the weights

$$\begin{aligned} \mu^\natural &= \mu_1\delta_1 + \ldots + \mu_m\delta_m + \nu_1\varepsilon_1 + \ldots + \nu_{n-1}\varepsilon_{n-1} + \nu_n\varepsilon_n, \\ \mu^\natural_- &= \mu_1\delta_1 + \ldots + \mu_m\delta_m + \nu_1\varepsilon_1 + \ldots + \nu_{n-1}\varepsilon_{n-1} - \nu_n\varepsilon_n. \end{aligned} \tag{2.4}$$

($\mu^\natural_-$ is only used for $\mathfrak{spo}(2m|2n)$ below.) It is sometimes convenient to identify $\mu^\natural$ and $\mu^\natural_-$ with the following tuples of integers:

$$\begin{aligned} \mu^\natural &= (\mu_1, \ldots, \mu_m; \nu_1, \ldots, \nu_n), \\ \mu^\natural_- &= (\mu_1, \ldots, \mu_m; \nu_1, \ldots, -\nu_n). \end{aligned}$$

Theorem 2.11. *Let $\mathfrak{g} = \mathfrak{spo}(2m|2n+1)$, and let $\lambda = \sum_{i=1}^m \lambda_i\delta_i + \sum_{j=1}^n \upsilon_j\varepsilon_j \in \mathfrak{h}^*$ be a $\Phi_{\bar{0}}^+$-dominant integral weight.*

(1) *Suppose that λ is a half-integer weight. Then $L(\lambda)$ is finite dimensional if and only if $\lambda_m \geq n$.*
(2) *Suppose that λ is an integer weight. Then $L(\lambda)$ is finite dimensional if and only if λ is of the form $\mu^\natural$, where μ is an $(m|n)$-hook partition.*

Proof. Part (1) is Proposition 2.9. So it remains to prove (2).

Let μ be an $(m|n)$-hook partition. For an integer M such that $M \geq \max\{m, \ell(\mu)\}$, consider the Lie superalgebra $\mathfrak{spo}(2M|2n+1)$ acting on its standard representation $\mathbb{C}^{2M|2n+1}$. It has highest weight δ_1. The module $\wedge^k(\mathbb{C}^{2M|2n+1})$ contains a highest weight vector of highest weight $\sum_{j=1}^{k}\delta_j$, for $k \leq M$. Write the conjugate partition $\mu' = (\nu_1, \nu_2, \ldots)$. Then the $\mathfrak{spo}(2M|2n+1)$-module $\wedge^{\nu_1}(\mathbb{C}^{2M|2n+1}) \otimes \wedge^{\nu_1}(\mathbb{C}^{2M|2n+1}) \otimes \cdots$ contains the irreducible $\mathfrak{spo}(2M|2n+1)$-module V of highest weight μ as a quotient, and hence V is finite dimensional. Now, we use a sequence of odd reflections to change the standard Borel subalgebra of $\mathfrak{spo}(2M|2n+1)$ into a Borel subalgebra that contains as a subalgebra the standard Borel subalgebra of $\mathfrak{spo}(2m|2n+1)$. (Here $\mathfrak{spo}(2m|2n+1)$ is regarded as a subalgebra of $\mathfrak{spo}(2M|2n+1)$.) A sequence of odd reflections that we may use for this purpose is the following: We first apply the n odd reflections with respect to $\delta_M - \varepsilon_1, \delta_M - \varepsilon_2, \ldots, \delta_M - \varepsilon_n$, then the n odd reflections with respect to $\delta_{M-1} - \varepsilon_1, \ldots, \delta_{M-1} - \varepsilon_n$ etc., until we finally apply the n odd reflections with respect to $\delta_{m+1} - \varepsilon_1, \delta_{m+1} - \varepsilon_2, \ldots, \delta_{m+1} - \varepsilon_n$. Thus, we have applied a total of $(M-m)n$ odd reflections. The subdiagram consisting of the first $(m+n-1)$ vertices of the new Dynkin diagram is of type $\mathfrak{gl}(m|n)$. Similar to Example 1.41, one shows that the highest weight of V with respect to the new Borel subalgebra is $\mu^\natural$. Hence by restriction we conclude that the $\mathfrak{spo}(2m|2n+1)$-module $L(\mu^\natural)$ is finite dimensional, proving the "if" direction of (2).

By Lemma 2.8, an integer $\Phi_{\bar{0}}^+$-dominant integral highest weight for a simple finite-dimensional $\mathfrak{g}$-module is necessarily of the form $\mu = \mu_1\delta_1 + \ldots + \mu_m\delta_m + \nu_1\varepsilon_1 + \ldots + \nu_n\varepsilon_n$, where $(\mu_1, \ldots, \mu_m)$ and $(\nu_1, \ldots, \nu_n)$ are partitions. To prove the remaining condition $\nu_1' \leq \mu_m$, it suffices to prove $\nu_1' \leq \mu_1$ in the case of $m = 1$ by noting that $\mathfrak{spo}(2|2n+1)$ is a subalgebra of $\mathfrak{spo}(2m|2n+1)$. Let V be a finite-dimensional simple $\mathfrak{spo}(2|2n+1)$-module of highest weight μ. Via the sequence of odd reflections with respect to $\delta_1 - \varepsilon_1, \delta_1 - \varepsilon_2, \ldots, \delta_1 - \varepsilon_n$, we change the standard Borel subalgebra to a Borel subalgebra with an odd non-isotropic simple root δ_1. Now if $\mu_1 \geq n$, then there is no condition on μ and we are done. Suppose now that $\mu_1 = k < n$; we shall prove by contradiction by assuming that $\nu_{k+1} \geq 1$ (and so $\nu_1' > \mu_1$). This implies that $\nu_i \geq 1$, for all $1 \leq i \leq k$, and hence, when applying the odd reflection with respect to $\delta_1 - \varepsilon_i$, the new highest weight is obtained from the old one by subtracting $\delta_1 - \varepsilon_i$, for $1 \leq i \leq k$. This also remains true when applying the odd reflection with respect to $\delta_1 - \varepsilon_{k+1}$. Hence the resulting new highest weight has negative δ_1-coefficient. But this implies that the new highest weight evaluated at the coroot $h_{2\delta_1}$ must be negative, contradicting the finite dimensionality of the module V. Hence μ must satisfy the hook condition, as claimed. □

2.1.5. Finite-dimensional simple modules of $\mathfrak{spo}(2m|2n)$. Let $n \geq 2$. Let $\mathfrak{g} = \mathfrak{spo}(2m|2n)$ with $\mathfrak{g}_{\bar{0}} = \mathfrak{sp}(2m) \oplus \mathfrak{so}(2n)$. In this subsection Φ^+ is the standard positive system (1.41) for $\mathfrak{g}$. The next lemma is the well-known dominance integral conditions for classical Lie algebras $\mathfrak{sp}(2m)$ and $\mathfrak{so}(2n)$.

Lemma 2.12. *Let $\mathfrak{g} = \mathfrak{spo}(2m|2n)$. Assume $n \geq 2$. A weight $\lambda = \sum_{i=1}^{m} \lambda_i \delta_i + \sum_{j=1}^{n} \upsilon_j \varepsilon_j \in \mathfrak{h}^*$ is $\Phi_{\bar{0}}^+$-dominant integral if and only if the following two conditions are satisfied:*

(DI.d-1) $\lambda_1 \geq \ldots \geq \lambda_m$ *with all* $\lambda_i \in \mathbb{Z}_+$.

(DI.d-2) $\upsilon_1 \geq \ldots \geq \upsilon_{n-1} \geq |\upsilon_n|$, *with either* (i) *all* $\upsilon_j \in \mathbb{Z}$ *or* (ii) *all* $\upsilon_j \in \frac{1}{2} + \mathbb{Z}$.

For $\mathfrak{g} = \mathfrak{spo}(2m|2n)$, a $\Phi_{\bar{0}}^+$-dominant integral weight $\lambda \in \mathfrak{h}^*$ is called an **integer weight** if it satisfies (DI.d-1), and it is called a **half-integer weight** if it satisfies (DI.d-2). The $L(\lambda)$'s for integer and half-integer weights λ again are quite different.

Proposition 2.13. *Let $\mathfrak{g} = \mathfrak{spo}(2m|2n)$ with $n \geq 2$. Let $\lambda = \sum_{i=1}^{m} \lambda_i \delta_i + \sum_{j=1}^{n} \upsilon_j \varepsilon_j$ be a half-integer $\Phi_{\bar{0}}^+$-dominant integral weight. Then the highest weight $\mathfrak{g}$-module $L(\lambda)$ is finite dimensional if and only if $\lambda_m \geq n$.*

Proof. Take a half-integer $\Phi_{\bar{0}}^+$-dominant integral weight λ with $\lambda_m \geq n$. We compute that $\rho_1 = n \sum_{i=1}^{m} \delta_i$, and so $\mu := \lambda - \rho_1$ is a $\Phi_{\bar{0}}^+$-dominant integral weight by Lemma 2.12. By virtue of the classical Weyl character formula, the virtual character of finite-dimensional $\mathfrak{g}$-modules (2.3) in Proposition 2.5 can be rewritten as $\mathrm{ch} L^0(\mu) \prod_{\beta \in \Phi_{\bar{1}}^+} (e^{\beta/2} + e^{-\beta/2})$. The highest weight in this expression is clearly $\mu + \rho_1 = \lambda$, and we conclude by Proposition 2.5 that $L(\lambda)$ is finite dimensional.

Now let λ be a half-integer $\Phi_{\bar{0}}^+$-dominant integral weight such that $L(\lambda)$ is finite dimensional. We apply exactly the same sequence of mn odd reflections as in the proof of Proposition 2.9 now to the standard fundamental system for $\mathfrak{spo}(2m|2n)$ in (1.41) in Section 1.3.4. The resulting fundamental system is that of the diagram in (1.42). Bearing in mind that λ is half-integer, as in the proof of Proposition 2.9, we can show that the highest weight with respect to this new Borel subalgebra is $\lambda - n \sum_{i=1}^{m} \delta_i + m \sum_{j=1}^{n} \varepsilon_j$. Since this weight must take non-negative integer value at $h_{2\delta_m}$ to ensure that $L(\lambda)$ is finite dimensional, we conclude that $\lambda_m \geq n$. □

Theorem 2.14. *Let $\mathfrak{g} = \mathfrak{spo}(2m|2n)$ with $n \geq 2$. Let $\lambda = \sum_{i=1}^{m} \lambda_i \delta_i + \sum_{j=1}^{n} \upsilon_j \varepsilon_j$ be a $\Phi_{\bar{0}}^+$-dominant integral weight.*

(1) *Suppose that λ is a half-integer weight. Then $L(\lambda)$ is finite dimensional if and only if $\lambda_m \geq n$.*

(2) *Suppose that λ is an integer weight. Then $L(\lambda)$ is finite dimensional if and only if λ is of the form $\mu^\natural$ or $\mu^\natural_-$, where μ is an $(m|n)$-hook partition.*

Proof. Part (1) is Proposition 2.13. Since the proof for (2) is very similar to the proof for Theorem 2.11(2), we shall only give a sketch below.

We recall from (1.41) that the standard Dynkin diagram of $\mathfrak{g} = \mathfrak{spo}(2m|2n)$ possesses a nontrivial diagram involution, which induces an involution of the Lie superalgebra $\mathfrak{g}$. Twisting the module structure with this involution interchanges the irreducible modules $L(\mu^\natural)$ and $L(\mu^\natural_-)$.

Given an $(m|n)$-hook partition μ, we choose $M \geq \max\{m, \ell(\mu)\}$ and consider the Lie superalgebra $\mathfrak{spo}(2M|2n)$. Then, arguing as in the proof of Theorem 2.11, the irreducible $\mathfrak{spo}(2M|2n)$-module of highest weight μ is finite dimensional. We apply the same sequence of odd reflections as in that proof to change the standard Borel subalgebra of $\mathfrak{spo}(2M|2n)$ to a Borel subalgebra that is compatible with the standard Borel subalgebra of $\mathfrak{spo}(2m|2n)$. Again, the subdiagram consisting of the first $(m+n-1)$ vertices of the new Dynkin diagram is of type $\mathfrak{gl}(m|n)$, and it is easy to see that the new highest weight equals $\mu^\natural$. By restriction to $\mathfrak{spo}(2m|2n)$, we see that $L(\mu^\natural)$ is finite dimensional, and so $L(\mu^\natural_-)$ is also finite dimensional by applying the Dynkin diagram involution.

Conversely, let $\mu = \mu_1\delta_1 + \ldots + \mu_m\delta_m + \nu_1\varepsilon_1 + \ldots + \nu_n\varepsilon_n$ be an integer $\Phi^+_{\bar{0}}$-dominant integral weight. By Lemma 2.12, $(\mu_1, \ldots, \mu_m)$ and $(\nu_1, \ldots, |\nu_n|)$ are partitions. To prove that $\nu'_1 \leq \mu_m$, it suffices to prove $\nu'_1 \leq \mu_1$ in the case that $m = 1$. Using the sequence of odd reflections with respect to $\delta_1 - \varepsilon_1, \delta_1 - \varepsilon_2, \ldots, \delta_1 - \varepsilon_n$, we change the standard Borel subalgebra to a Borel subalgebra with an even simple root $2\delta_1$. Now the same argument as in the proof of Theorem 2.11(2) shows that if μ does not satisfy the hook condition, then μ must take a negative value at $h_{2\delta_1}$, contradicting the finite dimensionality of $L(\mu)$. □

Example 2.15. Let $\xi, \bar{\xi}$ be two odd indeterminates, and let $x = \{x_1, \cdots, x_m\}$, $\bar{x} = \{\bar{x}_1, \cdots, \bar{x}_m\}$ be $2m$ even indeterminates with $m > 0$. We identify the span of these indeterminates with $\mathbb{C}^{2|2m}$ so that the **exterior algebra**

$$\wedge(\mathbb{C}^{2|2m}) := \wedge(\mathbb{C}^2) \otimes S(\mathbb{C}^{2m})$$

is identified with the polynomial superalgebra generated by these indeterminates, which we denote by $\mathbb{C}[\xi, \bar{\xi}, x, \bar{x}]$. We denote by Π the fundamental system of $\mathfrak{osp}(2|2m)$ given by $\{\varepsilon_1 - \delta_1, \delta_1 - \delta_2, \ldots, \delta_{n-1} - \delta_n, 2\delta_n\}$ with Dynkin diagram as in (1.42) with $n = 1$. Let e_α and f_α, for $\alpha \in \Pi$, denote the Chevalley generators. Define the following first-order differential operators on $\mathbb{C}[\xi, \bar{\xi}, x, \bar{x}]$:

$$\breve{e}_{\varepsilon_1-\delta_1} = \xi\frac{\partial}{\partial x_1} + \bar{x}_1\frac{\partial}{\partial \bar{\xi}}, \quad \breve{f}_{\varepsilon_1-\delta_1} = x_1\frac{\partial}{\partial \xi} - \bar{\xi}\frac{\partial}{\partial \bar{x}_1},$$

$$\breve{e}_{\delta_i-\delta_{i+1}} = x_i\frac{\partial}{\partial x_{i+1}} - \bar{x}_{i+1}\frac{\partial}{\partial \bar{x}_i}, \quad \breve{f}_{\delta_i-\delta_{i+1}} = x_{i+1}\frac{\partial}{\partial x_i} - \bar{x}_i\frac{\partial}{\partial \bar{x}_{i+1}}, \quad i = 1, \cdots, m-1,$$

$$\breve{e}_{2\delta_m} = x_m\frac{\partial}{\partial \bar{x}_m}, \quad \breve{f}_{2\delta_m} = \bar{x}_m\frac{\partial}{\partial x_m}.$$

The map $e_\alpha \mapsto \breve{e}_\alpha$ and $f_\alpha \mapsto \breve{f}_\alpha$, for $\alpha \in \Pi$, defines a representation of $\mathfrak{osp}(2|2m)$ on $\mathbb{C}[\xi, \bar{\xi}, x, \bar{x}]$. Indeed, this gives an explicit realization of the irreducible representations $L(\varepsilon_1 + k\delta_1) \cong \wedge^{k+1}(\mathbb{C}^{2|2m})$, for all $k \in \mathbb{Z}_+$ (see Exercise 2.12).

Example 2.16. In order to construct the symmetric tensors of the natural $\mathfrak{osp}(2|2m)$-module we consider $x, \bar{x}$, two even, and $\xi = \{\xi_1, \cdots, \xi_m\}$, $\bar{\xi} = \{\bar{\xi}_1, \cdots, \bar{\xi}_m\}$, $2m$ odd indeterminates. We identify the span of these indeterminates with $\mathbb{C}^{2|2m}$ so that the **supersymmetric algebra**

$$\mathcal{S}(\mathbb{C}^{2|2m}) := S(\mathbb{C}^2) \otimes \wedge(\mathbb{C}^{2m})$$

is identified with $\mathbb{C}[x, \bar{x}, \xi, \bar{\xi}]$. We continue to use the notation from Example 2.15 and define the following first-order differential operators on $\mathbb{C}[x, \bar{x}, \xi, \bar{\xi}]$:

$$\hat{e}_{\varepsilon_1 - \delta_1} = x\frac{\partial}{\partial \xi_1} + \bar{\xi}_1 \frac{\partial}{\partial \bar{x}}, \quad \hat{f}_{\varepsilon_1 - \delta_1} = \xi_1 \frac{\partial}{\partial x} - \bar{x}\frac{\partial}{\partial \bar{\xi}_1},$$

$$\hat{e}_{\delta_i - \delta_{i+1}} = \xi_i \frac{\partial}{\partial \xi_{i+1}} - \bar{\xi}_{i+1}\frac{\partial}{\partial \bar{\xi}_i}, \quad \hat{f}_{\delta_i - \delta_{i+1}} = \xi_{i+1}\frac{\partial}{\partial \xi_i} - \bar{\xi}_i \frac{\partial}{\partial \bar{\xi}_{i+1}},$$

$$\hat{e}_{2\delta_m} = \xi_m \frac{\partial}{\partial \bar{\xi}_m}, \quad \hat{f}_{2\delta_m} = \bar{\xi}_m \frac{\partial}{\partial \xi_m}.$$

The map $e_\alpha \mapsto \hat{e}_\alpha$ and $f_\alpha \mapsto \hat{f}_\alpha$, for $\alpha \in \Pi$, defines a representation of $\mathfrak{osp}(2|2m)$ on $\mathbb{C}[x, \bar{x}, \xi, \bar{\xi}]$. Just like for the orthogonal Lie algebra, the symmetric tensors are not irreducible. The computation of the composition factors of the symmetric tensor here is more involved than in the case of orthogonal Lie algebras (see Exercises 2.13–2.16).

2.1.6. Finite-dimensional simple modules of $\mathfrak{q}(n)$. Now suppose that $\mathfrak{g}$ is the queer Lie superalgebra $\mathfrak{q}(n)$. Let $\mathfrak{h}$ be the standard Cartan subalgebra of block diagonal matrices and let $\mathfrak{b} = \mathfrak{h} \oplus \mathfrak{n}^+$ be the standard Borel subalgebra of block upper triangular matrices; cf. Section 1.2.6. Recall that $\{H_i = E_{\bar{i},\bar{i}} + E_{ii} \mid 1 \le i \le n\}$ forms the standard basis for $\mathfrak{h}_{\bar{0}}$, with dual basis $\{\varepsilon_i \mid 1 \le i \le n\}$ for $\mathfrak{h}_{\bar{0}}^*$. We have a triangular decomposition $\mathfrak{g} = \mathfrak{n}^- \oplus \mathfrak{h} \oplus \mathfrak{n}^+$.

Recall from Lemma 1.42 that W_λ is a finite-dimensional irreducible $\mathfrak{h}$-module, for $\lambda \in \mathfrak{h}_{\bar{0}}^*$. Regarding W_λ as a $\mathfrak{b}$-module by letting $\mathfrak{n}^+$ act trivially, we define the Verma module of $\mathfrak{g}$ to be

$$\Delta(\lambda) = \mathrm{Ind}_{\mathfrak{b}}^{\mathfrak{g}} W_\lambda.$$

The highest weight space of $\Delta(\lambda)$ can be naturally identified with W_λ and has dimension $2^{(\ell(\lambda)+\delta(\lambda))/2}$. Any nonzero vector in the highest weight space W_λ generates the whole $\mathfrak{g}$-module $\Delta(\lambda)$. Following standard arguments for semisimple Lie algebras, any $\mathfrak{g}$-submodule of $\Delta(\lambda)$ has a weight space decomposition, and then the Verma module $\Delta(\lambda)$ contains a unique maximal proper submodule $J(\lambda)$ (which is the sum of all proper submodules). Hence, $\Delta(\lambda)$ has a unique irreducible quotient $\mathfrak{g}$-module $L(\lambda) = \Delta(\lambda)/J(\lambda)$.

Assume that the $\mathfrak{g}$-module $L(\lambda)$ is finite dimensional, and write $\lambda = \sum_{i=1}^n \lambda_i \varepsilon_i$. Since $\mathfrak{g}_{\bar{0}} \cong \mathfrak{gl}(n)$, the $\mathfrak{g}_{\bar{0}}$-dominance integral condition imposes a necessary condition on λ: $\lambda_i - \lambda_{i+1} \in \mathbb{Z}_+$, for all $1 \le i \le n-1$. However, this condition on λ is not sufficient for $L(\lambda)$ to be finite dimensional, as we see in the following lemma.

Lemma 2.17. *Let $\mathfrak{g} = \mathfrak{q}(2)$ and $\lambda = m\varepsilon_1 + m\varepsilon_2 \in \mathfrak{h}_{\bar{0}}^*$ with $m \neq 0$. Then $L(\lambda)$ is infinite dimensional.*

Proof. Recall the notations $\widetilde{E}_{ij}, \overline{E}_{ij}$, etc. from Sections 1.1.4 and 1.2.6. Then $\widetilde{E}_{12}$, $\widetilde{E}_{21}$, and $H_1 - H_2 = \widetilde{E}_{11} - \widetilde{E}_{22}$ form a standard $\mathfrak{sl}(2)$-triple $\mathfrak{s}$. Take a highest weight vector v_λ in $L(\lambda)$. For $\lambda = m\varepsilon_1 + m\varepsilon_2$, we have

$$\widetilde{E}_{12} v_\lambda = 0 = (H_1 - H_2) v_\lambda.$$

To prove the lemma, it suffices to show that $\widetilde{E}_{21} v_\lambda \neq 0$ in $L(\lambda)$. Indeed, by the $\mathfrak{sl}(2)$ representation theory, $\widetilde{E}_{21} v_\lambda \neq 0$ implies that the $\mathfrak{s}$-submodule generated by v_λ must be the infinite-dimensional Verma module of zero highest weight.

To that end, recalling $\overline{H}_i = E_{\bar{i}i} + E_{i\bar{i}}$, we compute that

$$\begin{aligned}\overline{E}_{12}\widetilde{E}_{21} v_\lambda &= \widetilde{E}_{21}\overline{E}_{12} v_\lambda + (\overline{H}_1 - \overline{H}_2) v_\lambda \\ &= (\overline{H}_1 - \overline{H}_2) v_\lambda,\end{aligned}$$

and $(\overline{H}_1 - \overline{H}_2)^2 v_\lambda = (\widetilde{E}_{11} + \widetilde{E}_{22}) v_\lambda = 2m v_\lambda \neq 0$. Hence, we have $(\overline{H}_1 - \overline{H}_2) v_\lambda \neq 0$ and $\widetilde{E}_{21} v_\lambda \neq 0$. This proves the lemma. □

Theorem 2.18. *The highest weight irreducible $\mathfrak{q}(n)$-module $L(\lambda)$ is finite dimensional if and only if $\lambda = \sum_{i=1}^n \lambda_i \varepsilon_i$ satisfies the conditions (i)-(ii):*

(i) $\lambda_i - \lambda_{i+1} \in \mathbb{Z}_+$; (ii) $\lambda_i = \lambda_{i+1}$ *implies that* $\lambda_i = 0$, *for* $1 \leq i \leq n-1$.

Proof. For $L(\lambda)$ to be finite dimensional, $\lambda = \sum_{i=1}^n \lambda_i \varepsilon_i$ must satisfy the usual dominance integral condition for $\mathfrak{q}(n)_{\bar{0}} \cong \mathfrak{gl}(n)$, which is Condition (i). Now λ also must satisfy Condition (ii) by applying Lemma 2.17 to various standard subalgebras of $\mathfrak{q}(n)$ that are isomorphic to $\mathfrak{q}(2)$. This proves the "only if" part.

Assume that λ satisfies Conditions (i)-(ii), and assume in addition that $\lambda_i \notin \mathbb{Z}$. This implies that $\lambda_i \neq 0$, for all i, and $\lambda_j > \lambda_{j+1}$, for $j = 1, \ldots, n-1$. We want to prove that $L(\lambda)$ is finite dimensional. Write $\mathfrak{g}_\varepsilon = \mathfrak{n}_\varepsilon^- + \mathfrak{h}_\varepsilon + \mathfrak{n}_\varepsilon^+$ for the standard triangular decomposition of $\mathfrak{g}_\varepsilon$, for $\varepsilon \in \mathbb{Z}_2$. The Verma $\mathfrak{g}$-module $\Delta(\lambda)$, regarded as a $\mathfrak{g}_{\bar{0}}$-module, has a $\mathfrak{g}_{\bar{0}}$-Verma flag, and we have

$$\mathrm{ch}\Delta(\lambda) = 2^{\lfloor (n+1)/2 \rfloor} \mathrm{ch}\left(\mathrm{Ind}_{\mathfrak{h}_{\bar{0}} + \mathfrak{n}_{\bar{0}}^+}^{\mathfrak{g}_{\bar{0}}} \mathbb{C}_\lambda \otimes \wedge(\mathfrak{n}_{\bar{1}}^-)\right),$$

where $\lfloor r \rfloor$ for a real number r denotes the largest integer no greater than r. As in the proof of Proposition 2.5, we apply the Bernstein functor $\mathcal{L}_0$ and its derived functor $\mathcal{L}_i$ (with $\mathfrak{g}_{\bar{0}} = \mathfrak{gl}(n)$) to $\Delta(\lambda)$ and compute the corresponding Euler characteristic (recall $\Phi_{\bar{1}}$ and $\Phi_{\bar{0}}$ are identical copies of the root system for $\mathfrak{gl}(n)$):

$$\begin{aligned}\sum_{i\geq 0}(-1)^i \mathrm{ch}\mathcal{L}_i(\Delta(\lambda)) =& 2^{\lfloor (n+1)/2\rfloor}\sum_{i\geq 0}(-1)^i \mathrm{ch}\mathcal{L}_i\left(\mathrm{Ind}_{\mathfrak{h}_{\bar{0}}+\mathfrak{n}_{\bar{0}}^+}^{\mathfrak{g}_{\bar{0}}}\mathbb{C}_\lambda\otimes\wedge(\mathfrak{n}_{\bar{1}}^-)\right)\\ =& 2^{\lfloor (n+1)/2\rfloor}\frac{\sum_{w\in W}(-1)^{\ell(w)}w\left(e^{\lambda+\rho_{\bar{0}}}\prod_{\alpha\in\Phi_{\bar{1}}^+}(1+e^{-\alpha})\right)}{\prod_{\alpha\in\Phi_{\bar{0}}^+}(e^{\alpha/2}-e^{-\alpha/2})}\\ =& 2^{\lfloor (n+1)/2\rfloor}\frac{\prod_{\alpha\in\Phi_{\bar{1}}^+}(1+e^{-\alpha})}{\prod_{\alpha\in\Phi_{\bar{0}}^+}(1-e^{-\alpha})}\sum_{w\in W}(-1)^{\ell(w)}e^{w(\lambda)}.\end{aligned}$$

Since the highest weight appearing in this last expression is clearly λ and the Euler characteristic is a virtual character of finite-dimensional $\mathfrak{g}$-modules (see the proof of Proposition 2.5), $L(\lambda)$ is a finite-dimensional module.

Now suppose that λ satisfies Conditions (i)-(ii) of the theorem and in addition that $\lambda_i \in \mathbb{Z}$ for each i. We write $\lambda = (\lambda_1,\dots,\lambda_k,0,\dots,0,\lambda_l,\dots,\lambda_n)$ such that $\lambda_1 > \cdots > \lambda_k > 0 > \lambda_l > \cdots > \lambda_n$. Denote $\lambda^+ = (\lambda_1,\dots,\lambda_k,0,\dots,0)$ and $\lambda^- = (0,\dots,0,\lambda_l,\dots,\lambda_n)$. By Theorem 3.49 on Sergeev duality in Chapter 3, $L(\mu)$ is finite dimensional for $\mu = \sum_i \mu_i\varepsilon_i$ corresponding to a strict partition $(\mu_1,\dots,\mu_n)$. Then, the dual module $L(\mu)^* \cong L(\mu^*)$ is finite dimensional, where we have denoted $\mu^* = -\sum_{i=1}^n \mu_{n+1-i}\varepsilon_i$ (see Exercise 2.18). It follows that both $L(\lambda^+)$ and $L(\lambda^-)$ are finite dimensional. Hence, as a quotient of the finite-dimensional $\mathfrak{g}$-module $L(\lambda^+)\otimes L(\lambda^-)$, $L(\lambda)$ must be finite dimensional. □

2.2. Harish-Chandra homomorphism and linkage

In this section, we study the center of the universal enveloping algebra $U(\mathfrak{g})$ when $\mathfrak{g}$ is a basic Lie superalgebra, in particular of type $\mathfrak{gl}$ and $\mathfrak{osp}$. The image of the Harish-Chandra homomorphism is described with the help of supersymmetric functions (see Appendix A). Then, we provide a characterization of when two central characters are equal and hence obtain a linkage principle for composition factors of a Verma module. The finite-dimensional typical irreducible representations of $\mathfrak{g}$ are classified, and a Weyl-type character formula for these modules is obtained.

2.2.1. Supersymmetrization. Let $\mathfrak{g}$ be a finite-dimensional Lie superalgebra. Denote by $Z(\mathfrak{g}) = Z(\mathfrak{g})_{\bar{0}} \oplus Z(\mathfrak{g})_{\bar{1}}$ the center of the enveloping superalgebra $U(\mathfrak{g})$, where

$$Z(\mathfrak{g})_i = \{z\in U(\mathfrak{g})_i \mid za = (-1)^{ik}az, \forall a\in\mathfrak{g}_k \text{ for } k\in\mathbb{Z}_2\}, \quad i\in\mathbb{Z}_2.$$

Recall $\mathcal{S}(\mathfrak{g})$ denotes the symmetric superalgebra of $\mathfrak{g}$. As a straightforward super generalization of a standard construction for Lie algebras (cf. Carter [**18**, Proposition 11.4]), we may define a supersymmetrization map

$$\gamma : \mathcal{S}(\mathfrak{g}) \longrightarrow U(\mathfrak{g})$$

and show that it is a $\mathfrak{g}$-module isomorphism. When restricting to the $\mathfrak{g}$-invariants, we have the following.

Proposition 2.19. *The supersymmetrization map $\gamma : \mathcal{S}(\mathfrak{g}) \to U(\mathfrak{g})$ induces a linear isomorphism $\gamma : \mathcal{S}(\mathfrak{g})^{\mathfrak{g}} \to Z(\mathfrak{g})$.*

Given a finite-dimensional representation (π, V) of a Lie superalgebra $\mathfrak{g}$, we denote its **supercharacter** by schV. That is,

$$\mathrm{sch}V = \sum_{\mu \in \mathfrak{h}^*} \mathrm{sdim}(V_\mu) e^\mu,$$

where we recall from Section 1.1.1 that $\mathrm{sdim}W = \dim W_{\bar{0}} - \dim W_{\bar{1}}$ stands for the superdimension of a vector superspace $W = W_{\bar{0}} \oplus W_{\bar{1}}$. If we regard e as the Euler number, a formal expansion of e^μ gives us

$$[\mathrm{sch}V](h) = \sum_{i=0}^{\infty} \frac{1}{i!} \mathrm{str}\left[\pi(h)^i\right], \quad \forall h \in \mathfrak{g}.$$

A straightforward super generalization of the arguments for Lie algebras (see Carter [**18**, pp. 212-3]) shows that $x \mapsto \mathrm{str}\left[\pi(x)^i\right]$ defines a $\mathfrak{g}$-invariant polynomial function on $\mathfrak{g}$. From this we conclude the following.

Proposition 2.20. *Let $\mathfrak{g}$ be a basic Lie superalgebra with Cartan subalgebra $\mathfrak{h}$. Let (π, V) be a finite-dimensional $\mathfrak{g}$-module. Then the polynomial functions* $\mathrm{str}\left[\pi(h)^i\right]$ *on $\mathfrak{h}$, for $i > 0$, can be identified with the homogeneous components of the supercharacter of π, and they arise as restrictions of $\mathfrak{g}$-invariant polynomial functions on $\mathfrak{g}$.*

2.2.2. Central characters. In this subsection, we let $\mathfrak{g}$ be a basic Lie superalgebra, and we let $\mathfrak{h}$ be its Cartan subalgebra.

Recall that $\mathfrak{g}$ is equipped with an even non-degenerate invariant supersymmetric bilinear form (see Section 1.2.2). Furthermore, the bilinear form $(\cdot,\cdot)$ on $\mathfrak{g}$ allows us to identify $\mathfrak{g}$ with its dual $\mathfrak{g}^*$ and $\mathfrak{h}$ with $\mathfrak{h}^*$ so that we have corresponding isomorphisms $\mathcal{S}(\mathfrak{g})^{\mathfrak{g}} \cong \mathcal{S}(\mathfrak{g}^*)^{\mathfrak{g}}$ and $S(\mathfrak{h}) \cong S(\mathfrak{h}^*)$. Via such identifications, the restriction map $\mathcal{S}(\mathfrak{g}^*) \to S(\mathfrak{h}^*)$ and the induced map

$$\eta : \mathcal{S}(\mathfrak{g}) \longrightarrow S(\mathfrak{h}) \tag{2.5}$$

are homomorphisms of algebras.

Let $\mathfrak{g} = \mathfrak{n}^- \oplus \mathfrak{h} \oplus \mathfrak{n}^+$ be a triangular decomposition. By the PBW Theorem 1.36, we have $U(\mathfrak{g}) = U(\mathfrak{h}) \oplus (\mathfrak{n}^- U(\mathfrak{g}) + U(\mathfrak{g})\mathfrak{n}^+)$, and we denote by $\phi : U(\mathfrak{g}) \to U(\mathfrak{h})$ the projection associated to this direct sum decomposition. Since $[\mathfrak{h}, \mathfrak{n}^\pm] \subseteq \mathfrak{n}^\pm$, by considering $\mathrm{ad}\,\mathfrak{h}(z) = 0$ for any element $z \in Z(\mathfrak{g})$, we conclude that z affords a unique expression of the form

$$z = h_z + \sum_i n_i^- h_i n_i^+, \quad \text{for } h_z, h_i \in U(\mathfrak{h}),\ n_i^\pm \in \mathfrak{n}^\pm U(\mathfrak{n}^\pm). \tag{2.6}$$

Hence $Z(\mathfrak{g})$ consists of only even elements. The restriction of ϕ to $Z(\mathfrak{g})$, denoted again by ϕ, defines an algebra homomorphism from $Z(\mathfrak{g})$ to $U(\mathfrak{h})$ that sends z to h_z.

As usual, we freely identify $S(\mathfrak{h})$ with the algebra of polynomial functions on $\mathfrak{h}^*$. Extending the standard pairing $\langle\cdot,\cdot\rangle : \mathfrak{h}^* \times \mathfrak{h} \to \mathbb{C}$, we can make sense of $\langle\lambda, p\rangle$ for $\lambda \in \mathfrak{h}^*$ and $p \in S(\mathfrak{h})$ by regarding it as the value at λ of the polynomial function p on $\mathfrak{h}^*$. For $\lambda \in \mathfrak{h}^*$, define a linear map

$$\begin{aligned} \chi_\lambda : Z(\mathfrak{g}) &\longrightarrow \mathbb{C} \\ \chi_\lambda(z) &= \langle\lambda, \phi(z)\rangle, \quad \forall z \in Z(\mathfrak{g}). \end{aligned} \tag{2.7}$$

Lemma 2.21. *Let $\mathfrak{g}$ be a basic Lie superalgebra. An element $z \in Z(\mathfrak{g})$ acts as the scalar $\chi_\lambda(z)$ on any highest weight $\mathfrak{g}$-module $V(\lambda)$ of highest weight λ.*

Proof. Our proof here is phrased so that it makes sense for $\mathfrak{q}(n)$ later on as well.

Let $z \in Z(\mathfrak{g})$. The highest weight space $V(\lambda)_\lambda$ is an irreducible $\mathfrak{h}$-module, which is one-dimensional for basic $\mathfrak{g}$ (but may fail to be one-dimensional in the $\mathfrak{q}(n)$ case). The even element z commutes with $\mathfrak{h}$, and hence by Schur's Lemma, z acts as a scalar $\xi(z)$ on $V(\lambda)_\lambda$. Now let v_λ be a highest weight vector of $V(\lambda)$, and write an arbitrary vector $u \in V(\lambda)$ as $u = xv_\lambda$ for $x \in U(\mathfrak{h}+\mathfrak{n}^-)$. Since z is central, we have $zu = xzv_\lambda = \xi(z)u$. It follows by (2.6) that

$$zv_\lambda = \big(h_z + \sum_i n_i^- h_i n_i^+\big)v_\lambda = \langle\lambda, h_z\rangle v_\lambda = \chi_\lambda(z)v_\lambda.$$

Hence $\xi(z) = \chi_\lambda(z)$. The lemma is proved. □

Therefore, $\chi_\lambda : Z(\mathfrak{g}) \to \mathbb{C}$ in (2.7) defines a one-dimensional representation of $Z(\mathfrak{g})$, which will be called the **central character** associated to a highest weight λ.

2.2.3. Harish-Chandra homomorphism for basic Lie superalgebras. In this subsection, we assume that $\mathfrak{g}$ is a basic Lie superalgebra.

Let Φ^+ be a positive system in the root system Φ for $\mathfrak{g}$. Recall the Weyl vector ρ from (1.35). Regarding $S(\mathfrak{h})$ as the algebra of polynomial functions on $\mathfrak{h}^*$, we define an automorphism τ of the algebra $S(\mathfrak{h})$ by

$$\tau(f)(\lambda) := f(\lambda - \rho), \quad \forall f \in S(\mathfrak{h}) \text{ and } \lambda \in \mathfrak{h}^*.$$

We then define the **Harish-Chandra homomorphism** to be the composition

$$\mathfrak{hc} = \tau\phi : Z(\mathfrak{g}) \longrightarrow S(\mathfrak{h}).$$

The non-degenerate invariant bilinear form $(\cdot,\cdot)$ on $\mathfrak{h}$ induces a W-equivariant algebra isomorphism $\theta : S(\mathfrak{h}) \to S(\mathfrak{h}^*)$ and hence an isomorphism $\theta : S(\mathfrak{h})^W \to S(\mathfrak{h}^*)^W$. By composition we obtain an algebra homomorphism

$$\mathfrak{hc}_* = \theta\mathfrak{hc} : Z(\mathfrak{g}) \longrightarrow S(\mathfrak{h}^*).$$

Proposition 2.22. *Let $\mathfrak{g}$ be a basic Lie superalgebra. We have*

(1) $\chi_\lambda = \chi_{w(\lambda+\rho)-\rho}$, *for all* $w \in W$ *and* $\lambda \in \mathfrak{h}^*$.

(2) $\mathfrak{hc}(Z(\mathfrak{g})) \subseteq S(\mathfrak{h})^W$ *and* $\mathfrak{hc}_*(Z(\mathfrak{g})) \subseteq S(\mathfrak{h}^*)^W$.

Proof. The character of the Verma module $\Delta(\mu)$ of highest weight $\mu \in \mathfrak{h}^*$ is given by $\mathrm{ch}\Delta(\mu) = e^{\mu+\rho}D^{-1}$, where

$$D := \frac{\prod_{\beta\in\Phi_{\bar{0}}^+}(e^{\beta/2} - e^{-\beta/2})}{\prod_{\alpha\in\Phi_{\bar{1}}^+}(e^{\alpha/2} + e^{-\alpha/2})}.$$

Since the character of a module is equal to the sum of the characters of its composition factors, we have

$$\mathrm{ch}\Delta(\lambda) = \sum_\mu b_{\mu\lambda}\mathrm{ch}L(\mu), \tag{2.8}$$

where $b_{\mu\lambda} \in \mathbb{Z}_+$ and $b_{\lambda\lambda} = 1$. Since $\Delta(\lambda)$ is a highest weight module, the μ's in the summation can be assumed to satisfy $\lambda - \mu \in \sum_{\alpha\in\Phi^+}\mathbb{Z}_+\alpha$ and also $\chi_\lambda = \chi_\mu$ by Lemma 2.21. We choose a total order $\geq$ on the set $\lambda - \sum_\alpha \mathbb{Z}_+\alpha$ with the property that $\nu \geq \mu$ if $\nu - \mu \in \sum_{\alpha\in\Phi^+}\mathbb{Z}_+\alpha$. Then the ordered sets $\{\mathrm{ch}\Delta(\mu)\}_{\mu,\geq}$ and $\{\mathrm{ch}L(\mu)\}_{\mu,\geq}$ are two ordered bases for the same space. Now (2.8) says that the matrix expressing $\{\mathrm{ch}\Delta(\mu)\}_{\mu,\geq}$ in terms of $\{\mathrm{ch}L(\mu)\}_{\mu,\geq}$ is upper triangular with 1 along the diagonal, and hence we have

$$\mathrm{ch}L(\lambda) = \sum_\mu a_{\mu\lambda}\mathrm{ch}\Delta(\mu),$$

where $a_{\mu\lambda} \in \mathbb{Z}$ with $a_{\lambda\lambda} = 1$, and

$$a_{\mu\lambda} = 0 \text{ unless } \lambda - \mu \in \sum_{\alpha\in\Phi^+}\mathbb{Z}_+\alpha \text{ and } \chi_\lambda = \chi_\mu. \tag{2.9}$$

Thus

$$D\mathrm{ch}L(\lambda) = \sum_\mu a_{\mu\lambda}e^{\mu+\rho}. \tag{2.10}$$

Assume for now that $\lambda \in \mathfrak{h}^*$ is chosen so that the irreducible $\mathfrak{g}$-module $L(\lambda)$ is finite dimensional. Then $L(\lambda)$ is a semisimple $\mathfrak{g}_{\bar{0}}$-module, and so $\mathrm{ch}L(\lambda)$ is W-invariant. On the other hand, D is W-anti-invariant by Theorem 1.18(4). Thus the right-hand side of (2.10) is W-anti-invariant, and hence can be written as

$$\sum_{\mu\in X} a_{\mu\lambda} \sum_{w\in W} (-1)^{\ell(w)} e^{w(\mu+\rho)}, \tag{2.11}$$

where X consists of $\Phi_{\bar{0}}^+$-dominant integral weights such that $a_{\mu\lambda} \neq 0$. We compute that $a_{w(\lambda+\rho)-\rho,\lambda} = \pm a_{\lambda\lambda} = \pm 1$, and hence by (2.9) we have $\chi_\lambda = \chi_{w(\lambda+\rho)-\rho}$, for all $w \in W$.

Now Part (1) of the proposition as a polynomial identity on $\lambda \in \mathfrak{h}^*$ follows from the following claim.

Claim. $S := \{\lambda \in \mathfrak{h}^* \mid L(\lambda) \text{ is finite dimensional}\}$ is Zariski dense in $\mathfrak{h}^*$.

Let us verify the claim. Given a $\Phi_{\bar{0}}^+$-dominant integral weight $\mu \in \mathfrak{h}^*$, the induced module $\mathrm{Ind}_{\mathfrak{g}_{\bar{0}}}^{\mathfrak{g}} L^0(\mu)$ is finite dimensional and has a highest weight $\mu + 2\rho_{\bar{1}}$. It is clear that the set of such weights $\mu + 2\rho_{\bar{1}}$ associated to all $\Phi_{\bar{0}}^+$-dominant integral $\mu \in \mathfrak{h}^*$ (which is a subset of S) is Zariski dense in $\mathfrak{h}^*$. Hence the claim follows.

By (1), we have that $\langle \lambda + \rho, \mathfrak{hc}(z)\rangle = \langle w(\lambda+\rho), \mathfrak{hc}(z)\rangle$, for all $w \in W$ and $z \in Z(\mathfrak{g})$. Thus $\mathfrak{hc}(z) \in S(\mathfrak{h})^W$, and equivalently, $\mathfrak{hc}_*(Z(\mathfrak{g})) \subseteq S(\mathfrak{h}^*)^W$ by definition of $\mathfrak{hc}_*$. This proves (2). □

By Proposition 2.22, we have

$$\mathfrak{hc} : Z(\mathfrak{g}) \longrightarrow S(\mathfrak{h})^W, \qquad \mathfrak{hc}_* = \theta\mathfrak{hc} : Z(\mathfrak{g}) \longrightarrow S(\mathfrak{h}^*)^W, \tag{2.12}$$

and we further define a homomorphism (see (2.5) for η)

$$\eta_* = \theta\eta : \mathcal{S}(\mathfrak{g})^{\mathfrak{g}} \longrightarrow S(\mathfrak{h}^*). \tag{2.13}$$

In the remainder of this section, we shall strengthen Proposition 2.22 for $\mathfrak{g}$ of type $\mathfrak{gl}$ and $\mathfrak{osp}$ by describing $\mathfrak{hc}_*(Z(\mathfrak{g}))$ precisely.

2.2.4. Invariant polynomials for $\mathfrak{gl}$ and $\mathfrak{osp}$. Suppose that $\mathfrak{g}$ is of type $\mathfrak{gl}$ or $\mathfrak{osp}$. The goal of this subsection is to establish Proposition 2.23 which partially describes the image $\mathrm{Im}(\eta_*)$ via supersymmetric polynomials; see Appendix A.2.1. (The word "partially" can and will be removed subsequently.) The description of this image plays a key role in the study of the Harish-Chandra homomorphism later on.

Associated to a positive system Φ^+, we have the subset $\bar{\Phi}_{\bar{1}}^+$ of positive isotropic odd roots defined in (1.34). Introduce the following element

$$P := \prod_{\alpha \in \bar{\Phi}_{\bar{1}}^+} \alpha \in S(\mathfrak{h}^*). \tag{2.14}$$

Below we shall treat $\mathfrak{g} = \mathfrak{gl}(m|n)$, $\mathfrak{spo}(2m|2n+1)$, and $\mathfrak{spo}(2m|2n)$ case-by-case.

Invariant polynomials for $\mathfrak{g} = \mathfrak{gl}(m|n)$. Recall that the Weyl group W for $\mathfrak{g} = \mathfrak{gl}(m|n)$ is $W = \mathfrak{S}_m \times \mathfrak{S}_n$, which is by definition the same as the Weyl group of its even subalgebra $\mathfrak{gl}(m) \oplus \mathfrak{gl}(n)$,. Applying Proposition 2.20 to the natural representation (π, V) of $\mathfrak{g} = \mathfrak{gl}(m|n)$ and recalling η_* from (2.13), we see that

$$\sigma_{m,n}^k := \sum_{i=1}^m \delta_i^k - \sum_{j=1}^n \varepsilon_j^k \in \mathrm{Im}(\eta_*), \quad \forall k \in \mathbb{Z}_+. \tag{2.15}$$

Here and further, we shall identify $S(\mathfrak{h}^*) = \mathbb{C}[\delta, \varepsilon]$, where we set

$$\delta = \{\delta_1, \ldots, \delta_m\}, \quad \varepsilon = \{\varepsilon_1, \ldots, \varepsilon_n\}.$$

Hence an element of $S(\mathfrak{h}^*)^W$ is precisely a polynomial that is symmetric in δ and symmetric in ε. Recalling from A.2.1 the definition of supersymmetric polynomials, we introduce the following subalgebra of $S(\mathfrak{h}^*)^W$:

$$S(\mathfrak{h}^*)^W_{\mathrm{sup}} = \{\text{supersymmetric polynomials in variables } \delta, \varepsilon\}. \tag{2.16}$$

By (A.33), the polynomials $\{\sigma^k_{m,n} \mid k \in \mathbb{Z}_+\}$ generate the algebra $S(\mathfrak{h}^*)^W_{\mathrm{sup}}$. Thus we have $S(\mathfrak{h}^*)^W_{\mathrm{sup}} \subseteq \mathrm{Im}(\eta_*)$.

The element P defined in (2.14), up to a possible sign that depends on the choice of positive systems for $\mathfrak{gl}(m|n)$, can be computed to be

$$P = \prod_{1\le i\le m, 1\le j\le n} (\delta_i - \varepsilon_j).$$

It follows by (A.30) that $fP \in S(\mathfrak{h}^*)^W_{\mathrm{sup}}$ for any $f \in S(\mathfrak{h}^*)^W$.

Invariant polynomials for $\mathfrak{g} = \mathfrak{spo}(2m|2n+1)$. Recall the Weyl group W for $\mathfrak{g} = \mathfrak{spo}(2m|2n+1)$, which is by definition the same as for $\mathfrak{sp}(2m) \oplus \mathfrak{so}(2n+1)$, is

$$W_{\mathrm{B}} := (\mathbb{Z}_2^m \rtimes \mathfrak{S}_m) \times (\mathbb{Z}_2^n \rtimes \mathfrak{S}_n), \tag{2.17}$$

and hence an element $f \in S(\mathfrak{h}^*)^W = \mathbb{C}[\delta, \varepsilon]^W$ is exactly a polynomial that is symmetric among δ_i^2's and symmetric among ε_j^2's. We introduce the following subalgebra of $S(\mathfrak{h}^*)^W$:

$$S(\mathfrak{h}^*)^W_{\mathrm{sup}} = \{\text{supersymmetric polynomials in } \delta_i^2, \varepsilon_j^2, 1 \le i \le m, 1 \le j \le n\}. \tag{2.18}$$

Applying Proposition 2.20 to the natural representation of $\mathfrak{spo}(2m|2n+1)$, we have that

$$\sigma^{2k}_{m,n} = \sum_{i=1}^m \delta_i^{2k} - \sum_{j=1}^n \varepsilon_j^{2k} \in \mathrm{Im}(\eta_*), \quad \forall k \in \mathbb{Z}_+.$$

Hence we have $S(\mathfrak{h}^*)^W_{\mathrm{sup}} \subseteq \mathrm{Im}(\eta_*)$, since $\sigma^{2k}_{m,n}$ for $k \ge 0$ generate $S(\mathfrak{h}^*)^W_{\mathrm{sup}}$ by (A.33).

In this case, we calculate that $\bar{\Phi}^+_{\bar{1}} = \{\delta_i \pm \varepsilon_j \mid 1 \le i \le m, 1 \le j \le n\}$ for the standard positive system Φ^+, and hence, P in (2.14) is given by

$$P = \prod_{1\le i\le m, 1\le j\le n} (\delta_i - \varepsilon_j)(\delta_i + \varepsilon_j) = \prod_{1\le i\le m, 1\le j\le n} (\delta_i^2 - \varepsilon_j^2).$$

(For other positive systems, P might differ from the above formula by an irrelevant sign.) By (A.30), we have that $fP \in S(\mathfrak{h}^*)^W_{\mathrm{sup}}$ for any $f \in S(\mathfrak{h}^*)^W$.

Invariant polynomials for $\mathfrak{g} = \mathfrak{spo}(2m|2n)$. We regard $\mathfrak{g} = \mathfrak{spo}(2m|2n)$ as a subalgebra of $\mathfrak{spo}(2m|2n+1)$ with the same Cartan subalgebra $\mathfrak{h}$. We continue to denote by W_{B} the Weyl group of $\mathfrak{spo}(2m|2n+1)$ and then identify the Weyl group W of $\mathfrak{g} = \mathfrak{spo}(2m|2n)$ with a subgroup of W_{B} of index 2. By the same consideration as for $\mathfrak{g} = \mathfrak{spo}(2m|2n+1)$ above using Proposition 2.20, we conclude that the image $\mathrm{Im}(\eta_*)$ contains $S(\mathfrak{h}^*)^{W_{\mathrm{B}}}_{\mathrm{sup}}$, which consists of all the supersymmetric polynomials in $\delta_i^2, \varepsilon_j^2$, $1 \le i \le m$, $1 \le j \le n$.

Noting now that $\Phi^+_{\bar{1}} = \{\delta_i \pm \varepsilon_j \mid 1 \le i \le m, 1 \le j \le n\}$ for the standard positive system Φ^+, we set

$$D_{\bar{1}} := \prod_{\alpha \in \Phi^+_{\bar{1}}} (e^{\alpha/2} - e^{-\alpha/2})$$
$$= \prod_{i,j} (e^{\delta_i/2-\varepsilon_j/2} - e^{-\delta_i/2+\varepsilon_j/2})(e^{\delta_i/2+\varepsilon_j/2} - e^{-\delta_i/2-\varepsilon_j/2}).$$

For a homogeneous polynomial $g \in \mathbb{Z}[\delta, \varepsilon]^{W_B}$, the expression

$$Q := \prod_{j=1}^{n} (e^{\varepsilon_j} - e^{-\varepsilon_j}) \cdot g(e^{\delta_1} - e^{-\delta_1}, \dots, e^{\delta_m} - e^{-\delta_m}; e^{\varepsilon_1} - e^{-\varepsilon_1}, \dots, e^{\varepsilon_n} - e^{-\varepsilon_n})$$

has $\varepsilon_1 \varepsilon_2 \cdots \varepsilon_n g(\delta_1, \dots, \delta_m; \varepsilon_1, \dots, \varepsilon_n)$ as its homogeneous term of lowest degree. Also observe that $D_{\bar{1}}$ has P as its lowest degree term, where the element P in (2.14) associated to the standard positive system Φ^+ of $\mathfrak{spo}(2m|2n)$ is given by

$$P = \prod_{1 \le i \le m, 1 \le j \le n} (\delta_i^2 - \varepsilon_j^2).$$

(For other positive systems, P might differ by a sign from the above formula.) Since Q is an integral polynomial in $e^{\pm\delta_i}$ and $e^{\pm\varepsilon_j}$ that is W-invariant, Q is actually a virtual $\mathfrak{g}_{\bar{0}}$-character. By Remark 2.6, $D_{\bar{1}}Q$ is a virtual supercharacter of finite-dimensional $\mathfrak{g}$-modules. Hence every homogeneous term of $D_{\bar{1}}Q$ lies in $\mathrm{Im}(\eta_*)$ by Proposition 2.20, and in particular so does the lowest degree term of $D_{\bar{1}}Q$, i.e.,

$$\varepsilon_1 \varepsilon_2 \cdots \varepsilon_n P g \in \mathrm{Im}(\eta_*).$$

Introduce the following subspace $S(\mathfrak{h}^*)^W_{\mathrm{sup}}$ (which can be easily shown to be a subalgebra) of $S(\mathfrak{h}^*)^W = \mathbb{C}[\delta, \varepsilon]^W$:

$$S(\mathfrak{h}^*)^W_{\mathrm{sup}} = S(\mathfrak{h}^*)^{W_B}_{\mathrm{sup}} \bigoplus \mathbb{C}[\delta, \varepsilon]^{W_B} \varepsilon_1 \varepsilon_2 \cdots \varepsilon_n P. \tag{2.19}$$

Summarizing the above discussions, we have shown that $S(\mathfrak{h}^*)^W_{\mathrm{sup}} \subseteq \mathrm{Im}(\eta_*)$.

It is a standard fact about the classical Weyl group W that any element $f \in S(\mathfrak{h}^*)^W = \mathbb{C}[\delta, \varepsilon]^W$ can be written as

$$f = h + \varepsilon_1 \varepsilon_2 \cdots \varepsilon_n g,$$

where $h, g \in \mathbb{C}[\delta, \varepsilon]^{W_B}$. Thus, we have $hP \in S(\mathfrak{h}^*)^W_{\mathrm{sup}}$ and $\varepsilon_1 \varepsilon_2 \cdots \varepsilon_n g P \in S(\mathfrak{h}^*)^W_{\mathrm{sup}}$, and hence, $fP \in S(\mathfrak{h}^*)^W_{\mathrm{sup}}$.

Summarizing all three cases, we have established the following.

Proposition 2.23. *Let $\mathfrak{g}$ be either $\mathfrak{gl}(m|n)$, $\mathfrak{spo}(2m|2n+1)$, or $\mathfrak{spo}(2m|2n)$, and let W be its Weyl group. Then, we have*

(1) $P \in S(\mathfrak{h}^*)^W$.

(2) $fP \in S(\mathfrak{h}^*)^W_{\mathrm{sup}}$, *for all* $f \in S(\mathfrak{h}^*)^W$.

(3) $S(\mathfrak{h}^*)^W_{\mathrm{sup}} \subseteq \mathrm{Im}(\eta_*)$.

2.2.5. Image of Harish-Chandra homomorphism for $\mathfrak{gl}$ and $\mathfrak{osp}$. Let $\mathfrak{g}$ be a basic Lie superalgebra for now. We have a non-degenerate invariant bilinear form $(\cdot,\cdot)$ on $\mathfrak{g}$, $\mathfrak{h}$, and $\mathfrak{h}^*$. We recall here that $(\alpha,\alpha)=0$, for any isotropic odd root α. The non-degenerate form $(\cdot,\cdot)$ induces an isomorphism $\mathfrak{h}\cong\mathfrak{h}^*$. Also recall the central characters χ_λ from (2.7).

Lemma 2.24. *Let $\mathfrak{g}$ be a basic Lie superalgebra. Let α be an isotropic odd root such that $(\lambda+\rho,\alpha)=0$. Then we have $\chi_\lambda=\chi_{\lambda+t\alpha}$, for all $t\in\mathbb{C}$.*

Proof. Fix a positive system Φ^+. We may assume that the isotropic odd root $\alpha\in\Phi^+$ (otherwise, use $-\alpha$ to replace α).

First suppose that α is an isotropic *simple* root in Φ^+. Then $(\rho,\alpha)=\frac{1}{2}(\alpha,\alpha)=0$, and hence $(\lambda,\alpha)=0$ by assumption. Let v_λ be a highest weight vector in the Verma module $\Delta(\lambda)$. Then,

$$e_\alpha f_\alpha v_\lambda=[e_\alpha,f_\alpha]v_\lambda=h_\alpha v_\lambda=(\lambda,\alpha)v_\lambda=0.$$

Also, we have $e_\gamma f_\alpha v_\lambda=0$ for simple roots $\gamma\neq\alpha$ by weight considerations. Hence, $f_\alpha v_\lambda$ is a singular vector in $\Delta(\lambda)$. This gives rise to a nontrivial $\mathfrak{g}$-module homomorphism $\Delta(\lambda-\alpha)\to\Delta(\lambda)$, whence $\chi_{\lambda-\alpha}=\chi_\lambda$.

Now suppose α is any isotropic root in Φ^+. By Lemma 1.29, there exists $w\in W$ such that $\beta=w(\alpha)$ is simple. Then $(\alpha,\alpha)=(\beta,\beta)=0$ by the W-invariance of $(\cdot,\cdot)$. By Proposition 2.22, we have

$$\chi_{\lambda-\alpha}=\chi_{w(\lambda-\alpha+\rho)-\rho}=\chi_\mu,$$

where we have denoted $\mu:=w(\lambda+\rho)-\rho-\beta$ and used $w(\alpha)=\beta$. Now by the W-invariance of $(\cdot,\cdot)$ again, we compute that

$$(\mu+\rho,\beta)=(w(\lambda+\rho),w(\alpha))-(\beta,\beta)=(\lambda+\rho,\alpha)-0=0.$$

Thus, we can apply the special case established in the preceding paragraph to the weight μ and simple root β to obtain that

$$\chi_{\lambda-\alpha}=\chi_\mu=\chi_{\mu+\beta}=\chi_{w(\lambda+\rho)-\rho}=\chi_\lambda,$$

where we have used Proposition 2.22 again in the last equality. This implies that $\chi_\lambda=\chi_{\lambda-t\alpha}$, for any $t\in\mathbb{Z}_+$ and any isotropic odd root α with $(\lambda+\rho,\alpha)=0$. Since $\mathbb{Z}_+$ is Zariski dense in $\mathbb{C}$, we conclude that the polynomial identity $\chi_\lambda=\chi_{\lambda+t\alpha}$ holds for all $t\in\mathbb{C}$. □

Now we specialize to the type $\mathfrak{gl}$ and $\mathfrak{osp}$.

Proposition 2.25. *Let $\mathfrak{g}$ be either $\mathfrak{gl}(m|n)$, $\mathfrak{spo}(2m|2n+1)$, or $\mathfrak{spo}(2m|2n)$. Then we have $\mathfrak{hc}_*(Z(\mathfrak{g}))\subseteq S(\mathfrak{h}^*)^W_{\text{sup}}$, where $S(\mathfrak{h}^*)^W_{\text{sup}}$ is given in* (2.16), (2.18), *and* (2.19).

Proof. Let us proceed case-by-case.

First let us consider the case of $\mathfrak{g} = \mathfrak{gl}(m|n)$. Assume that $\lambda + \rho = \sum_{i=1}^{m} a_i \delta_i + \sum_{j=1}^{n} b_j \varepsilon_j$ satisfies $(\lambda+\rho, \alpha) = 0$ for some isotropic root $\alpha = \delta_i - \varepsilon_j$, which means that $a_i = -b_j$. Let $z \in Z(\mathfrak{g})$. Then by Lemma 2.24 we have

$$\langle \lambda+\rho, \mathfrak{hc}(z)\rangle = \chi_\lambda(z) = \chi_{\lambda - t\alpha}(z) = \langle \lambda - t\alpha + \rho, \mathfrak{hc}(z)\rangle, \quad \forall t \in \mathbb{C}.$$

This implies that the specialization $\mathfrak{hc}_*(z)|_{\delta_i = -\varepsilon_j = t}$ is independent of t, where we recall from (2.12) that $\mathfrak{hc}_* = \theta\mathfrak{hc}$. In addition, we have $\mathfrak{hc}_*(z) \in S(\mathfrak{h}^*)^W$ by Proposition 2.22, whence $\mathfrak{hc}_*(z) \in S(\mathfrak{h}^*)^W_{\text{sup}}$ by the definition (2.16).

The case for $\mathfrak{g} = \mathfrak{spo}(2m|2n+1)$ is analogous. Let $\lambda \in \mathfrak{h}^*$ be such that $(\lambda + \rho, \alpha) = 0$ for some isotropic root $\alpha = \delta_i \pm \varepsilon_j$. We show by the same argument as above that the specialization $\mathfrak{hc}_*(z)|_{\delta_i = \mp\varepsilon_j = t}$ is independent of t. In addition, it follows by Proposition 2.22 that $\mathfrak{hc}_*(z) \in S(\mathfrak{h}^*)^W$ for $W = W_{\mathrm{B}}$ (see (2.17)), which means that $\mathfrak{hc}_*(z)$ is symmetric in δ_i^2 for all i and symmetric in ε_j^2 for all j. Then we conclude that $\mathfrak{hc}_*(z) \in S(\mathfrak{h}^*)^W_{\text{sup}}$ by the definition (2.18).

Now suppose that $\mathfrak{g} = \mathfrak{spo}(2m|2n)$. Let W be the Weyl group of $\mathfrak{g}$, which is regarded as a subgroup of W_{B} of index 2 as usual. Given $z \in Z(\mathfrak{g})$, we can write $\mathfrak{hc}_*(z) \in S(\mathfrak{h}^*)^W$ as

$$\mathfrak{hc}_*(z) = f + \varepsilon_1 \cdots \varepsilon_n g, \qquad \text{for } f, g \in S(\mathfrak{h}^*)^{W_{\mathrm{B}}} = \mathbb{C}[\delta^2, \varepsilon^2]^{\mathfrak{S}_m \times \mathfrak{S}_n}.$$

Since $\chi_\lambda = \chi_{\lambda - t\alpha}$, for any λ and any isotropic root $\alpha = \delta_i - \varepsilon_j$ with $(\lambda+\rho, \alpha) = 0$, the specialization $\mathfrak{hc}_*(z)|_{\delta_i = \varepsilon_j = t} = f_1 + g_1$ is independent of t, where we denote

$$f_1 := f|_{\delta_i = \varepsilon_j = t}, \qquad g_1 := t\varepsilon_1 \cdots \varepsilon_{j-1}\varepsilon_{j+1} \cdots \varepsilon_n g|_{\delta_i = \varepsilon_j = t}.$$

As a polynomial of t, f_1 has even degree while g_1 has odd degree. Thus both f_1 and g_1 must be independent of t. Hence, recalling (2.18), we must have that

$$f \in S(\mathfrak{h}^*)^{W_{\mathrm{B}}}_{\text{sup}}, \qquad g|_{\delta_i = \varepsilon_j = t} = 0.$$

Since the polynomial g is symmetric in $\delta_1^2, \ldots, \delta_m^2$ and symmetric in $\varepsilon_1^2, \ldots, \varepsilon_n^2$, we conclude that g is divisible by $(\delta_i^2 - \varepsilon_j^2)$, for all i and j, and hence divisible by $P = \prod_{i,j}(\delta_i^2 - \varepsilon_j^2)$. Thus we have $g = g_0 P$ for some $g_0 \in S(\mathfrak{h}^*)^{W_{\mathrm{B}}}$.

The inclusion $\mathfrak{hc}_*(Z(\mathfrak{g})) \subseteq S(\mathfrak{h}^*)^W_{\text{sup}}$ now follows by the definition (2.19). □

Theorem 2.26. *Let $\mathfrak{g} = \mathfrak{gl}(m|n)$, $\mathfrak{spo}(2m|2n+1)$, or $\mathfrak{spo}(2m|2n)$, and let W be its Weyl group. Then we have*

$$\mathfrak{hc}_*(Z(\mathfrak{g})) = \mathrm{Im}(\eta_*) = S(\mathfrak{h}^*)^W_{\text{sup}},$$

where $S(\mathfrak{h}^*)^W_{\text{sup}}$ is defined in (2.16), (2.18), *and* (2.19), *respectively.*

Proof. By Propositions 2.23 and 2.25, we have a map $\mathfrak{hc}_* : Z(\mathfrak{g}) \to \mathrm{Im}(\eta_*)$ and that $\mathfrak{hc}_*(Z(\mathfrak{g})) \subseteq S(\mathfrak{h}^*)^W_{\text{sup}} \subseteq \mathrm{Im}(\eta_*)$. So it suffices to show that $\mathfrak{hc}_*(Z(\mathfrak{g})) = \mathrm{Im}(\eta_*)$

in order to complete the proof. The commutative diagram (where gr denotes the associated graded)

$$(2.20)\qquad \begin{array}{ccc} \mathrm{gr}\,Z(\mathfrak{g}) & \xrightarrow[\cong]{\mathrm{gr}\,\gamma^{-1}} & \mathcal{S}(\mathfrak{g})^{\mathfrak{g}} \\ & \searrow^{\mathrm{gr}\,\mathfrak{hc}_*} & \downarrow{\scriptstyle \eta_*} \\ & & S(\mathfrak{h}^*) \end{array}$$

implies that $\mathrm{gr}\,\mathfrak{hc}_* : \mathrm{gr}\,Z(\mathfrak{g}) \to \mathrm{Im}(\eta_*)$ is surjective.

Now the surjectivity of the map $\mathfrak{hc}_* : Z(\mathfrak{g}) \to \mathrm{Im}(\eta_*)$ follows by a standard filtered algebra argument. Indeed, consider $\mathfrak{hc}_*(Z(\mathfrak{g}))$ and $\mathrm{Im}(\eta_*)$ as filtered spaces:

$$\mathbb{C} \cong \mathfrak{hc}_*(Z(\mathfrak{g}))_0 \subseteq \mathfrak{hc}_*(Z(\mathfrak{g}))_1 \subseteq \cdots, \qquad \mathbb{C} \cong \mathrm{Im}(\eta_*)_0 \subseteq \mathrm{Im}(\eta_*)_1 \subseteq \cdots.$$

We shall prove by induction on k that $\mathfrak{hc}_*(Z(\mathfrak{g}))_k = \mathrm{Im}(\eta_*)_k$, for all k. Clearly we have $\mathfrak{hc}_*(Z(\mathfrak{g}))_0 = \mathrm{Im}(\eta_*)_0$. Now assume that $\mathfrak{hc}_*(Z(\mathfrak{g}))_{k-1} = \mathrm{Im}(\eta_*)_{k-1}$. Let us take $a = \eta_*(x) \in \mathrm{Im}(\eta_*)_k$ for $x \in \mathcal{S}(\mathfrak{g})^{\mathfrak{g}}$. Then $a = \mathrm{gr}\,(\mathfrak{hc}_*\gamma)(x)$, and

$$\mathfrak{hc}_*\gamma(x) - a \in \mathrm{Im}(\eta_*)_{k-1} = \mathfrak{hc}_*(Z(\mathfrak{g}))_{k-1}.$$

Therefore, we conclude that $a \in \mathfrak{hc}_*(Z(\mathfrak{g}))_k$. □

Define the subalgebra $S(\mathfrak{h})^W_{\mathrm{sup}}$ of $S(\mathfrak{h})$ to be the preimage of $S(\mathfrak{h}^*)^W_{\mathrm{sup}}$ under the isomorphism $\theta : S(\mathfrak{h}) \to S(\mathfrak{h}^*)$. Various maps in this section can now be put together in the following diagram, which only becomes commutative when we pass from $Z(\mathfrak{g})$ to its associated graded (note that the other algebras are already graded).

$$(2.21)\qquad \begin{array}{ccc} Z(\mathfrak{g}) & \xrightarrow[\cong]{\gamma^{-1}} & \mathcal{S}(\mathfrak{g})^{\mathfrak{g}} \\ {\scriptstyle \mathfrak{hc}_*}\downarrow & \swarrow^{\eta_*}_{\cong} & \downarrow{\scriptstyle \eta} \\ S(\mathfrak{h}^*)^W_{\mathrm{sup}} & \xleftarrow[\theta]{\cong} & S(\mathfrak{h})^W_{\mathrm{sup}} \end{array}$$

Remark 2.27. The surjective homomorphism $\mathfrak{hc} : Z(\mathfrak{g}) \to S(\mathfrak{h})^W_{\mathrm{sup}}$ is actually injective as well, and hence all the homomorphisms in the diagram (2.21) are isomorphisms. Actually Sergeev [**111**, Corollary 1.1] showed that $\eta : \mathcal{S}(\mathfrak{g})^{\mathfrak{g}} \to S(\mathfrak{h})$ is injective, and the injectivity of $\mathfrak{hc}$ follows from this by a filtered algebra argument. Except for its intrinsic value, the injectivity does not seem to play any crucial role in the representation theory of $\mathfrak{g}$.

The following corollary is immediate from Proposition 2.23 and Theorem 2.26, and will be used later on.

Corollary 2.28. *Let $\mathfrak{g}$ be one of the Lie superalgebras $\mathfrak{gl}(m|n)$, $\mathfrak{spo}(2m|2n+1)$, or $\mathfrak{spo}(2m|2n)$. Then for any $f \in S(\mathfrak{h})^W$, we have $fP \in \mathfrak{hc}_*(Z(\mathfrak{g}))$.*

2.2.6. Linkage for $\mathfrak{gl}$ and $\mathfrak{osp}$. In this subsection, we let $\mathfrak{g}$ denote the Lie superalgebra $\mathfrak{gl}(m|n)$, $\mathfrak{spo}(2m|2n)$, or $\mathfrak{spo}(2m|2n+1)$. Using the description of the images of Harish-Chandra homomorphisms, we obtain a necessary and sufficient condition for two central characters of $\mathfrak{g}$ to be equal.

Let $\mathfrak{h}$ be its standard Cartan subalgebra of diagonal matrices. Let $\{\delta_i, \varepsilon_j | 1 \leq i \leq m, 1 \leq j \leq n\}$ be the standard basis of $\mathfrak{h}^*$. We fix a positive system Φ^+ of the root system Φ for $\mathfrak{g}$. Recall the subset $\bar{\Phi}^+_{\bar{1}} \subseteq \Phi^+$ defined in (1.34).

Definition 2.29. The **degree of atypicality** of an element $\lambda \in \mathfrak{h}^*$, denoted by $\#\lambda$, is the maximum number of mutually orthogonal roots $\alpha \in \bar{\Phi}^+_{\bar{1}}$ such that $(\lambda+\rho, \alpha) = 0$. An element $\lambda \in \mathfrak{h}^*$ is said to be **typical** (relative to Φ^+) if $\#\lambda = 0$ and is **atypical** otherwise.

We define a relation $\sim$ on $\mathfrak{h}^*$ by declaring

$$\lambda \sim \mu, \qquad \text{for } \lambda, \mu \in \mathfrak{h}^*,$$

if there exist *mutually orthogonal* isotropic odd roots $\alpha_1, \alpha_2, \ldots, \alpha_\ell$, complex numbers $c_1, c_2, \ldots, c_\ell$, and an element $w \in W$ satisfying that

$$\mu+\rho = w\Big(\lambda+\rho-\sum_{a=1}^{\ell} c_a\alpha_a\Big), \qquad (\lambda+\rho, \alpha_j) = 0,\ j = 1, \ldots, \ell. \tag{2.22}$$

The weights λ and μ are said to be **linked** if $\lambda \sim \mu$. It follows from Theorem 2.30 below that linkage is an equivalence relation.

Theorem 2.30. *Let $\mathfrak{g}$ be $\mathfrak{gl}(m|n)$, $\mathfrak{spo}(2m|2n)$, or $\mathfrak{spo}(2m|2n+1)$, and let $\mathfrak{h}$ be its Cartan subalgebra. Let $\lambda, \mu \in \mathfrak{h}^*$. Then λ is linked to μ if and only if $\chi_\lambda = \chi_\mu$.*

Proof. Assume first that $\lambda \sim \mu$, i.e., (2.22), holds. Since the α_i's are orthogonal to one another, the second condition in (2.22) implies that we can repeatedly apply Lemma 2.24 to conclude that

$$\chi_\lambda = \chi_{\lambda-c_1\alpha_1} = \chi_{\lambda-c_1\alpha_1-c_2\alpha_2} = \cdots = \chi_{\lambda-\sum_{a=1}^{\ell} c_a\alpha_a}.$$

Since $\mu+\rho = w(\lambda+\rho-\sum_{a=1}^{\ell} c_a\alpha_a)$ by (2.22) again, it follows by Proposition 2.22 that $\chi_{\lambda-\sum_{a=1}^{\ell} c_a\alpha_a} = \chi_\mu$, whence $\chi_\lambda = \chi_\mu$.

Now assume that $\chi_\lambda = \chi_\mu$. Let us write $\lambda = \sum_{i=1}^m \lambda_i\delta_i + \sum_{j=1}^n \nu_j\varepsilon_j$ and $\mu = \sum_{i=1}^m \mu_i\delta_i + \sum_{j=1}^n \eta_j\varepsilon_j$. Recall from Theorem 2.26 that, for $\mathfrak{g} = \mathfrak{gl}(m|n)$, the polynomials $\sigma^k_{m,n}$ given in (2.15) lie in $\mathfrak{hc}_*(Z(\mathfrak{g}))$, for all $k \in \mathbb{Z}_+$, while in the case of $\mathfrak{g} = \mathfrak{spo}(2m|2n)$ or $\mathfrak{spo}(2m|2n+1)$, the polynomials $\sigma^k_{m,n}$ lie in $\mathfrak{hc}_*(Z(\mathfrak{g}))$, for all even $k \in \mathbb{Z}_+$. Let $z^k_{m,n}$ be an element in $Z(\mathfrak{g})$ with $\mathfrak{hc}_*(z^k_{m,n}) = \sigma^k_{m,n}$, where k is assumed to be even for type $\mathfrak{spo}$. Then for $\mathfrak{g}$ in all three cases,

$$\chi_\lambda(z^k_{m,n}) = \sum_{i=1}^{m} (\lambda+\rho, \delta_i)^k - \sum_{j=1}^{n} (\lambda+\rho, \varepsilon_j)^k.$$

We now proceed case-by-case.

(1) First, suppose that $\mathfrak{g} = \mathfrak{gl}(m|n)$. Then, for all $k \geq 0$, we have

$$\chi_\lambda(z_{m,n}^k) = \chi_\mu(z_{m,n}^k).$$

This implies that, for all $k \geq 0$,

$$\sum_{i=1}^{m} (\lambda+\rho,\delta_i)^k - \sum_{j=1}^{n} (\lambda+\rho,\varepsilon_j)^k = \sum_{i=1}^{m} (\mu+\rho,\delta_i)^k - \sum_{j=1}^{n} (\mu+\rho,\varepsilon_j)^k.$$

Using exponential generating functions in a formal indeterminate t, this is equivalent to

$$\sum_{i=1}^{m} e^{(\lambda+\rho,\delta_i)t} - \sum_{j=1}^{n} e^{(\lambda+\rho,\varepsilon_j)t} = \sum_{i=1}^{m} e^{(\mu+\rho,\delta_i)t} - \sum_{j=1}^{n} e^{(\mu+\rho,\varepsilon_j)t}. \tag{2.23}$$

Pick a maximal number of pairs (i_a, j_a) such that $(\lambda+\rho,\delta_{i_a}) = (\lambda+\rho,\varepsilon_{j_a})$ and all i_a and all j_a are distinct, for $1 \leq a \leq \ell$. Similarly, pick a maximal number of pairs (i'_b, j'_b) such that $(\mu+\rho,\delta_{i'_b}) = (\mu+\rho,\varepsilon_{j'_b})$ and all i'_b and all j'_b are distinct, for $1 \leq b \leq r$. Note that the functions e^{at} are linearly independent for distinct a. In the identity obtained from (2.23) by canceling the terms corresponding to all the pairs (i_a, j_a) and (i'_b, j'_b), the survived $(\lambda+\rho,\delta_i)$'s must match bijectively with the survived $(\mu+\rho,\delta_i)$'s, while the $(\lambda+\rho,\varepsilon_j)$'s and $(\mu+\rho,\varepsilon_j)$'s must match bijectively. So $\ell = r$. Now we can find a pair of permutations $w = (\sigma_1, \sigma_2) \in \mathfrak{S}_m \times \mathfrak{S}_n$ that extends the above bijections such that $w^{-1}(\mu+\rho) - \lambda - \rho$ is a linear combination of these mutually orthogonal isotropic odd roots $\delta_{i_a} - \varepsilon_{j_a}$, and hence $\lambda \sim \mu$.

(2) Now suppose that $\mathfrak{g} = \mathfrak{spo}(2m|2n+1)$. Then, for all even $k \geq 0$, we have

$$\chi_\lambda(z_{m,n}^k) = \chi_\mu(z_{m,n}^k).$$

This is equivalent to the following generating function identity:

$$\sum_{i=1}^{m} e^{(\lambda+\rho,\delta_i)^2 t} - \sum_{j=1}^{n} e^{(\lambda+\rho,\varepsilon_j)^2 t} = \sum_{i=1}^{m} e^{(\mu+\rho,\delta_i)^2 t} - \sum_{j=1}^{n} e^{(\mu+\rho,\varepsilon_j)^2 t}. \tag{2.24}$$

Pick a maximal number of pairs (i_a, j_a) such that $(\lambda+\rho,\delta_{i_a}) = s_a(\lambda+\rho,\varepsilon_{j_a})$ for some sign $s_a \in \{\pm\}$ and all i_a and all j_a are distinct, for $1 \leq a \leq \ell$. Similarly, pick a maximal number of pairs (i'_b, j'_b) such that $(\mu+\rho,\delta_{i'_b})^2 = (\mu+\rho,\varepsilon_{j'_b})^2$ and all i'_b and all j'_b are distinct, for $1 \leq b \leq r$. In the identity obtained from (2.24) by canceling the terms corresponding to all the pairs (i_a, j_a) and (i'_b, j'_b), the survived $(\lambda+\rho,\delta_i)$'s must match bijectively with the survived $(\mu+\rho,\delta_i)$'s up to signs, while the $(\lambda+\rho,\varepsilon_j)$'s and $(\mu+\rho,\varepsilon_j)$'s must match bijectively up to signs. So $\ell = r$. Then, there exists a pair of signed permutations $w = (w_1, w_2) \in W = W_{\mathrm{B}} = (\mathbb{Z}_2^m \rtimes \mathfrak{S}_m) \times (\mathbb{Z}_2^n \rtimes \mathfrak{S}_n)$ that extends the above bijections such that

$$w^{-1}(\mu+\rho) - (\lambda+\rho) = \sum_{a=1}^{\ell} c_a(\delta_{i_a} - s_a\varepsilon_{j_a}), \quad c_a \in \mathbb{C}. \tag{2.25}$$

This implies that $\lambda \sim \mu$.

(3) Finally, consider the case when $\mathfrak{g} = \mathfrak{spo}(2m|2n)$. For all even $k \geq 0$, we have $\chi_\lambda(z^k_{m,n}) = \chi_\mu(z^k_{m,n})$. This is equivalent again to the generating function identity (2.24). Following the case of $\mathfrak{spo}(2m|2n+1)$, we make a similar choice of pairs (i_a, j_a), for $1 \leq a \leq \ell$, and (i'_b, j'_b), for $1 \leq b \leq \ell$, which leads to a choice of $w = (w_1, w_2) \in W_{\mathrm{B}}$ and a sequence of isotropic odd roots satisfying (2.25). But we are not done yet, as the Weyl group W of $\mathfrak{spo}(2m|2n)$ is not W_{B}, but a subgroup of W_{B} of index 2. We will finish the job by considering two cases separately depending on whether λ is typical or atypical.

Assume first that λ is atypical (and hence so is μ), and we have $\ell \geq 1$. If $w = (w_1, w_2) \in W$, we are done. Otherwise, let $\tau_1 \in W_{\mathrm{B}}$ be the element changing the sign for δ_{i_1} while fixing all ε_j and δ_i $(i \neq i_1)$, and let $\sigma_1 \in W_{\mathrm{B}}$ be the element changing the sign for ε_{j_1} while fixing all δ_i and ε_j $(j \neq j_1)$. Note that $\tau_1 \in W$ and $\sigma_1 w^{-1} \in W$, and $\tau_1^2 = \sigma_1^2 = 1$. By the definition of the pair (i_1, j_1) and using (2.25), we have

$$\begin{aligned}
&\tau_1\sigma_1 w^{-1}(\mu+\rho) - (\lambda+\rho) \\
&= \tau_1\sigma_1\big(w^{-1}(\mu+\rho) - (\lambda+\rho)\big) + \big(\tau_1\sigma_1(\lambda+\rho) - (\lambda+\rho)\big) \\
&= -\big(2(\lambda+\rho, \delta_{i_1}) + c_1\big)(\delta_{i_1} - s_1\varepsilon_{j_1}) + \sum_{a=2}^{\ell} c_a(\delta_{i_a} - s_a\varepsilon_{j_a}).
\end{aligned}$$

Hence we have obtained a set of mutually orthogonal isotropic odd roots, a corresponding set of complex numbers, and the Weyl group element $w\sigma_1\tau_1 \in W$ satisfying (2.22). Thus, $\lambda \sim \mu$.

By the description of the center $Z(\mathfrak{g})$ in Theorem 2.26 and (2.19), we have additional elements in $Z(\mathfrak{g})$ to use. In particular, we have

$$\big\langle \varepsilon_1\varepsilon_2\cdots\varepsilon_n P, \theta^{-1}(\lambda+\rho)\big\rangle = \big\langle \varepsilon_1\varepsilon_2\cdots\varepsilon_n P, \theta^{-1}(\mu+\rho)\big\rangle. \tag{2.26}$$

Assume now that λ is typical (and so is μ), and hence $\ell = 0$. Then the $(m+n)$-tuple $\big((\lambda+\rho, \delta_i)^2, (\lambda+\rho, \varepsilon_j)^2\big)_{i,j}$ coincides with $\big((\mu+\rho, \delta_i)^2, (\mu+\rho, \varepsilon_j)^2\big)_{i,j}$, up to a permutation in $\mathfrak{S}_m \times \mathfrak{S}_n$. Recall $P = \prod_{i,j}(\delta_i^2 - \varepsilon_j^2)$. Note that $\langle P, \theta^{-1}(\lambda+\rho)\rangle = \langle P, \theta^{-1}(\mu+\rho)\rangle \neq 0$, since λ and μ are typical. By canceling this nonzero factor in (2.26), we obtain that

$$\big\langle \varepsilon_1\varepsilon_2\cdots\varepsilon_n, \theta^{-1}(\lambda+\rho)\big\rangle = \big\langle \varepsilon_1\varepsilon_2\cdots\varepsilon_n, \theta^{-1}(\mu+\rho)\big\rangle. \tag{2.27}$$

Note that in this case (2.22) reads that $\mu+\rho = w(\lambda+\rho)$, where we recall $w = (w_1, w_2)$. If $(\varepsilon_i, \lambda+\rho) \neq 0$ for all i, we conclude from (2.27) that w_2 changes the signs for an even number of ε_j's, and hence $w = (w_1, w_2)$ lies in the Weyl group W. If $(\varepsilon_i, \lambda+\rho) = 0$ for some i and $w \notin W$, then $w' = (w_1, w_2\sigma) \in W$ satisfies that $\mu+\rho = w'(\lambda+\rho)$, where σ denotes the sign change at ε_i. Hence $\lambda \sim \mu$ in this case as well.

The theorem is proved. □

Theorem 2.30 has the following implication, which will be referred to as the **linkage principle** for Lie superalgebra $\mathfrak{g}$ of type $\mathfrak{gl}$ and $\mathfrak{osp}$.

Proposition 2.31. *Let $\mathfrak{g}$ be a Lie superalgebra of type $\mathfrak{gl}$ and $\mathfrak{osp}$. For a composition factor $L(\mu)$ in a Verma module $\Delta(\lambda)$ of $\mathfrak{g}$, μ must satisfy the following conditions: $\lambda - \mu \in \mathbb{Z}_+\Phi^+$ and $\mu \sim \lambda$.*

Corollary 2.32. *If $\chi_\lambda = \chi_\mu$, then the degrees of atypicality for λ and μ coincide.*

Proof. This can be read off from the proof of Theorem 2.30. □

Corollary 2.33. *A finite-dimensional $\mathfrak{spo}(2m|1)$-module is completely reducible.*

Proof. Recall from Theorem 2.11 that the irreducible $\mathfrak{spo}(2m|1)$-module $L(\lambda)$ is finite dimensional if and only if the sequence $(\lambda_1,\ldots,\lambda_m)$ associated to the highest weight $\lambda = \sum_{i=1}^m \lambda_i\delta_i$ is a partition. Let λ and μ be highest weights for two irreducible finite-dimensional $\mathfrak{spo}(2m|1)$-modules. Then λ and μ are $\Phi_{\bar{0}}^+$-dominant integral, and hence $\lambda \sim \mu$ if and only if $\lambda = \mu$. By Theorem 2.30 this implies that $\chi_\lambda = \chi_\mu$ if and only if $\lambda = \mu$, and hence every finite-dimensional $\mathfrak{spo}(2m|1)$-module is completely reducible. □

2.2.7. Typical finite-dimensional irreducible characters. In this subsection we assume that $\mathfrak{g}$ is of type $\mathfrak{gl}$ or $\mathfrak{osp}$. As an application of the linkage principle, we obtain a character formula for the typical finite-dimensional irreducible $\mathfrak{g}$-modules.

We fix a positive system Φ^+ of $\mathfrak{g}$ and denote by $\mathfrak{g} = \mathfrak{n}^- \oplus \mathfrak{h} \oplus \mathfrak{n}^+$ the associated triangular decomposition. Recall the subset $\bar{\Phi}_{\bar{1}}^+ \subseteq \Phi^+$ defined in (1.34). Clearly, $\lambda \in \mathfrak{h}^*$ is typical (cf. Definition 2.29) if and only if $(\lambda+\rho,\alpha) \neq 0$, for all $\alpha \in \bar{\Phi}_{\bar{1}}^+$. Alternatively, λ is typical if and only if $\langle \lambda+\rho, P'\rangle \neq 0$, where $P' := \theta^{-1}(P) \in S(\mathfrak{h})$ in terms of P in (2.14). The following is a partial converse of Proposition 2.22 and follows immediately from Theorem 2.30. Below we shall provide a second proof based on the weaker Corollary 2.28.

Lemma 2.34. *Let $\mathfrak{g}$ be of type $\mathfrak{gl}$ or $\mathfrak{osp}$. Let $\lambda,\mu \in \mathfrak{h}^*$ with λ typical. Suppose that $\chi_\lambda = \chi_\mu$. Then, there exists $w \in W$ such that $w(\lambda+\rho) = \mu+\rho$, and μ is also typical.*

Proof. Suppose that there is no $w \in W$ such that $w(\lambda+\rho) = \mu+\rho$. Then we have $W(\lambda+\rho)\cap W(\mu+\rho) = \emptyset$. We can choose an element $f \in S(\mathfrak{h})$ that takes the value 1 on the finite set $W(\lambda+\rho)$ and 0 on the finite set $W(\mu+\rho)$. Averaging over W if necessary, we may also assume that $f \in S(\mathfrak{h})^W$. Now by Corollary 2.28 we have $\theta(f)P \in \mathfrak{hc}_*(Z(\mathfrak{g}))$. Thus there exists $z \in Z(\mathfrak{g})$ so that $\theta(f)P = \mathfrak{hc}_*(z)$ and equivalently $fP' = \mathfrak{hc}(z)$. Hence,

$$\chi_\lambda(z) = \langle \lambda+\rho, \mathfrak{hc}(z)\rangle = \langle \lambda+\rho, fP'\rangle = \langle \lambda+\rho, f\rangle\langle \lambda+\rho, P'\rangle = \langle \lambda+\rho, P'\rangle \neq 0.$$

Similarly, we have

$$\chi_\mu(z) = \langle \mu+\rho, \mathfrak{hc}(z)\rangle = \langle \mu+\rho, fP'\rangle = \langle \mu+\rho, f\rangle\langle \mu+\rho, P'\rangle = 0.$$

This contradicts $\chi_\lambda = \chi_\mu$. Hence there exists $w \in W$ such that $w(\lambda+\rho) = \mu+\rho$.

It follows by definition that $\bar{\Phi}_{\bar{1}} = \bar{\Phi}_{\bar{1}}^+ \cup -\bar{\Phi}_{\bar{1}}^+$ is W-invariant. Hence, for any $\alpha \in \bar{\Phi}_{\bar{1}}^+$, we have

$$(\mu+\rho,\alpha) = (w(\lambda+\rho),\alpha) = (\lambda+\rho, w^{-1}\alpha) \neq 0.$$

Therefore, μ is typical. □

Recall that the finite-dimensional irreducible $\mathfrak{g}$-modules have been classified in terms of highest weights in Section 2.1.

Theorem 2.35. *Let $\mathfrak{g}$ be of type $\mathfrak{gl}$ or $\mathfrak{osp}$. Let $\lambda \in \mathfrak{h}^*$ be a typical weight such that $L(\lambda)$ is a finite-dimensional irreducible representation of $\mathfrak{g}$. Then*

$$chL(\lambda) = \frac{\prod_{\alpha\in\Phi_{\bar{1}}^+}(1+e^{-\alpha})}{\prod_{\beta\in\Phi_{\bar{0}}^+}(1-e^{-\beta})} \sum_{w\in W} (-1)^{\ell(w)} e^{w(\lambda+\rho)-\rho}.$$

Proof. Recall from the proof of Proposition 2.22 that

$$\text{(2.28)} \qquad D chL(\lambda) = \sum_{\mu\in X} a_{\mu\lambda} \sum_{w\in W} (-1)^{\ell(w)} e^{w(\mu+\rho)},$$

where X consists of $\Phi_{\bar{0}}^+$-dominant weights such that $a_{\mu\lambda} \neq 0$.

Since λ is typical and $\chi_\lambda = \chi_\mu$ for all $\mu \in X$, it follows by Lemma 2.34 that $X = \{\lambda\}$. Recalling that $a_{\lambda\lambda} = 1$, we can rewrite (2.28) as

$$D chL(\lambda) = \sum_{w\in W} (-1)^{\ell(w)} e^{w(\lambda+\rho)},$$

from which the theorem follows. □

Example 2.36. Let $\mathfrak{g} = \mathfrak{gl}(1|1)$. Then $\lambda = a\delta + b\varepsilon$ is typical if and only if $a+b \neq 0$.

Example 2.37. The character formula in Theorem 2.35 applies to all irreducible finite-dimensional $\mathfrak{spo}(2m|1)$-modules, as their highest weights are always typical.

2.3. Harish-Chandra homomorphism and linkage for $\mathfrak{q}(n)$

This section is the counterpart of Section 2.2 for the queer Lie superalgebra $\mathfrak{q}(n)$. The image of the Harish-Chandra homomorphism for $\mathfrak{q}(n)$ is described with the help of symmetric functions such as Schur Q-functions (see Appendix A). Then, we obtain a linkage principle on composition factors of a Verma module of $\mathfrak{q}(n)$. The finite-dimensional typical irreducible representations of $\mathfrak{q}(n)$ are classified, and a Weyl-type character formula for these modules is obtained.

2.3.1. Central characters for $\mathfrak{q}(n)$. In this section, let $\mathfrak{g} = \mathfrak{q}(n)$ be the queer Lie superalgebra, and let $\mathfrak{h} = \mathfrak{h}_{\bar{0}} + \mathfrak{h}_{\bar{1}}$ be its (nonabelian) Cartan subalgebra. As this subsection is parallel to that of the basic Lie superalgebra case in Section 2.2.2, we will be brief and only point out the differences from therein.

Recall that $\mathfrak{g}$ is equipped with a non-degenerate odd invariant supersymmetric bilinear form $(\cdot,\cdot)$, which allows us to identify $\mathfrak{g}$ with its dual $\mathfrak{g}^*$, $\mathfrak{h}$ with $\mathfrak{h}^*$, and $\mathfrak{h}_{\bar{0}}$ with $\mathfrak{h}_{\bar{1}}^*$. Via such identifications, the restriction map $\mathcal{S}(\mathfrak{g}^*) \to \mathcal{S}(\mathfrak{h}^*)$ and the induced map $\eta : \mathcal{S}(\mathfrak{g}) \to \mathcal{S}(\mathfrak{h})$ are homomorphisms of algebras. Let $\mathfrak{g} = \mathfrak{n}^- \oplus \mathfrak{h} \oplus \mathfrak{n}^+$ be a triangular decomposition. As before, we have a projection $\phi : U(\mathfrak{g}) \to U(\mathfrak{h})$, and any element $z \in Z(\mathfrak{g})$ affords a unique expression of the form (2.6). Moreover, it follows from $\operatorname{ad}\mathfrak{h}(z) = 0$ that $h_z \in U(\mathfrak{h}_{\bar{0}})$. Hence $Z(\mathfrak{g})$ consists of only even elements. The restriction of ϕ to $Z(\mathfrak{g})$, also denoted by ϕ, defines an algebra homomorphism from $Z(\mathfrak{g})$ to $U(\mathfrak{h}_{\bar{0}})$, which sends z to h_z. For $\lambda \in \mathfrak{h}_{\bar{0}}^*$, define a linear map $\chi_\lambda : Z(\mathfrak{g}) \to \mathbb{C}$ by $\chi_\lambda(z) = \langle \lambda, \phi(z) \rangle$.

The proof of Lemma 2.21 also works for the following lemma.

Lemma 2.38. *Let $\mathfrak{g} = \mathfrak{q}(n)$. An element $z \in Z(\mathfrak{g})$ acts as the scalar $\chi_\lambda(z)$ on any highest weight $\mathfrak{g}$-module $V(\lambda)$ of highest weight λ.*

Therefore $\chi_\lambda : Z(\mathfrak{g}) \to \mathbb{C}$ defines a one-dimensional $Z(\mathfrak{g})$-module, which will be called the **central character** of $\mathfrak{g} = \mathfrak{q}(n)$ associated to the highest weight λ.

2.3.2. Harish-Chandra homomorphism for $\mathfrak{q}(n)$. Let $\mathfrak{g} = \mathfrak{n}^- \oplus \mathfrak{h} \oplus \mathfrak{n}^+$ be the standard triangular decomposition. Bearing in mind that $\rho = 0$ for $\mathfrak{q}(n)$, we call

$$\mathfrak{hc} = \phi : Z(\mathfrak{g}) \to S(\mathfrak{h}_{\bar{0}})$$

the **Harish-Chandra homomorphism** for $\mathfrak{q}(n)$.

Recall the elements $H_i, \overline{H}_i, \widetilde{E}_{ij}, \overline{E}_{ij}$ in $\mathfrak{q}(n)$ etc. from Sections 1.1.4 and 1.2.6. A **singular vector** in the Verma module $\Delta(\lambda)$ of $\mathfrak{q}(n)$ is a nonzero vector v such that $\widetilde{E}_{ij}v = \overline{E}_{ij}v = 0$, for $1 \le i < j \le n$.

Lemma 2.39. *Let $\mathfrak{g} = \mathfrak{q}(n)$. Let $\lambda \in \mathfrak{h}_{\bar{0}}^*$ be such that $\lambda_i - \lambda_{i+1} = k \in \mathbb{N}$ for some $1 \le i \le n-1$. Then*

$$u := \overline{E}_{i,i+1}\widetilde{E}_{i+1,i}^{k+1} v_\lambda$$

is a singular vector in the Verma module $\Delta(\lambda)$ of weight $s_i(\lambda) = \lambda - k(\varepsilon_i - \varepsilon_{i+1})$.

Proof. We have the following identities in $\Delta(\lambda)$, for $\lambda \in \mathfrak{h}_{\bar{0}}^*$ and $m \ge 0$:

$$(\overline{H}_i - \overline{H}_{i+1})\widetilde{E}_{i+1,i}^m v_\lambda = \widetilde{E}_{i+1,i}^m(\overline{H}_i - \overline{H}_{i+1})v_\lambda - 2m\overline{E}_{i+1,i}\widetilde{E}_{i+1,i}^{m-1}v_\lambda, \tag{2.29}$$

$$\overline{E}_{i,i+1}\widetilde{E}_{i+1,i}^{m+1}v_\lambda = (m+1)\widetilde{E}_{i+1,i}^m(\overline{H}_i - \overline{H}_{i+1})v_\lambda - m(m+1)\overline{E}_{i+1,i}\widetilde{E}_{i+1,i}^{m-1}v_\lambda. \tag{2.30}$$

Indeed, (2.29) can be directly verified by induction on m, and then (2.30) follows by induction on m using (2.29).

It follows by (2.30) for $m = k$ that $u \neq 0$. Now note that

$$\overline{E}_{i,i+1}u = (\overline{E}_{i,i+1})^2 \widetilde{E}_{i+1,i}^{k+1} v_\lambda = 0,$$
$$\widetilde{E}_{i,i+1}u = \overline{E}_{i,i+1}\widetilde{E}_{i,i+1}\widetilde{E}_{i+1,i}^{k+1} v_\lambda = 0,$$

by a standard $\mathfrak{sl}(2)$-calculation. By weight considerations in $\Delta(\lambda)$, we must have

$$\overline{E}_{j,j+1}u = \widetilde{E}_{j,j+1}u = 0, \qquad j \neq i.$$

Thus, we conclude that u is a singular vector in $\Delta(\lambda)$, and the weight of u is clearly equal to $s_i(\lambda) = \lambda - k(\varepsilon_i - \varepsilon_{i+1})$. □

Recall that the Weyl group W for $\mathfrak{g} = \mathfrak{q}(n)$ is the symmetric group $\mathfrak{S}_n$.

Proposition 2.40. *Let $\mathfrak{g} = \mathfrak{q}(n)$. We have $\chi_\lambda = \chi_{w(\lambda)}$, for all $w \in W$ and $\lambda \in \mathfrak{h}_{\bar{0}}^*$. Also, we have $\mathfrak{hc}(Z(\mathfrak{g})) \subseteq S(\mathfrak{h}_{\bar{0}})^W$.*

Proof. By Lemma 2.39, we have a nonzero homomorphism $\Delta(s_i(\lambda)) \to \Delta(\lambda)$, for $\lambda \in \mathfrak{h}_{\bar{0}}^*$ with $\lambda_i - \lambda_{i+1} \in \mathbb{N}$, which implies by Lemma 2.21 that $\chi_\lambda = \chi_{s_i(\lambda)}$. Hence, $\chi_\lambda = \chi_{w(\lambda)}$, for all $w \in W$ and for all integral weights $\lambda = \sum_{i=1}^n \lambda_i \varepsilon_i$ satisfying $\lambda_1 > \ldots > \lambda_n$. Since the set of such weights is Zariski dense in $\mathfrak{h}_{\bar{0}}^*$, it follows that $\chi_\lambda = \chi_{w(\lambda)}$, for all $w \in W$ and $\lambda \in \mathfrak{h}_{\bar{0}}^*$. That is, $\langle \lambda, \mathfrak{hc}(z) \rangle = \langle w(\lambda), \mathfrak{hc}(z) \rangle$, for all $z \in Z(\mathfrak{g})$, $w \in W$ and $\lambda \in \mathfrak{h}_{\bar{0}}^*$. Hence, we conclude that $\mathfrak{hc}(Z(\mathfrak{g})) \subseteq S(\mathfrak{h}_{\bar{0}})^W$, thanks to the W-invariance of $(\cdot,\cdot)$. □

Remark 2.41. One can imitate the proof of Proposition 2.22 to give an alternative proof of Proposition 2.40, bypassing Lemma 2.39. On the other hand, a second proof of Proposition 2.22 can be given in the spirit of the proof of Proposition 2.40 with some extra work involving odd reflections, as the Weyl group W is not generated by the even simple reflections.

The odd non-degenerate invariant bilinear form $(\cdot,\cdot)$ on $\mathfrak{h}$ induces an (even) isomorphism $\mathfrak{h}_{\bar{0}} \cong \Pi\mathfrak{h}_{\bar{1}}^*$, where Π denotes the parity reversing functor from Section 1.1.1. This in turn induces (even) isomorphisms $\theta : S(\mathfrak{h}_{\bar{0}}) \to S(\Pi\mathfrak{h}_{\bar{1}}^*)$ and $\theta : S(\mathfrak{h}_{\bar{0}})^W \to S(\Pi\mathfrak{h}_{\bar{1}}^*)^W$. We further define $\mathfrak{hc}_* = \theta\mathfrak{hc} : Z(\mathfrak{g}) \to S(\Pi\mathfrak{h}_{\bar{1}}^*)^W$ and $\eta_* = \theta\eta : S(\mathfrak{g})^\mathfrak{g} \to S(\Pi\mathfrak{h}_{\bar{1}}^*)$.

Let $\bar{\varepsilon}_1, \bar{\varepsilon}_2, \ldots, \bar{\varepsilon}_n \in \Pi\mathfrak{h}_{\bar{1}}^*$ be determined by

$$\bar{\varepsilon}_i(\overline{H}_j) = \delta_{ij}.$$

That is, $\theta(H_i) = \bar{\varepsilon}_i$, for each i. Let (π, V) be the natural representation of $\mathfrak{g} = \mathfrak{q}(n)$ on the vector superspace $\mathbb{C}^{n|n}$. We recall the odd trace operator otr : $\mathfrak{g} \to \mathbb{C}$ from (1.32). A straightforward super generalization of the arguments for Lie algebras as given in Carter [**18**, pp. 212-213] shows that $x \mapsto \mathrm{otr}\big(\pi(x)^{2k-1}\big)$, for $k \in \mathbb{N}$, defines a $\mathfrak{g}$-invariant homogeneous polynomial on $\mathfrak{g}$ of degree k (note that $\mathrm{otr}\big(\pi(x)^{2k}\big) = 0$ by definition of the odd trace). Restricting to $\mathfrak{h}_{\bar{1}}$ we obtain a homogeneous polynomial on $\Pi\mathfrak{h}_{\bar{1}}$ of degree $2k-1$. Then a direct calculation shows that, for $x \in \mathfrak{h}_{\bar{1}}$ and $k \geq 1$, $\operatorname{otr}\big(\pi(x)^{2k-1}\big)$ equals the value at x of the polynomial

$$p_{2k-1}(\overline{\varepsilon}) = \sum_{i=1}^{n} \overline{\varepsilon}_i^{2k-1}, \quad k \in \mathbb{N}.$$

Recall that a ring Γ_n of symmetric functions in n variables $\overline{\varepsilon}_1, \overline{\varepsilon}_2, \ldots, \overline{\varepsilon}_n$ is defined in Appendix A.3. By (A.53), Γ_n has a basis that consists of Schur Q-functions $Q_\lambda(\overline{\varepsilon}_1, \overline{\varepsilon}_2, \ldots, \overline{\varepsilon}_n)$, for $\lambda \in \mathcal{SP}$, such that $\ell(\lambda) \leq n$ (where $\mathcal{SP}$ denotes the set of strict partitions). By (A.45), the algebra $\Gamma_{n,\mathbb{C}} := \mathbb{C} \otimes_{\mathbb{Z}} \Gamma_n$ is generated by the odd degree power sums $p_{2k-1}(\overline{\varepsilon})$ for $k \geq 1$.

Summarizing the above discussions, we have established the following.

Proposition 2.42. *Let $\mathfrak{g} = \mathfrak{q}(n)$. We have $\Gamma_{n,\mathbb{C}} \subseteq \operatorname{Im}(\eta_*)$.*

We define

$$P := \prod_{1 \leq i < j \leq n} (\overline{\varepsilon}_i + \overline{\varepsilon}_j) \in S(\Pi\mathfrak{h}_{\bar{1}}^*). \tag{2.31}$$

The following lemma is a reformulation of (A.59) in Appendix A.

Lemma 2.43. *Let $f \in S(\Pi\mathfrak{h}_{\bar{1}}^*)^W$. Then we have $fP \in \operatorname{Im}(\eta_*)$.*

Lemma 2.44. *Assume that a weight $\lambda \in \mathfrak{h}_{\bar{0}}^*$ satisfies that $\lambda_i = -\lambda_{i+1} = k \in \mathbb{C}$, for some $1 \leq i \leq n-1$.*

(1) *If $k = 0$, then $u := \widetilde{E}_{i+1,i} v_\lambda$ for any highest weight vector v_λ is a singular vector in $\Delta(\lambda)$ of weight $\lambda - \varepsilon_i + \varepsilon_{i+1}$.*

(2) *If $k \neq 0$, then there exists a highest weight vector v_λ in the Verma module $\Delta(\lambda)$ such that $(\overline{H}_i - \overline{H}_{i+1})v_\lambda \neq 0$ and $(\overline{H}_i + \overline{H}_{i+1})v_\lambda = 0$. Moreover,*

$$u := \widetilde{E}_{i+1,i}(\overline{H}_i - \overline{H}_{i+1})v_\lambda - 2k\overline{E}_{i+1,i}v_\lambda$$

is a singular vector in $\Delta(\lambda)$ of weight $\lambda - \varepsilon_i + \varepsilon_{i+1}$.

Proof. (1) Assume that $k = 0$. Note that $\overline{H}_i$ and $\overline{H}_{i+1}$ act trivially on the highest weight space W_λ of $\Delta(\lambda)$, and hence $\overline{H}_i v_\lambda = \overline{H}_{i+1} v_\lambda = 0$ for any highest weight vector v_λ. Then, $\widetilde{E}_{i,i+1} u = (H_i - H_{i+1})v_\lambda = 0$, and $\overline{E}_{i,i+1} u = (\overline{H}_i - \overline{H}_{i+1})v_\lambda = 0$. Also it follows from weight consideration that $\widetilde{E}_{j,j+1} u = \overline{E}_{j,j+1} u = 0$ for $j \neq i$. Hence, u is a singular vector.

(2) Assume that $k \neq 0$. We again identify the highest weight space of the Verma module $\Delta(\lambda)$ as the $\mathfrak{h}$-module W_λ (see Section 2.1.6 and also see Lemma 1.42). We compute that

$$\lambda([\overline{H}_i \pm \overline{H}_{i+1}, \overline{H}_i \pm \overline{H}_{i+1}]) = 2\lambda(H_i + H_{i+1}) = \lambda_i + \lambda_{i+1} = 0.$$

Hence, by Lemma 1.42, there exists a highest weight vector $v_\lambda \in W_\lambda$ such that $w_\lambda := (\overline{H}_i - \overline{H}_{i+1})v_\lambda \neq 0$ and $(\overline{H}_i + \overline{H}_{i+1})v_\lambda = 0$. It follows that

$$(\overline{H}_i + \overline{H}_{i+1})w_\lambda = \lambda(H_i - H_{i+1})v_\lambda = 2kv_\lambda,$$
$$(\overline{H}_i - \overline{H}_{i+1})w_\lambda = 0.$$

We compute that

$$\widetilde{E}_{i,i+1}\widetilde{E}_{i+1,i}w_\lambda = \lambda(H_i - H_{i+1})w_\lambda = 2kw_\lambda,$$
$$\overline{E}_{i,i+1}\widetilde{E}_{i+1,i}w_\lambda = (\overline{H}_i - \overline{H}_{i+1})w_\lambda = 0,$$
$$\widetilde{E}_{i,i+1}\overline{E}_{i+1,i}v_\lambda = (\overline{H}_i - \overline{H}_{i+1})v_\lambda = w_\lambda,$$
$$\overline{E}_{i,i+1}\overline{E}_{i+1,i}v_\lambda = (\overline{H}_i + \overline{H}_{i+1})v_\lambda = 0.$$

It follows that $\widetilde{E}_{i,i+1}u = \overline{E}_{i,i+1}u = 0$. Also we have by weight consideration that $\widetilde{E}_{j,j+1}u = \overline{E}_{j,j+1}u = 0$ for $j \neq i$. Hence, u is a singular vector in $\Delta(\lambda)$ whose weight is clearly equal to $\lambda - \varepsilon_i + \varepsilon_{i+1}$. □

Lemma 2.45. *Let $\lambda = \sum_{i=1}^n \lambda_i \varepsilon_i \in \mathfrak{h}_{\bar{0}}^*$. Assume that $\lambda_i = -\lambda_j$ for some $1 \leq i \neq j \leq n$. Then, $\chi_\lambda = \chi_{\lambda - t\alpha}$ for $\alpha = \varepsilon_i - \varepsilon_j$ and any $t \in \mathbb{C}$.*

Proof. We first claim that $\chi_\lambda = \chi_{\lambda-\varepsilon_i+\varepsilon_j}$. By Proposition 2.40 we are reduced to proving the claim when $j = i+1$, and the claim in this case follows by Lemmas 2.38 and 2.44. By a repeated application of the claim, we have $\chi_\lambda = \chi_{\lambda - t\alpha}$ for any $t \in \mathbb{Z}_+$. Since $\mathbb{Z}_+$ is Zariski dense in $\mathbb{C}$, the lemma follows. □

The following theorem is the $\mathfrak{q}(n)$-counterpart of Theorem 2.26.

Theorem 2.46. *Let $\mathfrak{g} = \mathfrak{q}(n)$. Then $\mathfrak{hc}_*(Z(\mathfrak{g})) = \mathrm{Im}(\eta_*) = \Gamma_{n,\mathbb{C}}$.*

Proof. Let $\lambda = \sum_{i=1}^n \lambda_i \varepsilon_i$ be an integral weight. Let $z \in Z(\mathfrak{g})$. By Proposition 2.40, we know that $\mathfrak{hc}_*(z) \in S(\Pi\mathfrak{h}_{\bar{1}}^*)^W$. By unraveling the definition of χ_λ, we obtain by Lemma 2.45 that $\mathfrak{hc}_*(z)|_{\bar{\varepsilon}_i = -\bar{\varepsilon}_j = t}$ is independent of t. Then by the characterization of Γ_n given in (A.60), we have $\mathfrak{hc}_*(z) \in \Gamma_{n,\mathbb{C}}$. Together with Proposition 2.42, we have shown that $\mathfrak{hc}_*(Z(\mathfrak{g})) \subseteq \Gamma_{n,\mathbb{C}} \subseteq \mathrm{Im}(\eta_*)$.

It remains to show that $\mathfrak{hc}_*(Z(\mathfrak{g})) = \mathrm{Im}(\eta_*)$. This follows by the corresponding identity on the associated graded and then a standard filtered algebra argument, exactly as in the proof of Theorem 2.26. We leave the details to the interested reader. □

We summarize various maps for the $\mathfrak{q}(n)$ case in the following diagram, which only becomes commutative when we pass from $Z(\mathfrak{g})$ to its associated graded (note that the other algebras are graded).

$$\begin{array}{ccc} Z(\mathfrak{g}) & \xrightarrow[\cong]{\gamma^{-1}} & \mathcal{S}(\mathfrak{g})^{\mathfrak{g}} \\ {\scriptstyle \mathfrak{hc}_*}\downarrow & \swarrow{\scriptstyle \eta_*,\ \cong} & \downarrow{\scriptstyle \eta} \\ S(\Pi\mathfrak{h}_{\bar{1}}^*)^W & \xleftarrow[\theta]{\cong} & S(\mathfrak{h}_{\bar{0}})^W \end{array} \tag{2.32}$$

Remark 2.47. Actually $\mathfrak{hc}$ and $\mathfrak{hc}_*$ are injective (see Sergeev [**111**]), and so we have an isomorphism of algebras $\mathfrak{hc}_* : Z(\mathfrak{g}) \to \Gamma_{n,\mathbb{C}}$. The injectivity of $\mathfrak{hc}_*$ does not play any role in the representation theory of $\mathfrak{q}(n)$ in the book.

2.3.3. Linkage for $\mathfrak{q}(n)$. For a root α of the form $\varepsilon_i - \varepsilon_j$, define

$$\overline{\alpha}^\vee := H_i + H_j \in \mathfrak{h}_{\bar{0}}.$$

We define a relation on $\mathfrak{h}_{\bar{0}}^*$

$$\lambda \sim \mu, \qquad \text{for } \lambda, \mu \in \mathfrak{h}_{\bar{0}}^*,$$

if there exist a collection of roots α_i, complex numbers c_i, for $1 \le i \le \ell$, and an element $w \in W$ such that

$$\mu = w\Big(\lambda - \sum_{a=1}^{\ell} c_a\alpha_a\Big), \quad \Big\langle \alpha_i, \overline{\alpha}_j^\vee \Big\rangle = 0, \quad \Big\langle \lambda, \overline{\alpha}_j^\vee \Big\rangle = 0, \, 1 \le i, j \le \ell. \tag{2.33}$$

If $\lambda \sim \mu$, we say that λ and μ are **linked**. The following theorem is a $\mathfrak{q}(n)$-counterpart of Theorem 2.30. It also implies that linkage for $\mathfrak{q}(n)$ is an equivalence relation.

Theorem 2.48. *Let $\mathfrak{h} = \mathfrak{h}_{\bar{0}} + \mathfrak{h}_{\bar{1}}$ be a Cartan subalgebra of $\mathfrak{q}(n)$, and let $\lambda, \mu \in \mathfrak{h}_{\bar{0}}^*$. Then λ is linked to μ if and only if $\chi_\lambda = \chi_\mu$.*

Proof. Assume first that $\lambda \sim \mu$. By a repeated application of Lemma 2.45, the second and third conditions in (2.33) for $\lambda \sim \mu$ imply that

$$\chi_\lambda = \chi_{\lambda - c_1\alpha_1} = \chi_{\lambda - c_1\alpha_1 - c_2\alpha_2} = \cdots = \chi_{\lambda - \sum_{a=1}^{\ell} c_a\alpha_a}.$$

Since $\mu = w(\lambda - \sum_{a=1}^{\ell} c_a\alpha_a)$ by (2.33) again, it follows by Proposition 2.40 that $\chi_{\lambda - \sum_{a=1}^{\ell} c_a\alpha_a} = \chi_\mu$, whence $\chi_\lambda = \chi_\mu$.

Now assume that $\chi_\lambda = \chi_\mu$. Write $\lambda = \sum_{i=1}^n \lambda_i\varepsilon_i$ and $\mu = \sum_{i=1}^n \mu_i\varepsilon_i$. Let $z_{2k+1} \in Z(\mathfrak{g})$ such that $\mathfrak{hc}(z_{2k+1}) = p_{2k+1}$, the $(2k+1)$st power sum in $H_1, \ldots, H_n$. From $\chi_\lambda(z_{2k+1}) = \chi_\mu(z_{2k+1})$ we obtain that $\langle \lambda, \mathfrak{hc}(z_{2k+1})\rangle = \langle \mu, \mathfrak{hc}(z_{2k+1})\rangle$. Hence we have

$$\sum_{i=1}^n \lambda_i^{2k+1} = \sum_{j=1}^n \mu_j^{2k+1}, \quad \forall k \in \mathbb{Z}_+,$$

which is equivalent to the following generating function identity in an indeterminate t:

$$\sum_{i=1}^{n} \sinh(\lambda_i t) = \sum_{j=1}^{n} \sinh(\mu_j t).$$

Here we recall that the function

$$\sinh(t) = \frac{e^t - e^{-t}}{2} = \sum_{k\geq 0} \frac{t^{2k+1}}{(2k+1)!}$$

satisfies $\sinh(-at) = -\sinh(at)$ for any scalar $a \in \mathbb{C}$.

Pick a maximal number of pairs (i_a, j_a) such that $\lambda_{i_a} = -\lambda_{j_a}$, with $1 \leq i_a < j_a \leq n$ and all i_a, j_a are distinct, for $1 \leq a \leq \ell$. Denote

$$I_\lambda = \{1 \leq i \leq n \mid i \neq i_a, i \neq j_a, \forall a\}.$$

Similarly, pick a maximal number of pairs (i'_b, j'_b) such that $\mu_{i'_b} = -\mu_{j'_b}$, with $1 \leq i'_b < j'_b \leq n$ and all i'_b, j'_b are distinct, for $1 \leq b \leq r$, and define I_μ accordingly. It follows by definition that

$$\sum_{i\in I_\lambda} \sinh(\lambda_i t) = \sum_{j\in I_\mu} \sinh(\mu_j t). \tag{2.34}$$

Let c be the maximum of the set $\{|\lambda_i|, |\mu_j| \mid i \in I_\lambda, j \in I_\mu\}$. Furthermore, assume without loss of generality that for some s we have $c = |\lambda_s|$ and λ_s appears in the set $\{\lambda_i | i \in I_\lambda\}$ with multiplicity ℓ. The identity (2.34) which is valid for all t implies that there exists μ_r with $r \in I_\mu$ such that $\mu_r = \lambda_s$ and also μ_r must appear in $\{\mu_j | j \in I_\mu\}$ with multiplicity ℓ as well. Canceling the corresponding hyperbolic sine functions from both sides of (2.34) we can proceed similarly as before and conclude that there is a bijection between I_λ and I_μ such that the corresponding λ_i and μ_j coincide. Now we can find a permutation $w \in \mathfrak{S}_n$ that extends the bijection between I_λ and I_μ such that $w^{-1}\mu - \lambda$ is a linear combination of the roots $\varepsilon_{i_a} - \varepsilon_{j_a}$. Thus, $\lambda \sim \mu$. □

Theorem 2.48 has the following implications. For a composition factor $L(\mu)$ in a Verma module $\Delta(\lambda)$ of $\mathfrak{q}(n)$, μ must satisfy the following conditions: $\lambda - \mu \in \mathbb{Z}_+\Phi^+$, and $\mu \sim \lambda$. This will be referred to as the **linkage principle** for $\mathfrak{q}(n)$.

Definition 2.49. Let $\mathfrak{g} = \mathfrak{q}(n)$ and let $\mathfrak{h}$ be its standard Cartan subalgebra. The **degree of atypicality** of a weight $\lambda \in \mathfrak{h}^*_{\bar{0}}$, denoted by $\#\lambda$, is the maximum number of pairs (i_a, j_a) with all i_a, j_a distinct such that $\langle \lambda, H_{i_a} + H_{j_a} \rangle = 0$. A weight $\lambda \in \mathfrak{h}^*_{\bar{0}}$ is called **typical** if $\#\lambda = 0$, and is **atypical** otherwise.

We have the following corollary from the proof of Theorem 2.48.

Corollary 2.50. *If $\chi_\lambda = \chi_\mu$ for $\lambda, \mu \in \mathfrak{h}^*_{\bar{0}}$, then the degrees of atypicality for λ and μ coincide.*

2.3.4. Typical finite-dimensional characters of $\mathfrak{q}(n)$. Note that $\lambda \in \mathfrak{h}_{\bar{0}}^*$ is typical (cf. Definition 2.49) if and only if

$$\prod_{1\le i<j\le n} \langle \lambda, H_i + H_j \rangle \neq 0.$$

Recall that the finite-dimensional irreducible $\mathfrak{q}(n)$-modules are classified in terms of highest weights in Theorem 2.18. The following is the $\mathfrak{q}(n)$-counterpart of Lemma 2.34.

Lemma 2.51. *Let $\mathfrak{g} = \mathfrak{q}(n)$ and $\lambda, \mu \in \mathfrak{h}_{\bar{0}}^*$ with λ typical. Suppose that $\chi_\lambda = \chi_\mu$. Then, there exists $w \in W$ such that $\mu = w(\lambda)$, and μ is typical.*

Proof. Recalling P from (2.31), we have $\bar{\varepsilon}_i + \bar{\varepsilon}_j = \theta(H_i + H_j)$, and

$$P = \theta\Big(\prod_{1\le i<j\le n} (H_i + H_j) \Big).$$

Clearly P is W-invariant. By Lemma 2.43 and Theorem 2.46, we have $\theta(f)P \in \mathfrak{hc}_*(Z(\mathfrak{g}))$, for $f \in S(\mathfrak{h}_{\bar{0}})^W$. Also recall $\rho = 0$.

With these ingredients in place, the argument in the proof of Lemma 2.34 goes through in this case as well. We leave the details to the interested reader. □

The following is a $\mathfrak{q}(n)$-analogue of Theorem 2.35.

Theorem 2.52. *Let $\mathfrak{g} = \mathfrak{q}(n)$ and let $\lambda \in \mathfrak{h}_{\bar{0}}^*$ be a typical weight such that the irreducible $\mathfrak{q}(n)$-module $L(\lambda)$ is finite dimensional. Then*

$$chL(\lambda) = 2^{\frac{\ell(\lambda)+\delta(\lambda)}{2}} \frac{\prod_{\beta\in\Phi_{\bar{1}}^+}(1+e^{-\beta})}{\prod_{\alpha\in\Phi_{\bar{0}}^+}(1-e^{-\alpha})} \sum_{w\in W} (-1)^{\ell(w)} e^{w(\lambda)}.$$

Proof. We follow the strategy of the proof of Theorem 2.35.

The character of the Verma module $\Delta(\lambda)$ is given by

$$\mathrm{ch}\Delta(\lambda) = 2^{\frac{\ell(\lambda)+\delta(\lambda)}{2}} e^{\lambda} \frac{\prod_{\beta\in\Phi_{\bar{1}}^+}(1+e^{-\beta})}{\prod_{\alpha\in\Phi_{\bar{0}}^+}(1-e^{-\alpha})}.$$

Observe that the highest weight spaces of $L(\mu)$ and $\Delta(\mu)$, for every $\mu \in W\lambda$, have dimension $2^{\frac{\ell(\lambda)+\delta(\lambda)}{2}}$, thanks to $\ell(\mu) = \ell(\lambda)$.

Imitating the proof of Proposition 2.22, we obtain that

$$\frac{\prod_{\alpha\in\Phi_{\bar{0}}^+}(1-e^{-\alpha})}{\prod_{\beta\in\Phi_{\bar{1}}^+}(1+e^{-\beta})} \mathrm{ch}L(\lambda) = \sum_{\mu\in X} a_{\mu\lambda} \sum_{w\in W} (-1)^{\ell(w)} e^{w(\mu)}, \tag{2.35}$$

where X consists of $\Phi_{\bar{0}}^+$-dominant integral weights such that $a_{\mu\lambda} \neq 0$.

Since λ is typical and $\chi_\lambda = \chi_\mu$ for all $\mu \in X$, it follows by Lemma 2.51 that $X = \{\lambda\}$. Recalling that $a_{\lambda\lambda} = 2^{\frac{\ell(\lambda)+\delta(\lambda)}{2}}$, we can rewrite (2.35) as

$$\frac{\prod_{\alpha \in \Phi_{\bar{0}}^+}(1-e^{-\alpha})}{\prod_{\beta \in \Phi_{\bar{1}}^+}(1+e^{-\beta})} \mathrm{ch} L(\lambda) = 2^{\frac{\ell(\lambda)+\delta(\lambda)}{2}} \sum_{w \in W} (-1)^{\ell(w)} e^{w(\lambda+\rho)},$$

from which the theorem follows. □

Remark 2.53. The assumption on λ in Theorem 2.52 forces $\lambda_i > \lambda_{i+1}$ for all i, and so $\ell(\lambda) = n$ or $\ell(\lambda) = n-1$.

2.4. Extremal weights of finite-dimensional simple modules

In this section, we study the extremal weights of a finite-dimensional irreducible module over a Lie superalgebra of type $\mathfrak{gl}$ and $\mathfrak{osp}$. A main complication arises from the existence of non-conjugate Borel subalgebras for Lie superalgebras. For the Lie superalgebra of type $\mathfrak{gl}$ the extremal weights for irreducible polynomial representations are determined. For the Lie superalgebras of type $\mathfrak{osp}$ the extremal weights of finite-dimensional irreducible representations of integer weights are determined. In all cases, the answers are given in terms of the hook diagrams.

2.4.1. Extremal weights for $\mathfrak{gl}(m|n)$. We start with some general remarks for any basic Lie superalgebra $\mathfrak{g}$. Let L be a finite-dimensional $\mathfrak{g}$-module. Then given any Borel subalgebra $\mathfrak{b}$ of $\mathfrak{g}$ associated to a positive system Φ^+, there exists a unique weight $\lambda^{\mathfrak{b}}$ for L such that $\lambda^{\mathfrak{b}} + \alpha$ is not a weight for L for any $\alpha \in \Phi^+$, and the weight space $L_{\lambda^{\mathfrak{b}}}$ is one-dimensional (cf. Proposition 1.39). The weight $\lambda^{\mathfrak{b}}$ is called the **$\mathfrak{b}$-extremal weight** for the $\mathfrak{g}$-module L. Two Borel subalgebras $\mathfrak{b}$ and $\mathfrak{b}'$ of $\mathfrak{g}$ are in general not conjugate. Thus, the $\mathfrak{b}$-extremal weight and the $\mathfrak{b}'$-extremal weight of L may in general not be conjugate by the Weyl group W, in contrast to the semisimple Lie algebra setting.

In this subsection we let $\mathfrak{g} = \mathfrak{gl}(m|n)$. Let $\mathfrak{h}$ be its standard Cartan subalgebra with standard Borel subalgebra $\mathfrak{b}^{\mathrm{st}}$. For an $(m|n)$-hook partition λ, we may regard $\lambda^\natural$ as an element in $\mathfrak{h}^*$ as in (2.4). We shall determine the $\mathfrak{b}$-extremal weight of $L(\mathfrak{g}, \mathfrak{b}^{\mathrm{st}}, \lambda^\natural)$ for any Borel subalgebra $\mathfrak{b}$. The class of $\mathfrak{gl}(m|n)$-modules $L(\mathfrak{g}, \mathfrak{b}^{\mathrm{st}}, \lambda^\natural)$, for all $(m|n)$-hook partitions λ, are called the **polynomial modules** of $\mathfrak{gl}(m|n)$, and they will feature significantly in Chapter 3 on Schur duality.

Recall the weights δ_i and ε_j from Section 1.2.3. Let $\mathfrak{b}$ be a Borel subalgebra of $\mathfrak{g}$ and let Φ^+ be its positive system. Assume that the $\varepsilon\delta$-sequence from Section 1.3 associated to $\mathfrak{b}$ is $\delta^{d_1}\varepsilon^{e_1}\delta^{d_2}\varepsilon^{e_2}\cdots\delta^{d_r}\varepsilon^{e_r}$ where the exponents denote the corresponding multiplicities (all d_i, e_i are positive except possibly $d_1 = 0$ or $e_r = 0$). There exist a permutation s of $\{1, \ldots, m\}$ and a permutation t of $\{1, \ldots, n\}$ such that the δ's and ε's appearing in the $\varepsilon\delta$-sequence from left to right are $\delta_{s(1)}, \delta_{s(2)}, \ldots, \delta_{s(m)}$, and $\varepsilon_{t(1)}, \varepsilon_{t(2)}, \ldots, \varepsilon_{t(n)}$, respectively.

Define

$$\mathsf{d}_u := \sum_{a=1}^{u} d_a \quad \text{and} \quad \mathsf{e}_u := \sum_{a=1}^{u} e_a \tag{2.36}$$

for $u = 1, \ldots, r$, and let $\mathsf{d}_0 = \mathsf{e}_0 = 0$. Note $\mathsf{d}_r = m, \mathsf{e}_r = n$. Define the **$\mathfrak{b}$-Frobenius coordinates** $(p_i|q_j)$ of an $(m|n)$-hook partition λ as follows. For $1 \leq i \leq m, 1 \leq j \leq n$, let

$$\begin{aligned} p_i &= \max\{\lambda_i - \mathsf{e}_u, 0\}, \quad \text{if } \mathsf{d}_u < i \leq \mathsf{d}_{u+1} \text{ for some } 0 \leq u \leq r-1, \\ q_j &= \max\{\lambda'_j - \mathsf{d}_{u+1}, 0\}, \quad \text{if } \mathsf{e}_u < j \leq \mathsf{e}_{u+1} \text{ for some } 0 \leq u \leq r-1. \end{aligned} \tag{2.37}$$

Associated to an $(m|n)$-hook Young diagram λ and a Borel subalgebra $\mathfrak{b}$, we define a weight $\lambda^{\mathfrak{b}} \in \mathfrak{h}^*$ in terms of the $\mathfrak{b}$-Frobenius coordinates $(p_i|q_i)$ to be

$$\lambda^{\mathfrak{b}} := \sum_{i=1}^{m} p_i \delta_{s(i)} + \sum_{j=1}^{n} q_j \varepsilon_{t(j)}.$$

It is elementary to read off the $\mathfrak{b}$-Frobenius coordinates of λ from the Young diagram of λ in general, as illustrated by the next example.

Example 2.54. Let $\mathfrak{b}$ be the Borel subalgebra of $\mathfrak{gl}(5|4)$ associated to the following fundamental system:

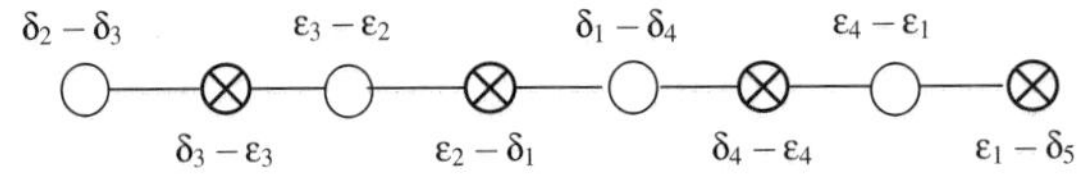

Consider the $(5|4)$-hook diagram $\lambda = (14, 11, 8, 8, 7, 4, 3, 2)$. The $\mathfrak{b}$-Frobenius coordinates for λ are:

$$p_1 = 14,\ p_2 = 11,\ p_3 = p_4 = 6,\ p_5 = 3; \quad q_1 = q_2 = 6,\ q_3 = 3,\ q_4 = 2.$$

These are read off from the Young diagram of λ by following the $\varepsilon\delta$ sequence $\delta\delta\varepsilon\varepsilon\delta\delta\varepsilon\varepsilon\delta$ (with δ for rows and ε for columns of λ) as follows:

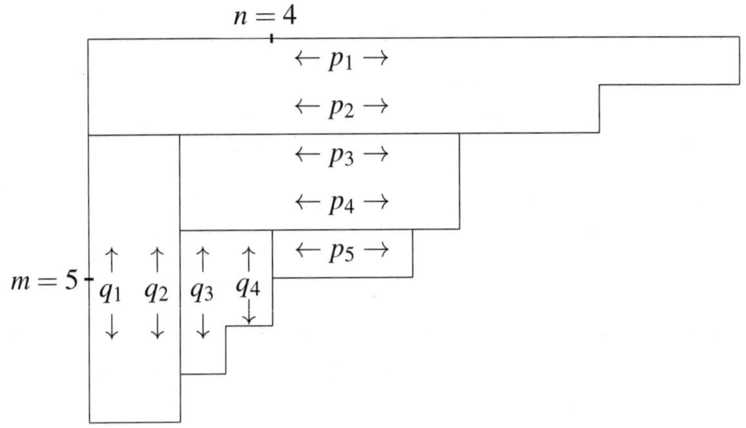

Then we obtain that

$$\begin{aligned}\lambda^{\mathfrak{b}} &= p_1\delta_2 + p_2\delta_3 + p_3\delta_1 + p_4\delta_4 + p_5\delta_5 + q_1\varepsilon_3 + q_2\varepsilon_2 + q_3\varepsilon_4 + q_4\varepsilon_1 \\ &= 6\delta_1 + 14\delta_2 + 11\delta_3 + 6\delta_4 + 3\delta_5 + 2\varepsilon_1 + 6\varepsilon_2 + 6\varepsilon_3 + 3\varepsilon_4.\end{aligned}$$

Theorem 2.55. *Let λ be an $(m|n)$-hook partition. Let $\mathfrak{b}$ be an arbitrary Borel subalgebra of $\mathfrak{gl}(m|n)$. Then, the $\mathfrak{b}$-extremal weight of the simple $\mathfrak{gl}(m|n)$-module $L(\mathfrak{g}, \mathfrak{b}^{st}, \lambda^{\natural})$ is $\lambda^{\mathfrak{b}}$.*

Proof. Set $V = L(\mathfrak{g}, \mathfrak{b}^{\mathrm{st}}, \lambda^{\natural})$. One distinguished feature for V is that all weights ν of V are polynomial in the sense that $\nu(E_{ii}) \in \mathbb{Z}_+$ for all $i \in I(m|n)$; see Chapter 3, Theorem 3.11.

The theorem holds for the standard Borel subalgebra $\mathfrak{b}^{\mathrm{st}}$, which corresponds to the sequence of m δ's followed by n ε's, thanks to $\lambda^{\mathfrak{b}^{\mathrm{st}}} = \lambda^{\natural}$. By Corollary 1.32, $\mathfrak{b}^{\mathrm{st}}$ can be converted to $\mathfrak{b}$ by a finite number of real and odd reflections. Thus, we are led to consider two Borel subalgebras $\mathfrak{b}_1$ and $\mathfrak{b}_2$, where $\mathfrak{b}_2$ is obtained from $\mathfrak{b}_1$ by applying a simple reflection corresponding to a simple root α. Let μ be the $\mathfrak{b}_2$-extremal weight for V. To complete the proof of the theorem, we will show that $\mu = \lambda^{\mathfrak{b}_2}$, assuming the theorem holds for $\mathfrak{b}_1$.

If α is an even root, then $r_\alpha(\lambda^{\mathfrak{b}_1}) = \lambda^{\mathfrak{b}_2}$ by definition, and the validity of the theorem for $\mathfrak{b}_1$ implies its validity for $\mathfrak{b}_2$.

Now assume that $\alpha = \pm(\delta_i - \varepsilon_j)$ is an odd simple root. The corresponding coroot is $h_\alpha = \pm(E_{\bar{i},\bar{i}} + E_{jj})$.

If $\lambda^{\mathfrak{b}_1}(h_\alpha) = 0$, then we must have $\lambda^{\mathfrak{b}_1}(E_{\bar{i},\bar{i}}) = \lambda^{\mathfrak{b}_1}(E_{jj}) = 0$ (as each has to be nonnegative), and $\lambda^{\mathfrak{b}_1} = \lambda^{\mathfrak{b}_2}$. Then by Lemma 1.40 we have $\mu = \lambda^{\mathfrak{b}_1} = \lambda^{\mathfrak{b}_2}$.

Suppose that $\lambda^{\mathfrak{b}_1}(h_\alpha) \neq 0$ and $\alpha = \delta_i - \varepsilon_j$. Diagrammatically we can represent $\lambda^{\mathfrak{b}_1}(E_{\bar{i},\bar{i}}) = a$ and $\lambda^{\mathfrak{b}_1}(E_{jj}) = b$ by Diagram (ii) below, and this forces that $a > 0$. By Lemma 1.40, we have $\mu = \lambda^{\mathfrak{b}_1} - \delta_i + \varepsilon_j$, and hence $\mu(E_{jj}) = b + 1$ and $\mu(E_{\bar{i},\bar{i}}) = a - 1$, which are represented by Diagram (i) below. The remaining parts of μ are the same as those for $\lambda^{\mathfrak{b}_1}$ and hence we conclude that $\mu = \lambda^{\mathfrak{b}_2}$, as claimed.

Suppose that $\lambda^{\mathfrak{b}_1}(h_\alpha) \neq 0$ and $\alpha = \varepsilon_j - \delta_i$. Diagrammatically we can represent $\lambda^{\mathfrak{b}_1}(E_{jj}) = b + 1$ and $\lambda^{\mathfrak{b}_1}(E_{\bar{i},\bar{i}}) = a - 1$ by Diagram (i) below, and this forces that $b \geq 0$ and $a \geq 1$. Lemma 1.40 implies that $\mu = \lambda^{\mathfrak{b}_1} - \varepsilon_j + \delta_i$ and hence $\mu(E_{\bar{i},\bar{i}}) = a$ and $\mu(E_{jj}) = b$ can be represented by Diagram (ii) below. Hence $\mu = \lambda^{\mathfrak{b}_2}$.

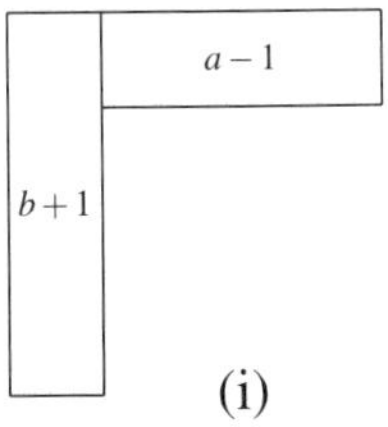

(i)

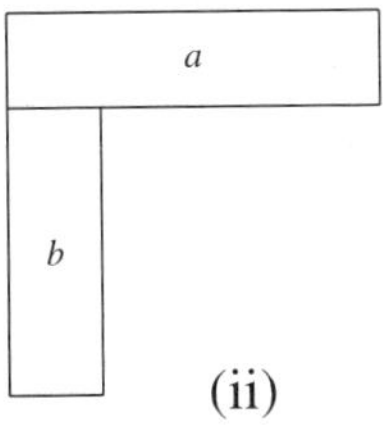

(ii)

This completes the proof of the theorem. □

Example 2.56. Let us describe the highest weights in Theorem 2.55 with respect to the three Borel subalgebras of special interest.

(1) As seen above, $\lambda^{\mathfrak{b}^{\mathrm{st}}} = \lambda^{\natural}$.

(2) If we take the opposite Borel subalgebra $\mathfrak{b}^{\mathrm{op}}$ corresponding to a sequence of n ε's followed by m δ's, then

$$\lambda^{\mathfrak{b}^{\mathrm{op}}} = \lambda'_1\varepsilon_1 + \ldots + \lambda'_n\varepsilon_n + \max\{\lambda_1 - n, 0\}\delta_1 + \ldots + \max\{\lambda_m - n, 0\}\delta_m.$$

(3) In the case when $|m-n| \le 1$, we may take a Borel subalgebra $\mathfrak{b}^{\mathrm{o}}$ whose simple roots are all odd (or equivalently, the corresponding $\varepsilon\delta$-sequence is alternating between ε and δ):

In this case, Theorem 2.55 states that the coefficients of δ and ε in $\lambda^{\mathfrak{b}^{\mathrm{o}}}$ are given by the **modified Frobenius coordinates** $(p_i|q_i)_{i\ge 1}$ of the partition λ (respectively, λ'), when the first simple root is of the form $\delta - \varepsilon$ (respectively, $\varepsilon - \delta$). Here by modified Frobenius coordinates we mean

$$p_i = \max\{\lambda_i - i + 1, 0\}, \quad q_i = \max\{\lambda'_i - i, 0\}$$

so that $\sum_i (p_i + q_i) = |\lambda|$. "Modified" here refers to a shift by 1 from the p_i coordinates defined in [**83**, Chapter 1, Page 3].

The modified Frobenius coordinates in this case can be read off from the Young diagram by alternatively reading off (and deleting thereafter) the number of boxes in rows and columns. As an illustration, if $\lambda = (7,5,4,3,1)$, then we have $(p_1,p_2,p_3|q_1,q_2,q_3) = (7,4,2|4,2,1)$.

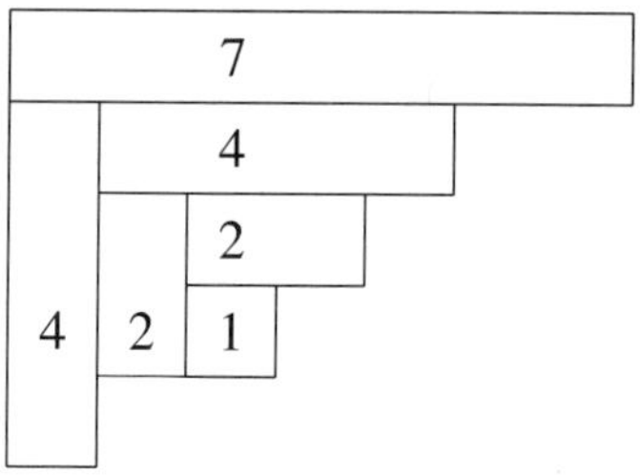

2.4.2. Extremal weights for $\mathfrak{spo}(2m|2n+1)$. Let us denote the weights of the natural $\mathfrak{spo}(2m|2n+1)$-module $\mathbb{C}^{2m|2n+1}$ by $\pm\delta_i, 0, \pm\varepsilon_j$ for $1 \le i \le m, 1 \le j \le n$. Recall that the **standard** Borel subalgebra $\mathfrak{b}^{\mathrm{st}}$ of $\mathfrak{spo}(2m|2n+1)$ is associated to the fundamental system (1.39) which we recall:

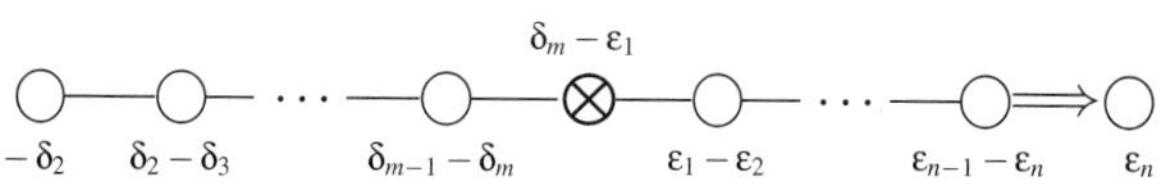

Let $\mathfrak{b}$ be a Borel subalgebra. As explained in Section 1.3.3, a Dynkin diagram for $\mathfrak{spo}(2m|2n+1)$ always has a type A end while the other end is a short (even or odd) root. For example, the above Dynkin diagram has its type A end labeled by the simple root $\delta_1 - \delta_2$. Starting from the type A end, the simple roots for $\mathfrak{b}$ of $\mathfrak{spo}(2m|2n+1)$ give rise to an $\varepsilon\delta$-sequence $\delta^{d_1}\varepsilon^{e_1}\delta^{d_2}\varepsilon^{e_2}\cdots\delta^{d_r}\varepsilon^{e_r}$ and sequences of ± 1's: $(\xi_i)_{1\le i\le m} \cup (\eta_j)_{1\le j\le n}$ (all the d_i and e_j are positive except possibly $d_1 = 0$ or $e_r = 0$). Furthermore, the Dynkin diagram contains a short *odd* root if and only if $e_r = 0$.

Recall the definitions of d_u and e_u from (2.36) and those of p_i and q_j from (2.37). Similar to the type A case there exist a permutation s of $\{1,\ldots,m\}$ and a permutation t of $\{1,\ldots,n\}$, so that the simple roots for $\mathfrak{b}$ are given by

$$\begin{aligned}
\xi_i\delta_{s(i)} - \xi_{i+1}\delta_{s(i+1)}, &\quad 1\le i\le m,\ i\notin\{\mathsf{d}_u | u=1,\ldots,r\};\\
\eta_j\varepsilon_{t(j)} - \eta_{j+1}\varepsilon_{t(j+1)}, &\quad 1\le j\le n,\ j\notin\{\mathsf{e}_u | u=1,\ldots,r\};\\
\xi_{\mathsf{d}_u}\delta_{s(\mathsf{d}_u)} - \eta_{1+\mathsf{e}_{u-1}}\varepsilon_{t(1+\mathsf{e}_{u-1})}, &\quad \text{for } 1\le u\le r \text{ if } e_r>0 \ (\text{or } 1\le u<r \text{ if } e_r=0);\\
\eta_{\mathsf{e}_u}\varepsilon_{t(\mathsf{e}_u)} - \xi_{1+\mathsf{d}_u}\delta_{s(1+\mathsf{d}_u)}, &\quad u=1,\ldots,r-1;\\
\eta_{\mathsf{e}_r}\varepsilon_{t(\mathsf{e}_r)}, &\quad \text{if } e_r>0 \qquad (\text{or } \xi_{\mathsf{d}_r}\delta_{s(\mathsf{d}_r)} \text{ if } e_r=0).
\end{aligned}$$

By Theorem 2.11, a complete list of finite-dimensional irreducible modules of $\mathfrak{spo}(2m|2n+1)$ of integer weights are the highest weight $\mathfrak{g}$-modules $L(\mathfrak{g},\mathfrak{b}^{\mathrm{st}},\lambda^{\natural})$ of $\mathfrak{b}^{\mathrm{st}}$-highest weights $\lambda^{\natural}$ in (2.4), for some $(m|n)$-hook partition λ.

Example 2.57. Consider the following Dynkin diagram of $\mathfrak{spo}(10|9)$ with simple roots attached:

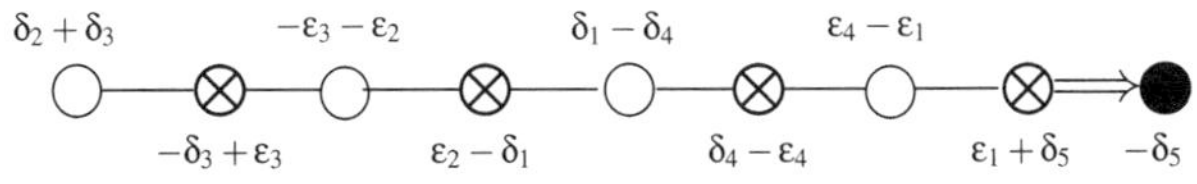

We read off a signed sequence with indices $\delta_2(-\delta_3)(-\varepsilon_3)\varepsilon_2\delta_1\delta_4\varepsilon_4\varepsilon_1(-\delta_5)$. In particular, we obtain a sequence $\delta\delta\varepsilon\varepsilon\delta\delta\varepsilon\varepsilon\delta$ by ignoring the signs and indices. In this case, $d_1 = d_2 = 2, d_3 = 1$, and $e_1 = e_2 = 2$. Furthermore, the sequences $(\xi_i)_{1\le i\le 5}$ and $(\eta_j)_{1\le j\le 4}$ are $(1,-1,1,1,-1)$ and $(-1,1,1,1)$, respectively.

Theorem 2.58. *Let λ be an $(m|n)$-hook partition. Let $\mathfrak{b}$ be a Borel subalgebra of $\mathfrak{spo}(2m|2n+1)$ and retain the above notation. Then, the $\mathfrak{b}$-highest weight of the simple $\mathfrak{spo}(2m|2n+1)$-module $L(\mathfrak{g},\mathfrak{b}^{st},\lambda^{\natural})$ is*

$$\lambda^{\mathfrak{b}} := \sum_{i=1}^{m} \xi_i p_i \delta_{s(i)} + \sum_{j=1}^{n} \eta_j q_j \varepsilon_{t(j)}.$$

Proof. The arguments here are similar to the ones given in the proof of Theorem 2.55 for $\mathfrak{gl}(m|n)$, and so we will be sketchy.

We first note that the theorem holds for the standard Borel subalgebra $\mathfrak{b}^{\mathrm{st}}$, which corresponds to the sequence of m δ's followed by n ε's with all signs ξ_i and η_j being positive, i.e., $\lambda^{\mathfrak{b}^{\mathrm{st}}} = \lambda^{\natural}$.

All Borels are linked to the standard Borel by a sequence of even and odd simple reflections. Let $\mathfrak{b}$ and $\mathfrak{b}'$ be two Borel subalgebras related by a simple reflection with respect to a simple root α, and assume the theorem holds for $\mathfrak{b}'$. If α is even, then clearly $r_\alpha(\lambda^{\mathfrak{b}'}) = \lambda^{\mathfrak{b}}$, and the theorem holds for $\mathfrak{b}$. In the case when α is odd, we have four cases to consider, by setting $\alpha = \pm\delta_i \pm \varepsilon_j$. Each case is verified by the same type of argument as in the proof of Theorem 2.55. □

Example 2.59. With respect to the Borel $\mathfrak{b}$ of $\mathfrak{spo}(10|9)$ as in Example 2.57, the $\mathfrak{b}$-extremal weight of $L(\mathfrak{g}, \mathfrak{b}^{\mathrm{st}}, \lambda^{\natural})$ for λ as given in Example 2.54 equals

$$\begin{aligned}\lambda^{\mathfrak{b}} &= p_1\delta_2 - p_2\delta_3 + p_3\delta_1 + p_4\delta_4 - p_5\delta_5 - q_1\varepsilon_3 + q_2\varepsilon_2 + q_3\varepsilon_4 + q_4\varepsilon_1 \\ &= 6\delta_1 + 14\delta_2 - 11\delta_3 + 6\delta_4 - 3\delta_5 + 2\varepsilon_1 + 6\varepsilon_2 - 6\varepsilon_3 + 3\varepsilon_4.\end{aligned}$$

Corollary 2.60. *Every finite-dimensional irreducible $\mathfrak{spo}(2m|2n+1)$-module of integer highest weight is self-dual.*

Proof. By a standard fact in highest weight theory, the dual module $L(\mathfrak{g}, \mathfrak{b}^{\mathrm{st}}, \lambda^{\natural})^*$ has $\mathfrak{b}^{\mathrm{st}}$-highest weight equal to the opposite of the $\mathfrak{b}^{\mathrm{st}}$-lowest weight of $L(\mathfrak{g}, \mathfrak{b}^{\mathrm{st}}, \lambda^{\natural})$. Denote by $\mathfrak{b}^{\mathrm{op}}$ the opposite Borel to the standard one $\mathfrak{b}^{\mathrm{st}}$. It follows by Theorem 2.58 that the $\mathfrak{b}^{\mathrm{op}}$-extremal weight (i.e., the $\mathfrak{b}^{\mathrm{st}}$-lowest weight) of the module $L(\mathfrak{g}, \mathfrak{b}^{\mathrm{st}}, \lambda^{\natural})$ is $-\lambda^{\natural}$. This proves the corollary. □

Consider the following Dynkin diagram of $\mathfrak{spo}(2m|2n+1)$ with simple roots attached, and let us denote its associated Borel subalgebra by $\mathfrak{b}^{\mathrm{sd}}$:

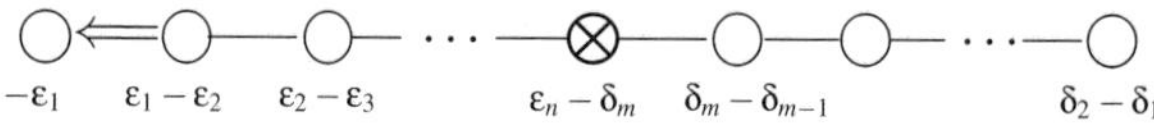

We have the following immediate corollary of Theorem 2.11 and Theorem 2.58.

Corollary 2.61. *An irreducible $\mathfrak{spo}(2m|2n+1)$-module of integer highest weight with respect to the Borel subalgebra $\mathfrak{b}^{sd}$ is finite dimensional if and only if the $\mathfrak{b}^{sd}$-highest weight is of the form*

$$\lambda^{\mathfrak{b}^{\mathrm{sd}}} = -\sum_{i=1}^{m} \lambda_i \delta_i - \sum_{j=1}^{n} \max\{\lambda'_j - m, 0\}\, \varepsilon_{n-j+1}, \tag{2.38}$$

where $\lambda = (\lambda_1, \lambda_2, \ldots)$ is an $(m|n)$-hook partition.

2.4.3. Extremal weights for $\mathfrak{spo}(2m|2n)$. Let $n \geq 2$. Let us denote the weights of the natural $\mathfrak{spo}(2m|2n)$-module $\mathbb{C}^{2m|2n}$ by $\pm\delta_i, \pm\varepsilon_j$ for $1 \leq i \leq m, 1 \leq j \leq n$. The **standard** Borel subalgebra $\mathfrak{b}^{\mathrm{st}}$ of $\mathfrak{spo}(2m|2n)$ is the one associated to the Dynkin diagram (1.41), which we recall for the reader's convenience:

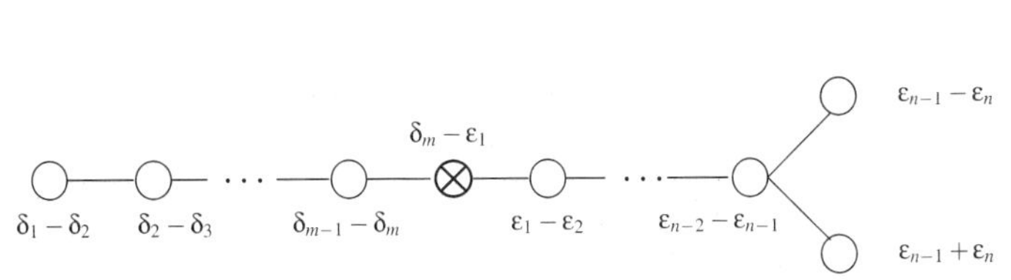

According to Section 1.3.4, there are two types of Dynkin diagrams and corresponding Borel subalgebras for $\mathfrak{spo}(2m|2n)$:

(i) Diagrams of $|$-shape, i.e., Dynkin diagrams with a long simple root $\pm 2\delta_i$.

(ii) Diagrams of $\curlyvee$-shape, i.e., Dynkin diagrams with no long simple root.

We will follow the notation for $\mathfrak{spo}(2m|2n+1)$ in Section 2.4.2 for fundamental systems in terms of signed $\varepsilon\delta$-sequences, so we have permutations s,t, and signs ξ_i,η_j. We fix the ambiguity on the choice of the last sign η_n associated to a Borel $\mathfrak{b}$ of $\curlyvee$-shape by demanding the total number of negative signs among $\eta_j (1 \le j \le n)$ to be always even.

Let λ be an $(m|n)$-hook partition, and let the $\mathfrak{b}$-Frobenius coordinates $(p_i|q_j)$ be as defined in Section 2.4.1. Introduce the following weights:

$$\lambda^{\mathfrak{b}} := \sum_{i=1}^{m} \xi_i p_i \delta_{s(i)} + \sum_{j=1}^{n} \eta_j q_j \varepsilon_{t(j)},$$

$$\lambda_-^{\mathfrak{b}} := \sum_{i=1}^{m} \xi_i p_i \delta_{s(i)} + \sum_{j=1}^{n-1} \eta_j q_j \varepsilon_{t(j)} - \eta_n q_n \varepsilon_{t(n)}.$$

The weight $\lambda_-^{\mathfrak{b}}$ will only be used for Borel $\mathfrak{b}$ of $\curlyvee$-shape. Note that $\lambda^{\mathfrak{b}^{\mathrm{st}}} = \lambda^{\natural}$ and we shall denote $\lambda_-^{\natural} := \lambda_-^{\mathfrak{b}^{\mathrm{st}}}$.

Given a Borel $\mathfrak{b}$ of $|$-shape, we define $\mathrm{s}(\mathfrak{b})$ to be the sign of $\prod_{j=1}^{n} \eta_j$.

Recall from Theorem 2.14 that a finite-dimensional irreducible $\mathfrak{spo}(2m|2n)$-module of integer highest weight with respect to the standard Borel subalgebra is of the form $L(\mathfrak{g},\mathfrak{b}^{\mathrm{st}},\lambda^{\natural})$ or $L(\mathfrak{g},\mathfrak{b}^{\mathrm{st}},\lambda_-^{\natural})$, where λ is an $(m|n)$-hook partition.

Theorem 2.62. *Let $n \geq 2$, and let λ be an $(m|n)$-hook partition.*

(1) *Assume $\mathfrak{b}$ is of $\curlyvee$-shape. Then,*
  (i) *$\lambda^{\mathfrak{b}}$ is the $\mathfrak{b}$-extremal weight for the module $L(\mathfrak{g},\mathfrak{b}^{st},\lambda^{\natural})$.*
  (ii) *$\lambda_-^{\mathfrak{b}}$ is the $\mathfrak{b}$-extremal weight for the module $L(\mathfrak{g},\mathfrak{b}^{st},\lambda_-^{\natural})$.*

(2) *Assume $\mathfrak{b}$ is of $|$-shape. Then,*
  (i) *$\lambda^{\mathfrak{b}}$ is the $\mathfrak{b}$-extremal weight for $L(\mathfrak{g},\mathfrak{b}^{st},\lambda^{\natural})$ if $s(\mathfrak{b}) = +$.*
  (ii) *$\lambda^{\mathfrak{b}}$ is the $\mathfrak{b}$-extremal weight for $L(\mathfrak{g},\mathfrak{b}^{st},\lambda_-^{\natural})$ if $s(\mathfrak{b}) = -$.*

Proof. Let $\mathfrak{b}$ and $\mathfrak{b}'$ be Borel subalgebras. Suppose that they are related by $\mathfrak{b}' = w(\mathfrak{b})$, for some Weyl group element w. Note that the set of weights for $L(\mathfrak{g},\mathfrak{b}^{\mathrm{st}},\lambda^{\natural})$ is W-invariant. By definition we observe that $\lambda^{\mathfrak{b}'} = w(\lambda^{\mathfrak{b}})$ and $\lambda_-^{\mathfrak{b}'} = w(\lambda_-^{\mathfrak{b}})$, and hence the validity of the theorem for $\mathfrak{b}$ implies its validity for $\mathfrak{b}'$. Thus it suffices to prove the theorem for a Borel subalgebra $\mathfrak{b}$ with even roots $\{\delta_i \pm \delta_j | i \leq j\} \cup \{\varepsilon_k \pm \varepsilon_l | k < l\}$. We shall make this assumption for the remainder of the proof.

Suppose that the Dynkin diagram of $\mathfrak{b}$ is of Y-shape. Thus the corresponding $\varepsilon\delta$-sequence ends with an ε, i.e., it is of the form

$$\cdots \delta \cdots \varepsilon \cdots \varepsilon.$$

The fundamental system of $\mathfrak{b}$ can be brought to the fundamental system corresponding to $\mathfrak{b}^{\mathrm{st}}$ by applying a sequence of odd reflections corresponding to odd isotropic root of the form $\varepsilon_i - \delta_j$ or $\delta_j - \varepsilon_i$. As Part (1) of the theorem is true for the standard Borel $\mathfrak{b}^{\mathrm{st}}$, exactly the same argument as in the proof of Theorem 2.55 proves (1).

Now suppose that $\mathfrak{b}$ is of |-shape. We first consider the Borel subalgebras $\mathfrak{b}_1$ and $\mathfrak{b}_2$ corresponding to the two fundamental systems below, respectively:

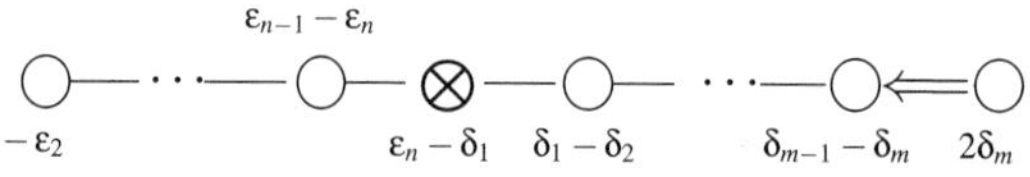

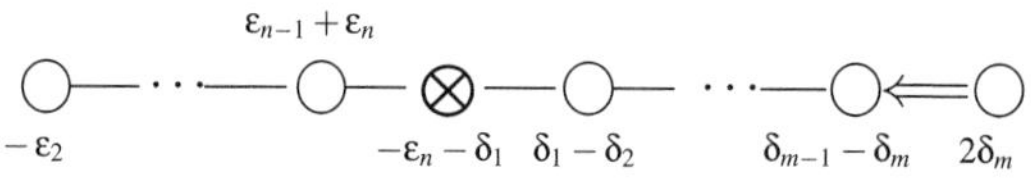

Note the sign sequence $(\eta_1, \ldots, \eta_n)$ associated to the ε_i's is $(1, \ldots, 1, 1)$ for $\mathfrak{b}_1$ and $(1, \ldots, 1, -1)$ for $\mathfrak{b}_2$. So we have $s(\mathfrak{b}_1) = +$ and $s(\mathfrak{b}_2) = -$.

First, using sequences of odd reflections, we can transform the standard fundamental system to the fundamental systems associated with $\mathfrak{b}_1$ and $\mathfrak{b}_2$. We list below two such sequences consisting of mn odd reflections that will take $\mathfrak{b}^{\mathrm{st}}$ to $\mathfrak{b}_1$ and $\mathfrak{b}_2$, respectively.

$$\{\delta_m - \varepsilon_1, \ldots, \delta_1 - \varepsilon_1; \delta_m - \varepsilon_2, \ldots, \delta_1 - \varepsilon_2; \ldots; \delta_m - \varepsilon_n, \ldots, \delta_1 - \varepsilon_n\},$$
$$\{\delta_m - \varepsilon_1, \ldots, \delta_1 - \varepsilon_1; \delta_m - \varepsilon_2, \ldots, \delta_1 - \varepsilon_2; \ldots; \delta_m + \varepsilon_n, \ldots, \delta_1 + \varepsilon_n\}.$$

We note that the above two sequences are identical except for the last n reflections (where $+$ signs replace $-$ signs). Starting from $\lambda_{\pm}^{\mathfrak{b}^{\mathrm{st}}}$ for the standard Borel $\mathfrak{b}^{\mathrm{st}}$ and repeatedly using Lemma 1.30 at each step, it is straightforward to show that Part (2) of the theorem is true for $\mathfrak{b}_1$ and $\mathfrak{b}_2$. Now if $s(\mathfrak{b}) = +$ (respectively, $s(\mathfrak{b}) = -$) then $\mathfrak{b}$ can be brought to $\mathfrak{b}_1$ (respectively $\mathfrak{b}_2$) by a sequence of odd reflections corresponding to odd isotropic roots of the form $\delta_j - \varepsilon_i$ or $\varepsilon_i - \delta_j$. Now the same type of argument as in the proof of Theorem 2.55 proves (2). □

Corollary 2.63. *Let $n \geq 0$ be even. Then every finite-dimensional irreducible $\mathfrak{spo}(2m|2n)$-module of integral highest weight is self-dual.*

Proof. The argument here is similar to the proof of Corollary 2.60. For $n = 0$, the corollary follows by the well-known fact that the longest element in the Weyl group for $\mathfrak{sp}(2m)$ is -1. Now assume $n \geq 2$ is even. Denote by $\mathfrak{b}^{\mathrm{op}}$ the opposite Borel to the standard one $\mathfrak{b}^{\mathrm{st}}$. It follows by Theorem 2.62 that the $\mathfrak{b}^{\mathrm{op}}$-extremal weight of the module $L(\mathfrak{g}, \mathfrak{b}^{\mathrm{st}}, \lambda^{\natural})$ (and respectively, $L(\mathfrak{g}, \mathfrak{b}^{\mathrm{st}}, \lambda^{\natural}_{-})$) is $-\lambda^{\natural}$ (and respectively, $-\lambda^{\natural}_{-}$). The corollary follows. □

Remark 2.64. The remaining $\mathfrak{b}$-extremal weights for the modules $L(\mathfrak{g}, \mathfrak{b}^{\mathrm{st}}, \lambda^{\natural})$ when $\mathrm{s}(\mathfrak{b}) = -$ or for the modules $L(\mathfrak{g}, \mathfrak{b}^{\mathrm{st}}, \lambda^{\natural}_{-})$ when $\mathrm{s}(\mathfrak{b}) = +$ do not seem to afford a uniform simple answer.

Consider the following Dynkin diagram of $\mathfrak{spo}(2m|2n)$ whose Borel subalgebra is denoted by $\mathfrak{b}^{\mathrm{sd}}$.

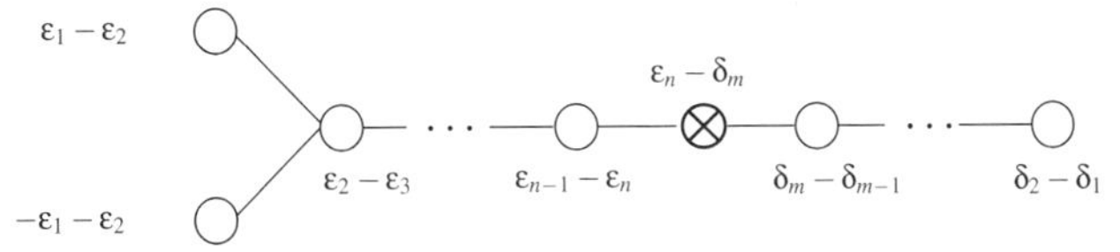

We record the following corollary of Theorems 2.14 and 2.62.

Corollary 2.65. *Let $n \geq 2$. An irreducible $\mathfrak{spo}(2m|2n)$-module of integral highest weight with respect to the Borel subalgebra $\mathfrak{b}^{\mathrm{sd}}$ is finite dimensional if and only if the highest weight is of the form*

$$-\sum_{i=1}^{m} \lambda_i \delta_i - \sum_{j=1}^{n-1} \max\{\lambda'_j - m, 0\}\, \varepsilon_{n-j+1} \pm \max\{\lambda'_n - m, 0\}\, \varepsilon_1, \tag{2.39}$$

where $\lambda = (\lambda_1, \lambda_2, \ldots)$ is an $(m|n)$-hook partition.

2.5. Exercises

Exercise 2.1. Let $m \geq 2$. Prove:

(1) A highest weight irreducible representation $L(\lambda)$ of $\mathfrak{gl}(m|m)$ remains an irreducible representation when restricted to $\mathfrak{sl}(m|m)$.

(2) Every highest weight irreducible representation of $\mathfrak{sl}(m|m)$ extends to a representation of $\mathfrak{gl}(m|m)$.

Hints: For (1) if v were an $\mathfrak{sl}(m|m)$-singular weight vector in $L(\lambda)$ and we can write $v = \sum_i v_i$, where the v_i's are $\mathfrak{gl}(m|m)$-weight vectors of distinct weights. But then each v_i would be $\mathfrak{gl}(m|m)$-singular. Use (1) to prove (2).

Exercise 2.2. We use the notation for $\mathfrak{osp}(1|2)$ from Example 1.25. Prove:

(1) The identities $[E, F^{2n}] = -nF^{2n-1}$ and $[E, F^{2n+1}] = F^{2n}(h - n)$ hold in $U(\mathfrak{osp}(1|2))$, for all $n \in \mathbb{Z}_+$.

(2) A complete list of inequivalent irreducible finite-dimensional representations of $\mathfrak{osp}(1|2)$ is given by $\{L(n\delta_1) \mid n \in \mathbb{Z}_+\}$.

(3) The character of $L(n\delta_1)$ is given by

$$\mathrm{ch}L(n\delta_1) = \frac{e^{(n+\frac{1}{2})\delta_1} - e^{-(n+\frac{1}{2})\delta_1}}{e^{\frac{1}{2}\delta_1} - e^{-\frac{1}{2}\delta_1}}.$$

Exercise 2.3. Let $\mathfrak{g}$ be a finite-dimensional simple Lie superalgebra equipped with an even supersymmetric invariant bilinear form $(\cdot,\cdot)$. Let $\{u_i|1 \le i \le k\}$ be a homogeneous basis for $\mathfrak{g}$ and let $\{u^i|1 \le i \le k\}$ be its dual basis with respect to $(\cdot,\cdot)$ so that $(u_i, u^j) = \delta_{ij}$. Prove:

(1) The **Casimir operator** $\Omega := \sum_{i=1}^{k}(-1)^{|u_i|}u_i u^i \in U(\mathfrak{g})$ is independent of the choice of the basis $\{u_i|1 \le i \le k\}$.

(2) $\Omega \in Z(\mathfrak{g})$ and Ω acts on a highest weight $\mathfrak{g}$-module of highest weight λ as the scalar $(\lambda + 2\rho, \lambda)$.

Exercise 2.4. Let $\mathfrak{g} = \mathfrak{osp}(1|2)$. Prove that the Casimir operator Ω acts as different scalars on inequivalent finite-dimensional irreducible $\mathfrak{g}$-modules and that every finite-dimensional $\mathfrak{g}$-module is completely reducible.

Exercise 2.5. Let $\mathfrak{g} = \mathfrak{gl}(1|1)$. Prove:

(1) The Verma $\mathfrak{g}$-module $\Delta(a\delta_1 + b\varepsilon_1)$ is reducible if and only if $a = -b$.

(2) There exists a non-split short exact sequence of $\mathfrak{g}$-modules:

$$0 \longrightarrow L((a-1)\delta_1 - (a-1)\varepsilon_1) \longrightarrow \Delta(a\delta_1 - a\varepsilon_1) \longrightarrow L(a\delta_1 - a\varepsilon_1) \longrightarrow 0.$$

(Hence the infinitely many simple $\mathfrak{g}$-modules $L(a\delta_1 - a\varepsilon_1)$ for all a in a congruence class $\mathbb{C}/\mathbb{Z}$ belong to one block.)

Exercise 2.6. Let $\mathfrak{g} = \mathfrak{sl}(1|2)$. Prove that the Casimir operator acts trivially on the polynomial module $L(\lambda)$, for $\lambda = (k+1)\delta_1 + (k-1)\varepsilon_1$, $k \in \mathbb{N}$.

Exercise 2.7. Let $\mathfrak{g}$ be a basic Lie superalgebra and let $\mathfrak{b}$ be a Borel subalgebra with Weyl vector $\rho_{\mathfrak{b}}$. Let $\mathfrak{b}'$ be a Borel subalgebra with $\mathfrak{b}_{\bar{0}} = \mathfrak{b}'_{\bar{0}}$. Suppose that $\mathfrak{b}'$ is obtained from $\mathfrak{b}$ by a sequence of odd reflections with respect to the odd roots $\alpha_1, \alpha_2, \ldots, \alpha_k$. Set $\lambda_0 = \lambda^{\mathfrak{b}}$ and define inductively for $i = 1, \ldots, k$

$$\lambda_i := \begin{cases} \lambda_{i-1}, & \text{if } (\lambda_{i-1}, \alpha_i) = 0, \\ \lambda_{i-1} - \alpha, & \text{if } (\lambda_{i-1}, \alpha_i) \neq 0. \end{cases}$$

Set $\lambda^{\mathfrak{b}'} = \lambda_k$. Prove:

(1) $\lambda^{\mathfrak{b}'}$ is well-defined, i.e., it is independent of the sequence of odd reflections.

(2) $(\lambda^{\mathfrak{b}} + \rho_{\mathfrak{b}}, \alpha) \neq 0$, $\forall \alpha \in \bar{\Phi}_{\bar{1}}$, if and only if $(\lambda^{\mathfrak{b}'} + \rho_{\mathfrak{b}'}, \alpha) \neq 0$, $\forall \alpha \in \bar{\Phi}_{\bar{1}}$ (here we recall the notation $\bar{\Phi}_{\bar{1}}$ from (1.18)).

Exercise 2.8. Let $\mathfrak{g} = \mathfrak{gl}(m|n)$ and let $\mathfrak{b}$ be its standard Borel subalgebra with positive system Φ^+. Let $\mathfrak{b}'$ be another Borel subalgebra of $\mathfrak{g}$ with positive system $'\Phi^+$. Assume that $\mathfrak{b}_{\bar{0}} = \mathfrak{b}'_{\bar{0}}$ and $-\Phi^+_{\bar{1}} = {}'\Phi^+_{\bar{1}}$. Give explicitly a sequence of mn odd reflections that transforms $\mathfrak{b}$ to $\mathfrak{b}'$.

Exercise 2.9. Suppose that two positive systems of a basic Lie superalgebra are related by an odd reflection α and they contain identical even positive roots. Prove that the respective Harish-Chandra homomorphisms are identical.

(Hints: It suffices to prove that the images of the Harish-Chandra homomorphisms evaluated at typical weights coincide, and note that the Weyl vectors corresponding to the two positive systems differ by α.)

Exercise 2.10. Let $\mathfrak{g}$ be a finite-dimensional reductive Lie algebra. Let $V(\lambda)$ be a highest weight module of $\mathfrak{g}$ of highest weight λ, and let E be a finite-dimensional representation of $\mathfrak{g}$. Denote the set of weights of E by P_E. Show that if $L(\mu)$ is a composition factor of $V(\lambda) \otimes E$ then $\mu + \rho \in W(\lambda + \rho + P_E)$.

Exercise 2.11. Let $\mathfrak{g} = \mathfrak{gl}(m|n)$ or $\mathfrak{osp}(2|2m)$. Let $\mathfrak{g} = \mathfrak{g}_{-1} \oplus \mathfrak{g} \oplus \mathfrak{g}_1$ be the $\mathbb{Z}$-gradation as in (2.1). For an integral weight λ let $K(\lambda)$ denote the (not necessarily finite-dimensional) Kac module. Prove:

(1) If there exists a singular vector of weight μ in $K(\lambda)$, then $\mu + \rho$ is of the form $w(\lambda - \sum_\alpha k_\alpha \alpha + \rho)$, $w \in W$, $k_\alpha \in \mathbb{Z}_+$, and $\alpha \in \Phi^+ \cap \mathfrak{g}_1$. (Hint: Use Exercise 2.10.)

(2) If there exists a singular vector of weight μ in $K(\lambda)$ and λ is typical, then $\mu + \rho = w(\lambda + \rho)$, for some $w \in W$.

(3) $K(\lambda)$ is irreducible, for λ typical.

Exercise 2.12. Let $\mathfrak{g} = \mathfrak{osp}(2|2m)$ with $m > 0$ and $\mathfrak{g}_{\bar{0}} = \mathfrak{so}(2) \oplus \mathfrak{sp}(2m)$. Prove:

(1) The map $e_\alpha \to \breve{e}_\alpha$ and $f_\alpha \to \breve{f}_\alpha$, for $\alpha \in \Pi$, given in Example 2.15, defines a representation of $\mathfrak{osp}(2|2m)$ on $\mathbb{C}[\xi, \bar{\xi}, x, \bar{x}]$.

(2) As a $\mathfrak{g}_{\bar{0}}$-module, we have $\mathbb{C}^{2|2m} \cong L^0(\varepsilon_1) \oplus L^0(-\varepsilon_1) \oplus L^0(\delta_1)$, with highest weight vectors ξ, $\bar{\xi}$, and x_1, respectively.

(3) For $k \geq 2$, the $\mathfrak{g}_{\bar{0}}$-module $\wedge^k(\mathbb{C}^{2|2m})$ decomposes as
$$L^0((k-1)\delta_1 + \varepsilon_1) \oplus L^0((k-1)\delta_1 - \varepsilon_1) \oplus L^0(k\delta_1) \oplus L^0((k-2)\delta_1),$$
with highest weight vectors ξx_1^{k-1}, $\bar{\xi} x_1^{k-1}$, x_1^k, and $x_1^{k-2}\xi\bar{\xi}$, respectively.

(4) The $\mathfrak{osp}(2|2m)$-module $\wedge^k(\mathbb{C}^{2|2m})$ is irreducible for all $k \geq 1$. (Hint: Use (4) and the formulas for $\breve{e}_{\varepsilon_1 - \delta_1}$ and $\breve{f}_{\varepsilon_1 - \delta_1}$.)

In Exercises 2.13–2.16 below, $\mathfrak{g} = \mathfrak{osp}(2|2m)$ with $m > 0$ and $\mathfrak{g}_{\bar{0}} = \mathfrak{so}(2) \oplus \mathfrak{sp}(2m)$, and we follow the notation of Example 2.16.

Exercise 2.13. Prove:

(1) The map $e_\alpha \to \hat{e}_\alpha$ and $f_\alpha \to \hat{f}_\alpha$, for $\alpha \in \Pi$, given in Example 2.16, defines a representation of $\mathfrak{osp}(2|2n)$.

(2) The **Laplace operator**

$$\Delta = \frac{\partial}{\partial x}\frac{\partial}{\partial \bar{x}} - \sum_{i=1}^{m} \frac{\partial}{\partial \xi_i}\frac{\partial}{\partial \bar{\xi}_i} : \mathcal{S}^k(\mathbb{C}^{2|2m}) \to \mathcal{S}^{k-2}(\mathbb{C}^{2|2m})$$

is surjective and commutes with the action of $\mathfrak{osp}(2|2m)$, for $k \geq 2$.

Exercise 2.14. Let $k \geq 2m+1$. Prove that $\dim \mathcal{S}^k(\mathbb{C}^{2|2m}) - \dim \mathcal{S}^{k-2}(\mathbb{C}^{2|2m}) = 2^{2m+1}$.

Exercise 2.15. Let $\Xi := \sum_{i=1}^m \xi_i \bar{\xi}_i$ and $k \geq 2m$. Set

$$\Gamma := \sum_{i=0}^{m} (-1)^i \binom{k-m}{i} \bar{x}^{k-m-i} x^{m-i} \Xi \in \mathcal{S}^k(\mathbb{C}^{2|2m}).$$

Prove:

(1) $\Delta(x^k) = 0$ and $\hat{e}_\alpha x^k = 0$, for $\alpha \in \Pi$.

(2) $\Gamma \neq 0$, $\Delta(\Gamma) = 0$, and $\hat{e}_\alpha \Gamma = 0$, for $\alpha \in \Pi$.

Exercise 2.16. Let $k \geq 2m+1$. Prove:

(1) $k\varepsilon_1$ and $(2m-k)\varepsilon_1$ are typical weights for $\mathfrak{osp}(2|2m)$.

(2) We have an isomorphism of $\mathfrak{osp}(2|2m)$-modules:

$$\ker \Delta \cong L(k\varepsilon_1) \oplus L((2m-k)\varepsilon_1).$$

What are the highest weights of these two irreducible modules with respect to the following new fundamental system?

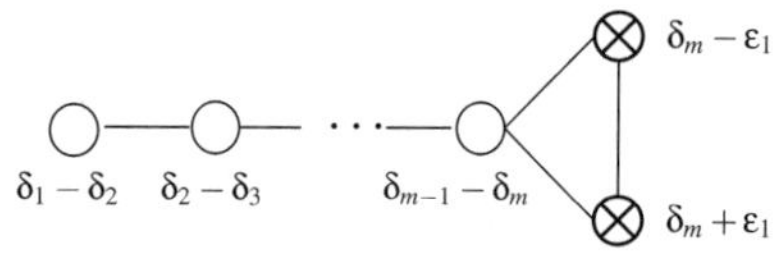

Exercise 2.17. Let $\lambda = \lambda_1\varepsilon_1 + \lambda_2\varepsilon_2$ with $\lambda_1 > \lambda_2$, $\lambda_1 \neq -\lambda_2$, and $\lambda_1, \lambda_2 \in \mathbb{Z}$. Write $\alpha = \varepsilon_1 - \varepsilon_2$. Prove:

(1) There is a short exact sequence of $\mathfrak{q}(2)$-modules:

$$0 \longrightarrow \Delta(\lambda_2\varepsilon_1 + \lambda_1\varepsilon_2) \longrightarrow \Delta(\lambda) \longrightarrow L(\lambda) \longrightarrow 0.$$

(2) We have the following decomposition of $L(\lambda)$ as $\mathfrak{gl}(2)$-modules:

$$L(\lambda) \cong \begin{cases} 2L^0(\lambda), & \text{if } \lambda_1 - \lambda_2 = 1, \\ 2L^0(\lambda) \oplus 2L^0(\lambda - \alpha), & \text{if } \lambda_1 - \lambda_2 > 1. \end{cases}$$

Exercise 2.18. Let $L(\mu)$ be a finite-dimensional irreducible $\mathfrak{q}(n)$-module of highest weight $\mu = \sum_{i=1}^n \mu_i \varepsilon_i$. Prove that $L(\mu^*) \cong L(\mu)^*$, where $\mu^* := -\sum_{i=1}^n \mu_{n+1-i}\varepsilon_i$. (Hint: If ν is a $\Phi_{\bar{0}}^+$-dominant integral weight of $L(\mu)$, then $w_0\nu - w_0\mu \in \sum_{\alpha \in \Phi_{\bar{0}}^+} \mathbb{Z}_+\alpha$, where w_0 is the longest element in W.)

Notes

Section 2.1. A systematic study of the representation theory of finite-dimensional (simple) Lie superalgebras was initiated by Kac [**62**]. The classification of finite-dimensional irreducible representations of the Lie superalgebras of types $\mathfrak{gl}$ and $\mathfrak{spo}$ (Propositions 2.2 and 2.2; Theorems 2.11 and 2.14) appeared first in Kac [**60**], where the finite dimensionality condition was formulated in terms of Dynkin labels instead of hook partitions used in this book. The proof using odd reflections in the most challenging $\mathfrak{spo}$ type given here follows Shu-Wang [**115**] for the cases of integer weights, and the proof using odd reflections and the Bernstein functor (see Santos [**104**]) for the cases of half-integer weights here is new. Yet another closely related approach to the classification is provided in Azam-Yamane-Yousofzadeh [**4**] by means of Weyl groupoids. Theorem 2.18 on the classification of finite-dimensional irreducible modules of $\mathfrak{q}(n)$ was formulated by Penkov [**93**, Theorem 4], who proved the "only if" direction. We are not aware of a written proof for the "if" direction in the literature. Our proof uses the Bernstein functor again.

Section 2.2. The Harish-Chandra homomorphisms for Lie superalgebras were first studied by Kac [**62**], who formulated results in terms of fractional fields. Proposition 2.22 and Lemma 2.24 (and their proofs) are taken from [**61**] and [**62**], respectively. The analogue of Chevalley theorem on invariant polynomials for simple Lie superalgebras has been established by Sergeev [**111**], which can be recovered as a part of Theorem 2.26 for type $\mathfrak{gl}$ and $\mathfrak{osp}$. Our approach to calculating the images of the Harish-Chandra homomorphisms in this book is different and new, and it uses the connection to the theory of supersymmetric functions for types $\mathfrak{gl}$ and $\mathfrak{spo}$ and the theory of Schur Q-functions for type $\mathfrak{q}$. There is another approach developed by Gorelik [**48**] following Kac [**63**] on the centers of the universal enveloping algebras (also see Alldridge [**1**]). A variant of Theorem 2.30 on linkage principle for $\mathfrak{gl}$ and $\mathfrak{osp}$ was originally stated as a conjecture in Kac [**62**, §6].

The notion of typicality for basic Lie superalgebras is due to Kac [**62**]. Corollary 2.28 (which is due to Kac [**62**, Theorem 2] by different arguments) was used by Kac to derive the irreducible character formula for finite-dimensional typical modules of basic classical Lie superalgebas (see Theorem 2.35). Also see Gorelik [**47**].

Section 2.3. This section is the counterpart to Section 2.2 for the queer Lie superalgebra $\mathfrak{q}(n)$. Theorem 2.46 on the image of the Harish-Chandra homomorphism for the queer Lie superalgebra was formulated without proof in a short announcement of Sergeev [**109**]. We fill in the details of a proof by presenting several explicit formulas of singular vectors in Verma modules. The notion of typicality for $\mathfrak{q}(n)$ was formulated by Penkov [**93**]. Theorem 2.52 on the typical character formula for $\mathfrak{q}(n)$ is due to Penkov [**93**, Theorem 2]. Theorem 2.48 on linkage principle for $\mathfrak{q}(n)$ or its variant has been regarded as folklore after [**109**].

Section 2.4. As we shall see in Chapter 3, the irreducible polynomial representations of $\mathfrak{gl}(m|n)$ were classified by Sergeev [**110**] (and independently later by Berele and Regev [**7**]). The results on extremal weights of the irreducible polynomial representations for $\mathfrak{gl}$ and finite-dimensional irreducible integer weight modules for $\mathfrak{spo}$ were due to Ngau Lam and the authors (for the $\mathfrak{spo}$ case see [**24**] and for the $\mathfrak{gl}$ case see [**32**]). The special case for type $\mathfrak{gl}$ when a fundamental system consists of only odd simple roots goes back to [**30**].

The irreducible character problem for finite-dimensional modules over Lie superalgebras has been a challenging one. For $\mathfrak{gl}(m|n)$, there have been complete solutions by completely different approaches due to Serganova [**107**], Brundan [**11**], Brundan-Stroppel [**15**], and Cheng-Lam [**23**] (also see Cheng-Wang-Zhang [**34**]). There is also a combinatorial dimension formula by Su-Zhang [**120**]. For $\mathfrak{q}(n)$, there has been complete solutions due to Penkov-Serganova [**96**] and Brundan [**12**]. There were numerous partial results in the literature over the years; see for example [**8, 33, 123, 124, 96, 104, 110, 133**]. Chapter 6 of this book offers a complete solution for a wider class of modules (including all finite-dimensional ones) of $\mathfrak{gl}(m|n)$ and $\mathfrak{osp}(k|2m)$.

Exercise 2.12 implies that the exterior powers of the natural $\mathfrak{osp}(2|2m)$-module are irreducible. For general $\mathfrak{osp}(k|2m)$ a similar argument as outlined there can be used to show the irreducibility of every exterior power of the natural module. As in the orthogonal Lie algebra case, (super)symmetric tensors are not irreducible. In contrast to the orthogonal Lie algebra, the determination of the composition factors in the symmetric powers is significantly more involved as seen in Exercises 2.13–2.16, which are taken from [**33**]. In [**33**] it is shown that the kernel of the Laplacian in the case of $\mathfrak{osp}(2|2m)$ can either have one, two, or three composition factors. The proof outlined in Exercise 2.11 for the irreducibility of Kac modules of a typical highest weight is taken from [**80**]. A different argument of this fact can be found in [**62**].

Chapter 3

Schur duality

Schur duality, which was popularized in Weyl's book *The Classical Groups*, is an interplay between representations of the general linear Lie group/algebra and representations of the symmetric group. On the combinatorial level, it explains their mutual connections to Schur functions and Young tableaux.

In this chapter, we describe Sergeev's superalgebra generalization of Schur duality. The first generalization is a duality between the general linear Lie superalgebra and the symmetric group. This allows us to classify the irreducible polynomial representations of the general linear Lie superalgebra and obtain their character formula in terms of supertableaux and also in terms of super Schur functions. The second generalization is a duality between the queer Lie superalgebra and a twisted (or spin) group algebra $\mathcal{H}_d$ of the hyperoctahedral group. The representation theory of the algebra $\mathcal{H}_d$ is systematically developed, and its connection with symmetric functions such as Schur Q-functions is explained in detail. The irreducible polynomial representations of the queer Lie superalgebra are classified in terms of strict partitions, and their characters are shown to be the Schur Q-functions up to some 2-powers.

3.1. Generalities for associative superalgebras

In this section, we classify the finite-dimensional simple associative superalgebras over $\mathbb{C}$. We also formulate the superalgebra generalizations of Schur's Lemma, the double centralizer property, and Wedderburn's Theorem. By studying the center of a finite group superalgebra, we obtain a numerical identity relating the number of simple supermodules to the number of split conjugacy classes.

In this section, a superalgebra is always understood to be a finite-dimensional associative superalgebra with unity 1, and a module is always understood to be finite-dimensional.

3.1.1. Classification of simple superalgebras. Let V be a vector superspace with even and odd subspaces of equal dimension. Given an odd automorphism P of V of order 2, we define the following subalgebra of the endomorphism superalgebra $\mathrm{End}(V)$:

$$Q(V) = \{x \in \mathrm{End}(V) \mid [x, P] = 0\}.$$

In the case when $V = \mathbb{C}^{n|n}$ and P is the linear transformation in the block matrix form (1.10), we write $Q(V)$ as $Q(n)$, which consists of $2n \times 2n$ matrices of the form:

$$\begin{pmatrix} a & b \\ b & a \end{pmatrix},$$

where a and b are arbitrary $n \times n$ matrices, for $n \geq 1$. Note that we have a superalgebra isomorphism $Q(V) \cong Q(n)$ by properly choosing coordinates in V, whenever $\dim V = (n|n)$.

The matrix superalgebra $M(m|n)$ is the algebra of matrices in the $m|n$-block form

$$\begin{pmatrix} a & b \\ c & d \end{pmatrix},$$

whose even subspace consists of the matrices with $b = 0$ and $c = 0$, and whose odd subspace consists of the matrices with $a = 0$ and $d = 0$. The matrix superalgebra $M(m|n)$ is isomorphic to the endomorphism superalgebra of $\mathbb{C}^{m|n}$.

Recall that by an ideal I and a module M of a superalgebra $\mathcal{A}$, we always mean that I and M are $\mathbb{Z}_2$-graded, i.e., $I = (I \cap \mathcal{A}_{\bar{0}}) \oplus (I \cap \mathcal{A}_{\bar{1}})$, and $M = M_{\bar{0}} \oplus M_{\bar{1}}$ such that $\mathcal{A}_i M_j \subseteq M_{i+j}$, for $i, j \in \mathbb{Z}_2$. For a superalgebra $\mathcal{A}$ and an $\mathcal{A}$-module M, we shall denote by $|\mathcal{A}|$ and $|M|$ the underlying (i.e., ungraded) algebra and module. Ideals and modules of $|\mathcal{A}|$ are understood in the usual (i.e., ungraded) sense.

We shall denote by $\mathcal{A}$-mod the category of (finite-dimensional) modules of the superalgebra $\mathcal{A}$ (with morphisms of degree one allowed). The underlying even subcategory $\mathcal{A}$-mod$_{\bar{0}}$, which consists of the same objects and only even morphisms of $\mathcal{A}$-mod, is an abelian category. Recall the parity reversing functor Π from Section 1.1.1. We define the **Grothendieck group** $[\mathcal{A}\text{-mod}]$ of the category $\mathcal{A}$-mod to be the $\mathbb{Z}$-module generated by all objects in $\mathcal{A}$-mod subject to the following two relations: (i) $[M] = [L] + [N]$; (ii) $[\Pi M] = -[M]$, for all L, M, N in $\mathcal{A}$-mod satisfying a short exact sequence $0 \to L \to M \to N \to 0$ with even morphisms.

Denote by $\mathcal{Z}(|\mathcal{A}|)$ the center of $|\mathcal{A}|$. Clearly, $\mathcal{Z}(|\mathcal{A}|) = \mathcal{Z}(|\mathcal{A}|)_{\bar{0}} \oplus \mathcal{Z}(|\mathcal{A}|)_{\bar{1}}$, where $\mathcal{Z}(|\mathcal{A}|)_i = \mathcal{Z}(|\mathcal{A}|) \cap \mathcal{A}_i$ for $i \in \mathbb{Z}_2$. Recall a superalgebra $\mathcal{A}$ is simple if $\mathcal{A}$ has no nontrivial ideal.

Theorem 3.1. *Let $\mathcal{A}$ be a finite-dimensional simple associative superalgebra.*

(1) *If* $\mathcal{Z}(|\mathcal{A}|)_{\bar{1}} = 0$, *then* $\mathcal{A}$ *is isomorphic to the matrix superalgebra* $M(m|n)$ *for some* m *and* n.

(2) *If* $\mathcal{Z}(|\mathcal{A}|)_{\bar{1}} \neq 0$, *then* $\mathcal{A}$ *is isomorphic to* $Q(n)$ *for some* n.

We shall need the following lemma. Let $p_i : \mathcal{A} \to \mathcal{A}_i$ be the projection for $i \in \mathbb{Z}_2$.

Lemma 3.2. *Let* $\mathcal{A}$ *be a simple superalgebra. If* J *is a proper ideal of* $|\mathcal{A}|$, *then the induced maps* $p_i|_J : J \to \mathcal{A}_i$ *are isomorphisms of vector spaces and, moreover,* $J \cap \mathcal{A}_i = 0$ *for* $i \in \mathbb{Z}_2$.

Proof. We claim that if I is a nonzero ideal of $\mathcal{A}_{\bar{0}}$, then

$$I + \mathcal{A}_{\bar{1}} I \mathcal{A}_{\bar{1}} = \mathcal{A}_{\bar{0}}, \tag{3.1}$$

and $\mathcal{A}_{\bar{1}} I + I \mathcal{A}_{\bar{1}} = \mathcal{A}_{\bar{1}}$. Indeed, the $\mathbb{Z}_2$-graded subspace $(I + \mathcal{A}_{\bar{1}} I \mathcal{A}_{\bar{1}}) + (\mathcal{A}_{\bar{1}} I + I \mathcal{A}_{\bar{1}})$ of $\mathcal{A}$ is closed under left and right multiplications by elements of $\mathcal{A}_{\bar{0}}$ and $\mathcal{A}_{\bar{1}}$; hence it is an ideal of $\mathcal{A}$. Now the claim follows from the simplicity of $\mathcal{A}$.

For a proper ideal J of $|\mathcal{A}|$, $J \cap \mathcal{A}_{\bar{0}}$ and $p_{\bar{0}}(J)$ are ideals of $\mathcal{A}_{\bar{0}}$, and $J \cap \mathcal{A}_{\bar{0}} \subseteq p_{\bar{0}}(J)$. Actually, we must have $J \cap \mathcal{A}_{\bar{0}} \subsetneq p_{\bar{0}}(J)$, for otherwise J would be an ideal of $\mathcal{A}$, contradicting the simplicity of $\mathcal{A}$. By inspection, $\mathcal{A}_{\bar{1}}(J \cap \mathcal{A}_{\bar{0}})\mathcal{A}_{\bar{1}} \subseteq J \cap \mathcal{A}_{\bar{0}}$ and $\mathcal{A}_{\bar{1}} p_{\bar{0}}(J) \mathcal{A}_{\bar{1}} \subseteq p_{\bar{0}}(J)$. It follows by (3.1) that

$$J \cap \mathcal{A}_{\bar{0}} = 0, \qquad p_{\bar{0}}(J) = \mathcal{A}_{\bar{0}}. \tag{3.2}$$

We have $\mathcal{A}_{\bar{1}}^2 = \mathcal{A}_{\bar{0}}$, for otherwise $\mathcal{A}_{\bar{1}}^2 + \mathcal{A}_{\bar{1}}$ would be a proper ideal of $\mathcal{A}$, which contradicts with the simplicity of $\mathcal{A}$. Hence, we have

$$J \cap \mathcal{A}_{\bar{1}} = 0, \qquad p_{\bar{1}}(J) = \mathcal{A}_{\bar{1}}, \tag{3.3}$$

by the following computations (which use (3.2) and that $1 \in \mathcal{A}_{\bar{0}}$):

$$\begin{aligned} J \cap \mathcal{A}_{\bar{1}} &= \mathcal{A}_{\bar{0}}(J \cap \mathcal{A}_{\bar{1}}) = \mathcal{A}_{\bar{1}}^2(J \cap \mathcal{A}_{\bar{1}}) \subseteq \mathcal{A}_{\bar{1}}(J \cap \mathcal{A}_{\bar{0}}) = 0, \\ p_{\bar{1}}(J) &\supseteq \mathcal{A}_{\bar{1}} p_{\bar{0}}(J) = \mathcal{A}_{\bar{1}} \mathcal{A}_{\bar{0}} = \mathcal{A}_{\bar{1}}. \end{aligned}$$

Now the lemma follows from (3.2) and (3.3). □

Proof of Theorem 3.1. We claim that $\mathcal{Z}(|\mathcal{A}|)_{\bar{0}} = \mathbb{C}$. To see this, let $z \in \mathcal{Z}(|\mathcal{A}|)_{\bar{0}}$. The map $\ell_z : A \to A$ defined by left multiplication by z has an eigenvalue $c \in \mathbb{C}$, and the kernel of $\ell_z - c$ is a nonzero 2-sided ideal of $\mathcal{A}$. The simplicity of $\mathcal{A}$ implies that $\ell_z - c = 0$, and hence $z = c \cdot 1$.

(1) Assume $\mathcal{Z}(|\mathcal{A}|)_{\bar{1}} = 0$. We claim that $|\mathcal{A}|$ is a simple algebra (and so $|\mathcal{A}|$ is a matrix algebra by a classical structure theorem of simple algebras). Indeed, assume $|\mathcal{A}|$ is a non-simple algebra with a proper ideal J. Denote $w = p_{\bar{0}}^{-1}(1) \in J$ and $u = p_{\bar{1}}(w) \in \mathcal{A}_{\bar{1}}$, where $u \neq 0$ by Lemma 3.2. Then, $w = 1 + u$. For any homogeneous element $x \in \mathcal{A}_i$ for some $i \in \mathbb{Z}_2$, we have $xw = x + xu$ and $wx = x + ux$, and so $p_i(xw) = x = p_i(wx)$. By Lemma 3.2, $xw = wx$ and so $xu = ux$; that is, $u \in \mathcal{Z}(|\mathcal{A}|)_{\bar{1}}$. It follows by assumption that $u = 0$, which is a contradiction.

Consider the automorphism $\sigma : |\mathcal{A}| \to |\mathcal{A}|$ given by $\sigma(a_i) = (-1)^i a_i$ for $a_i \in \mathcal{A}_i$ and $i \in \mathbb{Z}_2$. There exists $u \in |\mathcal{A}|$ such that $\sigma(a) = uau^{-1}$ for all $a \in |\mathcal{A}|$, since every automorphism of a matrix algebra is inner. We have $u \in \mathcal{A}_{\bar{0}}$ because $\sigma(u) = u$. It follows from $\sigma^2 = 1$ that $u^2 \in \mathcal{Z}(|\mathcal{A}|)_{\bar{0}} = \mathbb{C}$. We can choose u such that $u^2 = 1$ and then, by a change of basis, take u to be a diagonal matrix with m 1's followed by n (-1)'s in the main diagonal. Hence, we conclude that $\mathcal{A} \cong M(m|n)$.

(2) For a nonzero element $w \in \mathcal{Z}(|\mathcal{A}|)_{\bar{1}}$, we have $w^2 \in \mathcal{Z}(|\mathcal{A}|)_{\bar{0}} = \mathbb{C}$. We must have $w^2 \neq 0$, for otherwise the annihilator of w is a proper ideal of the superalgebra $\mathcal{A}$, contradicting the simplicity of $\mathcal{A}$. Then we may assume $w^2 = 1$. We have $\mathcal{Z}(|\mathcal{A}|)_{\bar{1}} = (\mathcal{Z}(|\mathcal{A}|)_{\bar{1}} w)w \subseteq \mathcal{Z}(|\mathcal{A}|)_{\bar{0}} w = \mathbb{C}w \subseteq \mathcal{Z}(|\mathcal{A}|)_{\bar{1}}$, and hence, $\mathcal{Z}(|\mathcal{A}|) = \mathbb{C} + \mathbb{C}w$. Moreover, $\mathcal{A}_{\bar{1}} = (\mathcal{A}_{\bar{1}} w)w \subseteq \mathcal{A}_{\bar{0}} w \subseteq \mathcal{A}_{\bar{1}}$, i.e., $\mathcal{A}_{\bar{1}} = \mathcal{A}_{\bar{0}} w$. The algebra $\mathcal{A}_{\bar{0}}$ must be simple, for otherwise a proper ideal I of $\mathcal{A}_{\bar{0}}$ would give rise to a proper ideal $I + Iw$ of $\mathcal{A}$ (which is again a contradiction). Then $\mathcal{A}_{\bar{0}}$ is a matrix algebra $M(n)$ for some n, and the superalgebra $\mathcal{A} = M(n) \oplus M(n)w$ is isomorphic to $Q(n)$. □

3.1.2. Wedderburn Theorem and Schur's Lemma. The basic results of finite-dimensional semisimple (unital associative) algebras over $\mathbb{C}$ admit natural super generalizations.

A module M of a superalgebra $\mathcal{A}$ is called **semisimple** if every $\mathcal{A}$-submodule of M is a direct summand of M. The following are straightforward generalizations of the classical structure results on semisimple modules and algebras, and they can be proved in the same way as in the non-super setting.

Theorem 3.3 (Super Wedderburn Theorem)**.** *The following statements are equivalent for a superalgebra $\mathcal{A}$:*

(1) *Every $\mathcal{A}$-module is semisimple.*

(2) *The left regular module $\mathcal{A}$ is a direct sum of minimal left ideals.*

(3) *$\mathcal{A}$ is a direct sum of simple superalgebras.*

A superalgebra $\mathcal{A}$ is called **semisimple** if it satisfies one of the three equivalent conditions (1)–(3) in Theorem 3.3.

By Theorems 3.1 and 3.3, a finite-dimensional semisimple superalgebra $\mathcal{A}$ is isomorphic to a direct sum of simple superalgebras as follows:

$$\mathcal{A} \cong \bigoplus_{i=1}^{m} M(r_i|s_i) \oplus \bigoplus_{j=1}^{q} Q(n_j), \tag{3.4}$$

where $m = m(\mathcal{A})$ and $q = q(\mathcal{A})$ are invariants of $\mathcal{A}$. Observe that $\mathbb{C}^{r|s}$ is a simple module (unique up to isomorphism) of the simple superalgebra $M(r|s)$, and $\mathbb{C}^{n|n}$ is a simple module (unique up to isomorphism) of the simple superalgebra $Q(n)$. A simple $\mathcal{A}$-module V is annihilated by all but one of the summands in (3.4). We say V is **of type** `M` if this summand is of the form $M(r_i|s_i)$ and **of type** `Q` if this summand is of the form $Q(n_j)$. These two types of simple modules are distinguished by a super analogue of Schur's Lemma.

Lemma 3.4 (Super Schur's Lemma). *If M and L are simple modules over a superalgebra $\mathcal{A}$, then*

$$\dim \mathrm{Hom}_{\mathcal{A}}(M,L) = \begin{cases} 1 & \text{if } M \cong L \text{ is of type } \mathtt{M}, \\ 2 & \text{if } M \cong L \text{ is of type } \mathtt{Q}, \\ 0 & \text{if } M \ncong L. \end{cases}$$

Proof. In the case when $M \cong L$ is of type $\mathtt{M}$, we can assume that $\mathcal{A} \cong M(r|s)$ for some r,s, and that $M = L = \mathbb{C}^{r|s}$. Then it is straightforward to check (as in the non-super case) that $\mathrm{Hom}_{M(r|s)}(\mathbb{C}^{r|s}, \mathbb{C}^{r|s}) = \mathbb{C}I$.

In the case when $M \cong L$ is of type $\mathtt{Q}$, we can assume that $\mathcal{A} \cong Q(n)$ for some n, and that $M = L = \mathbb{C}^{n|n}$. Then one checks that $\mathrm{Hom}_{Q(n)}(\mathbb{C}^{n|n}, \mathbb{C}^{n|n}) = \mathbb{C}I + \mathbb{C}\mathfrak{P}$, where $\mathfrak{P}$ is the linear transformation in the block matrix form (1.14).

The case where $M \ncong L$ is clear. □

3.1.3. Double centralizer property for superalgebras. Given two associative superalgebras $\mathcal{A}$ and $\mathcal{B}$, the tensor product $\mathcal{A} \otimes \mathcal{B}$ is naturally a superalgebra, with multiplication defined by

$$(a \otimes b)(a' \otimes b') = (-1)^{|b| \cdot |a'|}(aa') \otimes (bb') \qquad (a, a' \in \mathcal{A},\ b, b' \in \mathcal{B}). \tag{3.5}$$

Note that we have a superalgebra isomorphism (see Exercise 3.10)

$$Q(m) \otimes Q(n) \cong M(mn|mn).$$

Hence, as a $Q(m) \otimes Q(n)$-module, the tensor product $\mathbb{C}^{m|m} \otimes \mathbb{C}^{n|n}$ is a direct sum of two isomorphic copies of a simple module (which is $\cong \mathbb{C}^{mn|mn}$), and we have $\mathrm{Hom}_{Q(n)}(\mathbb{C}^{n|n}, \mathbb{C}^{mn|mn}) \cong \mathbb{C}^{m|m}$ as a $Q(m)$-module.

Let $\mathcal{A}$ and $\mathcal{B}$ be semisimple superalgebras. Let M be a simple $\mathcal{A}$-module of type $\mathtt{Q}$ and let N be a simple $\mathcal{B}$-module of type $\mathtt{Q}$. Then, the $\mathcal{A} \otimes \mathcal{B}$-module $M \otimes N$ is a direct sum of two isomorphic copies of a simple module of type $\mathtt{M}$, denoted by $2^{-1} M \otimes N$. Moreover, $\mathrm{Hom}_{\mathcal{B}}(N, 2^{-1} M \otimes N)$ is naturally an $\mathcal{A}$-module, which is isomorphic to the $\mathcal{A}$-module M.

Proposition 3.5. *Suppose that W is a finite-dimensional vector superspace, and $\mathcal{B}$ is a semisimple subalgebra of $\mathrm{End}(W)$. Let $\mathcal{A} = \mathrm{End}_{\mathcal{B}}(W)$. Then, $\mathrm{End}_{\mathcal{A}}(W) = \mathcal{B}$.*

As an $\mathcal{A} \otimes \mathcal{B}$-module, W is multiplicity-free, i.e.,

$$W \cong \sum_i 2^{-\delta_i} U_i \otimes V_i,$$

where $\delta_i \in \{0,1\}$, $\{U_i\}$ are pairwise non-isomorphic simple $\mathcal{A}$-modules, and $\{V_i\}$ are pairwise non-isomorphic simple $\mathcal{B}$-modules. Moreover, U_i and V_i for a given i are of the same type, and they are of type $\mathtt{M}$ if and only if $\delta_i = 0$.

Proof. Assume that V_a are all the pairwise non-isomorphic simple $\mathcal{B}$-modules of type M, and V_b are all the pairwise non-isomorphic simple $\mathcal{B}$-modules of type Q. Then the hom-spaces $U_a := \mathrm{Hom}_{\mathcal{B}}(V_a, W)$ and $U_b := \mathrm{Hom}_{\mathcal{B}}(V_b, W)$ are naturally $\mathcal{A}$-modules. By the super Schur's Lemma, we know $\dim U_b = (k_b|k_b)$ for some k_b. By the semisimplicity assumption on $\mathcal{B}$, we have an isomorphism of $\mathcal{B}$-modules

$$W \cong \bigoplus_a U_a \otimes V_a \oplus \bigoplus_b 2^{-1} U_b \otimes V_b.$$

By applying the super Schur's Lemma, we obtain the following superalgebra isomorphisms:

$$\begin{aligned}
\mathcal{A} = \mathrm{End}_{\mathcal{B}}(W) &\cong \bigoplus_a \mathrm{End}_{\mathcal{B}}(U_a \otimes V_a) \oplus \bigoplus_b \mathrm{End}_{\mathcal{B}}(2^{-1} U_b \otimes V_b) \\
&\cong \bigoplus_a \mathrm{End}(U_a) \otimes I_{V_a} \oplus \bigoplus_b (\mathrm{End}(2^{-1} U_b) \otimes Q(1)) \otimes I_{V_b} \\
&\cong \bigoplus_a \mathrm{End}(U_a) \otimes I_{V_a} \oplus \bigoplus_b Q(U_b) \otimes I_{V_b}.
\end{aligned}$$

Hence, $\mathcal{A}$ is a semisimple superalgebra, U_a are all the pairwise non-isomorphic simple $\mathcal{A}$-modules of type M, and U_b are all the pairwise non-isomorphic simple $\mathcal{A}$-modules of type Q.

Since $\mathcal{A}$ is now a semisimple superalgebra, we can reverse the roles of $\mathcal{A}$ and $\mathcal{B}$ in the above computation and obtain the following superalgebra isomorphisms:

$$\mathrm{End}_{\mathcal{A}}(W) \cong \bigoplus_a I_{U_a} \otimes \mathrm{End}(V_a) \oplus \bigoplus_b I_{U_b} \otimes Q(V_b) \cong \mathcal{B}.$$

The proposition is proved. □

3.1.4. Split conjugacy classes in a finite supergroup. For a finite group G, the group algebra $\mathbb{C}G$ is a semisimple algebra. It follows by using the Wedderburn Theorem and comparing two different bases for the center of $\mathbb{C}G$ that the number of simple G-modules coincides with the number of conjugacy classes of G.

Now assume that the finite group G contains a subgroup G_0 of index 2. We call elements in G_0 even and elements in $G \backslash G_0$ odd. Then the group algebra of G is naturally a superalgebra and we shall denote it by $\mathbb{C}[G, G_0] = \mathbb{C}G_0 \oplus \mathbb{C}[G \backslash G_0]$ to make clear its $\mathbb{Z}_2$-grading and its dependence on G_0. Just as for the usual group algebras, it is standard to show that $\mathbb{C}[G, G_0]$ is a semisimple superalgebra.

Since elements in a given conjugacy class of G share the same parity (i.e., $\mathbb{Z}_2$-grading), it makes sense to talk about even and odd conjugacy classes.

Proposition 3.6. (1) *The number of simple $\mathbb{C}[G, G_0]$-modules coincides with the number of even conjugacy classes of G.*

(2) *The number of simple $\mathbb{C}[G, G_0]$-modules of type* Q *coincides with the number of odd conjugacy classes of G.*

Proof. Set $\mathcal{A} = \mathbb{C}[G, G_0]$. By (3.4), $\mathcal{A}$ can be decomposed as a direct sum of simple superalgebras:

$$\mathcal{A} \cong \bigoplus_{i=1}^{m} M(r_i|s_i) \oplus \bigoplus_{j=1}^{q} Q(n_j).$$

The center of the algebra $|(M(r|s)|$ is $\cong \mathbb{C}I$, and the center of the algebra $|Q(n)|$ is of dimension $(1|1)$, spanned by I_{2n} and $\begin{pmatrix} 0 & I_n \\ I_n & 0 \end{pmatrix}$. Hence, $\dim \mathcal{Z}(|\mathcal{A}|)_{\bar{0}} = m+q$ and $\dim \mathcal{Z}(|\mathcal{A}|)_{\bar{1}} = q$.

On the other hand, note that $|\mathcal{A}|$ is the usual group algebra $\mathbb{C}G$. Hence the class sums for the even (respectively, odd) conjugacy classes of G form a basis of $\mathcal{Z}(|\mathcal{A}|)_{\bar{0}}$ (respectively, $\mathcal{Z}(|\mathcal{A}|)_{\bar{1}}$). The proposition is proved. □

For a finite group $\widetilde{G}$ that contains a subgroup $\widetilde{G}_0$ of index 2, let us further assume that $\widetilde{G}$ contains a central element z that is even and of order 2. Setting $G = \widetilde{G}/\{1, z\}$, we have a short exact sequence of groups

$$1 \longrightarrow \{1, z\} \longrightarrow \widetilde{G} \xrightarrow{\theta} G \longrightarrow 1. \tag{3.6}$$

Let $\mathcal{C}$ be a conjugacy class of G. Depending on whether an element $\tilde{x}$ in $\theta^{-1}(\mathcal{C})$ is conjugate to $z\tilde{x}$, $\theta^{-1}(\mathcal{C})$ is either a single conjugacy class of $\widetilde{G}$ or it splits into two conjugacy classes of $\widetilde{G}$ (see Exercise 3.2). In the latter case, $\mathcal{C}$ is called a **split** conjugacy class, and either conjugacy class in $\theta^{-1}(\mathcal{C})$ will also be called **split**. An element $x \in G$ is called **split** if the conjugacy class of x is split. If we denote $\theta^{-1}(x) = \{\tilde{x}, z\tilde{x}\}$, then x is split if and only if $\tilde{x}$ is not conjugate to $z\tilde{x}$.

Denote by $\mathbb{C}G^-$ the quotient superalgebra of $\mathbb{C}[\widetilde{G}, \widetilde{G}_0]$ by the ideal generated by $z+1$. Now z acts as $\pm I$ on any simple $\widetilde{G}$-module by Schur's Lemma. Thus, we have an isomorphism of left $\mathbb{C}[\widetilde{G}, \widetilde{G}_0]$-modules

$$\mathbb{C}[\widetilde{G}, \widetilde{G}_0] \cong \mathbb{C}[G, G_0] \bigoplus \mathbb{C}G^-, \tag{3.7}$$

according to the eigenvalues of z. Since z is central, (3.7) is an isomorphism of superalgebras as well. Hence, $\mathbb{C}G^-$ is naturally a semisimple superalgebra. A **spin** $\mathbb{C}[\widetilde{G}, \widetilde{G}_0]$-module M means a $\mathbb{Z}_2$-graded module over the superalgebra $\mathbb{C}[\widetilde{G}, \widetilde{G}_0]$ on which z acts by $-I$, and we will sometimes simply refer to it as a spin $\widetilde{G}$-module (with $\mathbb{Z}_2$-grading implicitly assumed). A spin $\mathbb{C}[\widetilde{G}, \widetilde{G}_0]$-module is then naturally identified as a $\mathbb{C}G^-$-module.

Example 3.7. A double cover $\widetilde{\mathfrak{S}}_n$ (nontrivial for $n \geq 4$) of the symmetric group $\mathfrak{S}_n$ was constructed by Schur (see [**57**]). The subgroup of $\mathfrak{S}_n$ of index 2 in this case is the alternating group A_n. The quotient algebra $\mathbb{C}\mathfrak{S}_n^- = \mathbb{C}\widetilde{\mathfrak{S}}_n/\langle z+1 \rangle$ by the ideal generated by $(z+1)$ is call the **spin symmetric group algebra**. The algebra $\mathbb{C}\mathfrak{S}_n^-$ is an algebra generated by $t_1, t_2, \ldots, t_{n-1}$ subject to the relations:

$$t_i^2 = 1, \quad t_i t_{i+1} t_i = t_{i+1} t_i t_{i+1}, \quad t_i t_j = -t_j t_i, \quad |i-j| > 1.$$

(A presentation for $\widetilde{\mathfrak{S}}_n$ can be obtained from the above formulas by keeping the first two relations and replacing the third one by $t_i t_j = z t_j t_i$.) The algebra $\mathbb{C}\mathfrak{S}_n^-$ is a superalgebra with each t_i being odd, for $1 \le i \le n-1$.

Another major example of a double cover $\widetilde{G}$ will be presented in Section 3.3.1.

Proposition 3.8. (1) *The number of simple spin $\mathbb{C}[\widetilde{G}, \widetilde{G}_0]$-modules equals the number of even split conjugacy classes of G.*

(2) *The number of simple spin $\mathbb{C}[\widetilde{G}, \widetilde{G}_0]$-modules of type $\mathtt{Q}$ equals the number of odd split conjugacy classes of G.*

Proof. Let us denote by $\mathcal{A} = \mathbb{C}[\widetilde{G}, \widetilde{G}_0]$, and further denote by $\bar{m}$ (respectively, $\bar{q}$) the number of simple spin $\mathcal{A}$-modules of type $\mathtt{M}$ (respectively, type $\mathtt{Q}$). The central element z acts on $\mathcal{Z}(|\mathcal{A}|)$ by multiplication with eigenvalues ± 1, and the (-1)-eigenspace $\mathcal{Z}^-(|\mathcal{A}|)$ can be naturally identified with $\mathcal{Z}(|\mathbb{C}G^-|)$; see (3.7). Hence, using (3.7) and a decomposition of $\mathbb{C}[\widetilde{G}, \widetilde{G}_0]$ as a direct sum of simples, we can show by a similar argument as for Proposition 3.6 that

$$\dim \mathcal{Z}^-(|\mathcal{A}|)_{\bar{0}} = \bar{m} + \bar{q}, \quad \dim \mathcal{Z}^-(|\mathcal{A}|)_{\bar{1}} = \bar{q}. \tag{3.8}$$

List all conjugacy classes of $\widetilde{G}$ as follows: $D_1, zD_1, \ldots, D_\ell, zD_\ell, D_{\ell+1}, \ldots, D_k$, where $D_i \cap zD_i = \emptyset$ for $1 \le i \le \ell$ and $zD_i = D_i$ for $\ell+1 \le i \le k$. The corresponding class sums, $d_1, zd_1, \ldots, d_\ell, zd_\ell, d_{\ell+1}, \ldots, d_k$, form a basis for $\mathcal{Z}(|\mathcal{A}|)$. It follows that a basis for $\mathcal{Z}^-(|\mathcal{A}|)$ is $\{d_i - zd_i \mid 1 \le i \le \ell\}$, and thus, $\dim \mathcal{Z}^-(|\mathcal{A}|) = \ell$, which is the number of split conjugacy classes of G. The proposition follows by a division of the split conjugacy classes by parity and a comparison with (3.8). □

3.2. Schur-Sergeev duality of type A

In this section, we formulate the Schur-Sergeev duality as a double centralizer property for the commuting actions of the Lie superalgebra $\mathfrak{gl}(V)$ and the symmetric group $\mathfrak{S}_d$ on the tensor space $V^{\otimes d}$, where $V = \mathbb{C}^{m|n}$. We obtain a multiplicity-free decomposition of $V^{\otimes d}$ as a $U(\mathfrak{gl}(V)) \otimes \mathbb{C}\mathfrak{S}_d$-module. The character formula for the irreducible $\mathfrak{gl}(V)$-modules arising this way is given in terms of super Schur polynomials. We then provide a Young diagrammatic interpretation of the degree of atypicality of the weight $\lambda^\natural$. We further show that the category of polynomial $\mathfrak{gl}(m|n)$-modules is semisimple.

3.2.1. Schur-Sergeev duality, I. Let $\mathfrak{g} = \mathfrak{gl}(m|n)$ and let $V = \mathbb{C}^{m|n}$ be the natural $\mathfrak{g}$-module. Then $V^{\otimes d}$ is naturally a $\mathfrak{g}$-module by letting

$$\begin{aligned}\Phi_d(g).(v_1 \otimes v_2 \otimes \ldots \otimes v_d) = & g.v_1 \otimes \ldots \otimes v_d + (-1)^{|g|\cdot|v_1|} v_1 \otimes g.v_2 \otimes \ldots \otimes v_d \\ & + \ldots + (-1)^{|g|\cdot(|v_1|+\ldots+|v_{d-1}|)} v_1 \otimes v_2 \otimes \ldots \otimes g.v_d,\end{aligned}$$

where $g \in \mathfrak{g}$ and $v_i \in V$ is assumed to be $\mathbb{Z}_2$-homogeneous for all i.

On the other hand, let

$$\begin{aligned} &\Psi_d((i,i+1)).v_1 \otimes \ldots \otimes v_i \otimes v_{i+1} \otimes \ldots \otimes v_d \\ &= (-1)^{|v_i|\cdot|v_{i+1}|} v_1 \otimes \ldots \otimes v_{i+1} \otimes v_i \otimes \ldots \otimes v_d, \quad 1 \le i \le d-1, \end{aligned} \tag{3.9}$$

where (i,j) denotes a transposition in $\mathfrak{S}_d$ and v_i, v_{i+1} are $\mathbb{Z}_2$-homogeneous.

Lemma 3.9. *The formula* (3.9) *defines a left action of the symmetric group* $\mathfrak{S}_d$ *on* $V^{\otimes d}$. *The actions of* $(\mathfrak{gl}(m|n), \Phi_d)$ *and* $(\mathfrak{S}_d, \Psi_d)$ *on* $V^{\otimes d}$ *commute with each other.*

Proof. It is straightforward to check that the formula (3.9) satisfies the Coxeter relations for $\mathfrak{S}_d$ (see Exercise 3.5), and hence it defines a left action of $\mathfrak{S}_d$.

The verification of the commuting action boils down to the case below when $d = 2$ and $i = 1$ (where we let $s = (1,2) \in \mathfrak{S}_2$, $g \in \mathfrak{gl}(m|n)$, and $v_1, v_2 \in V$):

$$\begin{aligned} &\Psi_d(s)\Phi_d(g)(v_1 \otimes v_2) \\ &\quad = \Psi_d(s)(g.v_1 \otimes v_2 + (-1)^{|g|\cdot|v_1|} v_1 \otimes g.v_2) \\ &\quad = (-1)^{(|g|+|v_1|)|v_2|} v_2 \otimes g.v_1 + (-1)^{|g|\cdot|v_1| + |v_1|(|g|+|v_2|)} g.v_2 \otimes v_1, \\ &\Phi_d(g)\Psi_d(s)(v_1 \otimes v_2) = (-1)^{|v_1|\cdot|v_2|}(g.v_2 \otimes v_1 + (-1)^{|g|\cdot|v_2|} v_2 \otimes g.v_1). \end{aligned}$$

Clearly, $\Psi_d(s)\Phi_d(g) = \Phi_d(g)\Psi_d(s)$. □

We are going to formulate a superalgebra analogue of Schur duality.

Theorem 3.10 (Schur-Sergeev duality, Part I). *Let* $\mathfrak{g} = \mathfrak{gl}(m|n)$. *The images of* Φ_d *and* Ψ_d, $\Phi_d(U(\mathfrak{g}))$ *and* $\Psi_d(\mathbb{C}\mathfrak{S}_d)$, *satisfy the double centralizer property, i.e.,*

$$\begin{aligned} \Phi_d(U(\mathfrak{g})) = &\mathrm{End}_{\mathfrak{S}_d}(V^{\otimes d}), \\ &\mathrm{End}_{U(\mathfrak{g})}(V^{\otimes d}) = \Psi_d(\mathbb{C}\mathfrak{S}_d). \end{aligned}$$

Proof. By Lemma 3.9, we have $\Phi_d(U(\mathfrak{g})) \subseteq \mathrm{End}_{\mathfrak{S}_d}(V^{\otimes d})$.

We shall proceed to prove that $\Phi_d(U(\mathfrak{g})) \supseteq \mathrm{End}_{\mathfrak{S}_d}(V^{\otimes d})$.

We have a natural isomorphism of vector superspaces $\mathrm{End}(V)^{\otimes d} \cong \mathrm{End}(V^{\otimes d})$, which is then made $\mathfrak{S}_d$-equivariant, as both superspaces admit natural $\mathfrak{S}_d$-actions. This isomorphism allows us to identify $\mathrm{End}_{\mathfrak{S}_d}(V^{\otimes d}) \equiv \mathrm{Sym}^d(\mathrm{End}(V))$, the space of $\mathfrak{S}_d$-invariants in $\mathrm{End}(V)^{\otimes d}$.

Denote by Y_k, $1 \le k \le d$, the $\mathbb{C}$-span of the super-symmetrization

$$\varpi(x_1, \ldots, x_k) := \sum_{\sigma \in \mathfrak{S}_d} \sigma.(x_1 \otimes \ldots \otimes x_k \otimes I^{\otimes d-k}),$$

for all $x_i \in \mathrm{End}(V)$. Note that $Y_d = \mathrm{Sym}^d(\mathrm{End}(V)) \equiv \mathrm{End}_{\mathfrak{S}_d}(V^{\otimes d})$.

Let $\tilde{x} = \Phi_d(x) = \sum_{i=1}^d I^{\otimes i-1} \otimes x \otimes I^{\otimes d-i}$, for $x \in \mathfrak{g} = \mathrm{End}(V)$, and denote by X_k, $1 \le k \le d$, the $\mathbb{C}$-span of $\tilde{x}_1 \ldots \tilde{x}_k$ for all $x_i \in \mathrm{End}(V)$.

Claim. We have $Y_k \subseteq X_k$ for $1 \le k \le d$.

Assuming the claim is true (in particular for $k = d$), then the theorem follows thanks to

$$\mathrm{End}_{\mathfrak{S}_d}(V^{\otimes d}) = Y_d \subseteq X_d \subseteq \Phi_d(U(\mathfrak{g})).$$

We will prove the claim by induction on k. The case $k = 1$ holds, thanks to $\varpi(x) = (d-1)!\tilde{x}$.

Assuming that $Y_{k-1} \subseteq X_{k-1}$, we have

$$\varpi(x_1, \ldots, x_{k-1}) \cdot \tilde{x}_k \in X_k. \tag{3.10}$$

On the other hand, we have

$$\begin{aligned}
&\varpi(x_1, \ldots, x_{k-1}) \cdot \tilde{x}_k \\
&= \sum_{\sigma \in \mathfrak{S}_d} \sigma.(x_1 \otimes \ldots \otimes x_{k-1} \otimes I^{\otimes d-k+1}) \cdot \tilde{x}_k \\
&= \sum_{j=1}^{d} \sum_{\sigma \in \mathfrak{S}_d} \sigma.\Big((x_1 \otimes \ldots \otimes x_{k-1} \otimes I^{\otimes d-k+1}) \cdot (I^{\otimes j-1} \otimes x_k \otimes I^{\otimes d-j})\Big),
\end{aligned}$$

which can be written as a sum $A_1 + A_2$, where

$$A_1 = \sum_{j=1}^{k-1} \varpi(x_1, \ldots, x_j x_k, \ldots, x_{k-1}) \in Y_{k-1},$$

and

$$\begin{aligned}
A_2 &= \sum_{j=k}^{d} \sum_{\sigma \in \mathfrak{S}_d} \sigma.(x_1 \otimes \ldots \otimes x_{k-1} \otimes I^{\otimes j-k} \otimes x_k \otimes I^{\otimes d-j}) \\
&= (d-k+1)\varpi(x_1, \ldots, x_{k-1}, x_k).
\end{aligned}$$

Note that $A_1 \in X_k$, since $Y_{k-1} \subseteq X_{k-1} \subseteq X_k$. Hence (3.10) implies that $A_2 \in X_k$, and so, $Y_k \subseteq X_k$. This proves the claim.

Hence, $\Phi_d(U(\mathfrak{g})) = \mathrm{End}_{\mathfrak{S}_d}(V^{\otimes d}) = \mathrm{End}_{\mathcal{B}}(V^{\otimes d})$, for $\mathcal{B} := \Psi_d(\mathbb{C}\mathfrak{S}_d)$.

Note that $\mathcal{B}$ is a semisimple algebra, and so the assumption of Proposition 3.5 is satisfied. Therefore, we have $\mathrm{End}_{U(\mathfrak{g})}(V^{\otimes d}) = \Psi_d(\mathbb{C}\mathfrak{S}_d)$. □

3.2.2. Schur-Sergeev duality, II. Recall from Definition 2.10 that an $(m|n)$-hook partition $\mu = (\mu_1, \mu_2, \ldots)$ is a partition with $\mu_{m+1} \leq n$. Given an $(m|n)$-hook partition μ, we denote by $\mu^+ = (\mu_{m+1}, \mu_{m+2}, \ldots)$ and write its conjugate partition, which is necessarily of length $\leq n$, as $\nu = (\mu^+)' = (\nu_1, \ldots, \nu_n)$. We recall the weight defined in (2.4):

$$\mu^\natural = \mu_1\delta_1 + \ldots + \mu_m\delta_m + \nu_1\varepsilon_1 + \ldots + \nu_n\varepsilon_n.$$

Denote by $\mathcal{P}_d(m|n)$ the set of all $(m|n)$-hook partitions of size d and let

$$\mathcal{P}(m|n) = \cup_{d \geq 0} \mathcal{P}_d(m|n) = \{\lambda \mid \lambda \text{ is a partition with } \lambda_{m+1} \leq n\}.$$

In particular, $\mathcal{P}_d(m) = \mathcal{P}_d(m|0)$ is the set of partitions of d of length at most m. Accordingly, we let

$$\mathcal{P}_d = \{\lambda \mid \lambda \text{ is a partition of } d\}.$$

We denote by $L(\lambda^\natural)$ for $\lambda \in \mathcal{P}(m|n)$ the simple $\mathfrak{g}$-module of highest weight $\lambda^\natural$ with respect to the standard Borel subalgebra. For a partition λ of d, we denote by S^λ the Specht module of $\mathfrak{S}_d$. It is well known that $\{S^\lambda \mid \lambda \in \mathcal{P}_d\}$ is a complete list of simple $\mathfrak{S}_d$-modules. For example, $S^{(d)}$ is the trivial representation $\mathbf{1}_d$, and $S^{(1^d)} = \mathrm{sgn}_d$ is the sign representation.

Theorem 3.11 (Schur-Sergeev duality, Part II). *As a $U(\mathfrak{gl}(m|n)) \otimes \mathbb{C}\mathfrak{S}_d$-module, we have*

$$(\mathbb{C}^{m|n})^{\otimes d} \cong \bigoplus_{\lambda \in \mathcal{P}_d(m|n)} L(\lambda^\natural) \otimes S^\lambda.$$

We will refer to a $U(\mathfrak{gl}(m|n)) \otimes \mathbb{C}\mathfrak{S}_d$-module as a $(\mathfrak{gl}(m|n), \mathfrak{S}_d)$-module. Similar conventions also apply to similar setups later on.

Proof. The proof of this theorem will occupy most of Section 3.2.2. Let $V = \mathbb{C}^{m|n}$ with standard basis $\{e_i | i \in I(m|n)\}$, and let $W = V^{\otimes d}$. Note that $\Psi_d(\mathbb{C}\mathfrak{S}_d)$ is a semisimple algebra and all its simple modules are of type $\mathtt{M}$. By Proposition 3.5 and Theorem 3.10, we have a multiplicity-free decomposition of the $(\mathfrak{gl}(m|n), \mathfrak{S}_d)$-module W of the following form:

$$W \equiv V^{\otimes d} \cong \bigoplus_{\lambda \in \mathfrak{P}_d(m|n)} L^{[\lambda]} \otimes S^\lambda,$$

where $L^{[\lambda]}$ is some simple $\mathfrak{gl}(m|n)$-module associated to λ, whose highest weight (with respect to the standard Borel) is to be determined. Also to be determined is the index set $\mathfrak{P}_d(m|n) = \{\lambda \vdash d \mid L^{[\lambda]} \neq 0\}$.

First we need to prepare some notation.

Let $\mathcal{CP}(m|n)$ be the set of pairs $\nu|\mu$ of compositions $\nu = (\nu_1, \ldots, \nu_m)$ and $\mu = (\mu_1, \ldots, \mu_n)$, and let

$$\mathcal{CP}_d(m|n) = \{\nu|\mu \in \mathcal{CP}(m|n) \mid \sum_i \nu_i + \sum_j \mu_j = d\}.$$

We have the following weight space decomposition with respect to the Cartan subalgebra of diagonal matrices $\mathfrak{h} \subset \mathfrak{gl}(m|n)$:

$$W = \bigoplus_{\nu|\mu \in \mathcal{CP}_d(m|n)} W_{\nu|\mu}, \tag{3.11}$$

where $W_{\nu|\mu}$ has a linear basis $e_{i_1} \otimes \ldots \otimes e_{i_d}$, with the indices satisfying the following equality of multisets:

$$\{i_1, \ldots, i_d\} = \{\underbrace{\overline{1}, \ldots, \overline{1}}_{\nu_1}, \ldots, \underbrace{\overline{m}, \ldots, \overline{m}}_{\nu_m}; \underbrace{1, \ldots, 1}_{\mu_1}, \ldots, \underbrace{n, \ldots, n}_{\mu_n}\}.$$

Introduce the following Young subgroup of $\mathfrak{S}_d$:

$$\mathfrak{S}_{\nu|\mu} = \mathfrak{S}_{\nu_1} \times \ldots \times \mathfrak{S}_{\nu_m} \times \mathfrak{S}_{\mu_1} \times \ldots \times \mathfrak{S}_{\mu_n}.$$

The span of the vector $e_{\nu|\mu} := e_{\overline{1}}^{\otimes \nu_1} \otimes \ldots \otimes e_{\overline{m}}^{\otimes \nu_m} \otimes e_1^{\otimes \mu_1} \otimes \ldots \otimes e_n^{\otimes \mu_n}$ can be identified with the one-dimensional $\mathfrak{S}_{\nu|\mu}$-module $\mathbf{1}_\nu \otimes \mathrm{sgn}_\mu$. Here $\mathbf{1}_\nu$ is the trivial $\mathfrak{S}_\nu$-module and sgn_μ is the $\mathfrak{S}_\mu$-module $\mathrm{sgn}_{\mu_1} \otimes \cdots \otimes \mathrm{sgn}_{\mu_n}$. Since the orbit $\mathfrak{S}_d e_{\nu|\mu}$ spans $W_{\nu|\mu}$, we have a surjective $\mathfrak{S}_d$-homomorphism from $\mathrm{Ind}_{\mathfrak{S}_{\nu|\mu}}^{\mathfrak{S}_d}(\mathbf{1}_\nu \otimes \mathrm{sgn}_\mu)$ onto $W_{\nu|\mu}$ by Frobenius reciprocity. By counting the dimensions we have an $\mathfrak{S}_d$-isomorphism:

$$W_{\nu|\mu} \cong \mathrm{Ind}_{\mathfrak{S}_{\nu|\mu}}^{\mathfrak{S}_d}(\mathbf{1}_\nu \otimes \mathrm{sgn}_\mu). \tag{3.12}$$

Let us denote the decomposition of the $\mathfrak{S}_d$-module $W_{\nu|\mu}$ into irreducibles by

$$W_{\nu|\mu} \cong \bigoplus_\lambda K_{\lambda,\nu|\mu} S^\lambda, \qquad \text{for } K_{\lambda,\nu|\mu} \in \mathbb{Z}_+.$$

Let λ be a partition that is identified with its Young diagram. Recall $I(m|n)$ is totally ordered by (1.3). A **supertableau** T of shape λ, or a **super λ-tableau** T, is an assignment of an element in $I(m|n)$ to each box of the Young diagram λ satisfying the following conditions:

(HT1) The numbers are weakly increasing along each row and column.

(HT2) The numbers from $\{\overline{1}, \ldots, \overline{m}\}$ are strictly increasing along each column.

(HT3) The numbers from $\{1, \ldots, n\}$ are strictly increasing along each row.

Such a supertableau T is said to have **content** $\nu|\mu \in \mathcal{CP}(m|n)$ if $\overline{i}$ $(1 \le i \le m)$ appears ν_i times and j $(1 \le j \le n)$ appears μ_j times. Denote by $\mathcal{HT}(\lambda, \nu|\mu)$ the set of super λ-tableaux of content $\nu|\mu$. In the case $n = 0$, a supertableau becomes a usual (semistandard) tableau.

Lemma 3.12. *We have* $K_{\lambda,\nu|\mu} = \#\mathcal{HT}(\lambda, \nu|\mu)$.

Proof. Recall that $\mathrm{Ind}_{\mathfrak{S}_{\nu|\mu}}^{\mathfrak{S}_d}(\mathbf{1}_\nu \otimes \mathrm{sgn}_\mu) \cong \bigoplus_\lambda K_{\lambda,\nu|\mu} S^\lambda$.

First assume that $\mu = \emptyset$, and we prove the formula by induction on the length $r = \ell(\nu)$. A (semistandard) tableau T of shape λ and content ν gives rise to a sequence of partitions $\emptyset = \lambda^0 \subset \lambda^1 \subset \ldots \subset \lambda^r = \lambda$ such that λ^i has the shape given by the parts of T with entries $\le i$, and λ^i/λ^{i-1} has ν_i boxes for each i. This sets up a bijection between $\mathcal{HT}(\lambda, \nu)$ and the set of such sequences of partitions. Denote $d_1 = d - \nu_r$ and $\widetilde{\nu} = (\nu_1, \ldots, \nu_{r-1})$. We have $\mathrm{Ind}_{\mathfrak{S}_{\widetilde{\nu}}}^{\mathfrak{S}_{d_1}} \mathbf{1}_{d_1} \cong \bigoplus_{\rho \vdash d_1} K_{\rho,\widetilde{\nu}} S^\rho$, where $K_{\rho,\widetilde{\nu}} = \#\mathcal{HT}(\rho, \widetilde{\nu})$ by the induction hypothesis. Now the induction step is simply the following representation theoretic version of Pieri's formula (A.22) (which can be seen using the Frobenius characteristic map and (A.26)): for a partition $\rho \vdash d_1$,

$$\mathrm{Ind}_{\mathfrak{S}_{d_1} \times \mathfrak{S}_{\nu_r}}^{\mathfrak{S}_d}(S^\rho \otimes \mathbf{1}_{\nu_r}) \cong \bigoplus_\lambda S^\lambda,$$

where λ is such that λ/ρ is a horizontal strip of ν_r boxes (that is, λ/ρ is a skew diagram of size ν_r whose columns all have length at most one).

Then using the above special case (for $\mu = \emptyset$) as the initial step, we complete the proof in the general case by induction on the length of μ, in which the induction step is exactly the conjugated Pieri's formula. □

Lemma 3.13. *Let* $\lambda \in \mathcal{P}_d$ *and* $\nu|\mu \in \mathcal{CP}_d(m|n)$. *Then* $K_{\lambda,\nu|\mu} = 0$ *unless* $\lambda \in \mathcal{P}_d(m|n)$.

Proof. By the identity $K_{\lambda,\nu|\mu} = \#\mathcal{HT}(\lambda,\nu|\mu)$, it suffices to prove that if a super λ-tableau T of content $\nu|\mu$ exists, then $\lambda_{m+1} \leq n$.

By applying the supertableau condition (HT2) to the first column of T, we see that the first entry $k \in I(m|n)$ in row $(m+1)$ satisfies $k > 0$. Applying the supertableau condition (HT3) to the $(m+1)$st row, we conclude that $\lambda_{m+1} \leq n$. □

Recall that $\mathfrak{P}_d(m|n) = \{\lambda \vdash d \mid L^{[\lambda]} \neq 0\}$. It follows by Lemma 3.13 that $\mathfrak{P}_d(m|n) \subset \mathcal{P}_d(m|n)$. On the other hand, given $\lambda \in \mathcal{P}_d(m|n)$, clearly a super λ-tableau exists; e.g., we can fill in the numbers $\overline{1},\ldots,\overline{m}$ on the first m rows of λ row-by-row downward, and then for the remaining subdiagram of λ, we fill in the numbers $1,\ldots,n$ column-by-column from left to right. This distinguished super λ-tableau will be denoted by T_λ. Hence, we have proved that

$$\mathfrak{P}_d(m|n) = \mathcal{P}_d(m|n).$$

For a given $\lambda \in \mathcal{P}_d(m|n)$, we have $L^{[\lambda]} = \bigoplus_{\nu|\mu \in \mathcal{CP}_d(m|n)} L^{[\lambda]}_{\nu|\mu}$. Among all the contents of super λ-tableaux, the one for T_λ corresponds to a highest weight relative to the standard Borel subalgebra of $\mathfrak{g}$ by the supertableau conditions (HT1-3). Hence we conclude that $L^{[\lambda]} = L(\lambda^\natural)$, the simple $\mathfrak{g}$-module of highest weight $\lambda^\natural$. This completes the proof of Theorem 3.11. □

Remark 3.14. (1) For $n = 0$, the Schur-Sergeev duality reduces to the usual Schur duality. If in addition $d = 2$, then $(\mathbb{C}^m)^{\otimes 2} = S^2(\mathbb{C}^m) \oplus \wedge^2(\mathbb{C}^m)$. This fits well with the well-known fact that, as $\mathfrak{gl}(m)$-modules, $S^2(\mathbb{C}^m)$ and $\wedge^2(\mathbb{C}^m)$ are, respectively, irreducible of highest weights $2\delta_1$ and $\delta_1 + \delta_2$.

(2) If $d \leq mn + m + n$, then $\mathcal{P}_d(m|n)$ is the set of all partitions of d, and every simple $\mathfrak{S}_d$-module appears in the Schur-Sergeev duality decomposition.

(3) For $d = 2$, the Schur-Sergeev duality reads that $(\mathbb{C}^{m|n})^{\otimes 2} = \mathcal{S}^2(\mathbb{C}^{m|n}) \oplus \Lambda^2(\mathbb{C}^{m|n})$, where we recall that $\mathcal{S}^2$ and Λ^2 are understood in the super sense. In particular, as an ordinary vector space,

$$\mathcal{S}^2(\mathbb{C}^{m|n}) = S^2(\mathbb{C}^m) \oplus (\mathbb{C}^m \otimes \mathbb{C}^n) \oplus \wedge^2(\mathbb{C}^n).$$

3.2.3. The character formula. Let $x = \{x_1, \dots, x_m\}$ and $y = \{y_1, \dots, y_n\}$ be two sets of independent variables. We shall compute the **character** of the $\mathfrak{gl}(m|n)$-module $L(\lambda^\natural)$:

$$\mathrm{ch} L(\lambda^\natural) := \mathrm{tr}\,|_{L(\lambda^\natural)}\, x_1^{E_{\bar{1}\bar{1}}} \dots x_m^{E_{\bar{m}\bar{m}}} y_1^{E_{11}} \dots y_n^{E_{nn}} = \sum_{\nu|\mu \in \mathcal{CP}(m|n)} \dim L(\lambda^\natural)_{\nu|\mu} \prod_{i,j} x_i^{\nu_i} y_j^{\mu_j}.$$

Recall that m_ν denotes the monomial symmetric function associated to a partition ν (cf. Appendix A), and that hs_ν denotes the super Schur function (A.37).

Theorem 3.15. *Let λ be an $(m|n)$-hook partition, i.e., a partition with $\lambda_{m+1} \le n$. Let $x = \{x_1, \dots, x_m\}$ and $y = \{y_1, \dots, y_n\}$. Then the following character formula holds:*

$$\mathrm{ch} L(\lambda^\natural) = \mathrm{hs}_\lambda(x; y).$$

Proof. It follows from Theorem 3.11 and (3.11) that

$$\bigoplus_{\nu|\mu \in \mathcal{CP}_d(m|n)} W_{\nu|\mu} \cong \bigoplus_{\lambda \in \mathcal{P}_d(m|n)} L(\lambda^\natural) \otimes S^\lambda.$$

Let us apply simultaneously the trace operator $\mathrm{tr}\,|_{L(\lambda^\natural)}\, x_1^{E_{\bar{1}\bar{1}}} \dots x_m^{E_{\bar{m}\bar{m}}} y_1^{E_{11}} \dots y_n^{E_{nn}}$ and the Frobenius characteristic map ch^{F} (see Appendix A.1.4) to both sides of the above isomorphism. Then, summing over all $d \ge 0$ and using (3.12) and (A.26), we have that

$$\sum_{\nu|\mu \in \mathcal{CP}(m|n)} m_\nu(x) m_\mu(y) h_\nu(z) e_\mu(z) = \sum_{\lambda \in \mathcal{P}(m|n)} \mathrm{ch} L(\lambda^\natural) s_\lambda(z),$$

where $z = \{z_1, z_2, \dots\}$ is infinite. By the identities in (A.11), we have

$$\sum_{\nu|\mu \in \mathcal{CP}(m|n)} m_\nu(x) m_\mu(y) h_\nu(z) e_\mu(z) = \prod_{k=1}^{\infty} \prod_{i=1}^{m} \prod_{j=1}^{n} \frac{1 + y_j z_k}{1 - x_i z_k} = \sum_{\lambda \in \mathcal{P}(m|n)} \mathrm{hs}_\lambda(x; y) s_\lambda(z).$$

Now the theorem follows by comparing the above two equations and noting the linear independence of $s_\lambda(z)$. □

Given a super λ-tableau T of content $\nu|\mu \in \mathcal{CP}(m|n)$, we denote by

$$(x|y)^T := x_1^{\nu_1} x_2^{\nu_2} \cdots x_m^{\nu_m} y_1^{\mu_1} \cdots y_n^{\mu_n}.$$

We have the following alternative character formula.

Theorem 3.16. *Let λ be an $(m|n)$-hook partition. Then,*

$$\mathrm{ch} L(\lambda^\natural) = \sum_T (x|y)^T, \tag{3.13}$$

where the summation is taken over all super λ-tableaux T. Also, we have

$$\mathrm{ch} L(\lambda^\natural) = \sum_{\nu,\mu} \#\mathcal{HT}(\lambda, \nu|\mu)\, m_\nu(x) m_\mu(y), \tag{3.14}$$

where the summation is over partitions ν and μ of length at most m and n, respectively, such that $|\nu| + |\mu| = d$.

Proof. The formula (3.13) is simply a reformulation of Theorem 3.15 by the definition of super Schur functions in (A.37) and the combinatorial formula for (skew) Schur functions (A.19).

Let us collect the same monomials together in the sum (3.13). Regarding $L(\lambda^\natural)$ as a module over $\mathfrak{gl}(m) \times \mathfrak{gl}(n)$, we observe that $\mathrm{ch}L(\lambda^\natural)$ is symmetric with respect to $x_1, \dots, x_m$ and symmetric with respect to $y_1, \dots, y_n$. Hence, we must have

$$\#\mathcal{HT}(\lambda, \widetilde{\nu}|\widetilde{\mu}) = \#\mathcal{HT}(\lambda, \nu|\mu)$$

if $\widetilde{\nu}$ and $\widetilde{\mu}$ are compositions obtained by rearranging the parts from ν and μ, respectively. Now (3.14) follows from this observation and (3.13). □

Corollary 3.17. *The following weight multiplicity formula for the module $L(\lambda^\natural)$ holds:*

$$\dim L(\lambda^\natural)_{\nu|\mu} = \#\mathcal{HT}(\lambda, \nu|\mu).$$

Example 3.18. The standard basis vectors for $\mathcal{S}^2(\mathbb{C}^{m|n})$ are $e_i \otimes e_j + (-1)^{|i||j|} e_j \otimes e_i$, where $i, j \in I(m|n)$ satisfy $i \le j < 0$, $i < 0 < j$, or $0 < i < j$ (cf. Remark 3.14 (4)). They are in bijection with the supertableaux of shape $\lambda = (2)$:

i	j

This is compatible with the isomorphism $\mathcal{S}^2(\mathbb{C}^{m|n}) \cong L(\mathfrak{gl}(m|n), 2\delta_1)$.

3.2.4. The classical Schur duality.

3.2.4. The classical Schur duality. We sketch below a more standard argument for the standard Schur duality on $W = (\mathbb{C}^n)^{\otimes d}$, which emphasizes the decomposition of W as a $\mathfrak{gl}(n)$-module instead of as an $\mathfrak{S}_d$-module.

Given a composition (or a partition) μ of d of length $\le n$, we denote by W_μ the μ-weight space of the $\mathfrak{gl}(n)$-module. Clearly W_μ has a basis

$$e_{i_1} \otimes \ldots \otimes e_{i_d}, \quad \text{where } \{i_1, \dots, i_d\} = \{\underbrace{1, \dots, 1}_{\mu_1}, \dots, \underbrace{n, \dots, n}_{\mu_n}\}. \tag{3.15}$$

As a $(\mathfrak{gl}(n), \mathfrak{S}_d)$-module,

$$W \cong \bigoplus_\lambda L(\lambda) \otimes U^\lambda,$$

where $U^\lambda := \mathrm{Hom}_{\mathfrak{gl}(n)}(L(\lambda), W) \cong W_\lambda^{\mathfrak{n}^+}$ (the space of highest weight vectors in W of weight λ). Only $\lambda \in \mathcal{P}_d(n)$ can be highest weights of $\mathfrak{gl}(n)$-modules that appear in the decomposition of $(\mathbb{C}^n)^{\otimes d}$, and every such λ indeed appears as $L(\lambda)$ is clearly a summand of the submodule $\wedge^{\lambda'_1}(\mathbb{C}^n) \otimes \wedge^{\lambda'_2}(\mathbb{C}^n) \otimes \cdots$ of W.

Note that $\mathcal{CP}_d(n)$ has two interpretations: one as the polynomial weights for $\mathfrak{gl}(n)$ and the other as compositions of d in at most n parts. A remarkable fact is that the partial order on weights induced by the positive roots of $\mathfrak{gl}(n)$ coincides with the dominance partial order $\ge$ on compositions.

Since $W = \bigoplus_\mu W_\mu = \bigoplus_{\mu,\lambda:\lambda\geq\mu} L(\lambda)_\mu \otimes U^\lambda$, we conclude that, as an $\mathfrak{S}_d$-module,

$$W_\mu \cong \bigoplus_{\lambda\geq\mu} \dim L(\lambda)_\mu U^\lambda. \tag{3.16}$$

On the other hand, $\mathfrak{S}_d$ acts on the basis (3.15) of W_μ transitively, and the stabilizer of the basis element $e_1^{\otimes\mu_1} \otimes \ldots \otimes e_n^{\otimes\mu_n}$ is the Young subgroup $\mathfrak{S}_\mu$. Therefore we have

$$W_\mu \cong \mathrm{Ind}_{\mathfrak{S}_\mu}^{\mathfrak{S}_d} \mathbf{1}_\mu = \bigoplus_{\lambda\geq\mu} K_{\lambda\mu} S^\lambda, \tag{3.17}$$

where $K_{\lambda\mu}$ is the Kostka number that satisfies $K_{\lambda\lambda} = 1$. It is well known that $K_{\lambda\mu}$ is equal to the number of semistandard λ-tableaux of content μ.

By the double centralizer property (see Proposition 3.5 and Theorem 3.10), U^λ has to be an irreducible $\mathfrak{S}_d$-module for each λ. We compare the interpretations (3.16) and (3.17) of W_μ in the special case when μ is dominant (i.e., a partition). One by one downward along the dominance order starting with $\mu = (d)$, this provides the identification $U^\mu = S^\mu$ for every μ, and moreover, we obtain the well-known equality $\dim L(\lambda)_\mu = K_{\lambda\mu}$.

3.2.5. Degree of atypicality of $\lambda^\natural$. In this section, we provide a Young diagrammatic interpretation of the degree of atypicality of the weight $\lambda^\natural$, for an $(m|n)$-hook partition λ.

Up to a shift by $-\frac{1}{2}(m+n+1)\mathbf{1}_{m|n}$ (which is irrelevant in all applications), the Weyl vector ρ for the standard positive system of $\mathfrak{gl}(m|n)$ from Lemma 1.22 can be written as

$$\rho = \sum_{i=1}^{m} (m-i+1)\delta_i - \sum_{j=1}^{n} j\varepsilon_j. \tag{3.18}$$

We introduce an integer i_λ, with $0 \leq i_\lambda \leq \min\{m,n\}$, to stand for the smallest nonnegative integer i such that the $(m-i, n-i)$-th box belongs to the diagram λ.

Example 3.19. Let $\lambda = (7,4,2,2,1,1)$ with $m=4$ and $n=5$. Then i_λ is the number of $(m-i, n-i)$ boxes that do not lie in the diagram of λ, for $i = 0, 1, \ldots, \min\{m,n\}$. Such boxes are marked with crosses in this example. So $i_\lambda = 2$.

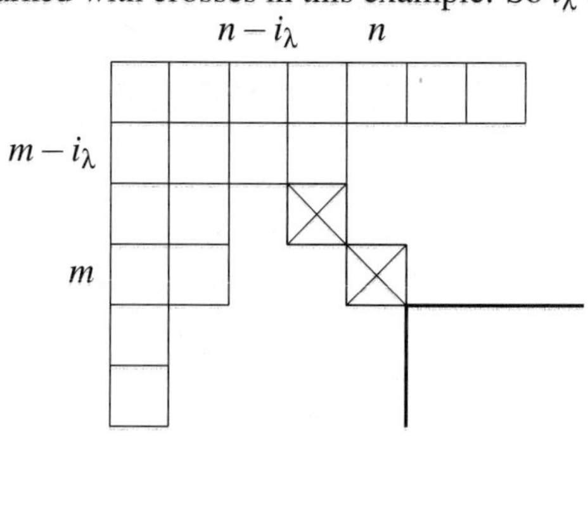

Lemma 3.20. *For $0 \le j \le i_\lambda - 1$, we have $\lambda_{m-j} \le n - i_\lambda \le \lambda_{m-i_\lambda}$.*

Proof. The lemma is clearly equivalent to the claim that $\lambda_{m-i_\lambda+1} \le n - i_\lambda \le \lambda_{m-i_\lambda}$. The latter is evident from the diagram in Example 3.19, as i_λ is equal to the number of boxes of coordinates $(m-i, n-i)$ that do not lie in the diagram of λ, for $i = 0, 1, \ldots, \min\{m, n\}$. □

Recall the degree of atypicality $\#\mu$ for a weight $\mu \in \mathfrak{h}^*$ from Definition 2.29.

Proposition 3.21. *Let λ be an $(m|n)$-hook partition. Then we have $i_\lambda = \#\lambda^\natural$.*

Proof. We compute by (3.18) that

$$\lambda^\natural + \rho = \sum_{i=1}^{m} (\lambda_i + m - i + 1)\delta_i - \sum_{j=1}^{n} (j - \nu_j)\varepsilon_j. \tag{3.19}$$

We observe that the sequence $\{\lambda_1 + m, \lambda_2 + m - 1, \ldots, \lambda_m + 1\}$ is strictly decreasing, while the sequence $\{1 - \nu_1, 2 - \nu_2, \ldots, n - \nu_n\}$ is strictly increasing.

Suppose that $i_\lambda = 0$. This is equivalent to saying that $\lambda_m \ge n$, by Lemma 3.20. Thus

$$\lambda_i + m - i + 1 \ge \lambda_m + 1 > j - \nu_j, \qquad \forall 1 \le j \le n, 1 \le i \le m.$$

It follows by Definition 2.29 that $\#\lambda^\natural = 0$.

Now suppose that $i_\lambda > 0$. Then, we have $\lambda_m = n - j_0 < n$ for some $0 < j_0 \le n$. This implies that $\nu_n = \cdots = \nu_{n-j_0+1} = 0$, and thus

$$\{n - j_0 + 1 - \nu_{n-j_0+1}, \ldots, n - \nu_n\} = \{n - j_0 + 1, \ldots, n\}.$$

It follows by Lemma 3.20 that, for $0 \le j \le i_\lambda - 1$,

$$n - j_0 + 1 = \lambda_m + 1 \le \lambda_{m-j} + j + 1 \le n - i_\lambda + j + 1 \le n.$$

Thus, in the set $\{n - j_0 + 1 - \nu_{n-j_0+1}, \ldots, n - \nu_n\} = \{n - j_0 + 1, \ldots, n\}$, there is a unique element that is equal to $\lambda_{m-j} + j + 1$, for $0 \le j \le i_\lambda - 1$. Hence, $\#\lambda^\natural \ge i_\lambda$.

Finally, for $i = 1, \ldots, m - i_\lambda$, Lemma 3.20 implies that

$$\lambda_i + m - i + 1 \ge \lambda_{m-i_\lambda} + i_\lambda + 1 \ge (n - i_\lambda) + i_\lambda + 1 = n + 1.$$

Thus any such $\lambda_i + m - i + 1$, for $i = 1, \ldots, m - i_\lambda$, cannot be equal to an element of the form $j - \nu_j$, for $j = 1, \ldots, n$, whence $\#\lambda^\natural = i_\lambda$. □

In light of (3.19), we define

$$H_{\lambda^\natural} = \{\lambda_i + m - i + 1 \mid 1 \le i \le m\},$$
$$T_{\lambda^\natural} = \{j - \nu_j \mid 1 \le j \le n\}, \qquad I_{\lambda^\natural} = H_{\lambda^\natural} \cap T_{\lambda^\natural}.$$

The following corollary can be read off from the proof of Proposition 3.21.

Corollary 3.22. *Let λ be an $(m|n)$-hook partition. Then $I_{\lambda^\natural}$ consists of precisely the smallest $\#\lambda^\natural$ numbers in $H_{\lambda^\natural}$, and also, $I_{\lambda^\natural}$ consists of precisely the largest $\#\lambda^\natural$ numbers in $\{1,2,\ldots,n\}$ that are not in $T_{\lambda^\natural} \setminus I_{\lambda^\natural}$.*

Example 3.23. Let $\lambda = (7,4,2,2,1,1)$ with $m = 4$ and $n = 5$ as in Example 3.19. Then $H_{\lambda^\natural} = \{11,7,4,3\}, T_{\lambda^\natural} = \{-1,2,3,4,5\}$, and $I_{\lambda^\natural} = \{3,4\}$. This agrees with Corollary 3.22.

3.2.6. Category of polynomial modules. In this subsection $\mathfrak{g} = \mathfrak{gl}(m|n)$ and $L(\nu)$ is the irreducible highest weight $\mathfrak{g}$-module of highest weight $\nu \in \mathfrak{h}^*$ with respect to the standard Borel subalgebra. Recall from Section 2.2.2 the notion of central characters $\chi_\nu : Z(\mathfrak{g}) \to \mathbb{C}$.

Recall that, given a Lie (super)algebra $\mathcal{G}$ and $\mathcal{G}$-modules M and N, the vector space $\mathrm{Ext}^1_{\mathcal{G}}(M,N)$ classifies (up to equivalence) the short exact sequences of $\mathcal{G}$-modules of the form

$$0 \longrightarrow N \longrightarrow E \longrightarrow M \longrightarrow 0. \tag{3.20}$$

Proposition 3.24. *Let λ,μ be $(m|n)$-hook diagrams. Then,*

(1) $\chi_{\lambda^\natural} = \chi_{\mu^\natural}$ *if and only if* $\lambda = \mu$.

(2) $\mathrm{Ext}^1_{\mathfrak{gl}(m|n)}\big(L(\lambda^\natural), L(\mu^\natural)\big) = 0$.

Proof. (1) Suppose that $\chi_{\lambda^\natural} = \chi_{\mu^\natural}$. Then by Theorem 2.30 we have

$$H_{\lambda^\natural} \setminus I_{\lambda^\natural} = H_{\mu^\natural} \setminus I_{\mu^\natural}, \qquad T_{\lambda^\natural} \setminus I_{\lambda^\natural} = T_{\mu^\natural} \setminus I_{\mu^\natural}.$$

Thus, it suffices to show that $\lambda^\natural$ can be reconstructed from the sets $H_{\lambda^\natural} \setminus I_{\lambda^\natural}$ and $T_{\lambda^\natural} \setminus I_{\lambda^\natural}$. To that end, note that $\lambda^\natural + \rho$ and $\lambda^\natural$ can be recovered from the sets $H_{\lambda^\natural}$ and $T_{\lambda^\natural}$, which in turn are determined by the three sets $I_{\lambda^\natural}$, $H_{\lambda^\natural} \setminus I_{\lambda^\natural}$ and $T_{\lambda^\natural} \setminus I_{\lambda^\natural}$. But by Corollary 3.22 the set $I_{\lambda^\natural}$ is determined from the set $T_{\lambda^\natural} \setminus I_{\lambda^\natural}$. This proves (1).

(2) Consider a short exact sequence of the form (3.20) with $M = L(\lambda^\natural)$ and $N = L(\mu^\natural)$. First assume $\lambda \neq \mu$. It follows by (1) that there exists a central element z such that $\chi_{\lambda^\natural}(z) \neq \chi_{\mu^\natural}(z)$. Then, $\ker(z - \chi_{\lambda^\natural}(z))$ is a nonzero proper submodule of E, which must be isomorphic to M as it cannot be isomorphic to N. Hence the short exact sequence splits. Now assume $\lambda = \mu$. Then by weight considerations, the two-dimensional $\mu^\natural$-weight subspace of E has to be the highest weight space. Thus E contains two simple highest weight submodules of highest weight $\mu^\natural$ that must intersect trivially. Hence the short exact sequence splits in this case as well. □

Definition 3.25. A weight $\mu = \sum_{i=1}^m a_i\delta_i + \sum_{j=1}^n b_j\varepsilon_j$ is called a **polynomial weight** for $\mathfrak{gl}(m|n)$ if all a_i, b_j are nonnegative integers. A $\mathfrak{gl}(m|n)$-module M is called a **polynomial module** if M is $\mathfrak{h}$-semisimple and every weight of M is a polynomial weight.

Proposition 3.26. *An irreducible highest weight $\mathfrak{gl}(m|n)$-module M is a polynomial module if and only if $M \cong L(\mu^\natural)$ for some $(m|n)$-hook partition μ.*

Proof. Assume that $M \cong L(\mu^\natural)$, for some $(m|n)$-hook partition μ. Denote by $d = |\mu|$. Then, by Theorem 3.11, $L(\mu^\natural)$ is a direct summand of $(\mathbb{C}^{m|n})^{\otimes d}$, which is clearly a polynomial module.

Now assume that M is an irreducible polynomial $\mathfrak{gl}(m|n)$-module, which by Definition 3.25(3) is isomorphic to $L(\lambda)$ for some $\lambda = \sum_{i=1}^m \lambda_i \delta_i + \sum_{j=1}^n b_j \varepsilon_j$, with all $\lambda_i, b_j \in \mathbb{Z}_+$. Since $L(\mathfrak{gl}(m) \oplus \mathfrak{gl}(n), \lambda)$ is a polynomial module, we must have $\lambda_i \geq \lambda_{i+1}$ and $b_j \geq b_{j+1}$, for all possible i, j.

We shall proceed to complete the proof by contradiction. Suppose that $\lambda \neq \mu^\natural$, for any $(m|n)$-hook partition μ. Then we have $\lambda_m = k-1$ and $b_k > 0$, for some $1 \leq k \leq n$. We apply the sequence of odd reflections corresponding to the odd roots $\delta_m - \varepsilon_1, \ldots, \delta_m - \varepsilon_{k-1}$ to obtain a new Borel subalgebra $\widetilde{\mathfrak{b}}$ from the standard one. By applying Lemma 1.40 repeatedly, we compute the $\widetilde{\mathfrak{b}}$-highest weight of M to be $\widetilde{\lambda} = \lambda - (k-1)\delta_m + \varepsilon_1 + \ldots + \varepsilon_{k-1}$. Observe by a repeated use of Lemma 1.30 that $\delta_m - \varepsilon_k$ is a simple root of $\widetilde{\mathfrak{b}}$. Let $w_{\widetilde{\lambda}}$ be a nonzero $\widetilde{\mathfrak{b}}$-highest weight vector of M. Then the vector $e_{-\delta_m + \varepsilon_k} w_{\widetilde{\lambda}}$ is nonzero in M, and its weight $\widetilde{\lambda} - \delta_m + \varepsilon_k$ has -1 as the coefficient for δ_m, thanks to $\lambda_m = k-1$ and $b_k > 0$. This contradicts the assumption that M is a polynomial module. □

Theorem 3.27. *The category of polynomial* $\mathfrak{gl}(m|n)$*-modules is a semisimple tensor category.*

Proof. Let M be a polynomial $\mathfrak{gl}(m|n)$-module. For $v \in M$ note that $U(\mathfrak{gl}(m|n)_{\bar{0}})v$ is finite-dimensional. Thus, $U(\mathfrak{gl}(m|n))v$ is also finite-dimensional, and hence it is a direct sum of irreducible polynomial modules by Propositions 3.24 and 3.26. It follows that M is a sum of irreducible polynomial modules, and hence it is a direct sum of irreducible polynomial modules.

It remains to show that the tensor product of any two irreducible polynomial $\mathfrak{gl}(m|n)$-modules is a direct sum of irreducible polynomial $\mathfrak{gl}(m|n)$-modules. To that end, note that by Theorem 3.11 any irreducible polynomial module is a direct summand of $(\mathbb{C}^{m|n})^{\otimes d}$, for some $d \geq 0$, and furthermore any such tensor power is a direct sum of irreducible polynomial modules. Now take $L(\lambda^\natural) \subseteq (\mathbb{C}^{m|n})^{\otimes d}$ and $L(\mu^\natural) \subseteq (\mathbb{C}^{m|n})^{\otimes l}$, for $(m|n)$-hook partitions λ and μ. Then

$$L(\lambda^\natural) \otimes L(\mu^\natural) \subseteq (\mathbb{C}^{m|n})^{\otimes k} \otimes (\mathbb{C}^{m|n})^{\otimes l} \cong (\mathbb{C}^{m|n})^{\otimes d+l}.$$

Since $(\mathbb{C}^{m|n})^{\otimes d+l}$ is a direct sum of irreducible polynomial modules, so is the submodule $L(\lambda^\natural) \otimes L(\mu^\natural)$. □

3.3. Representation theory of the algebra $\mathcal{H}_n$

In this section, we develop systematically the representation theory of an algebra $\mathcal{H}_n$, which is equivalent to the spin representation theory of a distinguished double cover $\widetilde{B}_n$ of the hyperoctahedral group B_n. We classify the split conjugacy classes in

$\widetilde{B}_n$. We then define a characteristic map using the character table for the simple spin modules of $\widetilde{B}_n$, analogous to the Frobenius characteristic map for the symmetric groups. The images of the irreducible spin characters of $\widetilde{B}_n$ under the characteristic map are shown to be Schur Q-functions (up to some 2-powers).

3.3.1. A double cover. Let Π_n be the finite group generated by a_i $(i = 1, \dots, n)$ and the central element z subject to the relations

$$a_i^2 = 1, \quad z^2 = 1, \quad a_i a_j = z a_j a_i \quad (i \neq j). \tag{3.21}$$

The symmetric group $\mathfrak{S}_n$ acts on Π_n by $\sigma(a_i) = a_{\sigma(i)}$, $\sigma \in \mathfrak{S}_n$. The semidirect product $\widetilde{B}_n := \Pi_n \rtimes \mathfrak{S}_n$ admits a natural finite group structure and will be called the **twisted hyperoctahedral group**. Explicitly, the multiplication in $\widetilde{B}_n$ is given by

$$(a, \sigma)(a', \sigma') = (a\sigma(a'), \sigma\sigma'), \qquad a, a' \in \Pi_n, \sigma, \sigma' \in \mathfrak{S}_n.$$

Since $\Pi_n/\{1, z\} \simeq \mathbb{Z}_2^n$, the group $\widetilde{B}_n$ is a double cover of the hyperoctahedral group $B_n := \mathbb{Z}_2^n \rtimes \mathfrak{S}_n$, and the order $|\widetilde{B}_n|$ is $2^{n+1} n!$. That is, we have a short exact sequence of groups

$$1 \longrightarrow \{1, z\} \longrightarrow \widetilde{B}_n \xrightarrow{\theta_n} B_n \longrightarrow 1,$$

where θ_n sends each a_i to the generator b_i of the ith copy of $\mathbb{Z}_2$ in B_n.

We define a $\mathbb{Z}_2$-grading on the group $\widetilde{B}_n$ by setting the degree of each a_i to be $\bar{1}$ and the degree of elements in $\mathfrak{S}_n$ to be $\bar{0}$. Hence the group $\widetilde{B}_n$ fits into the general setting of $\widetilde{G}$ in Section 3.1.4. The group B_n inherits a $\mathbb{Z}_2$-grading from $\widetilde{B}_n$ via the homomorphism θ_n. This induces **parity** epimorphisms $p : \widetilde{B}_n \to \mathbb{Z}_2$ and $p : B_n \to \mathbb{Z}_2$.

The conjugacy classes of the group B_n (a special case of a wreath product) can be described as follows, cf. Macdonald [**83**, I, Appendix B]. It is convenient to identify $\mathbb{Z}_2$ as $\{+, -\}$ with $+$ being the identity element. Given a cycle $t = (i_1, \dots, i_m)$, we call the set $\{i_1, \dots, i_m\}$ the **support** of t, denoted by $\mathrm{supp}(t)$. The subgroup $\mathbb{Z}_2^n$ of B_n consists of elements of the form $b_I := \prod_{i \in I} b_i$ for $I \subset \{1, \dots, n\}$. Each element $b_I \sigma \in B_n$ can be written as a product of the form (unique up to reordering)

$$b_I \sigma = (b_{I_1} \sigma_1)(b_{I_2} \sigma_2) \dots (b_{I_k} \sigma_k),$$

where $\sigma \in \mathfrak{S}_n$ is a product of disjoint cycles $\sigma = \sigma_1 \dots \sigma_k$, and $I_a \subset \mathrm{supp}(\sigma_a)$ for each $1 \le a \le k$; $b_{I_a} \sigma_a$ is called a **signed cycle** of $b_I \sigma$. The **cycle-product** of each signed cycle $b_{I_a} \sigma_a$ is defined to be the element $\prod_{i \in I_a} b_i \in \mathbb{Z}_2$ (which can be conveniently thought of as a sign $+$ or $-$). Let m_i^+ (respectively, m_i^-) be the number of i-cycles of $b_I \sigma$ with associated cycle-product being $+$ (respectively, $-$). Then $\rho^+ = (i^{m_i^+})_{i \ge 1}$ and $\rho^- = (i^{m_i^-})_{i \ge 1}$ are partitions such that $|\rho^+| + |\rho^-| = n$. The pair of partitions (ρ^+, ρ^-) will be called the **type** of the element $b_I \sigma$.

The following is the basic fact on the conjugacy classes of B_n, cf. [**83**, I, Appendix B]. We leave the proof to the reader (Exercise 3.9).

Lemma 3.28. *Two elements of B_n are conjugate if and only if their types are the same.*

We shall denote by C_{ρ^+,ρ^-} the conjugacy class of type (ρ^+,ρ^-). Note that if $b_I\sigma \in C_{\rho^+,\rho^-}$, then C_{ρ^+,ρ^-} is even (respectively, odd) if $|I|$ is even (respectively, odd).

Example 3.29. Let $\tau = (1,2,3,4)(5,6,7)(8,9), \sigma = (1,3,8,6)(2,7,9)(4,5) \in \mathfrak{S}_{10}$. It is straightforward to check that both $x = ((+,+,+,-,+,+,+,-,+,-),\tau)$ and $y = ((+,-,-,-,+,-,-,-,+,-),\sigma)$ in B_{10} have the same type

$$(\rho^+,\rho^-) = ((3),(4,2,1)).$$

Thus, x is conjugate to y in B_{10}.

3.3.2. Split conjugacy classes in $\widetilde{B}_n$. A partition $\lambda = (\lambda_1,\dots,\lambda_\ell)$ of length ℓ is called **strict** if $\lambda_1 > \lambda_2 > \dots > \lambda_\ell$, and it is called **odd** if each part λ_i is odd. We denote by $\mathcal{SP}_n$ the set of all strict partitions of n, and by $\mathcal{OP}_n$ the set of all odd partitions of n. Moreover, we denote

$$\mathcal{SP} = \bigcup_{n\geq 0} \mathcal{SP}_n, \qquad \mathcal{OP} = \bigcup_{n\geq 0} \mathcal{OP}_n.$$

Recall $\mathcal{P}_n$ denotes the set of all partitions of n. Let

$$\begin{aligned} \mathcal{P}_n^+ &= \{\lambda \in \mathcal{P}_n \mid \ell(\lambda) \text{ is even}\}, \\ \mathcal{P}_n^- &= \{\lambda \in \mathcal{P}_n \mid \ell(\lambda) \text{ is odd}\}. \end{aligned}$$

Given $\sigma \in \mathfrak{S}_n$ of cycle type μ, we denote by

$$d(\sigma) = n - \ell(\mu).$$

For an (ordered) subset $I = \{i_1,i_2,\dots,i_m\}$ of $\{1,\dots,n\}$, we denote

$$a_I = a_{i_1 i_2 \dots i_m} = a_{i_1} a_{i_2} \dots a_{i_m}.$$

It follows that $p(a_I) \equiv |I| \mod 2$. If $I \cap J = \emptyset$, then $a_I a_J = z^{|I||J|} a_J a_I$. Also we can easily show by induction that

$$a_{i_1 i_2 \dots i_m} = z^{d(s)} a_{s(i_1)s(i_2)\dots s(i_m)} \tag{3.22}$$

for a permutation s such that s fixes the letters other than $i_1,i_2,\dots,i_m$.

We can write a general element of $\widetilde{B}_n$ as

$$z^k a_I s = z^k (a_{I_1} s_1) \dots (a_{I_q} s_q),$$

where $s = s_1 \dots s_q$ is a cycle decomposition of s and $I_j \subset \operatorname{supp}(s_j)$ for each j. We denote by J^c the complement of a subset $J \subseteq \{1,\dots,n\}$.

Lemma 3.30. *Let $a_I s = (a_{I_1} s_1) \dots (a_{I_q} s_q)$ be an element of $\widetilde{B}_n$ in its cycle decomposition. Let $J \subseteq supp(s_1) \cap I_1^c$. Then*

$$(a_J s_1)(a_I s)(a_J s_1)^{-1} = z^{d(s_1)+|J||I|} a_I s.$$

Proof. Observe that $a_I^2 = z^{(|I|-1)|I|/2}$ for any subset I. For $k > 1$ we have

$$(a_J s_1)(a_{I_k} s_k)(a_J s_1)^{-1} = z^{|J||I_k|} a_{I_k} s_k.$$

Therefore it remains to see that

$$\begin{aligned}(a_J s_1)(a_{I_1} s_1)(a_J s_1)^{-1} &= z^{(|J|-1)|J|/2} a_J a_{s_1(I_1 \cup J)} s_1 \\ &= z^{(|J|-1)|J|/2+d(s_1)} a_J a_{(I_1 \cup J)} s_1 \\ &= z^{(|J|-1)|J|/2+d(s_1)+|J||I_1|} a_J^2 a_{I_1} s_1 \\ &= z^{|J||I_1|+d(s_1)} a_{I_1} s_1,\end{aligned}$$

where we have used the fact that $\mathrm{supp}(s_1) \supseteq I_1 \cup J$ and (3.22). □

Theorem 3.31. *The conjugacy class C_{ρ^+,ρ^-} in B_n splits if and only if*

(1) for even C_{ρ^+,ρ^-}, we have $\rho^+ \in \mathcal{OP}_n$ and $\rho^- = \emptyset$,

(2) for odd C_{ρ^+,ρ^-}, we have $\rho^+ = \emptyset$ and $\rho^- \in \mathcal{SP}_n^-$.

Proof. Assume that C_{ρ^+,ρ^-} is an even conjugacy class such that $\rho^+ \notin \mathcal{OP}$. Then $\theta_n^{-1}(C_{\rho^+,\rho^-})$ contains an element $a_I s$ with a signed cycle decomposition of the form

$$a_I s = s_1 (a_{I_2} s_2) \dots (a_{I_p} s_p),$$

where $s_1 = (1, 2, \dots, r)$ for $r = 2k$ even, $I_1 = \emptyset$ and $|I|$ is even. Consider the element $x = a_{12\dots r}(1, 2, \dots, r) \in \widetilde{B}_n$. By Lemma 3.30 we have

$$x(a_I s)x^{-1} = z^{(2k-1)+2k\cdot|I|} a_I s = z a_I s.$$

Therefore, if an even conjugacy class C_{ρ^+,ρ^-} splits, then $\rho^+ \in \mathcal{OP}$.

Assume that C_{ρ^+,ρ^-} is an even conjugacy class such that $\rho^- \neq \emptyset$. Then ρ^- contains at least two parts, and $\theta_n^{-1}(C_{\rho^+,\rho^-})$ contains an element of the form

$$a_I s = (a_{i_1} s_1)(a_{i_2} s_2)(a_{I_3} s_3) \dots (a_{I_p} s_p),$$

where $i_1 \in \mathrm{supp}(s_1), i_2 \in \mathrm{supp}(s_2)$. Then

$$\begin{aligned}&(a_{i_1} s_1)^{-1} a_I s (a_{i_1} s_1) \\ &= (a_{i_1} s_1)^{-1} (a_{i_1} s_1 a_{i_2} s_2)(a_{i_1} s_1)(a_{I_3} s_3) \dots (a_{I_p} s_p) \\ &= a_{i_2} s_2 (a_{i_1} s_1)(a_{I_3} s_3) \dots (a_{I_p} s_p) \\ &= z(a_{i_1} s_1)(a_{i_2} s_2)(a_{I_3} s_3) \dots (a_{I_p} s_p) \\ &= z a_I s.\end{aligned}$$

Hence, if an even conjugacy class C_{ρ^+,ρ^-} splits, then $\rho^- = \emptyset$. Together with the above, we have shown that an even split conjugacy class should satisfy (1).

Now assume that C_{ρ^+,ρ^-} is an odd conjugacy class such that $\rho^+ \neq \emptyset$. Then, $\theta_n^{-1}(C_{\rho^+,\rho^-})$ contains an element $a_I s$ with signed cycle decomposition of the form

$$a_I s = (s_1)(a_{I_2} s_2)\ldots(a_{I_q} s_q),$$

where $I_1 = \emptyset$ and $|I|$ is odd. Let $J = \mathrm{supp}(s_1)$. Then, by Lemma 3.30,

$$(a_J s_1)(a_I s)(a_J s_1)^{-1} = z^{(|J|-1)+|I||J|} a_I s = z a_I s,$$

since $|I|$ is odd. Hence if C_{ρ^+,ρ^-} is an odd split conjugacy class then $\rho^+ = \emptyset$.

Next, assume that C_{ρ^+,ρ^-} is an odd conjugacy class such that ρ^- contains two identical parts. Then $\theta_n^{-1}(C_{\rho^+,\rho^-})$ contains an element of the form

$$a_I s = (a_{i_1}(i_1, i_2, \ldots, i_k))(a_{j_1}(j_1, j_2, \ldots, j_k))\ldots(a_{I_q} s_q).$$

Consider the element $t = (i_1, j_1)\ldots(i_k, j_k)$. We have

$$t(a_I s)t^{-1} = a_{t(I)} s = a_{j_1}(j_1, \ldots, j_k) a_{i_1}(i_1, \ldots, i_k)\ldots = z a_I s.$$

Hence, if an odd conjugacy class C_{ρ^+,ρ^-} splits then $\rho^- \in \mathcal{SP}$. Together with the above, we have shown that an odd split conjugacy class satisfies (2).

Assume that (1) holds. Then $\theta_n^{-1}(C_{\rho^+,\emptyset})$ contains an element $s = s_1 \ldots s_q \in \mathfrak{S}_n$ with each s_i being an odd cycle. Suppose on the contrary the conjugacy class $C_{\rho^+,\emptyset}$ does not split, that is, $(a_J t)s(a_J t)^{-1} = zs$ for some element $a_J t$. Then $(a_J t)s = zs(a_J t)$, and so $z a_{s(J)} = a_J$, which implies that $\mathrm{supp}(s) \subseteq J$. On the other hand, $a_{s(J)} = z^{d(s)} a_J = a_J$ by (3.22), since s is a product of disjoint odd cycles. This contradiction implies that the conjugacy class $C_{\rho^+,\emptyset}$ splits.

Now assume that we are given an odd conjugacy class $C_{\emptyset,\rho^-}$ with ρ^- strict as specified in (2). Thus $\theta_n^{-1}(C_{\emptyset,\rho^-})$ contains an element $a_I s = (a_{i_1} s_1)\ldots(a_{i_q} s_q)$, where q is odd and $i_k \in \mathrm{supp}(s_k)$. Suppose on the contrary that the conjugacy class $C_{\emptyset,\rho^-}$ does not split, that is, $(a_J t)(a_I s)(a_J t)^{-1} = z(a_I s)$ for some element $a_J t$. It follows that t commutes with s and hence $t = s_1^{r_1} \ldots s_q^{r_q}$ for $0 \le r_i < \mathrm{order}(s_i)$, since the cycle type of s is a strict partition. Write $a_J t = (a_{J_1} t_1)\ldots(a_{J_q} t_q)$ with $t_m = s_m^{r_m}$. As in the proof of Lemma 3.30 we have

$$\begin{aligned}
&(a_J t)(a_{i_1} s_1)(a_J t)^{-1} \\
&= (a_{J_1} t_1)(a_{J_2} t_2)\cdots(a_{J_q} t_q)(a_{i_1} s_1)((a_{J_2} t_2)\cdots(a_{J_q} t_q))^{-1}(a_{J_1} t_1)^{-1} \\
&= z^{|J|-|J_1|}(a_{J_1} t_1)(a_{i_1} s_1)(a_{J_1} t_1)^{-1},
\end{aligned}$$

which must equal $a_{i_1} s_1$ up to a power of z. Set $(a_{J_1} t_1)(a_{i_1} s_1)(a_{J_1} t_1)^{-1} = z^* a_{i_1} s_1$ where $*$ is 0 or 1. We claim that $*$ is always 0. Note that $a_{J_1} a_{t_1(i_1)} = z^* a_{i_1} a_{s_1(J_1)}$, and so J_1 differs from $s_1(J_1)$ by one element. Without loss of generality, we let $i_1 = 1$, $s_1 = (1, 2, \ldots, k)$, $J_1 = \{1, 2, \ldots, r\}$ with $0 \le r < k$. Then we have

$$a_1 \ldots a_r \cdot a_{r_1+1} = a_{J_1} a_{t_1(i_1)} = z^* a_{i_1} a_{s_1(J_1)} = z^* a_1 \cdot a_2 \ldots a_{r+1},$$

which implies that $r_1 = r$ and $* = 0$. Therefore, $(a_J t)(a_{i_1} s_1)(a_J t)^{-1} = z^{|J|-|J_1|} a_{i_1} s_1$, and similarly we have

$$\begin{aligned}(a_J t)(a_I s)(a_J t)^{-1} &= (a_J t)(a_{I_1} s_1)\dots(a_{I_q} s_q)(a_J t)^{-1} \\ &= z^{q|J|-(|J_1|+\dots+|J_q|)} a_I s = z^{(q-1)|J|} a_I s = a_I s,\end{aligned}$$

since q is odd. This is a contradiction. □

For $\alpha \in \mathcal{OP}_n$ we let $\mathcal{C}_\alpha^+$ be the split conjugacy class in $\widetilde{B}_n$ which lies in $\theta_n^{-1}(C_{\alpha,\emptyset})$ and contains a permutation in $\mathfrak{S}_n$ of cycle type α. Then $z\mathcal{C}_\alpha^+$ is the other conjugacy class in $\theta_n^{-1}(C_{\alpha,\emptyset})$, which will be denoted by $\mathcal{C}_\alpha^-$. Recall from Appendix A that z_α denotes the order of the centralizer of an element of type α in $\mathfrak{S}_n$. The order of the centralizer of an element of a given cycle type is known explicitly for B_n (and actually for any wreath product, cf. Macdonald [**83**, I, Appendix B]). The next lemma follows from this classical fact.

Lemma 3.32. *Let $\alpha \in \mathcal{OP}_n$. The order of the centralizer of an element in the conjugacy class $\mathcal{C}_\alpha^+$ of $\widetilde{B}_n$ is $2^{1+\ell(\alpha)} z_\alpha$. Thus, the order of the conjugacy class $\mathcal{C}_\alpha^+$ is $|\mathcal{C}_\alpha^+| = n! 2^{n-\ell(\alpha)} z_\alpha^{-1}$.*

3.3.3. A ring structure on R^-.

Let us introduce a basic example of superalgebras.

Definition 3.33. The **Clifford algebra** $\mathcal{C}_n$ is the $\mathbb{C}$-algebra generated by c_i, for $1 \le i \le n$, subject to relations

$$c_i^2 = 1, \qquad c_i c_j = -c_j c_i \ (i \neq j). \tag{3.23}$$

Letting $|c_i| = \bar{1}$, $\forall i$, $\mathcal{C}_n$ becomes a superalgebra also called the **Clifford superalgebra**.

The symmetric group $\mathfrak{S}_n$ acts as automorphisms on the algebra $\mathcal{C}_n$ naturally. We will refer to the semi-direct product $\mathcal{H}_n := \mathcal{C}_n \rtimes \mathbb{C}\mathfrak{S}_n$ as the **Hecke-Clifford algebra**, where

$$\sigma c_i = c_{\sigma(i)} \sigma, \quad \forall \sigma \in \mathfrak{S}_n. \tag{3.24}$$

Note that the algebra $\mathcal{H}_n$ is naturally a superalgebra by letting each $\sigma \in \mathfrak{S}_n$ be even and each c_i be odd.

Recall the group Π_n from (3.21). The quotient algebra $\mathbb{C}\Pi_n/\langle z+1\rangle$ of the group algebra $\mathbb{C}\Pi_n$ by the ideal generated by $z+1$ is isomorphic to the Clifford superalgebra $\mathcal{C}_n$ with the identification $\bar{a}_i = c_i, 1 \le i \le n$. Hence, we have an isomorphism of superalgebras

$$\mathbb{C}\widetilde{B}_n/\langle z+1\rangle \cong \mathcal{H}_n. \tag{3.25}$$

Recall our convention from Section 3.1.1 that a module of a superalgebra is always understood to be $\mathbb{Z}_2$-graded. We shall denote by $\mathcal{H}_n$-mod the category of modules of the superalgebra $\mathcal{H}_n$ (with morphisms of degree one allowed). Thanks to the superalgebra isomorphism (3.25), $\mathcal{H}_n$-mod is equivalent to the category of spin $\widetilde{B}_n$-modules. (Recall from Section 3.1.4 that a spin $\widetilde{B}_n$-module means a $\mathbb{Z}_2$-graded $\widetilde{B}_n$-module M on which $z \in \widetilde{B}_n$ acts as $-I$.) We shall not distinguish these two isomorphic categories below, and the latter one has the advantage that one can apply the standard arguments from the theory of finite groups directly.

Denote by $R_n^- = [\mathcal{H}_n\text{-mod}]$ the Grothendieck group of $\mathcal{H}_n$-mod (cf. Section 3.1.1). As in the usual (ungraded) case, we may replace the isomorphism classes of modules by their characters and then regard R_n^- as the free abelian group with a basis consisting of the characters of the simple spin $\widetilde{B}_n$-modules. Let

$$R^- := \bigoplus_{n=0}^{\infty} R_n^-, \qquad R_{\mathbb{Q}}^- := \bigoplus_{n=0}^{\infty} \mathbb{Q} \otimes_{\mathbb{Z}} R_n^-,$$

where it is understood that $R_0^- = \mathbb{Z}$. We shall define a ring structure on R^- as follows.

Denote by $\widetilde{B}_{m,n}$ the subgroup of $\widetilde{B}_{m+n}$ generated by $\mathfrak{S}_m \times \mathfrak{S}_n$ and Π_{m+n}. Then $\widetilde{B}_{m,n}$ can be identified with the quotient group $\widetilde{B}_m \times \widetilde{B}_n / \{(1,1),(z,z)\}$, where $\widetilde{B}_m \times \widetilde{B}_n$ denotes the product group in the super sense, i.e., elements from $\widetilde{B}_m$ and $\widetilde{B}_n$ supercommute with each other.

Given $\varphi \in R_m^-$ and $\psi \in R_n^-$, we define the spin character $\varphi \hat{\times} \psi$ of $\widetilde{B}_{m,n}$ by letting

$$\varphi \hat{\times} \psi \overline{(x,y)} = \varphi(x)\psi(y),$$

where $\overline{(x,y)}$ is the image of (x,y) in $\widetilde{B}_{m,n} \cong (\widetilde{B}_m \times \widetilde{B}_n)/\{(1,1),(z,z)\}$. We define a product on R^- by

$$\varphi \cdot \psi = \mathrm{Ind}_{\widetilde{B}_{m,n}}^{\widetilde{B}_{m+n}} (\varphi \hat{\times} \psi),$$

where $\varphi \in R_m^-$, $\psi \in R_n^-$ for all m,n, and Ind denotes the induced character. It follows from the properties of the induced characters that the multiplication on R^- is commutative and associative.

Remark 3.34. Equivalently, the multiplication in R^- can be described as follows. Let $\mathcal{H}_{m,n}$ be the subalgebra of $\mathcal{H}_{m+n}$ generated by $\mathcal{C}_{m+n}$ and $\mathfrak{S}_m \times \mathfrak{S}_n$. For $M \in \mathcal{H}_m$-mod and $N \in \mathcal{H}_n$-mod, $M \otimes N$ is naturally an $\mathcal{H}_{m,n}$-module, and we define the product

$$[M] \cdot [N] = [\mathcal{H}_{m+n} \otimes_{\mathcal{H}_{m,n}} (M \otimes N)],$$

and then extend it by $\mathbb{Z}$-bilinearity.

For $\varphi \in R_n^-$, we shall write $\varphi_\alpha = \varphi(x)$ for $x \in \mathcal{C}_\alpha^+, \alpha \in \mathcal{OP}_n$; hence $\varphi(y) = -\varphi_\alpha$ for $y \in \mathcal{C}_\alpha^-$. Given two partitions α, β, we let $\alpha \cup \beta$ denote the partition obtained by collecting the parts of α and β together and rearranging them in descending order.

Lemma 3.35. *Let* $\varphi \in R_m^-$, $\psi \in R_n^-$, *and* $\gamma \in \mathcal{OP}_{m+n}$. *Then*

$$(\varphi \cdot \psi)_\gamma = \sum_{\substack{\alpha \in \mathcal{OP}_m, \beta \in \mathcal{OP}_n \\ \alpha \cup \beta = \gamma}} \frac{z_\gamma}{z_\alpha z_\beta} \varphi_\alpha \psi_\beta.$$

Proof. It can be checked directly that

$$\mathcal{C}_\gamma^+ \cap \widetilde{B}_{m,n} = \bigcup_{\alpha, \beta \in \mathcal{OP}, \alpha \cup \beta = \gamma} \mathcal{C}_\alpha^+ \hat{\times} \mathcal{C}_\beta^+.$$

It follows from this, Lemma 3.32, and the standard induced character formula for finite groups that

$$\begin{aligned}
(\varphi \cdot \psi)_\gamma &= \frac{|\widetilde{B}_{m+n}|}{|\widetilde{B}_{m,n}| \cdot |\mathcal{C}_\gamma^+|} \sum_{w \in \mathcal{C}_\gamma^+} (\varphi \hat{\times} \psi)(w) \\
&= \frac{2^{1+\ell(\gamma)} z_\gamma}{2^{m+n+1} m! n!} \sum_{\alpha, \beta \in \mathcal{OP}, \alpha \cup \beta = \gamma} \varphi_\alpha \psi_\beta |\mathcal{C}_\alpha^+| \cdot |\mathcal{C}_\beta^+| \\
&= \frac{z_\gamma}{m! n! 2^{m+n-\ell(\gamma)}} \sum_{\alpha, \beta \in \mathcal{OP}, \alpha \cup \beta = \gamma} 2^{m-\ell(\alpha)} z_\alpha^{-1} m! 2^{n-\ell(\beta)} z_\beta^{-1} n! \varphi_\alpha \psi_\beta \\
&= \sum_{\alpha, \beta \in \mathcal{OP}, \alpha \cup \beta = \gamma} \frac{z_\gamma}{z_\alpha z_\beta} \varphi_\alpha \psi_\beta,
\end{aligned}$$

where we have used $\ell(\gamma) = \ell(\alpha) + \ell(\beta)$. □

3.3.4. The characteristic map. Recall from (A.45) in Appendix A that the ring $\Gamma_\mathbb{Q} := \mathbb{Q} \otimes_\mathbb{Z} \Gamma$ has a basis given by the power-sum symmetric functions p_μ for $\mu \in \mathcal{OP}$. Moreover, $\Gamma_\mathbb{Q}$ is equipped with a bilinear form $\langle \cdot, \cdot \rangle$ given in (A.55).

We define the (spin) **characteristic map**

$$\mathrm{ch} : R_\mathbb{Q}^- \longrightarrow \Gamma_\mathbb{Q}$$

to be the linear map given by

$$\mathrm{ch}(\varphi) = \sum_{\alpha \in \mathcal{OP}_n} z_\alpha^{-1} \varphi_\alpha p_\alpha, \qquad \text{for } \varphi \in R_n^-, n \geq 0.$$

Proposition 3.36. *The characteristic map* $\mathrm{ch} : \mathbb{R}_\mathbb{Q}^- \to \Gamma_\mathbb{Q}$ *is an algebra homomorphism.*

Proof. For $\varphi \in R_m^-$, $\psi \in R_n^-$, we have by Lemma 3.35 that

$$\begin{aligned}
\mathrm{ch}(\varphi \cdot \psi) &= \sum_{\gamma \in \mathcal{OP}} z_\gamma^{-1} (\varphi \cdot \psi)_\gamma p_\gamma \\
&= \sum_\gamma \sum_{\alpha, \beta \in \mathcal{OP}, \alpha \cup \beta = \gamma} z_\gamma^{-1} \frac{z_\gamma}{z_\alpha z_\beta} \varphi_\alpha \psi_\beta p_\gamma = \mathrm{ch}(\varphi) \mathrm{ch}(\psi),
\end{aligned}$$

where we have used $p_\gamma = p_\alpha p_\beta$. □

Denote by

$$\langle [M], [N] \rangle = \dim \mathrm{Hom}_{\widetilde{B}_n}(M, N)$$

for spin $\widetilde{B}_n$-modules M, N. This defines a bilinear form $\langle \cdot, \cdot \rangle$ on R_n^- by $\mathbb{Z}$-bilinearity. The $\mathrm{Hom}_{\widetilde{B}_n}$ here can be either understood as in the category of $\widetilde{B}_n$-modules (with degree one morphisms allowed) or in the category of (non-graded) $\widetilde{B}_n$-modules, and they give the same dimension. In light of super Schur's Lemma 3.4, this is reduced to a straightforward verification in the case when $M \cong N$ is simple of type $\mathtt{Q}$.

Then R^-, and hence also $R_{\mathbb{Q}}^-$, carry a symmetric bilinear form, still denoted by $\langle \cdot, \cdot \rangle$, which is induced from the ones on R_n^-, for all n, such that R_n^- and R_m^- are orthogonal whenever $n \neq m$.

Lemma 3.37. *For $\varphi, \psi \in R_n^-$, we have*

$$\langle \varphi, \psi \rangle = \sum_{\alpha \in \mathcal{OP}_n} 2^{-\ell(\alpha)} z_\alpha^{-1} \varphi_\alpha \psi_\alpha.$$

Proof. Note that $x \in \mathcal{C}_\alpha^+$ implies that $x^{-1} \in \mathcal{C}_\alpha^+$. Also note that $\varphi(x) = 0$ unless x is even and split. By Lemma 3.32 and applying the standard formula for the bilinear form on characters of a finite group, we have

$$\begin{aligned}
\langle \varphi, \psi \rangle &= \frac{1}{|\widetilde{B}_n|} \sum_{x \in \widetilde{B}_n} \varphi(x^{-1}) \psi(x) \\
&= \frac{1}{2^{n+1} n!} \sum_{\alpha \in \mathcal{OP}_n} \left(|\mathcal{C}_\alpha^+| \varphi_\alpha \psi_\alpha + |\mathcal{C}_\alpha^-| \varphi_\alpha \psi_\alpha \right) \\
&= \frac{1}{2^n n!} \sum_{\alpha \in \mathcal{OP}_n} 2^{n-\ell(\alpha)} z_\alpha^{-1} n! \varphi_\alpha \psi_\alpha \\
&= \sum_{\alpha \in \mathcal{OP}_n} 2^{-\ell(\alpha)} z_\alpha^{-1} \varphi_\alpha \psi_\alpha.
\end{aligned}$$

□

Proposition 3.38. *The characteristic map* $\mathrm{ch} : R_{\mathbb{Q}}^- \to \Gamma_{\mathbb{Q}}$ *is an isometry, i.e., it preserves the bilinear forms* $\langle \cdot, \cdot \rangle$ *on* $R_{\mathbb{Q}}^-$ *and* $\Gamma_{\mathbb{Q}}$.

Proof. This follows from a direct computation using Lemma 3.37 and (A.55) from Appendix A: for $\varphi, \psi \in R_n^-$,

$$\begin{aligned}
\langle \mathrm{ch}(\varphi), \mathrm{ch}(\psi) \rangle &= \sum_{\alpha, \beta \in \mathcal{OP}_n} z_\alpha^{-1} z_\beta^{-1} \langle p_\alpha, p_\beta \rangle \varphi_\alpha \psi_\beta \\
&= \sum_{\alpha \in \mathcal{OP}_n} z_\alpha^{-2} 2^{-\ell(\alpha)} z_\alpha \varphi_\alpha \psi_\alpha = \langle \varphi, \psi \rangle.
\end{aligned}$$

□

3.3.5. The basic spin module. The algebra $\mathcal{H}_n$ acts on the Clifford superalgebra $\mathcal{C}_n$ by the formulas

$$c_i.(c_{i_1}c_{i_2}\ldots) = c_i c_{i_1} c_{i_2}\ldots, \qquad \sigma.(c_{i_1}c_{i_2}\ldots) = c_{\sigma(i_1)}c_{\sigma(i_2)}\cdots,$$

for $\sigma \in \mathfrak{S}_n$. The $\mathcal{H}_n$-module $\mathcal{C}_n$ is called the **basic spin module** of $\widetilde{B}_n$. Let $\sigma = \sigma_1 \ldots \sigma_\ell \in \mathfrak{S}_n$ be a product of disjoint cycles with the cycle length of σ_i being μ_i, for $i = 1, \ldots, \ell$. If I is a union of some of the $\mathrm{supp}(\sigma_i)$'s, say $I = \mathrm{supp}(\sigma_{i_1}) \cup \ldots \cup \mathrm{supp}(\sigma_{i_s})$, then $\sigma(c_I) = (-1)^{\mu_{i_1}+\ldots+\mu_{i_s}-s} c_I$. If I is not such a union, then $\sigma(c_I)$ is not a scalar multiple of c_I.

Lemma 3.39. *The value of the character ξ^n of the basic spin $\widetilde{B}_n$-module at the conjugacy class $\mathcal{C}_\alpha^+$ is given by*

$$\xi^n_\alpha = 2^{\ell(\alpha)}, \qquad \alpha \in \mathcal{OP}_n. \tag{3.26}$$

Proof. Let $\alpha = (\alpha_1, \alpha_2, \ldots) \in \mathcal{OP}_n$ and $\ell(\alpha) = \ell$. Let $\sigma = \sigma_1 \ldots \sigma_\ell$ be an element in $\mathfrak{S}_n$ of cycle type α. The elements $c_I := \prod_{i \in I} c_i$ (which are defined up to a nonessential sign) for $I \subset \{1, \ldots, n\}$ form a basis of the basic spin module $\mathcal{C}_n$. Observe that $\sigma c_I = c_I$ if I is a union of a subset of the supports $\mathrm{supp}(\sigma_p)$ for $1 \le p \le \ell(\alpha)$; otherwise σc_I is equal to $\pm c_J$ for some $J \neq I$. Hence the character value ξ^n_α, which is the trace of σ on $\mathcal{C}_n$, is equal to $2^{\ell(\alpha)}$. □

Below we will freely use the statements in Appendix A.3 on Schur Q-functions Q_λ. Recall the symmetric function q_n defined by the generating function (A.42), which is

$$\sum_{n \ge 0} q_n t^n = \prod_{i \ge 1} \frac{1 + x_i t}{1 - x_i t}.$$

Lemma 3.40. *Let $n \ge 1$. We have*

(1) $\mathrm{ch}(\xi^n) = q_n$*;*

(2) $\langle \xi^n, \xi^n \rangle = 2$*;*

(3) *the basic spin $\widetilde{B}_n$-module $\mathcal{C}_n$ is simple of type* Q.

Proof. (1) It follows by the definition of ch, Lemma 3.39, and (A.47) that

$$\mathrm{ch}(\xi^n) = \sum_{\alpha \in \mathcal{OP}_n} 2^{\ell(\alpha)} z_\alpha^{-1} p_\alpha = q_n.$$

(2) Note that $p_k(1,0,0,\ldots) = 1$ for each $k \ge 1$, and hence $p_\alpha(1,0,0,\ldots) = 1$. Also, it follows by definition that $q_n(1,0,0,\ldots) = 2$, for $n \ge 1$. Thus, specializing the identity $\sum_{\alpha \in \mathcal{OP}_n} 2^{\ell(\alpha)} z_\alpha^{-1} p_\alpha = q_n$ at $(x_1, x_2, x_3, \ldots) = (1,0,0,\ldots)$, we obtain $\sum_{\alpha \in \mathcal{OP}_n} 2^{\ell(\alpha)} z_\alpha^{-1} = 2$. We compute, by Lemmas 3.37 and 3.39, that

$$\langle \xi^n, \xi^n \rangle = \sum_{\alpha \in \mathcal{OP}_n} 2^{-\ell(\alpha)} z_\alpha^{-1} (2^{\ell(\alpha)})^2 = \sum_{\alpha \in \mathcal{OP}_n} 2^{\ell(\alpha)} z_\alpha^{-1} = 2.$$

(3) As the $\widetilde{B}_n$-module $\mathcal{C}_n$ is semisimple, to prove (3) it suffices to exhibit an odd automorphism of the $\widetilde{B}_n$-module $\mathcal{C}_n$ by (2) and super Schur's Lemma 3.4. Indeed, the right multiplication with the element $\frac{1}{\sqrt{n}}(c_1+\ldots+c_n)$ provides such an automorphism of order 2. $\square$

Proposition 3.41. *The characteristic map* ch : $R_{\mathbb{Q}}^- \to \Gamma_{\mathbb{Q}}$ *is an isomorphism of graded vector spaces.*

Proof. Since $\Gamma_{\mathbb{Q}}$ is generated by q_r for $r \geq 1$ and ch is an algebra homomorphism by Proposition 3.36, ch is surjective by Lemma 3.40. By Proposition 3.8 and Theorem 3.31, the dimension of $\mathbb{Q}\otimes R_n^-$ is $|\mathcal{SP}_n|$, which is the same as $\dim \Gamma_{\mathbb{Q}}^n$, for each n. Hence ch is an isomorphism. $\square$

3.3.6. The irreducible characters. Using the algebra structure on R^-, we define the elements ξ^λ for a strict partition λ by the following recursive relations:

$$\xi^{(\lambda_1,\lambda_2)} = \xi^{\lambda_1}\xi^{\lambda_2} + 2\sum_{i=1}^{\lambda_2}(-1)^i\xi^{\lambda_1+i}\xi^{\lambda_2-i}, \tag{3.27}$$

$$\xi^{\lambda} = \sum_{j=2}^{k}(-1)^j\xi^{(\lambda_1,\lambda_j)}\xi^{(\lambda_2,\ldots,\hat{\lambda}_j,\ldots,\lambda_k)}, \quad \text{for } k=\ell(\lambda) \text{ even}, \tag{3.28}$$

$$\xi^{\lambda} = \sum_{j=1}^{k}(-1)^{j-1}\xi^{\lambda_j}\xi^{(\lambda_1,\ldots,\hat{\lambda}_j,\ldots,\lambda_k)}, \quad \text{for } k=\ell(\lambda) \text{ odd}. \tag{3.29}$$

We emphasize that these are precisely the same recursive relations for the Schur Q-functions Q_λ (see (A.51)).

Lemma 3.42. *We have* $\mathrm{ch}(\xi^\lambda) = Q_\lambda$ *and* $\langle \xi^\lambda, \xi^\mu\rangle = 2^{\ell(\lambda)}\delta_{\lambda\mu}$, *for* $\lambda,\mu \in \mathcal{SP}$.

Proof. By Lemma 3.40, we have $\mathrm{ch}(\xi^n) = q_n = Q_{(n)}$. The general case of the first identity follows since ch is a ring homomorphism, and, in addition, ξ^λ and Q_λ are obtained from ξ_n's and q_n's, respectively, by the same recursive relations.

The second identity follows from the fact that ch is an isometry and the formula $\langle Q_\lambda, Q_\mu\rangle = 2^{\ell(\lambda)}\delta_{\lambda\mu}$ from (A.57). $\square$

Recalling the notation $\delta(\lambda)$ from (1.52) we define

$$\zeta^\lambda := 2^{-\frac{\ell(\lambda)-\delta(\lambda)}{2}}\xi^\lambda, \qquad \text{for } \lambda \in \mathcal{SP}_n.$$

Lemma 3.43. *The element* ζ^λ *lies in* R_n^-, *for* $\lambda \in \mathcal{SP}_n$.

Proof. We proceed by induction on $\ell(\lambda)$. For $\ell(\lambda)=1$, it is clear. Since $\mathcal{C}_n$ is a simple spin $\widetilde{B}_n$-module of type $\mathtt{Q}$ by Lemma 3.40, the induced module with character $\xi^m\xi^n$ is a sum of two isomorphic copies of a genuine module (the two odd automorphisms of ξ^m and ξ^n give rise to an even automorphism of order 2); that is,

$\frac{1}{2}\xi^m\xi^n \in R^-$. Hence, for $\ell(\lambda)=2$, $\frac{1}{2}\xi^{(\lambda_1,\lambda_2)}$ lies in R^- by (3.27). The general case follows easily by induction using the recursive relations (3.28) and (3.29). □

Corollary 3.44. *For strict partitions λ,μ, we have*

$$\begin{aligned}\langle\zeta^\lambda,\zeta^\lambda\rangle &= \begin{cases}1 & \text{for } \ell(\lambda) \text{ even}\\ 2 & \text{for } \ell(\lambda) \text{ odd},\end{cases}\\ \langle\zeta^\lambda,\zeta^\mu\rangle &= 0, \quad \text{for } \lambda\neq\mu.\end{aligned}$$

Corollary 3.45. *For each $\lambda\in\mathcal{SP}_n$, we have*

$$Q_\lambda = 2^{\frac{\ell(\lambda)-\delta(\lambda)}{2}}\sum_{\alpha\in\mathcal{OP}_n} z_\alpha^{-1}\zeta^\lambda_\alpha p_\alpha.$$

Also, for each $\alpha\in\mathcal{OP}_n$, we have

$$p_\alpha = \sum_{\lambda\in\mathcal{SP}_n} 2^{-\frac{\ell(\lambda)+\delta(\lambda)}{2}-\ell(\alpha)}\zeta^\lambda_\alpha Q_\lambda.$$

Proof. The first identity follows from Lemma 3.42 and the definitions of ζ^λ and ch.

Write $p_\alpha = \sum_{\lambda\in\mathcal{SP}_n} a^\lambda_\alpha Q_\lambda$ for some scalars a^λ_α. Recall from (A.55) and (A.57) that $\langle Q_\lambda, Q_\mu\rangle = 2^{\ell(\lambda)}\delta_{\lambda\mu}$ and $\langle p_\alpha,p_\beta\rangle = 2^{-\ell(\alpha)}z_\alpha\delta_{\alpha\beta}$. Then

$$a^\lambda_\alpha = 2^{-\ell(\lambda)}\langle p_\alpha, Q_\lambda\rangle = 2^{-\ell(\lambda)}2^{\frac{\ell(\lambda)-\delta(\lambda)}{2}}z_\alpha^{-1}\zeta^\lambda_\alpha\langle p_\alpha,p_\alpha\rangle = 2^{-\frac{\ell(\lambda)+\delta(\lambda)}{2}-\ell(\alpha)}\zeta^\lambda_\alpha,$$

where the second equality above uses the first identity of the corollary. □

Theorem 3.46. *Let $\lambda\in\mathcal{SP}_n$ and $\ell(\lambda)=\ell$. Then ζ^λ is the character of a simple spin $\widetilde{B}_n$-module (which is to be denoted by D^λ). Moreover, the degree of ζ^λ is equal to*

$$2^{n-\frac{\ell-\delta(\lambda)}{2}}\frac{n!}{\lambda_1!\ldots\lambda_\ell!}\prod_{i<j}\frac{\lambda_i-\lambda_j}{\lambda_i+\lambda_j}.$$

Proof. Since $\zeta^\lambda\in R^-$ by Lemma 3.43 and $\langle\zeta^\lambda,\zeta^\lambda\rangle = 1$ for $\ell(\lambda)$ even by Corollary 3.44, ζ^λ or $-\zeta^\lambda$, for $\ell(\lambda)$ even, is a simple character of type $\mathtt{M}$. By Corollary 3.44, these simple characters are distinct. A simple count using Proposition 3.8 and Theorem 3.31 implies that these are all simple characters of type $\mathtt{M}$. Now since $\langle\zeta^\lambda,\zeta^\lambda\rangle = 2$ for $\ell(\lambda)$ odd by Corollary 3.44, we have two possibilities: (a) either ζ^λ or $-\zeta^\lambda$ is simple of type $\mathtt{Q}$, or (b) ζ^λ is of the form $\pm\zeta^\mu\pm\zeta^\nu$ with both ζ^μ and ζ^ν being simple of type $\mathtt{M}$, which means both $\ell(\mu)$ and $\ell(\nu)$ are even. Case (b) cannot occur; otherwise, it would contradict the linear independence of ζ^λ for all $\lambda\in\mathcal{SP}_n$. Hence these must be all type $\mathtt{Q}$ simple characters by a counting argument using Proposition 3.8 and Theorem 3.31 again.

To show that ζ^λ rather than $-\zeta^\lambda$ for each $\lambda \in \mathcal{SP}_n$ is a character of a simple module, it suffices to know that $\zeta^\lambda_{(1^n)}$ or $\xi^\lambda_{(1^n)}$ is positive. To that end, we claim that

$$\xi^\lambda_{(1^n)} = 2^n \frac{n!}{\lambda_1! \dots \lambda_\ell!} \prod_{i<j} \frac{\lambda_i - \lambda_j}{\lambda_i + \lambda_j},$$

which is equivalent to the degree formula in the theorem.

The claim can be proved by induction on $\ell(\lambda)$ using the relations (3.27)–(3.29). The initial case with $\ell(\lambda) = 1$ is taken care of by Lemma 3.39, and the case with $\ell(\lambda) = 2$ can be checked directly by using (3.27). As the induction is elementary though lengthy (see [**57**]), we will simply remark here that the sought-for identity by using (3.28) and (3.29) precisely corresponds to the Laplacian-type expansion of the classical Pfaffian identity (cf. Macdonald [**83**, III.8, Ex. 5]):

$$\mathrm{Pf}\left(\frac{t_i - t_j}{t_i + t_j}\right)_{1 \le i,j \le 2n} = \prod_{1 \le i < j \le 2n} \frac{t_i - t_j}{t_i + t_j}.$$

We refer to [**57**, proof of Proposition 4.13] for detail. □

3.4. Schur-Sergeev duality for $\mathfrak{q}(n)$

In this section, we formulate a double centralizer property for the actions of the Lie superalgebra $\mathfrak{q}(n)$ and the algebra $\mathcal{H}_d$ on the tensor space $(\mathbb{C}^{n|n})^{\otimes d}$. We obtain a multiplicity-free decomposition of $(\mathbb{C}^{n|n})^{\otimes d}$ as a $U(\mathfrak{q}(n)) \otimes \mathcal{H}_d$-module. The characters of the simple $\mathfrak{q}(n)$-modules arising this way are shown to be Schur Q-functions (up to some 2-powers).

3.4.1. A double centralizer property.

Recall from Section 3.2 (by setting $m = n$) that we have a representation $(\Phi_d, V^{\otimes d})$ of $\mathfrak{gl}(n|n)$, and hence of its subalgebra $\mathfrak{q}(n)$; and we also have a representation $(\Psi_d, V^{\otimes d})$ of the symmetric group $\mathfrak{S}_d$. Moreover, the actions of $\mathfrak{gl}(n|n)$ and the symmetric group $\mathfrak{S}_d$ on $V^{\otimes d}$ commute with each other.

Note in addition that the Clifford superalgebra $\mathcal{C}_d$ acts on $V^{\otimes d}$, and we denote this action also by Ψ_d:

$$\Psi_d(c_i).(v_1 \otimes \dots \otimes v_d) = (-1)^{(|v_1| + \dots + |v_{i-1}|)} v_1 \otimes \dots \otimes v_{i-1} \otimes P v_i \otimes \dots \otimes v_d,$$

where P is given in (1.10), each $v_i \in V$ is assumed to be $\mathbb{Z}_2$-homogeneous, and $1 \le i \le n$.

Lemma 3.47. *Let $V = \mathbb{C}^{n|n}$. The actions of $\mathfrak{S}_d$ and $\mathcal{C}_d$ above give rise to a representation $(\Psi_d, V^{\otimes d})$ of $\mathcal{H}_d$. Moreover, the actions of $\mathfrak{q}(n)$ and $\mathcal{H}_d$ on $V^{\otimes d}$ commute with each other.*

Proof. To see that (3.24) holds, it suffices to check for $\sigma = (i, i+1), 1 \le i \le d-1$. We compute that

$$\begin{aligned}
&\Psi_d((i,i+1))\Psi_d(c_i).(v_1 \otimes \ldots \otimes v_d) \\
&\quad = \Psi_d((i,i+1))(-1)^{(|v_1|+\ldots+|v_{i-1}|)} v_1 \otimes \ldots \otimes P v_i \otimes v_{i+1} \otimes \ldots \otimes v_d \\
&\quad = (-1)^{(|v_i|+1)|v_{i+1}|}(-1)^{(|v_1|+\ldots+|v_{i-1}|)} v_1 \otimes \ldots \otimes v_{i+1} \otimes P v_i \otimes \ldots \otimes v_d, \\
&\Psi_d(c_{i+1})\Psi_d((i,i+1)).(v_1 \otimes \ldots \otimes v_d) \\
&\quad = \Psi_d(c_{i+1})(-1)^{(|v_i|\cdot|v_{i+1}|)} v_1 \otimes \ldots \otimes v_{i+1} \otimes v_i \otimes \ldots \otimes v_d \\
&\quad = (-1)^{(|v_i|\cdot|v_{i+1}|)}(-1)^{(|v_1|+\ldots+|v_{i-1}|+|v_{i+1}|)} v_1 \otimes \ldots \otimes v_{i+1} \otimes P v_i \otimes \ldots \otimes v_d.
\end{aligned}$$

Hence, we have $\Psi_d((i,i+1))\Psi_d(c_i) = \Psi_d(c_{i+1})\Psi_d((i,i+1))$. This further implies that $\Psi_d((i,i+1))\Psi_d(c_{i+1}) = \Psi_d(c_i)\Psi_d((i,i+1))$. A similar calculation shows that $\Psi_d((j,j+1))\Psi_d(c_i) = \Psi_d(c_i)\Psi_d((j,j+1))$ for $j \neq i, i-1$.

By the definitions of $\mathfrak{q}(n)$ and of $\Psi_d(c_i)$ via P, the action of $\mathfrak{q}(n)$ commutes with the action of c_i for $1 \le i \le d$. Since $\mathfrak{gl}(n|n)$ commutes with $\mathfrak{S}_d$, so does the subalgebra $\mathfrak{q}(n)$ of $\mathfrak{gl}(n|n)$. Hence, the action of $\mathfrak{q}(n)$ commutes with the action of $\mathcal{H}_d$ on $V^{\otimes d}$. □

Theorem 3.48. *The images $\Phi_d(U(\mathfrak{q}(n)))$ and $\Psi_d(\mathcal{H}_d)$ satisfy the double centralizer property, i.e.,*

$$\begin{aligned}
\Phi_d(U(\mathfrak{q}(n))) =& \mathrm{End}_{\mathcal{H}_d}(V^{\otimes d}), \\
& \mathrm{End}_{\mathfrak{q}(n)}(V^{\otimes d}) = \Psi_d(\mathcal{H}_d).
\end{aligned}$$

Proof. Let $\mathfrak{g} = \mathfrak{q}(n)$. We denote by $Q(V)$ the associative subalgebra of endomorphisms on V which (super)commute with the linear operator P. By Lemma 3.47, we have

$$\Phi_d(U(\mathfrak{g})) \subseteq \mathrm{End}_{\mathcal{H}_d}(V^{\otimes d}). \tag{3.30}$$

We shall proceed to prove that $\Phi_d(U(\mathfrak{g})) \supseteq \mathrm{End}_{\mathcal{H}_d}(V^{\otimes d})$.

By examining the action of $\mathcal{C}_d$ on $V^{\otimes d}$, we see that the natural isomorphism $\mathrm{End}(V)^{\otimes d} \cong \mathrm{End}(V^{\otimes d})$ allows us to identify $\mathrm{End}_{\mathcal{C}_d}(V^{\otimes d}) \equiv Q(V)^{\otimes d}$. As $\mathcal{H}_d = \mathcal{C}_d \rtimes \mathfrak{S}_d$, this further leads to the identification $\mathrm{End}_{\mathcal{H}_d}(V^{\otimes d}) \equiv \mathrm{Sym}^d(Q(V))$, the space of $\mathfrak{S}_d$-invariants in $Q(V)^{\otimes d}$.

Denote by Y_k, $1 \le k \le d$, the $\mathbb{C}$-span of the super-symmetrization

$$\varpi(x_1, \ldots, x_k) := \sum_{\sigma \in \mathfrak{S}_d} \sigma.(x_1 \otimes \ldots \otimes x_k \otimes I^{\otimes d-k}),$$

for all $x_i \in Q(V)$. Note that $Y_d = \mathrm{Sym}^d(Q(V)) \equiv \mathrm{End}_{\mathcal{H}_d}(V^{\otimes d})$.

Let $\tilde{x} = \Phi_d(x) = \sum_{i=1}^d I^{\otimes i-1} \otimes x \otimes I^{\otimes d-i}$, for $x \in Q(V)$, and denote by X_k, $1 \le k \le d$, the $\mathbb{C}$-span of $\tilde{x}_1 \ldots \tilde{x}_k$ for all $x_i \in Q(V)$.

By the same argument as in the proof of Theorem 3.10 (Schur-Sergeev duality for $\mathfrak{gl}(m|n)$), we have $Y_k \subseteq X_k$ for $1 \leq k \leq d$. This implies that

$$\mathrm{End}_{\mathcal{H}_d}(V^{\otimes d}) = Y_d \subseteq X_d \subseteq \Phi_d(U(\mathfrak{g})). \tag{3.31}$$

Combining (3.31) with (3.30), we conclude that $\Phi_d(U(\mathfrak{g})) = \mathrm{End}_{\mathcal{H}_d}(V^{\otimes d}) = \mathrm{End}_{\mathcal{B}}(V^{\otimes d})$, for $\mathcal{B} := \Psi_d(\mathcal{H}_d)$.

Note that the spin group algebra $\mathcal{H}_d$, and hence also $\mathcal{B}$, are semisimple superalgebras, and so the assumption of Proposition 3.5 is satisfied. Therefore, we have $\mathrm{End}_{U(\mathfrak{g})}(V^{\otimes d}) = \Psi_d(\mathcal{H}_d)$. □

3.4.2. The Sergeev duality. Recall from (1.52) that, for a partition λ of d with length $\ell(\lambda)$,

$$\delta(\lambda) = \begin{cases} 0, & \text{if } \ell(\lambda) \text{ is even,} \\ 1, & \text{if } \ell(\lambda) \text{ is odd.} \end{cases}$$

Recall further that D^λ stands for the simple $\mathcal{H}_d$-module with character ζ^λ (see Theorem 3.46), and $L(\lambda)$ for the simple $\mathfrak{q}(n)$-module with highest weight λ (see Section 2.1.6).

Theorem 3.49. *Let $V = \mathbb{C}^{n|n}$. As a $U(\mathfrak{q}(n)) \otimes \mathcal{H}_d$-module, we have*

$$V^{\otimes d} \cong \bigoplus_{\lambda \in \mathcal{SP}_d, \ell(\lambda) \leq n} 2^{-\delta(\lambda)} L(\lambda) \otimes D^\lambda. \tag{3.32}$$

Here, 2^{-1} has the same meaning as in Proposition 3.5.

Proof. Let $W = V^{\otimes d}$. Recall D^λ is of type $\mathtt{M}$ if and only if $\delta(\lambda) = 0$. It follows from Proposition 3.5, Theorem 3.48, and the semisimplicity of the superalgebra $\mathcal{H}_d$ that we have a multiplicity-free decomposition of the $(\mathfrak{q}(n), \mathcal{H}_d)$-module W:

$$W \cong \bigoplus_{\lambda \in \Omega_d(n)} 2^{-\delta(\lambda)} L^{[\lambda]} \otimes D^\lambda,$$

where $L^{[\lambda]}$ is some simple $\mathfrak{q}(n)$-module associated to λ whose highest weight (with respect to the standard Borel) is to be determined. Also to be determined is the index set $\Omega_d(n) = \{\lambda \in \mathcal{SP}_d \mid L^{[\lambda]} \neq 0\}$.

We shall identify a weight $\mu = \sum_{i=1}^n \mu_i \varepsilon_i$ occuring in W with a composition $\mu = (\mu_1, \ldots, \mu_n) \in \mathcal{CP}_d(n)$. We have the following weight space decomposition:

$$W = \bigoplus_{\mu \in \mathcal{CP}_d(n)} W_\mu, \tag{3.33}$$

where W_μ has a linear basis $e_{i_1} \otimes \ldots \otimes e_{i_d}$, with the indices satisfying the following equality of multisets:

$$\{i_1, \ldots, i_d\} = \{\underbrace{1, \ldots, 1, \overline{1}, \ldots, \overline{1}}_{\mu_1}, \ldots, \underbrace{n, \ldots, n, \overline{n}, \ldots, \overline{n}}_{\mu_n}\}.$$

We have an $\mathcal{H}_d$-module isomorphism:

(3.34) $$W_\mu \cong \mathrm{Ind}_{\mathbb{C}\mathfrak{S}_\mu}^{\mathcal{H}_d} \mathbf{1}_\mu = \mathcal{H}_d \otimes_{\mathbb{C}\mathfrak{S}_\mu} \mathbf{1}_\mu.$$

Recall the integers $\widehat{K}_{\lambda\mu}$ for a composition μ and a strict partition λ defined via the following symmetric function identity in (A.53) :

(3.35) $$q_\mu = \sum_{\lambda\in\mathcal{SP},\lambda\geq\mu} \widehat{K}_{\lambda\mu} Q_\lambda,$$

where $\widehat{K}_{\lambda\lambda} = 1$. To complete the proof of the theorem, we shall need the following.

Lemma 3.50. *Let $\mu = (\mu_1, \ldots, \mu_n)$ be a composition of d. We have the following decomposition of W_μ as an $\mathcal{H}_d$-module:*

$$W_\mu \cong \bigoplus_{\lambda\in\mathcal{SP},\lambda\geq\mu} 2^{\frac{\ell(\lambda)-\delta(\lambda)}{2}} \widehat{K}_{\lambda\mu} D^\lambda.$$

In particular, $\widehat{K}_{\lambda\mu} \in \mathbb{Z}_+$.

Proof. Decompose the $\mathcal{H}_d$-module W_μ into irreducibles:

(3.36) $$W_\mu \cong \bigoplus_{\lambda\in\mathcal{SP}_d} \tilde{K}_{\lambda\mu} D^\lambda, \qquad \text{for } \tilde{K}_{\lambda\mu} \in \mathbb{Z}_+.$$

Recall the characteristic map $\mathrm{ch}: R^- \to \Gamma_{\mathbb{Q}}$ from Section 3.3.4, where we have denoted $R^- = \bigoplus_{n\geq 0}[\mathcal{H}_n\text{-mod}]$. By Lemmas 3.42 and 3.43, for $\lambda \in \mathcal{SP}_d$,

(3.37) $$\mathrm{ch}([D^\lambda]) = 2^{-\frac{l(\lambda)-\delta(\lambda)}{2}} Q_\lambda.$$

It follows by the ring structure on R^- and Remark 3.34 that, for any composition $\mu = (\mu_1, \mu_2, \ldots)$ of d, $\mathrm{Ind}_{\mathbb{C}\mathfrak{S}_\mu}^{\mathcal{H}_d} \mathbf{1}_\mu$ is equal to the product of the basic spin characters $\xi^{\mu_1} \cdot \xi^{\mu_2} \ldots$, and, hence by Lemma 3.40,

(3.38) $$\mathrm{ch}(W_\mu) = \mathrm{ch}\big(\mathrm{Ind}_{\mathbb{C}\mathfrak{S}_\mu}^{\mathcal{H}_d} \mathbf{1}_\mu\big) = q_\mu.$$

Applying the characteristic map to both sides of (3.36), and using (3.37), (3.38), and (3.34), we obtain that

$$q_\mu = \sum_\lambda 2^{-\frac{\ell(\lambda)-\delta(\lambda)}{2}} \tilde{K}_{\lambda\mu} Q_\lambda.$$

It follows by a comparison of this identity with (3.35) that $\tilde{K}_{\lambda\mu} = 2^{\frac{\ell(\lambda)-\delta(\lambda)}{2}} \widehat{K}_{\lambda\mu}$. □

We return to the proof of Theorem 3.49. Since the simple $\mathfrak{q}(n)$-module $L^{[\lambda]}$, for $\lambda \in \mathfrak{Q}_d(n)$, has a weight space decomposition $L^{[\lambda]} = \bigoplus_{\mu\in\mathcal{CP}_d(n),\mu\leq\lambda} L_\mu^{[\lambda]}$, we must have $\ell(\lambda) \leq n$ and so $\lambda \in \mathcal{SP}_d(n)$. Now let $\lambda \in \mathcal{SP}_d$ with $\ell(\lambda) \leq n$. Clearly λ is a weight of $L^{[\lambda]}$ and, moreover, corresponds to a highest weight. Hence, we conclude that $L^{[\lambda]} = L(\lambda)$, the simple $\mathfrak{g}$-module of highest weight λ, and that $\mathfrak{Q}_d(n) = \{\lambda \in \mathcal{SP}_d \mid \ell(\lambda) \leq n\}$. This completes the proof of Theorem 3.49. □

3.4.3. The irreducible character formula. A **character** of a $\mathfrak{q}(n)$-module with weight space decomposition $M = \oplus_\mu M_\mu$ is by definition

$$\mathrm{ch} M = \mathrm{tr}\,|_M\, x_1^{H_1}\ldots x_n^{H_n} = \sum_{\mu=(\mu_1,\ldots,\mu_n)} \dim M_\mu \cdot x_1^{\mu_1}\ldots x_n^{\mu_n}.$$

Theorem 3.51. *Let λ be a strict partition of length $\leq n$. The character of the simple $\mathfrak{q}(n)$-module $L(\lambda)$ is given by* $\mathrm{ch}L(\lambda) = 2^{-\frac{\ell(\lambda)-\delta(\lambda)}{2}} Q_\lambda(x_1,\ldots,x_n)$.

Proof. By (3.33) and (3.34), we have $V^{\otimes d} \cong \oplus_{\mu\in\mathcal{CP}_d(n)} \mathrm{Ind}_{\mathfrak{S}_\mu}^{\mathcal{H}_d} \mathbf{1}_\mu$. Using (3.38) and applying the characteristic map ch and the trace operator $\mathrm{tr} x_1^{H_1}\ldots x_n^{H_n}$ simultaneously, which we will denote by ch^2, we obtain that

$$\mathrm{ch}^2(V^{\otimes d}) = \sum_{\mu\in\mathcal{P}_d,\ell(\mu)\leq n} q_\mu(z) m_\mu(x),$$

which can be written using (A.54) and (A.57) as

$$\mathrm{ch}^2(V^{\otimes d}) = \prod_{1\leq i\leq n, 1\leq j} \frac{1+x_i z_j}{1-x_i z_j} = \sum_{\lambda\in\mathcal{SP}} 2^{-\ell(\lambda)} Q_\lambda(x_1,\ldots,x_n) Q_\lambda(z),$$

where $z = \{z_1, z_2, \ldots\}$ is infinite. On the other hand, by applying ch^2 to (3.32) and using (3.37), we obtain that

$$\mathrm{ch}^2(V^{\otimes d}) = \sum_{\lambda\in\mathcal{SP}_d,\ell(\lambda)\leq n} 2^{-\delta(\lambda)} \mathrm{ch}L(\lambda)\cdot 2^{-\frac{\ell(\lambda)-\delta(\lambda)}{2}} Q_\lambda(z).$$

Now the theorem follows by comparing the above two identities and noting the linear independence of the $Q_\lambda(z)$'s. □

3.5. Exercises

Exercise 3.1. (1) Show that the ungraded algebra $|Q(n)|$ for the superalgebra $Q(n)$ is isomorphic to $M(n)\oplus M(n)$.

(2) Assume that a semisimple superalgebra $\mathcal{A}$ has m (respectively, q) non-isomorphic simple modules of type $\mathtt{M}$ (respectively, of type $\mathtt{Q}$). Show that the algebra $|\mathcal{A}|$ is semisimple and that the number of non-isomorphic simple $|\mathcal{A}|$-modules is equal to $m+2q$.

Exercise 3.2. Let $\widetilde{G}$ be a double cover of a finite group G with the projection $\theta : \widetilde{G}\to G$; see (3.6). Let $\mathcal{C}$ be a conjugacy class of G. Show that $\theta^{-1}(\mathcal{C})$ is either a single conjugacy class of $\widetilde{G}$ or splits into two conjugacy classes of $\widetilde{G}$.

Exercise 3.3. Define an involution α on a superalgebra $\mathcal{A}$ by letting $\alpha(a_k) = (-1)^k a_k$, for $a_k\in\mathcal{A}_k, k=0,1$. Given an $\mathcal{A}$-module N, we define another $\mathcal{A}$-module N' with the same underlying vector superspace as N but with the action of $\mathcal{A}$ twisted by α. Show that $N\cong N'$ as $\mathcal{A}$-modules.

Exercise 3.4. Let $\mathcal{A}$ be a superalgebra and N an $|\mathcal{A}|$-module. Regarding α in Exercise 3.3 as an involution on $|\mathcal{A}|$, we have an $|\mathcal{A}|$-module N' obtained from N by a twisted action via α. Prove:

(1) If M is a simple $\mathcal{A}$-module of type $\mathtt{M}$, then $|M| \cong |M|'$ as $|\mathcal{A}|$-modules and $|M|$ is a simple $|\mathcal{A}|$-module.

(2) If Q is a simple $\mathcal{A}$-module of type $\mathtt{Q}$, then there exists a simple $|\mathcal{A}|$-module N such that $Q \cong N \oplus N'$ and $N \ncong N'$ as $|\mathcal{A}|$-modules.

Exercise 3.5. Prove that the formula (3.9) satisfies the Coxeter relations for the symmetric group $\mathfrak{S}_d$.

Exercise 3.6. (1) Let λ be an $(m|n)$-hook partition with $m=0$. Show that $\lambda^\natural = \lambda'$, $\ell(\lambda') \le n$, and $\mathrm{hs}_\lambda(0;y) = s_{\lambda'}(y)$, for $y = \{y_1,\dots,y_n\}$.

(2) Show that the Schur-Sergeev duality (Theorem 3.11) for $m=0$ reduces to a version of $(\mathfrak{gl}(n),\mathfrak{S}_d)$-Schur duality twisted by the sign representation of $\mathfrak{S}_d$, i.e., as a $(\mathfrak{gl}(n),\mathfrak{S}_d)$-module, $(\mathbb{C}^{0|n})^{\otimes d} \cong \bigoplus_{\mu\in\mathcal{P}_d(n)} L(\mu)\otimes S^{\mu'}$. (Note here the well-known fact that $S^\mu \otimes \mathrm{sgn} \cong S^{\mu'}$.)

Exercise 3.7. For an $(m|n)$-hook partition λ, show that the weight $\lambda^\natural$ is typical (i.e., the degree of atypicality $\#\lambda^\natural = 0$) if and only if $\lambda_m \ge n$.

Exercise 3.8. Let λ be an $(m|n)$-hook partition such that $\lambda_m \ge n$, and let the Young diagram λ be depicted below as a union of 3 regions: an $m\times n$ rectangle diagram and two subdiagrams μ, ν:

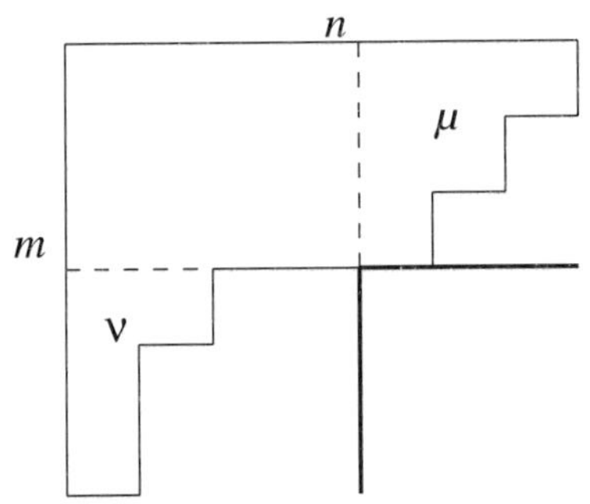

(1) Prove that the Kac $\mathfrak{gl}(m|n)$-module $K(\lambda^\natural)$ is irreducible.

(2) Let $x = \{x_1,\dots,x_m\}$ and $y = \{y_1,\dots,y_n\}$. Prove the following factorization identity:

$$\mathrm{hs}_\lambda(x;y) = \prod_{i=1}^{m}\prod_{j=1}^{n}(x_i+y_j)s_\mu(x)s_{\nu'}(y).$$

Exercise 3.9. Prove Lemma 3.28 which states that two elements of the hyperoctahedral group B_n are conjugate if and only if their types are the same.

Exercise 3.10. Prove the following isomorphisms of superalgebras (see (3.5)):

(1) $Q(m) \otimes Q(n) \cong M(mn|mn)$.

(2) $Q(m) \otimes M(r|s) \cong Q(m(r+s))$.

(3) $M(m|n) \otimes M(r|s) \cong M(mr+ns|ms+nr)$.

Exercise 3.11. Prove the following isomorphisms for Clifford superalgebras (see Definition 3.33):

(1) $\mathcal{C}_m \otimes \mathcal{C}_n \cong \mathcal{C}_{m+n}$.

(2) $\mathcal{C}_1 \cong Q(1), \quad \mathcal{C}_2 \cong M(1|1)$.

(3) $\mathcal{C}_{2m+1} \cong Q(2^m), \quad \mathcal{C}_{2m} \cong M(2^{m-1}|2^{m-1})$.

Hence, $\mathcal{C}_n$ is a simple superalgebra, and it is of type $\mathtt{M}$ if and only if n is even.

Exercise 3.12. Recall the spin symmetric group algebra from Example 3.7. Prove that sending

$$c_i \mapsto c_i,\ 1 \le i \le n; \qquad t_j \mapsto \sqrt{-2}^{-1} s_j(c_j - c_{j+1}),\ 1 \le j \le n-1$$

defines a superalgebra isomorphism $\mathbb{C}\mathfrak{S}_n^- \otimes \mathcal{C}_n \longrightarrow \mathcal{H}_n$.

Exercise 3.13. Recall a homomorphism $\varphi : \Lambda \longrightarrow \Gamma$ from (A.62) in Appendix A. For $n \ge 0$, we define functors $\Omega_n : \mathfrak{S}_n\text{-mod} \longrightarrow \mathcal{H}_n\text{-mod}, \Omega_n(M) = \mathrm{Ind}_{\mathbb{C}\mathfrak{S}_n}^{\mathcal{H}_n} M$. This induces a functor $\Omega = \oplus_n \Omega_n$, which gives rise to a $\mathbb{Z}$-linear map $\Omega : R \longrightarrow R^-$, by letting $\Omega([M]) = [\Omega_n(M)]$ for $M \in \mathfrak{S}_n$-mod. Prove that $\Omega : R_{\mathbb{Q}} \to R_{\mathbb{Q}}^-$ and $\varphi : \Lambda_{\mathbb{Q}} \longrightarrow \Gamma_{\mathbb{Q}}$ are homomorphisms of (Hopf) algebras. Moreover, the following is a commutative diagram of (Hopf) algebras:

$$\begin{array}{ccc} R_{\mathbb{Q}} & \xrightarrow{\ \Omega\ } & R_{\mathbb{Q}}^- \\ {\scriptstyle \mathrm{ch}^{\mathrm{F}}}\downarrow{\scriptstyle\cong} & & {\scriptstyle \mathrm{ch}}\downarrow{\scriptstyle\cong} \\ \Lambda_{\mathbb{Q}} & \xrightarrow{\ \varphi\ } & \Gamma_{\mathbb{Q}} \end{array} \tag{3.39}$$

Exercise 3.14. Show that the $\mathcal{H}_d$-module W_μ defined in (3.33) and (3.34) is isomorphic to $\Omega(M^\mu)$, where $M^\mu = \mathrm{Ind}_{\mathfrak{S}_\mu}^{\mathfrak{S}_d} \mathbf{1}_\mu$ is the permutation module of $\mathfrak{S}_d$. (This way, Lemma 3.50 fits well with (3.35) and the diagram (3.39) as well as Exercise 3.15 below.)

Exercise 3.15. (1) Prove using (A.53), (A.54), and (A.57) that, for $\lambda \in \mathcal{SP}$,

$$Q_\lambda = \sum_{\mu \in \mathcal{P}, \mu \le \lambda} 2^{\ell(\lambda)} \widehat{K}_{\lambda\mu} m_\mu.$$

(2) Prove that the μ-weight multiplicity of the simple $\mathfrak{q}(n)$-module $L(\lambda)$, for $\lambda \in \mathcal{SP}_d(n)$ and $\mu \in \mathcal{CP}_d(n)$, is given by

$$\dim L(\lambda)_\mu = 2^{\frac{\ell(\lambda)+\delta(\lambda)}{2}} \widehat{K}_{\lambda\mu}.$$

Exercise 3.16. For $\lambda \in \mathcal{SP}_d$, write $Q_\lambda(x) = \sum_{\mu \in \mathcal{P}} 2^{\ell(\lambda)} g_{\lambda\mu} s_\mu(x)$, for $g_{\lambda\mu} \in \mathbb{Q}$. Prove:

(1) $\dim \operatorname{Hom}_{\mathcal{H}_d}(D^\lambda, \Omega(S^\mu)) = 2^{\frac{\ell(\lambda)+\delta(\lambda)}{2}} g_{\lambda\mu}$.

(2) As a $\mathfrak{gl}(n)$-module, the $\mathfrak{q}(n)$-module $L(\lambda)$ decomposes into

$$L(\lambda) \cong \bigoplus_{\mu \in \mathcal{P}, \mu \leq \lambda, \ell(\mu) \leq n} 2^{\frac{\ell(\lambda)+\delta(\lambda)}{2}} g_{\lambda\mu} L^0(\mu).$$

Here $L^0(\mu)$ denotes the $\mathfrak{gl}(n)$-module of highest weight μ.

(3) $g_{\lambda\mu} = 0$ unless $\lambda \geq \mu$; $\quad g_{\lambda\lambda} = 1$.

(4) $g_{\lambda\mu} \in \mathbb{Z}_+$.

Exercise 3.17. The Jucys-Murphy elements J_k $(1 \leq k \leq n)$ in $\mathcal{H}_n$ are defined to be $J_k = \sum_{1 \leq j < k}(1 + c_j c_k)(j,k)$. Prove:

(1) $J_i J_k = J_k J_i, \quad$ for $1 \leq i \neq k \leq n$.

(2) $c_i J_i = -J_i c_i, \quad c_i J_k = J_k c_i, \quad$ for $1 \leq i \neq k \leq n$.

(3) $s_i J_i = J_{i+1} s_i - (1 + c_i c_{i+1}), \quad$ for $1 \leq i \leq n-1$.

(4) $s_i J_k = J_k s_i, \quad$ for $k \neq i, i+1$.

Exercise 3.18. Let λ be a partition of d. Derive the following branching rule (e.g. using Pieri's formula (A.22) in Appendix A):

$$\operatorname{Res}_{\mathfrak{S}_{d-1}} S^\lambda = \bigoplus_{\mu \in \mathcal{P}_{d-1}, \mu \leftarrow \lambda} S^\mu.$$

Here $\mu \leftarrow \lambda$ denotes that μ is obtained from λ by deleting a removable box in the Young diagram of λ, and equivalently, λ is obtained from μ by inserting an addable box in the diagram of μ.

Exercise 3.19. Let $\mu \in \mathcal{P}_d(m|n)$ be an $(m|n)$-hook partition of d. Prove the following isomorphism of $\mathfrak{gl}(m|n)$-modules using Schur-Sergeev duality:

$$\mathbb{C}^{m|n} \otimes L(\mu^\natural) \cong \bigoplus_{\lambda \in \mathcal{P}_{d+1}(m|n), \mu \leftarrow \lambda} L(\lambda^\natural).$$

Notes

Section 3.1. The classification of finite-dimensional simple associative superalgebras was due to Wall [**126**] over a general field, and it is somewhat simplified over $\mathbb{C}$ in Józefiak [**56**]. The basics on representation theory of superalgebras and finite supergroups, including Wedderburn's Theorem, Schur's Lemma, and the role of split conjugacy classes, have been developed in [**56**]. Our exposition follows [**56**] closely.

Section 3.2. The superalgebra generalization of Schur duality between the general linear Lie superalgebra $\mathfrak{gl}(m|n)$ and the symmetric group $\mathfrak{S}_d$ was due to Sergeev [**110**] and Berele-Regev [**7**] independently, and the character of irreducible polynomial representations of $\mathfrak{gl}(m|n)$ was given in terms of super Schur polynomials. This generalization is intimately related to the combinatorics of supertableaux. Proposition 3.21 on the diagrammatic interpretation of the degree of atypicality of a polynomial dominant weight of $\mathfrak{gl}(m|n)$ was announced in Cheng-Wang [**32**], and Corollary 3.22 appeared in Moens and van der Jeugt [**88**, Proposition 2.1]. The semisimplicity of the category of polynomial modules of $\mathfrak{gl}(m|n)$ was expected by experts. Our proof here, which is based on a comparison of central characters, is adapted from Cheng-Kwon [**19**].

Section 3.3. The algebra $\mathcal{H}_d$ is a twisted group algebra for a double cover $\widetilde{B}_d$ of the hyperoctahedral group. The classification of the split conjugacy classes for $\widetilde{B}_d$ (Theorem 3.31) was due to Read [**99**]. We follow Józefiak [**58**] to develop systematically a superalgebra approach toward the characteristic map and spin representation theory for $\widetilde{B}_d$.

Section 3.4. The Sergeev duality is a version of Schur duality between the queer Lie superalgebra $\mathfrak{q}(n)$ and an algebra $\mathcal{H}_d$, and it was outlined in [**110**]. This leads to a character formula in terms of Schur Q-functions for the irreducible polynomial $\mathfrak{q}(n)$-modules [**110**]. Here we present complete proofs which use extensively the results from Sections 3.1 and 3.3.

Various results on symmetric functions including super Schur functions and Schur Q-functions relevant to Chapter 3 are collected in Appendix A.

Exercises 3.1, 3.3, and 3.4 are taken from Józefiak [**56**]. Exercise 3.12 is a theorem of Sergeev [**112**] and Yamaguchi [**132**]. Exercises 3.7 and 3.8(2) are due to Berele-Regev [**7**]. Exercises 3.13 and 3.16 are taken from Wan-Wang [**127, 128**], while Exercise 3.16(3) and (4) go back to Stembridge [**118**]. The Jucys-Murphy elements of $\mathcal{H}_n$ in Exercise 3.17 were introduced by Nazarov [**92**].

Chapter 4

Classical invariant theory

In this chapter, we give an introduction to the invariant theory for a group G, which is one of the classical groups $\mathrm{GL}(V), \mathrm{Sp}(V)$, or $\mathrm{O}(V)$, where V is a finite-dimensional vector space. Traditionally, the G-invariants of a polynomial (or symmetric) algebra on U are more widely studied, where U is a direct sum of copies of V or its dual. It has gradually become clear that the algebra of G-invariants in an exterior algebra of U can be developed in a parallel fashion to a large extent. In our presentation, we treat the polynomial and exterior algebras in the united framework of supersymmetric algebras. We formulate and establish the First Fundamental Theorem (FFT) in a supersymmetric algebra setting, which states that the subalgebra of G-invariants is generated by a finite set of basic invariants of degree two. We also develop a tensor version of FFT for each of the classical groups, which is used in the proof of the FFT for supersymmetric algebras. In the case of type A, the tensor FFT is derived from Schur duality, and it is indeed equivalent to Schur duality.

The FFT for supersymmetric algebras of classical groups, Theorem 4.19, will be used in the development of Howe duality in Chapter 5. A reader can also choose to accept Theorem 4.19 and then skip this chapter.

4.1. FFT for the general linear Lie group

In this section, we shall formulate and prove both the tensor and the polynomial versions of the First Fundamental Theorem (FFT) of classical invariant theory for the general linear Lie group. For a vector space U we shall denote the polynomial algebra on U by $\mathcal{P}(U)$.

4.1.1. General invariant theory. Let U and V be finite-dimensional modules of a classical (or more generally a reductive) Lie group G. Then, as G-modules, $U^{\otimes d}$ and $\mathcal{P}(U)$ with induced G-actions are completely reducible. The basic question of classical invariant theory is to describe

(1) the space of tensor G-invariants $(U^{\otimes d})^G$;

(2) the algebra of polynomial G-invariants $\mathcal{P}(U)^G$;

(3) the algebra of G-invariants in the tensor algebra $\mathcal{P}(U) \otimes \wedge(V)$.

These different versions of G-invariants turn out to be closely related to one another.

Theorem 4.1. *Let G be a classical (or reductive) group, and let U,V be finite-dimensional rational G-modules. Then, the algebra of G-invariants in $\mathcal{P}(U) \otimes \wedge(V)$ is finitely generated as a $\mathbb{C}$-algebra.*

Proof. Set $\mathcal{R} := \mathcal{P}(U) \otimes \wedge(V)$. Under the assumptions of the theorem, the G-module $\mathcal{R}$ is completely reducible, and hence the subalgebra $\mathcal{J} := (\mathcal{P}(U) \otimes \wedge(V))^G$ is a direct summand of the G-module $\mathcal{R}$. Let us denote by $\varpi : \mathcal{R} \longrightarrow \mathcal{J}$ the natural projection that is G-equivariant.

The algebra $\mathcal{R}$ is naturally $\mathbb{Z}_+$-graded by the total degree, denoted by deg, on the polynomial subalgebra and the exterior subalgebra. Denote by $\mathcal{J}_+$ the subspace of $\mathcal{J}$ consisting of G-invariant elements in $\mathcal{R}$ with zero constant terms, and denote by $\langle \mathcal{J}_+ \rangle$ the (two-sided) ideal of $\mathcal{R}$ generated by $\mathcal{J}_+$. Then $\langle \mathcal{J}_+ \rangle$, when viewed as a submodule of the finitely generated $\mathcal{P}(U)$-module $\mathcal{R} = \mathcal{P}(U) \otimes \wedge(V)$, is finitely generated over $\mathcal{P}(U)$ by the Hilbert basis theorem. In particular, $\langle \mathcal{J}_+ \rangle$ as an ideal of $\mathcal{R}$ is finitely generated. We can further take a set of generators $\varphi_1, \ldots, \varphi_n$ of the ideal $\langle \mathcal{J}_+ \rangle$ to be homogeneous elements, say $\deg \varphi_i = d_i \geq 1$ for each i.

To complete the proof of the theorem, we shall show that $\varphi_1, \ldots, \varphi_n$ generate $\mathcal{J}$ as a $\mathbb{C}$-algebra. Indeed, let $\varphi \in \mathcal{J}_+$ be a homogeneous element. Write $\varphi = \sum_i f_i \varphi_i$ for some $f_i \in \mathcal{R}$. Then,

$$\varphi = \varpi(\varphi) = \varpi(\sum_i f_i \varphi_i) = \sum_i \varpi(f_i) \varphi_i,$$

where $\deg \varpi(f_i) \leq \deg f_i < \deg \varphi$. By induction on the degree, we can assume that $\varpi(f_i)$ lies in the $\mathbb{C}$-subalgebra generated by $\varphi_1, \ldots, \varphi_n$. □

For a general G-module U, it remains an open problem to describe a reasonable set of generators for the algebra $\mathcal{P}(U)^G$. However, in the case when G is one of the classical groups acting on the natural representation V, and U is a direct sum of copies of V and copies of V^*, the problem turns out to have an elegant solution, known as the **First Fundamental Theorem (FFT)** of classical invariant theory.

4.1.2. Tensor and multilinear FFT for $GL(V)$. Now let V be a finite-dimensional vector space, and let $G = \mathrm{GL}(V)$ be the general linear Lie group on V. Let $U = V^{\otimes d} \otimes (V^*)^{\otimes k}$, the space of mixed tensors of type (d,k). Denote the representation of G on such a U by $\rho_{d,k}$.

First observe that $\lambda I \in G$, $\lambda \neq 0$, acts on $V^{\otimes d} \otimes (V^*)^{\otimes k}$ by $\rho_{d,k}(\lambda I) = \lambda^{d-k} I$. Hence, $(V^{\otimes d} \otimes (V^*)^{\otimes k})^G = 0$ unless $k = d$.

Assume now that $k = d$, and we consider the G-representation $\rho_{d,d}$. For any finite-dimensional G-module W, we have a canonical identification as G-modules:

$$W \otimes W^* \cong \mathrm{End}(W).$$

Set $W = V^{\otimes d}$. We have $(V^{\otimes d})^* = (V^*)^{\otimes d}$ as G-modules. Then we have a canonical identification as G-modules:

$$V^{\otimes d} \otimes (V^*)^{\otimes d} \cong \mathrm{End}(V^{\otimes d}).$$

It follows that

$$\left(V^{\otimes d} \otimes (V^*)^{\otimes d}\right)^G \cong \mathrm{End}_G(V^{\otimes d}). \tag{4.1}$$

Recall the following commuting actions from Lemma 3.9 in Section 3.2 where we replace the action of $\mathfrak{gl}(V)$ by $G = \mathrm{GL}(V)$:

$$G \overset{\Phi_d}{\curvearrowright} V^{\otimes d} \overset{\Psi_d}{\curvearrowleft} \mathfrak{S}_d.$$

By Schur duality (Theorem 3.10), we have

$$\mathrm{End}_G(V^{\otimes d}) = \Psi_d(\mathbb{C}\mathfrak{S}_d). \tag{4.2}$$

Via (4.1) and (4.2), $\Psi_d(\sigma)$ for each permutation $\sigma \in \mathfrak{S}_d$ transfers to a G-invariant $\Theta_\sigma \in \left(V^{\otimes d} \otimes (V^*)^{\otimes d}\right)^G$, which can be written down explicitly in terms of dual bases. Let $e_1, \ldots, e_N$ be a basis for V, and let $e_1^*, \ldots, e_N^*$ be the dual basis of V^*. We have

$$\Theta_\sigma = \sum_{1 \leq i_1, \ldots, i_d \leq N} e_{i_{\sigma(1)}} \otimes \ldots \otimes e_{i_{\sigma(d)}} \otimes e_{i_1}^* \otimes \ldots \otimes e_{i_d}^*. \tag{4.3}$$

For example, Θ_σ for $\sigma = 1$ corresponds to the identity $I_{V^{\otimes d}}$ via (4.1).

Summarizing, we have proved the following.

Theorem 4.2 (Tensor FFT for $\mathrm{GL}(V)$). *Let $G = \mathrm{GL}(V)$. There are no non-zero G-invariants in the tensor space $V^{\otimes d} \otimes (V^*)^{\otimes k}$, for $k \neq d$. The space of G-invariants in $V^{\otimes d} \otimes (V^*)^{\otimes d}$ is spanned by the Θ_σ for $\sigma \in \mathfrak{S}_d$.*

For future applications, it will be convenient to formulate a multilinear version of FFT, which is simply a dual version to the tensor FFT for $\mathrm{GL}(V)$ above. We first recall a standard fact from multilinear algebra, which follows from the universal property of tensor product.

Lemma 4.3. *Let $V_1, \dots, V_p$ be finite-dimensional vector spaces. Then, the dual space $(V_1 \otimes \cdots \otimes V_p)^*$ can be naturally identified with the space of multilinear functions on $V_1 \oplus \cdots \oplus V_p$.*

By Lemma 4.3, we shall regard the dual space $\left(V^{\otimes d} \otimes (V^*)^{\otimes k}\right)^*$ as the space of multilinear functions $f : V^d \oplus V^{*k} \to \mathbb{C}$. There is no nonzero G-invariant multilinear function for $k \neq d$, since $\lambda I \in G$ with $\lambda \neq 0$ sends $(v, \varphi) = (v_1, \dots, v_d, \varphi_1, \dots, \varphi_k)$ to $(\lambda v_1, \dots, \lambda v_d, \lambda^{-1}\varphi_1, \dots, \lambda^{-1}\varphi_k)$; hence, if f is a G-invariant multilinear function, then $f(\lambda I.(v, \varphi)) = \lambda^{d-k} f(v, \varphi)$.

Let $k = d$. Define the **contraction** $\langle j | i^* \rangle$, for $1 \leq i, j \leq d$, by

$$\langle j | i^* \rangle (v_1, \dots, v_d, \varphi_1, \dots, \varphi_k) := \varphi_i(v_j). \tag{4.4}$$

Given $\sigma \in \mathfrak{S}_d$, we let

$$f_\sigma := \langle 1 | \sigma(1)^* \rangle \langle 2 | \sigma(2)^* \rangle \cdots \langle d | \sigma(d)^* \rangle.$$

The tensor FFT for $\mathrm{GL}(V)$, Theorem 4.2, can now be reformulated as follows.

Theorem 4.4 (Multilinear FFT for $\mathrm{GL}(V)$). *Let $G = \mathrm{GL}(V)$. Then there are no nonzero G-invariant multilinear functions on $V^d \oplus V^{*k}$, for $k \neq d$. The space of G-invariant multilinear functions on $V^d \oplus V^{*d}$ is spanned by the functions f_σ for $\sigma \in \mathfrak{S}_d$.*

Proof. Thanks to $V^{**} \cong V$, we identify $\left((V^*)^{\otimes d} \otimes V^{\otimes d}\right) \cong \left(V^{\otimes d} \otimes (V^*)^{\otimes d}\right)^*$. Thus, we have by (4.1) and (4.2) that

$$\left((V^*)^{\otimes d} \otimes V^{\otimes d}\right)^G \cong \left(V^{\otimes d} \otimes (V^*)^{\otimes d}\right)^G \cong \Psi_d(\mathbb{C}\mathfrak{S}_d).$$

By Lemma 4.3, $\left((V^*)^{\otimes d} \otimes V^{\otimes d}\right)^G$ can be identified with the space of G-invariant multilinear functions on $V^d \oplus V^{*d}$. Hence, the theorem follows as a dual version of the tensor FFT for $\mathrm{GL}(V)$ (see Theorem 4.2), where f_σ is the counterpart of Θ_σ in (4.3). □

4.1.3. Formulation of the polynomial FFT for *GL*(V). Again let $G = \mathrm{GL}(V)$. Then G acts naturally on the dual space V^* and also on the direct sums V^k and V^{*m}, for $k, m \geq 0$. We have natural identifications of G-modules:

$$V^k \xrightarrow{\sim} \mathrm{Hom}(\mathbb{C}^k, V), \qquad V^{*m} \xrightarrow{\sim} \mathrm{Hom}(V, \mathbb{C}^m).$$

This leads to a G-equivariant isomorphism between polynomial algebras

$$\mathcal{P}(V^k \oplus V^{*m}) \xrightarrow{\sim} \mathcal{P}(V_{m,k}),$$

where we have denoted

$$V_{m,k} := \mathrm{Hom}(\mathbb{C}^k, V) \oplus \mathrm{Hom}(V, \mathbb{C}^m).$$

We identify the hom-space $\mathrm{Hom}(\mathbb{C}^k, \mathbb{C}^m)$ with the space M_{mk} of $m \times k$ matrices over $\mathbb{C}$ and define the composition map

$$\tau : V_{m,k} \to M_{mk}, \quad \tau(x,y) = yx.$$

Note that τ is G-equivariant, where we let G act on M_{mk} trivially. Hence, it induces an algebra homomorphism

$$\tau^* : \mathcal{P}(M_{mk}) \longrightarrow \mathcal{P}(V_{m,k})^G, \quad \tau^*(f) = f \circ \tau.$$

The image of τ^* consists of G-invariants, since by the G-equivariance of τ we have $\tau^*(f)(g.x) = f(\tau(g.x)) = f(\tau(x)) = \tau^*(f)(x)$, for $g \in G, x \in V_{m,k}$.

Denote by x_{ij} the (i,j)th matrix coefficient on M_{mk}, where $1 \le i \le m, 1 \le j \le k$. It follows by a direct computation that $\tau^*(x_{ij})$ coincides with the contraction $\langle j|i^*\rangle$ (which is defined as in (4.4) with obvious modifications of indices), or equivalently,

$$\tau^*(x_{ij})(v_1, \dots, v_k, v_1^*, \dots, v_m^*) = v_i^*(v_j).$$

Theorem 4.5 (Polynomial FFT for $\mathrm{GL}(V)$). *The algebra of* $\mathrm{GL}(V)$*-invariants in* $\mathcal{P}(V^k \oplus V^{*m})$ *is generated by the contractions* $\langle j|i^*\rangle$, *for* $1 \le i \le m, 1 \le j \le k$.

The proof of Theorem 4.5 will be given in Section 4.1.4 below.

Remark 4.6. The above theorem can be equivalently reformulated as the surjectivity of the homomorphism τ^*. One shows that the image of τ consists of all the matrices $Z \in M_{mk}$ such that $\mathrm{rank}(Z) \le \min\{m, k, \dim V\}$ (Exercise 4.1). Under the assumption that $\dim V \ge \min\{m,k\}$, the map τ is surjective. Therefore τ^* is injective, and so τ^* is actually an isomorphism. Equivalently, the algebra of polynomial G-invariants in $\mathcal{P}(V^k \oplus V^{*m})$ is a polynomial algebra generated by the mk contractions $\langle j|i^*\rangle$.

4.1.4. Polarization and restitution. Let $\mathcal{P}^p(W)$ be the set of degree p homogeneous polynomials on a finite-dimensional vector space W, and let $f \in \mathcal{P}^p(W)$. We define a family of polynomials $f_{r_1 \dots r_p} \in \mathcal{P}(W^p)$, which are multi-homogeneous of degree $(r_1, \dots, r_p)$, by the following formula:

$$f(t_1 w_1 + \ldots + t_p w_p) = \sum_{r_1, \dots, r_p} f_{r_1 \dots r_p}(w_1, \dots, w_p) t_1^{r_1} \cdots t_p^{r_p}, \tag{4.5}$$

where $t_i \in \mathbb{C}, w_i \in W$, and the sum is over $r_i \ge 0$ with $r_1 + \ldots + r_p = p$. In particular, $f_{11\dots1}$ is multilinear.

Definition 4.7. The multilinear function $f_{11\dots1} \in \mathcal{P}(W^p)$ is called the **polarization** of the polynomial function $f \in \mathcal{P}^p(W)$. We shall denote $\mathfrak{P}f = f_{11\dots1}$. On the other hand, given a multilinear function $F : W^p \to \mathbb{C}$, the polynomial function $\mathfrak{R}F \in \mathcal{P}(W)$, defined by $\mathfrak{R}F(w) := F(w, \dots, w)$ for $w \in W$, is called the **restitution** of F.

Let us denote by $\mathcal{P}(W^p)_{(1,1,\dots,1)}$ the space of multilinear functions on W^p.

Proposition 4.8. *The polarization map* $\mathfrak{P} : \mathcal{P}^p(W) \to \mathcal{P}(W^p)_{(1,1,\ldots,1)}$, $f \mapsto \mathfrak{P}f$, *and the restitution map* $\mathfrak{R} : \mathcal{P}(W^p)_{(1,1,\ldots,1)} \to \mathcal{P}^p(W)$, $F \mapsto \mathfrak{R}F$, *satisfy the following properties:*

(1) *Both* $\mathfrak{P}$ *and* $\mathfrak{R}$ *are* $\mathrm{GL}(W)$*-equivariant linear maps.*

(2) $\mathfrak{P}f$ *is* $\mathfrak{S}_p$*-invariant, for* $f \in \mathcal{P}^p(W)$.

(3) $\mathfrak{R}\mathfrak{P}(f) = p!f$, *for* $f \in \mathcal{P}^p(W)$. *In particular,* $\mathfrak{R}$ *is surjective.*

Proof. Parts (1) and (2) follow directly from the definitions.

Setting $w_1 = \ldots = w_p = w$ in (4.5), we obtain that

$$(t_1 + \ldots + t_p)^p f(w) = \sum_{r_1,\ldots,r_p} f_{r_1\ldots r_p}(w,\ldots,w) t_1^{r_1} \cdots t_p^{r_p}.$$

Comparing the coefficients of $t_1 \cdots t_p$, we conclude that

$$p!f(w) = f_{1,1,\ldots,1}(w,\ldots,w) = \mathfrak{R}\mathfrak{P}f(w),$$

whence (3). □

Corollary 4.9. *Let* $G \subseteq \mathrm{GL}(W)$ *be a classical group. Then, by restriction of the restitution map to the subspace of* G*-invariants, we have a surjective linear map*

$$\mathfrak{R} : \mathcal{P}(W^p)^G_{(1,1,\ldots,1)} \longrightarrow \mathcal{P}^p(W)^G.$$

Proof. Since G is a classical group, all the representations of G involved are completely reducible. By Proposition 4.8, $\mathfrak{R} : \mathcal{P}(W^p)_{(1,1,\ldots,1)} \to \mathcal{P}^p(W)$ is surjective and G-equivariant, and so $\mathfrak{R}$ sends a given G-isotypic component of $\mathcal{P}(W^p)_{(1,1,\ldots,1)}$ onto the corresponding isotypic component of $\mathcal{P}^p(W)$. The corollary follows by considering the isotypic component of the trivial module. □

Now we are ready to derive the polynomial FFT for $\mathrm{GL}(V)$ from the multilinear FFT for $\mathrm{GL}(V)$.

Proof of Theorem 4.5. Set $G = \mathrm{GL}(V)$. It suffices to prove the following.

Claim. $\mathcal{P}^p(V^k \oplus V^{*m})^G = 0$, for p odd. The space $\mathcal{P}^{2d}(V^k \oplus V^{*m})^G$ is spanned by $\langle j_1|i_1^*\rangle\langle j_2|i_2^*\rangle \cdots \langle j_d|i_d^*\rangle$, where $1 \le i_a \le m, 1 \le j_a \le k$ for each $a = 1,\ldots,d$.

Let us set $W = V^k \oplus V^{*m}$, and let p be any positive integer to start with. Keeping Lemma 4.3 in mind, we observe that $\mathcal{P}(W^p)^G_{(1,1,\ldots,1)}$ is a direct sum of spaces of G-invariant multilinear functions on direct sums of p copies among V and V^* in various order. By Theorem 4.4, we have $\mathcal{P}(W^p)^G_{(1,1,\ldots,1)} = 0$, for p odd; moreover, $\mathcal{P}(W^{2d})^G_{(1,1,\ldots,1)}$ is spanned by the functions of the form $\langle j_1|i_1^*\rangle\langle j_2|i_2^*\rangle \cdots \langle j_d|i_d^*\rangle$. Now the claim follows by applying Corollary 4.9. □

4.2. Polynomial FFT for classical groups

In this section, we formulate and establish the polynomial FFT for the orthogonal and symplectic Lie groups in a uniform manner (see Theorem 4.13). A theorem of Weyl allows us to reduce the polynomial FFT to a special basic case. This basic case is then proved directly by induction.

4.2.1. A reduction theorem of Weyl. Given $0 \le s \le k$, we have a natural inclusion $V^s \subset V^k$ by identifying $(v_1, \dots, v_s)$ with $(v_1, \dots, v_s, 0, \dots, 0)$. We also have the natural projection from V^k to V^s which sends $(v_1, \dots, v_s, \dots, v_k)$ to $(v_1, \dots, v_s)$, which allows us to view $\mathcal{P}(V^s)$ as a subalgebra of $\mathcal{P}(V^k)$. Note that V^k is naturally a left $\mathrm{GL}(k)$-module, with the action π given by the multiplication by g^{-1} on the right: $\pi(g)(v_1, \dots, v_k) = (v_1, \dots, v_k)g^{-1}$. The induced left $\mathrm{GL}(k)$-action, denoted by π', on $\mathcal{P}(V^k)$ is given by $(\pi'(g)f)(v_1, \dots, v_k) = f((v_1, \dots, v_k)g)$.

Though the goal of this subsection is Theorem 4.11 (due to Hermann Weyl), it is natural to work in a more general setting first. Given a subset S of a module M over a group G, we denote by $\langle S \rangle_G$ the G-submodule of M generated by S.

Theorem 4.10. *Let $N = \dim V$. For $k \ge N$, let L be a $\mathrm{GL}(k)$-submodule of $\mathcal{P}(V^k)$. Then, as a $\mathrm{GL}(k)$-module, L is generated by the intersection $L \cap \mathcal{P}(V^N)$, i.e., $L = \langle L \cap \mathcal{P}(V^N) \rangle_{\mathrm{GL}(k)}$.*

Proof. We will assume $k > N$, since the case $k = N$ is trivial.

Since $\mathcal{P}(V^k)$ is a complete reducible $\mathrm{GL}(k)$-module, we may assume without loss of generality that L is an irreducible $\mathrm{GL}(k)$-module. Denote by $\mathtt{U}(k)$ the subgroup of $\mathrm{GL}(k)$ consisting of upper triangular $k \times k$-matrices with all diagonal entries being 1. Then $\mathcal{P}(V^k)^{\mathtt{U}(k)}$ is the space of highest weight vectors in $\mathcal{P}(V^k)$.

Claim. We have $\mathcal{P}(V^k)^{\mathtt{U}(k)} \subseteq \mathcal{P}(V^N)$.

Note by the standard highest weight theory that $L^{\mathtt{U}(k)} \neq 0$. Granting the claim, we have $L^{\mathtt{U}(k)} \subseteq \mathcal{P}(V^k)^{\mathtt{U}(k)} \subseteq \mathcal{P}(V^N)$, and hence, $L \cap \mathcal{P}(V^N) \neq 0$. Then, it follows by the irreducibility of L that $L = \langle L \cap \mathcal{P}(V^N) \rangle_{\mathrm{GL}(k)}$.

So it remains to prove the claim. Recall $\dim V = N$, and introduce the following Zariski-open subset in V^k:

$$Z = \{(v = (v_1, \dots, v_k) \in V^k \mid v_1, \dots, v_N \text{ are linearly independent in } V\}.$$

Given $v = (v_1, \dots, v_k) \in Z$, we can find $\alpha_1, \dots, \alpha_{k-1} \in \mathbb{C}$ such that

$$\alpha_1 v_1 + \dots + \alpha_{k-1} v_{k-1} + v_k = 0.$$

Hence, there exists $u \in \mathrm{U}(k)$ such that $\pi(u)(v_1,\ldots,v_{k-1},v_k) = (v_1,\ldots,v_{k-1},0)$. For example, the following element

$$u = \begin{pmatrix} 1 & 0 & \cdots & 0 & \alpha_1 \\ 0 & 1 & \cdots & 0 & \alpha_2 \\ \vdots & \ddots & \ddots & \ddots & \vdots \\ \vdots & \ddots & \ddots & \ddots & \alpha_{k-1} \\ 0 & \cdots & \cdots & 0 & 1 \end{pmatrix}^{-1}$$

will do the job.

Let $f \in \mathcal{P}(V^k)^{\mathrm{U}(k)}$. Then, $f(v) = f(v_1,\ldots,v_{k-1},0)$, for $v \in Z$. By induction on $k > N$, we see that $f(v) = f(v_1,\ldots,v_N,0,\ldots,0)$, for $v \in Z$. Since the polynomial function f is determined by its restriction to the Zariski-open subset Z, $f(v) = f(v_1,\ldots,v_N,0,\ldots,0)$, for all $v \in V$. This completes the proof of the claim and hence of the theorem. □

Theorem 4.11 (Weyl). *Let G be an arbitrary subgroup of $\mathrm{GL}(V)$. Let $N = \dim V$ and assume $k \geq N$. Then, $\mathcal{P}(V^k)^G$ is generated by $\langle \mathcal{P}(V^N)^G \rangle_{\mathrm{GL}(k)}$. In particular, if a subspace S generates the algebra $\mathcal{P}(V^N)^G$, then $\langle S \rangle_{\mathrm{GL}(k)}$ generates the algebra $\mathcal{P}(V^k)^G$.*

Proof. Thanks to the commuting actions of $G \subseteq \mathrm{GL}(V)$ and of $\mathrm{GL}(k)$ on $\mathcal{P}(V^k)$, we see that $\mathcal{P}(V^k)^G$ is a $\mathrm{GL}(k)$-module. The first part of the theorem follows by specializing Theorem 4.10 to the case $L = \mathcal{P}(V^k)^G$ and noting that $\mathcal{P}(V^N)^G = \mathcal{P}(V^k)^G \cap \mathcal{P}(V^N)$.

Denote by $\mathcal{A}$ the subalgebra of $\mathcal{P}(V^k)$ generated by $\langle S \rangle_{\mathrm{GL}(k)}$. Then, $\mathcal{A}$ is contained in the algebra $\mathcal{P}(V^k)^G$, thanks to $\langle S \rangle_{\mathrm{GL}(k)} \subseteq \mathcal{P}(V^k)^G$. On the other hand, $\mathcal{A} \supseteq S$, and so the algebra $\mathcal{A}$ contains the algebra $\mathcal{P}(V^N)^G$ that S generates. Since it follows by definition that $\mathcal{A}$ is $\mathrm{GL}(k)$-stable, $\mathcal{A}$ contains $\langle \mathcal{P}(V^N)^G \rangle_{\mathrm{GL}(k)}$, and so we can apply the first part of the theorem to conclude that $\mathcal{A}$ contains $\mathcal{P}(V^k)^G$. Hence, $\mathcal{A} = \mathcal{P}(V^k)^G$. □

The following proposition will be useful later on.

Proposition 4.12. *Let G be a subgroup of $\mathrm{GL}(V)$, and let $N = \dim V$. Assume that S is a subspace of $\mathcal{P}(V^N)^G$ that generates the algebra $\mathcal{P}(V^N)^G$. Then, for $k \leq N$, the algebra of invariants $\mathcal{P}(V^k)^G$ is generated by the subset S_k that consists of the restrictions to V^k of elements in S. Moreover, $S_k = S \cap \mathcal{P}(V^k)$ if S is $\mathrm{GL}(N)$-stable.*

Proof. The inclusion $V^k \to V^N$ and the projection $V^N \to V^k$ induce the restriction map $\mathrm{res} : \mathcal{P}(V^N) \to \mathcal{P}(V^k)$ and the inclusion $\mathcal{P}(V^k) \hookrightarrow \mathcal{P}(V^N)$, respectively. Since either of the compositions

$$\mathcal{P}(V^k) \hookrightarrow \mathcal{P}(V^N) \overset{\mathrm{res}}{\to} \mathcal{P}(V^k), \qquad \mathcal{P}(V^k)^G \to \mathcal{P}(V^N)^G \overset{\mathrm{res}}{\to} \mathcal{P}(V^k)^G$$

is the identity map, res : $\mathcal{P}(V^N)^G \to \mathcal{P}(V^k)^G$ is surjective. Hence, $\mathcal{P}(V^k)^G$ is generated by the restrictions S_k.

Now assume that S is $\mathrm{GL}(N)$-stable. In particular, T_N acts on S semisimply, where T_k denotes the diagonal subgroup of $\mathrm{GL}(k)$, for $k \leq N$. Note that $T_N = T_k \times T_{N-k}$. Both S_k and $S \cap \mathcal{P}(V^k)$ can easily be seen to be equal to the subspace of S where T_{N-k} acts trivially, and hence they must be equal. □

4.2.2. The symplectic and orthogonal groups. Let V be a vector space equipped with a non-degenerate symmetric or skew-symmetric bilinear form $\omega : V \times V \to \mathbb{C}$. We denote by $G(V, \omega)$ the subgroup of $\mathrm{GL}(V)$ that preserves the form ω, that is,

$$G(V,\omega) = \{g \in \mathrm{GL}(V) \mid \omega(gv_1, gv_2) = \omega(v_1, v_2), \forall v_1, v_2 \in V\}.$$

The group $G(V, \omega)$ is called an **orthogonal group** and is denoted by $\mathrm{O}(V)$ if ω is symmetric, and it is called a **symplectic group** and is denoted by $\mathrm{Sp}(V)$ if ω is skew-symmetric. The Lie algebra of the group $G(V, \omega)$, called the **orthogonal and symplectic Lie algebras** respectively, is

$$\mathfrak{g}(V,\omega) = \{x \in \mathfrak{gl}(V) \mid \omega(xv_1, v_2) = -\omega(v_1, xv_2), \forall v_1, v_2 \in V\}.$$

Let $V = \mathbb{C}^N$, which will mostly be viewed as the space of column vectors. A non-degenerate symmetric (respectively, skew-symmetric) bilinear form relative to the standard basis $\{e_1, \ldots, e_N\}$ in $\mathbb{C}^N$ is determined by its associated non-singular symmetric (respectively, skew-symmetric) matrix J. That is, $\omega(v_1, v_2) = v_1^t J v_2$, for $v_1, v_2 \in \mathbb{C}^N$. Note that N is necessarily even in the skew-symmetric case. When a form ω on $V = \mathbb{C}^N$ has its associated matrix J, we have

$$G(V,\omega) = \{X \in \mathrm{GL}(N) \mid X^t J X = J\}.$$

Different choices of non-singular symmetric (respectively, skew-symmetric) $N \times N$ matrices J lead to isomorphic orthogonal (respectively, symplectic) groups in different matrix forms. Some useful choices for the non-singular symmetric matrices are the identity matrix I_N, the following anti-diagonal matrix

$$J_N := \begin{pmatrix} 0 & \cdots & 0 & 1 \\ 0 & \cdots & 1 & 0 \\ \vdots & \vdots & \vdots & \vdots \\ 1 & \cdots & 0 & 0 \end{pmatrix}, \tag{4.6}$$

and

$$\begin{pmatrix} 0 & I_\ell \\ I_\ell & 0 \end{pmatrix} \ (\text{for } N = 2\ell), \qquad \begin{pmatrix} 0 & I_\ell & 0 \\ I_\ell & 0 & 0 \\ 0 & 0 & 1 \end{pmatrix} \ (\text{for } N = 2\ell + 1).$$

When we choose ω to be the standard symmetric form on $V = \mathbb{C}^N$ corresponding to the identity matrix I_N, $\mathrm{O}(V)$ is simply the group $\mathrm{O}(N)$ of all $N \times N$ orthogonal matrices X, i.e., $X^t X = I_N$.

On the other hand, some common choices for non-singular skew-symmetric matrices with $N = 2\ell$ used to define the symplectic Lie group and algebra are

$$\begin{pmatrix} 0 & I_\ell \\ -I_\ell & 0 \end{pmatrix} \quad \text{and} \quad \begin{pmatrix} \Omega & 0 & \cdots & 0 \\ 0 & \Omega & \cdots & 0 \\ \vdots & \vdots & \ddots & \vdots \\ 0 & 0 & \cdots & \Omega \end{pmatrix},$$

where $\Omega = \begin{pmatrix} 0 & 1 \\ -1 & 0 \end{pmatrix}$. Another common choice of a non-singular skew-symmetric matrix is

$$\begin{pmatrix} 0 & J_\ell \\ -J_\ell & 0 \end{pmatrix}. \tag{4.7}$$

The choices of (4.6) and (4.7) play prominent roles in the later Sections 5.3.2 and 5.3.1, respectively.

4.2.3. Formulation of the polynomial FFT. Let $V = \mathbb{C}^N$ be equipped with a symmetric (respectively, skew-symmetric) form ω^+ (respectively, ω^-) whose associated matrix is J_+ (respectively, J_-). It will be convenient to write $G^\pm(V)$ for $G(V, \omega^\pm)$, as this allows us to treat the $+$ and $-$ cases uniformly.

Given $k \in \mathbb{N}$, we denote by

$$\begin{aligned} \mathrm{SM}_k^+ &= \{k \times k \text{ complex symmetric matrices}\}, \\ \mathrm{SM}_k^- &= \{k \times k \text{ complex skew-symmetric matrices}\}. \end{aligned}$$

We have a natural identification of $G^\pm(V)$-modules between V^k and the space M_{Nk} of $N \times k$ complex matrices (on which $G^\pm(V)$ acts by left multiplication). Define maps

$$\begin{aligned} \tau_+ : V^k \equiv M_{Nk} &\longrightarrow \mathrm{SM}_k^+, \qquad \tau_+(X) = X^t J_+ X, \\ \tau_- : V^k \equiv M_{Nk} &\longrightarrow \mathrm{SM}_k^-, \qquad \tau_-(X) = X^t J_- X. \end{aligned}$$

It can be checked that $\tau_\pm(X)^t = \pm\tau_\pm(X)$, so $\tau_\pm$ are well-defined.

The map $\tau_\pm$ is $G^\pm(V)$-equivariant, where we let $G^\pm(V)$ act on $\mathrm{SM}_k^\pm$ trivially. Indeed, for $g \in G^\pm(V), X \in M_{Nk}$, we have

$$\tau_\pm(gX) = (gX)^t J_\pm(gX) = X^t g^t J_\pm g X = X^t J_\pm X = \tau_\pm(X).$$

Hence, the pullback via $\tau_\pm$ gives rise to an algebra homomorphism

$$\tau_\pm^* : \mathcal{P}(\mathrm{SM}_k^\pm) \longrightarrow \mathcal{P}(V^k)^{G^\pm(V)}, \qquad \tau_\pm^*(f) = f \circ \tau_\pm,$$

where $\mathcal{P}(V^k)^{G^\pm(V)}$ denotes the algebra of $G^\pm(V)$-invariant polynomials on V^k. Indeed, it follows from the $G^\pm(V)$-equivariance of $\tau_\pm$ that the image of $\tau_\pm^*$ lies in the $G^\pm(V)$-invariant subalgebra of $\mathcal{P}(V^k)$.

Write $X = (v_1, \ldots, v_k)$ with each $v_i \in V = \mathbb{C}^N$. It follows by definition that

$$(X^t J_\pm X)_{ij} = \omega^\pm(v_i, v_j).$$

Denote by the same notation the restriction to $\mathrm{SM}_k^\pm$ the matrix coefficients x_{ij} on M_{kk}, for $1 \le i, j \le k$. Introduce the $G^\pm(V)$-invariant functions $(i|j)$ on V^k, for $1 \le i, j \le k$, by letting

$$(i|j)(v_1, \ldots, v_k) := \omega^\pm(v_i, v_j).$$

Note that these functions are not independent in general, and

$$(i|j) = \pm(j|i)$$

in the $\pm$ case, respectively. It follows by definitions that

$$\tau_\pm^*(x_{ij})(v_1, \ldots, v_k) = (i|j).$$

Theorem 4.13 (Polynomial FFT for $G^\pm(V)$). *The homomorphism $\tau_\pm^* : \mathcal{P}(\mathrm{SM}_k^\pm) \to \mathcal{P}(V^k)^{G^\pm(V)}$ is surjective. Equivalently, $\mathcal{P}(V^k)^{G^\pm(V)}$ is generated by the functions $(i|j)$, where $1 \le i \le j \le k$ in the $+$ case and $1 \le i < j \le k$ in the $-$ case.*

The proof of Theorem 4.13 will be given in the subsequent subsections.

Remark 4.14. One can show that the image of $\tau_\pm$ consists of all the matrices $Z \in \mathrm{SM}_k^\pm$ such that $\mathrm{rank}(Z) \le \min\{k, \dim V\}$. In particular, if $\dim V \ge k$, then $\tau_\pm$ is surjective. Thus, under the assumption that $\dim V \ge k$, $\tau_\pm^*$ is injective and so $\tau_\pm^*$ is actually an isomorphism. Equivalently, the algebra of $G^\pm(V)$-invariant polynomials in V^k is a free polynomial algebra generated by the $k(k\pm 1)/2$ polynomials as listed in Theorem 4.13.

Remark 4.15. The non-degenerate form $\omega^\pm$ induces a canonical isomorphism of $G^\pm(V)$-modules: $V^* \cong V$. Via this isomorphism, we can obtain from Theorem 4.13 a (seemingly more general) polynomial FFT for the algebra of $G^\pm(V)$-invariant polynomials in $(V^*)^m \oplus V^k$.

4.2.4. From basic to general polynomial FFT. Let us now refer to the statement in the polynomial FFT for $G^\pm(V)$ as formulated in Theorem 4.13 as FFT(k), for $k \ge 1$. Set $N = \dim V$ as before.

Assuming a basic case FFT(N) holds for now, let us complete the proof of FFT(k) for all k. Denote by S_k the subspace spanned by the functions $(i|j)$, where $1 \le i, j \le k$.

FFT(N) $\Longrightarrow$ FFT(k) for $k < N$. (Indeed, the same argument below shows that FFT(ℓ) $\Longrightarrow$ FFT(k) for all $\ell \ge k \ge 1$.) The subspace S_N is $\mathrm{GL}(N)$-stable. Moreover, $S_N \cap \mathcal{P}(V^k)$ is simply S_k. Now FFT(k) follows by applying FFT(N) and Proposition 4.12 for $G = G^\pm(V)$ and $S = S_N$.

$\mathrm{FFT}(N) \Longrightarrow \mathrm{FFT}(k)$ for $k > N$. According to $\mathrm{FFT}(N)$, S_N generates the algebra $\mathcal{P}(V^N)^G$. Observe that $\langle S_N \rangle_{\mathrm{GL}(k)} = S_k$. Hence, by Weyl's Theorem 4.11, the algebra $\mathcal{P}(V^k)^G$ is generated by S_k, whence $\mathrm{FFT}(k)$.

This completes the proof of Theorem 4.13, modulo the proof of the basic case $\mathrm{FFT}(N)$. We shall prove the basic case $\mathrm{FFT}(N)$ as Theorem 4.16 in the next subsection.

4.2.5. The basic case. In this subsection, we shall prove the following basic case of the polynomial FFT for $G^{\pm}(V)$ (that is, Theorem 4.13 for $k = \dim V$).

Theorem 4.16. *Let $N = \dim V$. Then, the algebra $\mathcal{P}(V^N)^{G^{\pm}(V)}$ is generated by the functions $(i|j)$, for $1 \le i, j \le N$.*

Proof. We proceed by induction on N. We treat the cases of orthogonal and symplectic groups separately, though the overall strategy of the proof is the same.

The orthogonal group case. Let $\{e_1, \ldots, e_N\}$ be the standard basis of $V = \mathbb{C}^N$. Take $(x, y) = \sum_{i=1}^N x_i y_i$, where $x = (x_1, \ldots, x_N), y = (y_1, \ldots, y_N)$ in $\mathbb{C}^N$, and so $\mathrm{O}(V) = \mathrm{O}(N)$. Denote

$$V_1 := \mathbb{C}^{N-1} \times \{0\} \subseteq V$$

and denote by $(\cdot,\cdot)_1$ the bilinear form on V_1 obtained by restriction from $(\cdot,\cdot)$ on V.

For $N = 1$, $O_1 = \{\pm 1\}$, Theorem 4.16 is clear.

Assume now that $N > 1$. We have an orthogonal decomposition $V = V_1 \oplus \mathbb{C}e_N$ with respect to $(\cdot,\cdot)$.

Let $f \in \mathcal{P}(V^N)^{\mathrm{O}(N)}$. Since $-I_N \in \mathrm{O}(N)$, $f(X) = f(-I_N \cdot X) = f(-X)$, f cannot contain a nonzero odd degree term. Without loss of generality, let us assume that f is homogeneous of even degree $2d$. Below, we will consider the restriction of f to the Zariski open subset of vectors $v = (v_1, \ldots, v_N)$ of V^N given by the inequality $(v_N, v_N) \neq 0$.

Let $\kappa \in \mathbb{C}$ such that $\kappa^2 = (v_N, v_N)$. Since $(v_N, v_N) = (\kappa e_N, \kappa e_N)$, there exists an orthogonal matrix $g \in \mathrm{O}(N)$ such that $gv_N = \kappa e_N$. Set

$$v_i' = \kappa^{-1} g v_i, \quad (i = 1, \ldots, N-1). \tag{4.8}$$

Then,

$$f(v) = f(gv) = \kappa^{2d} f(v_1', \ldots, v_{N-1}', e_N) = (v_N, v_N)^d f(v_1', \ldots, v_{N-1}', e_N).$$

Let

$$v_i' = v_i'' + t_i e_N \quad (i = 1, \ldots, N-1), \tag{4.9}$$

be the orthogonal decomposition in $V = V_1 \oplus \mathbb{C}e_N$ for $v_i'' \in V_1$. Using (4.8), we compute that

$$t_i = (v_i', e_N) = (\kappa^{-1} g v_i, \kappa^{-1} g v_N) = (v_i, v_N)/(v_N, v_N). \tag{4.10}$$

Writing $t^I = t_1^{i_1} \cdots t_{N-1}^{i_{N-1}}$ for multi-indices $I = (i_1, \ldots, i_{N-1})$ and regarding t_i as independent variables, we have a decomposition of the following form:

$$f(v_1', \ldots, v_{N-1}', e_N) = \sum_I \hat{f}_I(v_1'', \ldots, v_{N-1}'') t^I, \tag{4.11}$$

where $\hat{f}_I$ are polynomial functions uniquely determined by f. Noting from (4.9) and (4.10) that t_i is invariant under the transformation $v_i' \mapsto h v_i'$ for $h \in \mathrm{O}(N-1)$, we conclude from (4.11) that $\hat{f}_I$ is $\mathrm{O}(N-1)$-invariant for each I.

By induction hypothesis, there exist polynomials φ_I in the functions $(i|j), 1 \leq i \leq j \leq N-1$, such that $\hat{f}_I = \varphi_I$ when evaluated on $(v_1'', \ldots, v_{N-1}'')$. Using (4.8), (4.9), and (4.10), we compute that

$$(v_i'', v_j'') = (v_i', v_j') - (t_i e_N, t_j e_N) = \frac{(v_i, v_j)(v_N, v_N) - (v_i, v_N)(v_j, v_N)}{(v_N, v_N)^2}. \tag{4.12}$$

Let us recall that we have been considering the restriction of f to the open subset defined by $(v_N, v_N) \neq 0$. Plugging (4.12) into $\hat{f}_I = \varphi_I$, we have $f(v) = (v_N, v_N)^{-p} F_1$ for some polynomial F_1 in $(v_i, v_j), i \leq j \leq N$, and some positive integer p whenever $(v_N, v_N) \neq 0$.

We will be done if we can show that the polynomial function $(v_N, v_N)^p$ divides F_1. Indeed, the same type of argument above allows us to conclude that $f(v) = (v_1, v_1)^{-q} F_2$ for some polynomial F_2 in $(v_i, v_j), i \leq j \leq N$, and some positive integer q whenever $(v_1, v_1) \neq 0$. In this way, we obtain a polynomial equation

$$(v_1, v_1)^q F_1 = (v_N, v_N)^p F_2,$$

which holds on a Zariski open subset of V^N given by the inequalities $(v_1, v_1) \neq 0$ and $(v_N, v_N) \neq 0$, and hence the polynomial equation must hold on V^N. This implies $(v_N, v_N)^p$ divides F_1, since the polynomials (v_N, v_N) and (v_1, v_1) are relatively prime.

The symplectic group case. Let $V = \mathbb{C}^N$ be equipped with the standard basis $\{e_1, \ldots, e_N\}$, where $N = 2\ell$ is even. For definiteness, let us take the symplectic group $\mathrm{Sp}(N) = \mathrm{Sp}(V)$ defined via the symplectic form ω on V:

$$\omega(x, y) = \sum_{i=1}^{\ell} (x_{2i-1} y_{2i} - x_{2i} y_{2i-1}),$$

where $x = (x_1, \ldots, x_N), y = (y_1, \ldots, y_N)$ in $\mathbb{C}^N$. We have an orthogonal decomposition $V = V_1 \oplus W$ with respect to the symplectic form ω, where

$$V_1 := \mathbb{C}^{N-2} \times \{0\} \subseteq V, \qquad W := \mathbb{C} e_{N-1} \oplus \mathbb{C} e_N.$$

We shall denote by ω_1 the bilinear form on V_1 induced from ω on V and regard $\mathrm{Sp}(N-2) = \mathrm{Sp}(V_1)$ as a natural subgroup of $\mathrm{Sp}(N)$ in this way.

Let $f \in \mathcal{P}(V^N)^{\mathrm{Sp}(N)}$. Since $-I_N \in \mathrm{Sp}(N)$, we may assume without loss of generality that f is homogeneous of even degree $2d$, just as for the $\mathrm{O}(N)$ case above. Below, we will consider the restriction of f to the Zariski open subset of vectors $v=(v_1,\ldots,v_N)$ of V^N given by the inequality $\omega(v_{N-1},v_N)\neq 0$.

Choose $\kappa\in\mathbb{C}$ such that $\kappa^2=\omega(v_{N-1},v_N)$. Since $\omega(e_{N-1},e_N)=1$ and so $\omega(v_{N-1},v_N)=\omega(\kappa e_{N-1},\kappa e_N)$, it is standard to show that there exists $g\in \mathrm{Sp}(N)$ such that

$$gv_{N-1}=\kappa e_{N-1},\qquad gv_N=\kappa e_N.$$

Set

$$v'_i=\kappa^{-1}gv_i,\quad (i=1,\ldots,N-2). \tag{4.13}$$

Then,

$$\begin{aligned}f(v)=f(gv)&=\kappa^{2d}f(v'_1,\ldots,v'_{N-2},e_{N-1},e_N)\\&=\omega(v_{N-1},v_N)^d f(v'_1,\ldots,v'_{N-2},e_{N-1},e_N).\end{aligned}$$

When $N=2$, we obtain a polynomial equation $f(v_1,v_2)=f(e_1,e_2)\omega(v_1,v_2)^d$, which holds on a Zariski open set $\omega(v_1,v_2)\neq 0$ and hence holds everywhere. This proves the theorem for $N=2$.

Now assume $N>2$. Let

$$v'_i=v''_i+s_ie_{N-1}+t_ie_N\quad (i=1,\ldots,N-2), \tag{4.14}$$

be the orthogonal decomposition in $V=V_1\oplus W$ for $v''_i\in V_1$. It follows by (4.13) that

$$\begin{cases} s_i &=\omega(v'_i,e_N)=\omega(\kappa^{-1}gv_i,\kappa^{-1}gv_N)=\omega(v_i,v_N)/\omega(v_{N-1},v_N),\\ t_i &=-\omega(v'_i,e_{N-1})=-\omega(v_i,v_{N-1})/\omega(v_{N-1},v_N).\end{cases} \tag{4.15}$$

Write $s^Jt^I=s_1^{j_1}\cdots s_{N-2}^{j_{N-2}}t_1^{i_1}\cdots t_{N-2}^{i_{N-2}}$ associated to the multi-indices $I=(i_1,\ldots,i_{N-2})$ and $J=(j_1,\ldots,j_{N-2})$. Regarding s_i and t_i as independent variables, we have a decomposition of the following form:

$$f(v'_1,\ldots,v'_{N-2},e_{N-1},e_N)=\sum_{I,J}\hat f_{IJ}(v''_1,\ldots,v''_{N-2})s^Jt^I, \tag{4.16}$$

where $\hat f_{IJ}$ are polynomial functions uniquely determined by f. Noting from (4.15) that s_i,t_i are invariant under the transformation $v'_i\mapsto gv'_i$ for $g\in\mathrm{Sp}(N-2)$, we conclude from (4.16) that $\hat f_{IJ}$ is $\mathrm{Sp}(N-2)$-invariant for every I,J.

By inductive assumption, there exist polynomials φ_{IJ} in the functions $(i|j)$, $1\le i\le j\le N-2$, such that $\hat f_{IJ}=\varphi_{IJ}$ when evaluated at $(v''_1,\ldots,v''_{N-2})$. Using (4.13), (4.14), and (4.15), we compute that

$$\begin{aligned}\omega(v''_i,v''_j)&=\omega(v'_i,v'_j)+t_is_j-s_it_j\\&=\frac{\omega(v_i,v_j)\omega(v_{N-1},v_N)-\omega(v_i,v_{N-1})\omega(v_j,v_N)+\omega(v_i,v_N)\omega(v_j,v_{N-1})}{\omega(v_{N-1},v_N)^2}.\end{aligned} \tag{4.17}$$

Recall that we have been considering the restriction of f to the open subset defined by $\omega(v_{N-1},v_N)\neq 0$. Plugging (4.17) into $\hat f_{IJ}=\varphi_{IJ}$, we have $f(v)=\omega(v_{N-1},v_N)^{-p}F$

for some polynomial F in $\omega(v_i, v_j)_{i\le j\le N}$ and some positive integer p whenever $\omega(v_{N-1}, v_N) \neq 0$. Repeating the same trick used in the orthogonal group case, we see that the polynomial function $\omega(v_{N-1}, v_N)^p$ divides F.

This completes the proof of the theorem. □

4.3. Tensor and supersymmetric FFT for classical groups

In this section, we will use the polynomial FFT for classical groups to derive the corresponding tensor FFT. Finally, the supersymmetric algebra version of FFT is derived from the tensor FFT.

4.3.1. Tensor FFT for classical groups. Let $G = G(V, \omega)$, where V is a vector space equipped with a non-degenerate symmetric or skew-symmetric bilinear form ω. Since $V \cong V^*$ as a G-module, it suffices to consider the G-invariants on the tensor space $V^{\otimes m}$ rather than a mixed tensor space $V^{\otimes m} \otimes (V^*)^{\otimes n}$, for $m, n \geq 0$. Observe that $-I_V \in G$ acts by the scalar multiple by $(-1)^m$ on $V^{\otimes m}$. Hence, $(V^{\otimes m})^G = 0$ unless $m = 2k$ is even.

Via the canonical identification of G-modules:

$$V \otimes V \cong V^* \otimes V \cong \mathrm{End}(V),$$

we obtain a natural G-module isomorphism:

$$T : V^{\otimes 2k} \xrightarrow{\sim} \mathrm{End}(V^{\otimes k}),$$

which maps $u = v_1 \otimes v_2 \otimes \ldots \otimes v_{2k}$ to $T(u)$ defined as follows:

$$T(u)(x_1 \otimes \ldots \otimes x_k) := \omega(x_1, v_2)\omega(x_2, v_4)\ldots\omega(x_k, v_{2k})v_1 \otimes v_3 \otimes \ldots \otimes v_{2k-1}.$$

Clearly the identity $I_{V^{\otimes k}}$ is G-invariant, and hence so is $T^{-1}(I_{V^{\otimes k}})$. We describe $T^{-1}(I_{V^{\otimes k}})$ explicitly as follows. Take a basis $\{f_i\}$ for V and its dual basis $\{f^i\}$ for V with respect to ω, i.e., $\omega(f_i, f^j) = \delta_{ij}$. Define

$$\theta = \sum_{i=1}^{N} f_i \otimes f^i, \quad \theta_k = \underbrace{\theta \otimes \ldots \otimes \theta}_{k}.$$

Then $T^{-1}(I_{V^{\otimes k}}) = \theta_k$.

Recall from Section 3.2 the following commuting actions:

$$GL(V) \overset{\Phi_{2k}}{\curvearrowright} V^{\otimes 2k} \overset{\Psi_{2k}}{\curvearrowleft} \mathfrak{S}_{2k}.$$

Since G is a subgroup of $GL(V)$ that commutes with $\mathfrak{S}_{2k}$,

$$\theta_\sigma := \Psi_{2k}(\sigma)(\theta_k)$$

is G-invariant for any $\sigma \in \mathfrak{S}_{2k}$.

Theorem 4.17 (Tensor FFT for $G(V,\omega)$). *Let $G = G(V,\omega)$, where ω is a non-degenerate symmetric or skew-symmetric bilinear form on V. Then $(V^{\otimes m})^G = 0$ for m odd. Moreover, $(V^{\otimes 2k})^G$ has a spanning set $\{\theta_\sigma \mid \sigma \in \mathfrak{S}_{2k}\}$.*

The tensor FFT for $G(V,\omega)$ affords an equivalent dual reformulation on the G-invariants of $(V^{*\otimes m}) \cong (V^{\otimes m})^*$. We let $(i|j)$ be the function such that

$$(i|j)(v_1,\dots,v_m) = \omega(v_i,v_j). \tag{4.18}$$

Denote by

$$\theta_k^* = (1|2)(3|4)\dots(2k-1|2k)$$

the G-invariant function on $(V^{\otimes m})$ that corresponds to the identity under the isomorphism $T^* : V^{*\otimes 2k} \xrightarrow{\sim} \mathrm{End}(V^{\otimes k})$. Denote by Ψ_{2k}^* the natural action of $\mathfrak{S}_{2k}$ on $(V^{*\otimes 2k})$, which clearly commutes with the natural action of G.

Theorem 4.18 (Tensor FFT for $G(V,\omega)$, dual version). *Let $G = G(V,\omega)$, where ω is non-degenerate symmetric or skew-symmetric. Then $(V^{*\otimes m})^G = 0$ for m odd. Moreover, $(V^{*\otimes 2k})^G$ has a spanning set $\{\Psi_{2k}^*(\sigma)(\theta_k^*) \mid \sigma \in \mathfrak{S}_{2k}\}$.*

Theorems 4.17 and 4.18 are equivalent, and we shall only present a detailed proof of the tensor FFT as formulated in Theorem 4.18. The strategy of the proof can be outlined as follows. First, we introduce an auxiliary polarization variable $X \in \mathrm{End}\, V$ to reformulate this tensor FFT problem as one involving the basic case of the polynomial FFT as in Theorem 4.16 (note $\mathrm{End}\, V \cong V^N$ for $N = \dim V$). Then, by applying Theorem 4.16, we transfer the problem at hand to an FFT problem for $\mathrm{GL}(V)$, for which Theorem 4.2 applies.

Proof of Theorem 4.18. Since $-I_V \in G$ acts on $V^{\otimes m}$ by a multiple of $(-1)^m$, we have $(V^{\otimes m})^G = 0$, for m odd.

Now let $m = 2k$ be even. We shall define an injective linear map

$$\Phi : V^{*\otimes m} \longrightarrow \mathcal{P}^m(\mathrm{End}\, V) \otimes V^{*\otimes m}, \quad \lambda \mapsto \Phi_\lambda,$$

where Φ_λ is defined as a function on $\mathrm{End}\, V \times V^{\otimes m}$, which is of polynomial degree m on $\mathrm{End}\, V$ and linear on $V^{\otimes m}$, by letting $\Phi_\lambda(X,w) = \langle \lambda, X^{\otimes m} w\rangle$, for $X \in \mathrm{End}\, V, w \in V^{\otimes m}$. The injectivity follows since λ is recovered as $\Phi_\lambda(I_V,-)$.

The space $\mathcal{P}^m(\mathrm{End}\, V) \otimes V^{*\otimes m}$ carries commuting actions of the groups G and $\mathrm{GL}(V)$ as follows: $g.f(X,w) = f(g^{-1}X,w)$ for $g \in G$; $h.f(X,w) = f(Xh,h^{-1}w)$ for $h \in \mathrm{GL}(V)$. Note that Φ is a G-equivariant map, i.e., $g.\Phi_\lambda = \Phi_{g.\lambda}$, for $g \in G$. Also note that Φ_λ is $\mathrm{GL}(V)$-invariant, i.e., $h.\Phi_\lambda = \Phi_\lambda$, for $h \in \mathrm{GL}(V)$. Hence, we obtain by restriction of Φ the following injective linear map (which can actually be shown to be an isomorphism, though this stronger statement is not needed below)

$$\Phi : (V^{*\otimes m})^G \longrightarrow \left((\mathcal{P}^m(\mathrm{End}\, V))^G \otimes V^{*\otimes m}\right)^{\mathrm{GL}(V)}.$$

For notational convenience below, let us fix $V = \mathbb{C}^N$ (regarded as a space of column vectors), and $\omega(x,y) = x^t Jy$, for $x,y \in V$. In this way, we identify End V as the space M_{NN} of $N \times N$ matrices.

The action of G on End V relevant to Φ above gives rise to a G-equivariant isomorphism End $V \cong V^N$. When applying the basic case of the polynomial FFT (Theorem 4.16), we obtain, for $\lambda \in (V^{*\otimes m})^G$, that

$$\Phi_\lambda(X,w) = F_\lambda(X^t JX, w),$$

where F_λ is a polynomial function on $SM_N^{\pm} \times V^{\otimes 2k}$, which is of polynomial degree k on $SM_N^{\pm}$ and linear on $V^{\otimes 2k}$ (here we recall $m = 2k$).

We have a natural isomorphism $SM_N^+ \cong S^2(V^*)$ in the case of symmetric ω, and $SM_N^- \cong \wedge^2(V^*)$ in the case of skew-symmetric ω. Recall that the action of $\mathrm{GL}(V)$ on End $V \cong M_{NN}$ relevant to Φ above is given by left multiplication on M_{NN}. This action induces an action of $\mathrm{GL}(V)$ on $SM_N^{\pm}$, which corresponds precisely to the $\mathrm{GL}(V)$-action on $S^2(V^*)$ (respectively, $\wedge^2(V^*)$) coming from the $\mathrm{GL}(V)$-module structure on V^*. Hence, we have a $\mathrm{GL}(V)$-invariant polynomial function $F_\lambda : S^2(V^*) \times V^{\otimes 2k} \to \mathbb{C}$ in the case of symmetric ω (or $F_\lambda : \wedge^2(V^*) \times V^{\otimes 2k} \to \mathbb{C}$ in the case of skew-symmetric ω), which is linear on $V^{\otimes 2k}$ and of polynomial degree k on $S^2(V^*)$ (or $\wedge^2(V^*)$). Thus, by the tensor FFT for $\mathrm{GL}(V)$ (see Theorem 4.2), F_λ can be written as a linear combination of the functions

$$(M,w) \to \Big(\sum_{i,j} u_1^i M_{ij} v_1^j\Big) \cdots \Big(\sum_{i,j} u_k^i M_{ij} v_k^j\Big),$$

where w is the tensor product of $u_1, v_1, \dots, u_k, v_k \in \mathbb{C}^N$ in some order, u_1^i's are the coordinates of u_1, and so on. Evaluating F_λ at $M = J$, we deduce that $\lambda = F_\lambda(J,-)$ is a linear combination of the $(i|j)$'s. □

4.3.2. From tensor FFT to supersymmetric FFT. Let $W = W_{\bar 0} \oplus W_{\bar 1}$ be a finite-dimensional vector superspace. Recall the supersymmetric algebra of W:

$$\mathcal{S}(W) = S(W_{\bar 0}) \otimes \wedge(W_{\bar 1}).$$

Let $G = \mathrm{GL}(V)$. We set

$$W_{\bar 0} = V^m \oplus V^{*n}, \qquad W_{\bar 1} = V^k \oplus V^{*\ell}. \tag{4.19}$$

As a $\mathrm{GL}(V)$-module, we have a decomposition $\mathcal{S}(W) = \oplus_{r\geq 0}\mathcal{S}^r(W)$ by the total degree, where $\mathcal{S}^r(W)$ as a $\mathrm{GL}(V)$-module is a direct sum of the following tensor spaces (which are also $\mathrm{GL}(V)$-submodules)

$$\begin{aligned} S^{\alpha\beta\gamma\delta} :=& S^{\alpha_1}(V) \otimes \ldots \otimes S^{\alpha_m}(V) \otimes S^{\beta_1}(V^*) \otimes \ldots \otimes S^{\beta_n}(V^*) \\ & \otimes \wedge^{\gamma_1}(V) \otimes \ldots \otimes \wedge^{\gamma_k}(V) \otimes \wedge^{\delta_1}(V^*) \otimes \ldots \otimes \wedge^{\delta_\ell}(V^*), \end{aligned} \tag{4.20}$$

where the summation is taken over all nonnegative integers $\alpha_a, \beta_b, \gamma_c$, and δ_d such that $|\alpha| + |\beta| + |\gamma| + |\delta| = r$. Here we have denoted $\alpha = (\alpha_1, \dots, \alpha_m)$, $|\alpha| = \sum_a \alpha_a$, similarly for $\beta, |\beta|$, and so on.

We now fix $\alpha, \beta, \gamma, \delta$, and want to describe a spanning set for $(S^{\alpha\beta\gamma\delta})^{\mathrm{GL}(V)}$. To that end, we regard the space (or the $\mathrm{GL}(V)$-module) $S^{\alpha\beta\gamma\delta}$ as a subspace of the following mixed tensor space (or $\mathrm{GL}(V)$-module)

$$(4.21)\qquad \begin{aligned} T^{\alpha\beta\gamma\delta} :=& V^{\otimes\alpha_1} \otimes \ldots \otimes V^{\otimes\alpha_m} \otimes (V^*)^{\otimes\beta_1} \otimes \ldots \otimes (V^*)^{\otimes\beta_n} \\ &\otimes V^{\otimes\gamma_1} \otimes \ldots \otimes V^{\otimes\gamma_k} \otimes (V^*)^{\otimes\delta_1} \otimes \ldots \otimes (V^*)^{\otimes\delta_\ell} \end{aligned}$$

(which is isomorphic to $V^{\otimes(|\alpha|+|\gamma|)} \otimes (V^*)^{\otimes(|\beta|+|\delta|)}$). Hence, the space $(S^{\alpha\beta\gamma\delta})^{\mathrm{GL}(V)}$ can be viewed as a subspace of $(T^{\alpha\beta\gamma\delta})^{\mathrm{GL}(V)}$. By the tensor FFT for $\mathrm{GL}(V)$ (Theorem 4.2), we will assume below that $|\alpha| + |\gamma| = |\beta| + |\delta| = N$ for some N, for otherwise $(T^{\alpha\beta\gamma\delta})^{\mathrm{GL}(V)} = 0$. By the same tensor FFT again, the $\mathrm{GL}(V)$-invariants of $T^{\alpha\beta\gamma\delta}$ are spanned by Θ_σ for $\sigma \in \mathfrak{S}_N$, and each Θ_σ as defined in (4.3) is a suitable tensor product of N degree-two invariants that arise from various pairings between V and V^* such that each copy of V and V^* in $T^{\alpha\beta\gamma\delta}$ is used exactly in one such pairing.

The subspace $(S^{\alpha\beta\gamma\delta})^{\mathrm{GL}(V)}$ is obtained from the space $(T^{\alpha\beta\gamma\delta})^{\mathrm{GL}(V)}$ by a (partial) supersymmetrization. Applying the $\mathfrak{S}_\alpha \times \mathfrak{S}_\beta \times \mathfrak{S}_\gamma^- \times \mathfrak{S}_\delta^-$-supersymmetrization operator $\varpi_{\alpha\beta\gamma\delta}$ to the Θ_σ's gives us a spanning set $\{\varpi_{\alpha\beta\gamma\delta}(\Theta_\sigma)\}$ for $(S^{\alpha\beta\gamma\delta})^{\mathrm{GL}(V)}$, where $\mathfrak{S}_\alpha = \mathfrak{S}_{\alpha_1} \times \ldots \times \mathfrak{S}_{\alpha_m}$ and by supersymmetrization we mean the symmetrization over the subgroup $\mathfrak{S}_\alpha \times \mathfrak{S}_\beta$ and the anti-symmetrization (denoted by the minus sign superscripts above) over the subgroup $\mathfrak{S}_\gamma \times \mathfrak{S}_\delta$. Denote the degree-two $\mathrm{GL}(V)$-invariants of $\mathcal{S}(W) = S(V^m \oplus V^{*n}) \otimes \wedge(V^k \oplus V^{*\ell})$, which are straightforward variants of the contractions $\langle j|i^*\rangle$ introduced in (4.4), by

$$(4.22)\qquad \langle a|b^*||\rangle, \quad \langle a|||d^*\rangle, \quad \langle|b^*|c|\rangle, \quad \langle||c|d^*\rangle,$$

where $1 \le a \le m$, $1 \le b \le n, 1 \le c \le k, 1 \le d \le \ell$. Here in the notation of (4.22) and similar ones below, we have used a vertical line $|$ to indicate the location of the chosen pairs of V and V^* in $\mathcal{S}(V^m \oplus V^{*n} \oplus V^k \oplus V^{*\ell})$. From the tensor product form of Θ_σ we derive that $\varpi_{\alpha\beta\gamma\delta}(\Theta_\sigma) \in (S^{\alpha\beta\gamma\delta})^{\mathrm{GL}(V)}$ is the corresponding product of N invariants among (4.22).

The cases for $G = \mathrm{O}(V)$ or $\mathrm{Sp}(V)$ can be pursued in an analogous manner. Set

$$(4.23)\qquad W_{\bar 0} = V^m, \qquad W_{\bar 1} = V^k.$$

We do not need V^* here thanks to $V^* \cong V$ as a G-module. We have a similar decomposition of the G-module $\mathcal{S}(W) = \oplus_{\alpha,\gamma} S^{\alpha\gamma}$, where $S^{\alpha\gamma}$ is modified from (4.20) with β and δ being zero. The space $S^{\alpha\gamma}$ is a subspace (and also a G-submodule) of the tensor space $T^{\alpha\gamma}$, where $T^{\alpha\gamma}$ is modified from (4.21) with β and δ being zero. Hence, the space $(S^{\alpha\gamma})^G$ can be viewed as a subspace of $(T^{\alpha\gamma})^G$. By the tensor FFT (Theorem 4.17), we may assume below that $|\alpha| + |\gamma|$ is even, say, equal to $2N$. Then by the same FFT again, the G-invariants of $T^{\alpha\gamma}$ are spanned by θ_σ for $\sigma \in \mathfrak{S}_{2N}$, and each θ_σ is a suitable tensor product of N degree two invariants that arise from pairings $V \otimes V$ so that each V in $T^{\alpha\gamma}$ is used exactly in one such pairing.

The subspace $(S^{\alpha\gamma})^G$ is obtained from the space $(T^{\alpha\gamma})^G$ by a (partial) supersymmetrization. Hence, applying the $\mathfrak{S}_\alpha \times \mathfrak{S}_\gamma^-$-supersymmetrization operator $\varpi_{\alpha\gamma}$ to the θ_σ's gives us a spanning set $\{\varpi_{\alpha\gamma}(\theta_\sigma)\}$ for $(S^{\alpha\gamma})^G$. Denote the degree-two G-invariants in $S^{\alpha\gamma}$, which are straightforward variants of $(i|j)$ introduced in (4.18), by

$$(a|c), \quad (a;a'|), \quad (|c;c'), \tag{4.24}$$

where $1 \le a \le a' \le m$ and $1 \le c \le c' \le k$. Note in addition that $(a;a|) = 0$ for $G = \mathrm{Sp}(V)$ and $(|c;c) = 0$ for $G = \mathrm{O}(V)$. Hence, we derive from the tensor product form of θ_σ that $\varpi_{\alpha\gamma}(\theta_\sigma)$ is a product of N invariants among $(a|c), (a;a'|), (|c;c')$.

Summarizing, we have established the following.

Theorem 4.19 (Supersymmetric FFT). *Let $G = \mathrm{GL}(V), \mathrm{O}(V)$, or $\mathrm{Sp}(V)$. Let W be as given in* (4.19) *for* $\mathrm{GL}(V)$ *and in* (4.23) *for* $\mathrm{O}(V)$ *or* $\mathrm{Sp}(V)$*, respectively. The algebra of G-invariants in the supersymmetric algebra $\mathcal{S}(W)$ is generated by its quadratic invariants given in* (4.22) *for* $\mathrm{GL}(V)$ *and in* (4.24) *for* $\mathrm{O}(V)$ *or* $\mathrm{Sp}(V)$*, respectively.*

Remark 4.20. Theorem 4.19 has two most important specializations:

(1) When $W_{\bar{1}} = 0$, the supersymmetric algebra on W reduces to the symmetric algebra $S(W_{\bar{0}})$. The corresponding FFT admits an equivalent reformulation as a polynomial FFT that describes the G-invariants in a polynomial algebra on a direct sum of copies of V (and V^*).

(2) When $W_{\bar{0}} = 0$, the supersymmetric algebra on W reduces to the exterior algebra $\wedge(W_{\bar{1}})$. The G-invariants in this case are known as **skew invariants**.

4.4. Exercises

Exercise 4.1. Prove that the image of

$$\tau : \mathrm{Hom}(\mathbb{C}^k, V) \oplus \mathrm{Hom}(V, \mathbb{C}^m) \to M_{mk}, \quad \tau(x,y) = yx,$$

consists of all the matrices $Z \in M_{mk}$ such that $\mathrm{rank}(Z) \le \min\{m, k, \dim V\}$.

Exercise 4.2. Consider the cyclic diagonal subgroup $C_n = \{\mathrm{diag}\,(\xi, \xi^{-1}) \mid \xi^n = 1\}$ of $\mathrm{SL}(2)$ and identify the symmetric algebra $S(\mathbb{C}^2)$ with $\mathbb{C}[x,y]$. Find a presentation in terms of generators and relations for the algebra of invariants $\mathbb{C}[x,y]^{C_n}$.

Exercise 4.3. Let V be a complex vector space of dimension n. For $A \in \mathrm{End}(V)$, consider its characteristic polynomial

$$\det(tI - A) = t^n + \sum_{i=1}^{n} (-1)^i e_i(A) t^{n-i}.$$

Also introduce the polynomial function $\mathrm{tr}_i : \mathrm{End}(V) \to \mathbb{C}$, $A \mapsto \mathrm{tr}\, A^i$, for $i \ge 1$. Prove that:

(1) The e_i and tr_i, for $1 \le i \le n$, are invariant polynomials on $\mathrm{End}(V)$.

(2) The algebra of invariants for the conjugation action of $\mathrm{GL}(V)$ on $\mathrm{End}(V)$ is isomorphic to a polynomial algebra generated by $e_1, \ldots, e_n$.

(3) The algebra of invariants for the conjugation action of $\mathrm{GL}(V)$ on $\mathrm{End}(V)$ is isomorphic to a polynomial algebra generated by $\mathrm{tr}_1, \ldots, \mathrm{tr}_n$.

(Hint: Establish connection to symmetric polynomials in n variables and note that the set of diagonalizable $n \times n$ matrices is Zariski dense in M_{nn}.)

Exercise 4.4. Let U and V be two finite-dimensional simple $\mathrm{GL}(n)$-modules, and assume that $U \cong V$ as $\mathrm{SL}(n)$-modules. Then $U \cong V \otimes \det^r$ for some $r \in \mathbb{Z}$ as $\mathrm{GL}(n)$-modules, where $\det : \mathrm{GL}(n) \to \mathbb{C}$ is the determinant $\mathrm{GL}(n)$-module.

Notes

Section 4.1. The materials here are mostly classical and standard. Weyl's book [**131**] on invariant theory was influential, where the terminology of First Fundamental Theorem (FFT) for invariants was introduced. See Howe [**51, 52**]; also see the notes of Kraft-Procesi [**74**]. Theorem 4.1 on finite generation of an invariant subalgebra of the supersymmetric algebra is somewhat novel, as it is usually formulated for a polynomial algebra only.

Section 4.2. The results here are classical. Our proof of Weyl's theorem follows Kraft-Procesi [**74**]. Weyl's theorem allows us to reduce the proof of the polynomial FFT for classical groups in general to a basic case. The basic case is then proved by induction on dimensions, and our presentation here does not differ much from Goodman-Wallach [**46**]. In [**46**], the general polynomial FFT was derived using the tensor FFT instead of Weyl's theorem, while in Weyl [**131**], the Cappelli identity was used essentially in the proof of the polynomial FFT.

Section 4.3. The tensor FFT for classical groups can be found in Weyl [**131**]. The quick proof of the tensor FFT for the orthogonal groups here is due to Atiyah-Bott-Patodi [**3**], and it works for the symplectic groups similarly (also see a presentation in Goodman-Wallach [**46**]). The supersymmetric FFT for classical groups, which is a common generalization of the polynomial FFT and skew FFT, is taken from Howe [**51**]. The supersymmetric FFT will be used in the next chapter on Howe duality for Lie superalgebras.

The exercises are pretty standard.

Howe duality

The goal of this chapter is to explain Howe duality for classical Lie groups, Lie algebras, and Lie superalgebras, and applications to irreducible character formulas for Lie superalgebras. Like Schur duality, Howe duality involves commuting actions of a pair of Lie group G and Lie superalgebra $\mathfrak{g}'$ on a supersymmetric algebra $\mathcal{S}(U)$, and $\mathcal{S}(U)$ is naturally an irreducible module over a Weyl-Clifford algebra $\mathfrak{WC}$. In the main examples, G is one of the three classical groups $\mathrm{GL}(V), \mathrm{Sp}(V)$, or $\mathrm{O}(V)$, $\mathfrak{g}'$ is a classical Lie algebra or superalgebra, and U is a direct sum of copies of the natural G-module and its dual. According to the FFT for classical invariant theory in Chapter 4, when applied to the G-action on the associated graded $\mathrm{gr}\,\mathfrak{WC}$, the basic invariants generating $(\mathrm{gr}\,\mathfrak{WC})^G$ turn out to form the associated graded space for a Lie (super)algebra $\mathfrak{g}'$. From this it follows that the algebra of G-invariants $\mathfrak{WC}^G$ is generated by $\mathfrak{g}'$.

The most important cases of $\mathcal{S}(U)$ are the two cases of a symmetric algebra and of an exterior algebra, corresponding respectively to $U_{\bar{1}} = 0$ and $U_{\bar{0}} = 0$. In these two cases, the Howe dualities only involve Lie groups and Lie algebras, and they may be regarded as Lie theoretic reformulations of the classical polynomial invariant and skew invariant theories, respectively, treated in Chapter 4.

Howe duality provides a realization of an important class of irreducible modules, called oscillator modules, of the Lie (super)algebra $\mathfrak{g}'$. The irreducible G-modules and irreducible oscillator $\mathfrak{g}'$-modules appearing in the decomposition of $\mathcal{S}(U)$ are paired in "duality", which is much stronger than a simple bijection. Character formulas for the irreducible oscillator $\mathfrak{g}'$-modules are then obtained via a comparison with Howe duality involving classical groups G and infinite-dimensional Lie algebras, which we develop in detail.

5.1. Weyl-Clifford algebra and classical Lie superalgebras

In this section, we introduce a mixture of Weyl algebra and Clifford algebra uniformly in the framework of superalgebras. We formulate the basic properties of this Weyl-Clifford algebra and its connections to classical Lie algebras and superalgebras. We then establish a general duality theorem that works particularly well in the setting of Weyl-Clifford algebra.

5.1.1. Weyl-Clifford algebra. Recall that for a superspace $U = U_{\bar{0}} \oplus U_{\bar{1}}$, we denote the supersymmetric algebra by

$$\mathcal{S}(U) = S(U_{\bar{0}}) \otimes \wedge(U_{\bar{1}}).$$

In a first round of reading of the constructions below, the reader is advised to understand first the two special cases with $U_{\bar{1}} = 0$ and then $U_{\bar{0}} = 0$, respectively.

The superspace $\mathcal{S}(U)$ is spanned by monomials of the form $u_1 \dots u_p w_1 \dots w_q$, for $u_1, \dots, u_p \in U_{\bar{0}}$ and $w_1, \dots, w_q \in U_{\bar{1}}$. Note that $u_i u_j = u_j u_i, u_i w_k = w_k u_i$, and $w_k w_\ell = -w_\ell w_k$. The space $\mathcal{S}(U)$ is $\mathbb{Z} \times \mathbb{Z}_2$-graded by letting the degree of a nonzero element $u_1 \dots u_p w_1 \dots w_q$ be $(p+q, \bar{q})$, where $\bar{q}$ denotes q modulo 2.

For $x \in U_{\bar{0}}$ (respectively, $x \in U_{\bar{1}}$), we define the left multiplication operator of degree $(1, \bar{0})$ (respectively, $(1, \bar{1})$)

$$M_x : \mathcal{S}(U) \longrightarrow \mathcal{S}(U), \quad y \mapsto xy.$$

We regard $U_{\bar{0}}^* \subseteq U^*$ by extending $u^* \in U_{\bar{0}}^*$ to an element in U^* by setting $\langle u^*, w \rangle = 0$ for all $w \in U_{\bar{1}}$. Similarly we have $U_{\bar{1}}^* \subseteq U^*$. In this way, we identify $U^* \equiv U_{\bar{0}}^* \oplus U_{\bar{1}}^*$ as a vector superspace. For $u^* \in U_{\bar{0}}^*$, we define the even derivation D_{u^*} of degree $(-1, \bar{0})$, and for $w^* \in U_{\bar{1}}^*$, we define the odd derivation D_{w^*} of degree $(-1, \bar{1})$ as follows:

$$D_{u^*} : \mathcal{S}(U) \longrightarrow \mathcal{S}(U), \quad D_{w^*} : \mathcal{S}(U) \longrightarrow \mathcal{S}(U),$$

$$u_1 \dots u_p w_1 \dots w_q \overset{D_{u^*}}{\longmapsto} \sum_{i=1}^{p} \langle u^*, u_i \rangle u_1 \dots \hat{u}_i \dots u_p w_1 \dots w_q,$$

$$u_1 \dots u_p w_1 \dots w_q \overset{D_{w^*}}{\longmapsto} \sum_{k=1}^{q} (-1)^{k-1} \langle w^*, w_k \rangle u_1 \dots u_p w_1 \dots \hat{w}_k \dots w_q,$$

where $u_i \in U_{\bar{0}}$ and $w_j \in U_{\bar{1}}$. Here and below $\hat{u}_i$ means the omittance of u_i, and so on.

The linear operators M_u and D_{u^*}, for $u \in U_{\bar{0}}, u^* \in U_{\bar{0}}^*$, in the superalgebra $\mathrm{End}(\mathcal{S}(U))$ defined above are even, while $M_w, D_{w^*} \in \mathrm{End}(\mathcal{S}(U))$, for $w \in U_{\bar{1}}, w^* \in U_{\bar{1}}^*$, are odd. We shall denote by $[\cdot,\cdot]$ the supercommutator among linear operators.

Lemma 5.1. *For $x^*, y^* \in U^*$ and $x, y \in U$, we have*

$$[D_{x^*}, D_{y^*}] = [M_x, M_y] = 0, \qquad [D_{x^*}, M_y] = \langle x^*, y \rangle 1.$$

Let us reformulate. Set

$$\boldsymbol{U} = U \oplus U^*.$$

Then $\boldsymbol{U}$ is a superspace $\boldsymbol{U} = \boldsymbol{U}_{\bar{0}} \oplus \boldsymbol{U}_{\bar{1}}$ with

$$\boldsymbol{U}_{\bar{0}} = U_{\bar{0}} \oplus U_{\bar{0}}^*, \quad \boldsymbol{U}_{\bar{1}} = U_{\bar{1}} \oplus U_{\bar{1}}^*. \tag{5.1}$$

Let $\langle \cdot, \cdot \rangle'$ be the even skew-supersymmetric bilinear form on the superspace $\boldsymbol{U}$ defined as follows: for $u_1, u_2 \in U_{\bar{0}}, u_1^*, u_2^* \in U_{\bar{0}}^*$, $w_1, w_2 \in U_{\bar{1}}$, and $w_1^*, w_2^* \in U_{\bar{1}}^*$, set

$$\begin{aligned}
\langle u_1 + u_1^*, u_2 + u_2^* \rangle' &= \langle u_1^*, u_2 \rangle - \langle u_2^*, u_1 \rangle, \\
\langle w_1 + w_1^*, w_2 + w_2^* \rangle' &= \langle w_1^*, w_2 \rangle + \langle w_2^*, w_1 \rangle, \\
\langle u_1 + u_1^*, w_1 + w_1^* \rangle' &= 0.
\end{aligned}$$

Define a linear map

$$\iota : \boldsymbol{U} \longrightarrow \mathrm{End}(\mathcal{S}(U))$$

by letting

$$\iota(x) = M_x, \; \iota(x^*) = D_{x^*}, \qquad \text{for } x \in U, \; x^* \in U^*.$$

It is straightforward to check that Lemma 5.1 affords the following equivalent reformulation.

Lemma 5.2. *We have*

$$[\iota(a), \iota(b)] = \langle a, b \rangle' 1, \qquad \text{for } a, b \in \boldsymbol{U}. \tag{5.2}$$

The subalgebra in $\mathrm{End}(\mathcal{S}(U))$ generated by $\iota(\boldsymbol{U})$ will be denoted by $\mathfrak{WC}(\boldsymbol{U})$ and referred to as the **Weyl-Clifford (super)algebra**. When $U_{\bar{1}} = 0$, $\mathfrak{WC}(\boldsymbol{U})$ reduces to the usual **Weyl algebra** $\mathfrak{W}(\boldsymbol{U}_{\bar{0}})$. Choose a basis $x_1, \ldots, x_m$ for the m-dimensional space $U_{\bar{0}}$, with dual basis $x_1^*, \ldots, x_m^*$, so that $S(U_{\bar{0}})$ is identified with $\mathbb{C}[x_1, \ldots, x_m]$. The operators M_{x_i} can then be identified with the multiplication operators x_i, while the derivations $D_{x_i^*}$ are identified with the differential operators $\frac{\partial}{\partial x_i} \equiv \partial_i$. Hence, $\mathfrak{W}(\boldsymbol{U}_{\bar{0}})$ is identified with the algebra of polynomial differential operators on $\mathbb{C}[x_1, \ldots, x_m]$. When $U_{\bar{0}} = 0$ and $\dim U_{\bar{1}} = n$, $\mathfrak{WC}(\boldsymbol{U})$ is isomorphic to the **Clifford superalgebra** $\mathcal{C}_{2n}$ (see Definition 3.33), which we also denote by $\mathfrak{C}(\boldsymbol{U}_{\bar{1}})$. In the same spirit as above, let $\eta_1, \ldots, \eta_n$ be a basis for $U_{\bar{1}}$, and let $\eta_1^*, \ldots, \eta_n^*$ be its dual basis, so that $\wedge(U_{\bar{1}}) \equiv \wedge(\eta_1, \ldots, \eta_n)$, the exterior superalgebra generated by $\eta_1, \ldots, \eta_n$. The multiplication operators M_{η_i} correspond to left multiplications by η_i and shall be denoted accordingly by η_i. The derivations $D_{\eta_i^*}$ correspond to the odd differential operators $\frac{\partial}{\partial \eta_i} \equiv \eth_i$. Hence, $\mathfrak{C}(\boldsymbol{U}_{\bar{1}})$ can be regarded as the superalgebra of differential operators on $\wedge(\eta_1, \ldots, \eta_n)$.

For general U, there is an algebra isomorphism $\mathfrak{WC}(\boldsymbol{U}) \cong \mathfrak{W}(\boldsymbol{U}_{\bar{0}}) \otimes \mathfrak{C}(\boldsymbol{U}_{\bar{1}})$ and an identification

$$\mathcal{S}(U) \equiv \mathbb{C}[x, \eta] := \mathbb{C}[x_1, \ldots, x_m] \otimes \wedge(\eta_1, \ldots, \eta_n).$$

The superalgebra $\mathfrak{WC}(\boldsymbol{U})$ is then identified with the superalgebra of polynomial differential operators on the superspace U, that is, the subalgebra of $\mathrm{End}(\mathcal{S}(U))$

generated by the multiplication operators x_i, η_j and the (super)derivations ∂_i, $\eth_j$, for $1 \le i \le m, 1 \le j \le n$.

5.1.2. A filtration on Weyl-Clifford algebra. The linear map $\iota|_U : U \to \mathfrak{WC}(\boldsymbol{U})$ induces an injective algebra homomorphism $\kappa : \mathcal{S}(U) \to \mathfrak{WC}(\boldsymbol{U})$, and similarly, $\iota|_{U^*} : U^* \to \mathfrak{WC}(\boldsymbol{U})$ induces an injective algebra homomorphism $\kappa^* : \mathcal{S}(U^*) \to \mathfrak{WC}(\boldsymbol{U})$.

Proposition 5.3. *The map $m : \mathcal{S}(U) \otimes \mathcal{S}(U^*) \to \mathfrak{WC}(\boldsymbol{U})$ given by $x \otimes y \mapsto \kappa(x)\kappa^*(y)$ is a linear isomorphism. Moreover, the algebra $\mathfrak{WC}(\boldsymbol{U})$ is isomorphic to the algebra generated by $\iota(\boldsymbol{U})$ subject to the relation* (5.2).

Proof. The two statements in the proposition are equivalent. It follows by definition and (5.2) that the map m is surjective. The injectivity of m is equivalent to the claim that the "monomials" of an ordered basis for U and U^* form a basis of $\mathfrak{WC}(\boldsymbol{U})$, and the claim can then be established by a straightforward inductive argument.

Alternatively, the proposition for $\mathfrak{WC}(\boldsymbol{U})$ reduces to the well-known counterparts for $\mathfrak{W}(\boldsymbol{U}_{\bar{0}})$ and $\mathfrak{C}(\boldsymbol{U}_{\bar{1}})$, thanks to $\mathfrak{WC}(\boldsymbol{U}) \cong \mathfrak{W}(\boldsymbol{U}_{\bar{0}}) \otimes \mathfrak{C}(\boldsymbol{U}_{\bar{1}})$ (note that $\mathfrak{W}(\boldsymbol{U}_{\bar{0}})$ and $\mathfrak{C}(\boldsymbol{U}_{\bar{1}})$ commute). □

We continue to identify $\mathfrak{WC}(\boldsymbol{U})$ with the superalgebra of polynomial differential operators generated by the multiplication operators x_i, η_j and the (super)derivations ∂_i, $\eth_j$, for $1 \le i \le m, 1 \le j \le n$. By Proposition 5.3, $\mathfrak{WC}(\boldsymbol{U})$ has a linear basis $\{x^\alpha \eta^\beta \partial^\gamma \eth^\delta \mid \alpha, \gamma \in \mathbb{Z}_+^m, \beta, \delta \in \mathbb{Z}_2^n\}$, where $x^\alpha = x_1^{\alpha_1} \cdots x_m^{\alpha_m}$ for $\alpha = (\alpha_1, \ldots, \alpha_m)$ and η^β, ∂^γ, $\eth^\delta$ are defined similarly. Define the degree

$$\deg x^\alpha \eta^\beta \partial^\gamma \eth^\delta = |\alpha| + |\beta| + |\gamma| + |\delta|,$$

where $|\alpha| = \alpha_1 + \ldots + \alpha_m$ and $|\beta|, |\gamma|, |\delta|$ are similarly defined.

The Weyl-Clifford algebra $\mathfrak{WC}(\boldsymbol{U})$ carries a natural filtration $\{\mathfrak{WC}(\boldsymbol{U})_k\}_{k \ge 0}$ by letting $\mathfrak{WC}(\boldsymbol{U})_k$ be spanned by the elements $x^\alpha \eta^\beta \partial^\gamma \eth^\delta$ of degree no greater than k. Clearly, $\mathfrak{WC}(\boldsymbol{U})_k \mathfrak{WC}(\boldsymbol{U})_\ell \subseteq \mathfrak{WC}(\boldsymbol{U})_{k+\ell}$, and so $\mathfrak{WC}(\boldsymbol{U})$ is a filtered algebra. Actually, $\mathfrak{W}(\boldsymbol{U}_{\bar{0}})$ and $\mathfrak{C}(\boldsymbol{U}_{\bar{1}})$ are filtered subalgebras of the filtered algebra $\mathfrak{WC}(\boldsymbol{U})$, and $\mathfrak{WC}(\boldsymbol{U}) = \mathfrak{W}(\boldsymbol{U}_{\bar{0}}) \otimes \mathfrak{C}(\boldsymbol{U}_{\bar{1}})$ as a tensor product of filtered algebras. As usual, the associated graded algebra for $\mathfrak{WC}(\boldsymbol{U})$ is defined to be

$$\operatorname{gr} \mathfrak{WC}(\boldsymbol{U}) = \bigoplus_{k=0}^{\infty} \mathfrak{WC}(\boldsymbol{U})_k / \mathfrak{WC}(\boldsymbol{U})_{k-1}.$$

For $T = \sum_{\alpha,\beta,\gamma,\delta} c_{\alpha\beta\gamma\delta} x^\alpha \eta^\beta \partial^\gamma \eth^\delta$ with $c_{\alpha\beta\gamma\delta} \in \mathbb{C}$ of filtered degree j, i.e., $T \in \mathfrak{WC}(\boldsymbol{U})_j \backslash \mathfrak{WC}(\boldsymbol{U})_{j-1}$, we define the **symbol** or the **Weyl symbol** of T to be

$$\sigma(T) = \sum_{|\alpha|+|\beta|+|\gamma|+|\delta|=j} c_{\alpha,\beta,\gamma,\delta} x^\alpha \eta^\beta y^\gamma \xi^\delta \in \mathcal{S}^j(\boldsymbol{U}),$$

where we denote by y_i and ξ_i the generators in $\mathcal{S}(\boldsymbol{U})$ that correspond to ∂_i and $\eth_i$, respectively. We define a **symbol map**, for $k \geq 0$,

$$\sigma : \mathfrak{WC}(\boldsymbol{U})_k \to \bigoplus_{j=0}^{k} \mathcal{S}^j(\boldsymbol{U}) \tag{5.3}$$

by sending T to $\sigma(T)$.

Proposition 5.4. *Let G be a classical subgroup of $\mathrm{GL}(U_{\bar{0}}) \times \mathrm{GL}(U_{\bar{1}})$. The symbol map σ is a G-module isomorphism from $\mathfrak{WC}(\boldsymbol{U})_k$ to $\bigoplus_{j=0}^{k} \mathcal{S}^j(\boldsymbol{U})$ for each k. As a G-module, the associated graded algebra $gr\,\mathfrak{WC}(\boldsymbol{U})$ is isomorphic to the supersymmetric algebra $\mathcal{S}(\boldsymbol{U})$.*

Proof. The G-module $\mathfrak{WC}(\boldsymbol{U})_k$ is semisimple and hence isomorphic to the associated graded $\bigoplus_{j=0}^{k} \mathfrak{WC}(\boldsymbol{U})_j / \mathfrak{WC}(\boldsymbol{U})_{j-1}$. For a fixed j, the linear span of

$$\{x^{\alpha}\eta^{\beta}\partial^{\gamma}\eth^{\delta} \mid |\alpha| + |\beta| + |\gamma| + |\delta| = j\}$$

as a G-module is isomorphic to $\mathfrak{WC}(\boldsymbol{U})_j / \mathfrak{WC}(\boldsymbol{U})_{j-1}$, which in turn is isomorphic to $\mathcal{S}^j(\boldsymbol{U})$. The proposition follows. □

5.1.3. Relation to classical Lie superalgebras. There is an intimate relation between the Weyl-Clifford algebra and classical Lie algebras/superalgebras, which we will now formulate. Introduce $\mathcal{G} = \mathcal{G}_{\bar{0}} \oplus \mathcal{G}_{\bar{1}}$, where

$$\mathcal{G}_{\bar{0}} = S^2(\iota(\boldsymbol{U}_{\bar{0}})) \oplus \wedge^2(\iota(\boldsymbol{U}_{\bar{1}})), \qquad \mathcal{G}_{\bar{1}} = \iota(\boldsymbol{U}_{\bar{0}}) \otimes \iota(\boldsymbol{U}_{\bar{1}}).$$

Proposition 5.5. (1) *The subspace $\mathcal{G}$ of $\mathfrak{WC}(\boldsymbol{U})$ is closed under the supercommutator and as Lie superalgebras $\mathcal{G} \cong \mathfrak{spo}(\boldsymbol{U})$. In particular, $\mathcal{G}_{\bar{0}}$ is a Lie algebra isomorphic to $\mathfrak{sp}(\boldsymbol{U}_{\bar{0}}) \oplus \mathfrak{so}(\boldsymbol{U}_{\bar{1}})$.*

(2) *Under the adjoint action, $\iota(\boldsymbol{U})$ is isomorphic to the natural representation of $\mathcal{G} \cong \mathfrak{spo}(\boldsymbol{U})$.*

Proof. We identify $S^2(\iota(\boldsymbol{U}_{\bar{0}}))$ with the space of anti-commutators $[\iota(\boldsymbol{U}_{\bar{0}}), \iota(\boldsymbol{U}_{\bar{0}})]_+$, that is, the linear span of $[a,b]_+ = ab + ba$, for $a, b \in \iota(\boldsymbol{U}_{\bar{0}})$; note by (5.2) that $[a,b] = ab - ba \in \mathbb{C}$. Similarly, we identify $\wedge^2(\iota(\boldsymbol{U}_{\bar{1}}))$ with $[\iota(\boldsymbol{U}_{\bar{1}}), \iota(\boldsymbol{U}_{\bar{1}})]$, the linear span of $[c,d] = cd - dc$, for $c, d \in \iota(\boldsymbol{U}_{\bar{1}})$; note by (5.2) that $[c,d]_+ \in \mathbb{C}$.

We claim that $S^2(\iota(\boldsymbol{U}_{\bar{0}}))$ is closed under the commutator, and moreover, we have $S^2(\iota(\boldsymbol{U}_{\bar{0}})) \cong \mathfrak{sp}(\boldsymbol{U}_{\bar{0}})$ as Lie algebras. To see this, let $x_1, \ldots, x_m$ be a basis for $U_{\bar{0}}$. Regard x_i and ∂_i as elements in $\mathfrak{WC}(\boldsymbol{U})$ as before. We may identify $S^2(\iota(\boldsymbol{U}_{\bar{0}}))$ with the space spanned by elements of the form

$$x_i\partial_j + \partial_j x_i, \quad x_i x_j, \quad \partial_i\partial_j, \quad 1 \leq i, j \leq m. \tag{5.4}$$

By a direct computation via this spanning set, one checks that $S^2(\iota(\boldsymbol{U}_{\bar{0}}))$ is closed under the commutator. Using that $[b, a_1] = \langle b, a_1 \rangle'$ by (5.2), one further computes that, for $a, b, a_1 \in \iota(\boldsymbol{U}_{\bar{0}})$,

$$[[a,b]_+, a_1] = a[b,a_1] + [b,a_1]a + [a,a_1]b + b[a,a_1] \\ = 2\langle b,a_1\rangle' a + 2\langle a,a_1\rangle' b. \tag{5.5}$$

Hence, $\iota(\boldsymbol{U}_{\bar{0}})$ is a representation of the Lie algebra $S^2(\iota(\boldsymbol{U}_{\bar{0}}))$ under the adjoint action. With respect to the ordered basis $\{x_1, \ldots, x_m, \partial_1, \ldots, \partial_m\}$ for $\iota(\boldsymbol{U}_{\bar{0}})$, we can write down the matrices of (5.4), the span of which is precisely given by the $2m \times 2m$ matrices of the $m|m$-block form

$$\begin{pmatrix} A & B \\ C & -A^t \end{pmatrix}, \quad B = B^t, C = C^t.$$

This shows that $S^2(\iota(\boldsymbol{U}_{\bar{0}})) \cong \mathfrak{sp}(\boldsymbol{U}_{\bar{0}})$ and that $\iota(\boldsymbol{U}_{\bar{0}})$ is its natural representation.

Similarly, $\wedge^2(\iota(\boldsymbol{U}_{\bar{1}}))$ is closed under the Lie bracket, $\wedge^2(\iota(\boldsymbol{U}_{\bar{1}})) \cong \mathfrak{so}(\boldsymbol{U}_{\bar{1}})$ as Lie algebras, and $\iota(\boldsymbol{U}_{\bar{1}})$ under the adjoint action of $\wedge^2(\iota(\boldsymbol{U}_{\bar{1}}))$ is its natural representation. Since the case for $\wedge^2(\iota(\boldsymbol{U}_{\bar{1}}))$ is completely parallel to the case for $S^2(\iota(\boldsymbol{U}_{\bar{0}}))$ above, we shall omit the details except recording the following identity analogous to (5.5):

$$[[c,d], c_1] = 2\langle d,c_1\rangle' c - 2\langle c,c_1\rangle' d, \quad \forall c,d,c_1 \in \iota(\boldsymbol{U}_{\bar{1}}). \tag{5.6}$$

On the other hand, by a direct computation, we have that, for $a, b \in \boldsymbol{U}_{\bar{0}}$ and $c, d \in \boldsymbol{U}_{\bar{1}}$,

$$[a \otimes c, b \otimes d]_+ = \frac{1}{2}([a,b] \otimes [c,d] + [a,b]_+ \otimes [c,d]_+).$$

Here we recall $[a,b] \in \mathbb{C}$ and $[c,d]_+ \in \mathbb{C}$. This implies that $[\mathcal{G}_{\bar{1}}, \mathcal{G}_{\bar{1}}]_+ \subseteq \mathcal{G}_{\bar{0}}$. Hence $\mathcal{G}$ is closed under the super-commutator and so is a Lie subalgebra of $\mathfrak{WC}(\boldsymbol{U})$.

We further observe that $\mathcal{G}$ preserves the bilinear form $\langle\cdot,\cdot\rangle'$ on $\boldsymbol{U}$ and hence is a subalgebra of $\mathfrak{spo}(\boldsymbol{U})$. Indeed, it follows by using (5.5) twice that, for $a, b, a_1, b_1 \in \iota(\boldsymbol{U}_{\bar{0}})$,

$$\langle [[a,b]_+, a_1], b_1 \rangle' = -\langle a_1, [[a,b]_+, b_1] \rangle'. \tag{5.7}$$

Similarly, it follows by (5.6) that, for $c, d, c_1, d_1 \in \iota(\boldsymbol{U}_{\bar{1}})$,

$$\langle [[c,d], c_1], d_1 \rangle' = -\langle c_1, [[c,d], d_1] \rangle'. \tag{5.8}$$

The remaining identity for the $\mathcal{G}$-invariance of $\langle\cdot,\cdot\rangle'$ can be verified similarly (see Exercise 5.1). Since $\mathcal{G}_{\bar{0}} = \mathfrak{sp}(\boldsymbol{U}_{\bar{0}}) \oplus \mathfrak{so}(\boldsymbol{U}_{\bar{1}})$ and $\mathcal{G}_{\bar{1}} \cong \boldsymbol{U}_{\bar{0}} \otimes \boldsymbol{U}_{\bar{1}}$, it follows that $\mathcal{G} \cong \mathfrak{spo}(\boldsymbol{U})$. Now (5.5) and (5.6) together with $[[a,b]_+, c] = [[c,d], a] = 0$ show that $\iota(\boldsymbol{U})$ is indeed the natural representation of $\mathcal{G}$ under the adjoint action. □

Proposition 5.6. *The adjoint action of the Lie superalgebra $\mathcal{G}$ on $\mathfrak{WC}(\boldsymbol{U})$ preserves the filtration on $\mathfrak{WC}(\boldsymbol{U})$. The action of $\mathcal{G}_{\bar{0}}$ can be lifted to an action by automorphisms of the group $\mathrm{Sp}(\boldsymbol{U}_{\bar{0}}) \times \mathrm{O}(\boldsymbol{U}_{\bar{1}})$ on $\mathfrak{WC}(\boldsymbol{U})$.*

Proof. The first statement follows from Proposition 5.5(2) and the definition of the filtration on $\mathfrak{WC}(\boldsymbol{U})$. Clearly the adjoint action of $\mathcal{G}_{\bar{0}}$ on $\iota(\boldsymbol{U})$ lifts to an action of the group $\mathrm{Sp}(\boldsymbol{U}_{\bar{0}}) \times \mathrm{O}(\boldsymbol{U}_{\bar{1}})$. As each finite-dimensional filtered subspace $\mathfrak{WC}(\boldsymbol{U})_k$ for $k \geq 1$ is preserved by the adjoint action of $\mathcal{G}_{\bar{0}}$, the $\mathcal{G}_{\bar{0}}$-module $\mathfrak{WC}(\boldsymbol{U})$ is semisimple and hence isomorphic to $\mathcal{S}(\boldsymbol{U})$ by a standard induction argument. The second statement follows. □

5.1.4. A general duality theorem. In this subsection, we establish a duality theorem in a general setting, which will be used subsequently.

Let G be one of the classical Lie groups $\mathrm{GL}(V)$, $\mathrm{Sp}(V)$, or $\mathrm{O}(V)$. Assume (ρ, L) is a rational representation of G of countable dimension, which in particular means that it is a direct sum of finite-dimensional simple G-modules. Assume that a subalgebra $\mathcal{R} \subseteq \mathrm{End}(L)$ satisfies:

(i) $\mathcal{R}$ acts irreducibly on L.

(ii) For $g \in G, T \in \mathcal{R}$, we have $\rho(g)T\rho(g)^{-1} \in \mathcal{R}$. That is, $\mathcal{R}$ is a G-module by conjugation.

(iii) As a G-module, $\mathcal{R}$ is a direct sum of finite-dimensional irreducible G-modules.

As we shall see, in the main applications developed in this chapter, the algebra $\mathcal{R}$ is taken to be the Weyl-Clifford algebra on a sum of copies of the natural G-module V, and Conditions (i)-(iii) are easily verified.

Define the algebra of G-invariants in $\mathcal{R}$ to be

$$\mathcal{R}^G = \{T \in \mathcal{R} \mid \rho(g)T = T\rho(g), \forall g \in G\}.$$

We write the G-module L as a direct sum of finite-dimensional simple G-modules $L(\lambda)$ of the isomorphism class λ as $L \cong \bigoplus_{\lambda \in \widehat{G}(\rho)} L(\lambda) \otimes M^\lambda$, where $\widehat{G}(\rho)$ denotes the set of isomorphism classes of simple G-modules appearing in (ρ, L), and

$$M^\lambda := \mathrm{Hom}_G(L(\lambda), L)$$

denotes the **multiplicity space** of λ.

Then M^λ is naturally a left $\mathcal{R}^G$-module by left multiplication. Indeed, for $T \in \mathcal{R}^G, f \in M^\lambda = \mathrm{Hom}_G(L(\lambda), L), v \in L(\lambda), g \in G$, we have

$$Tf(gv) = T\rho(g)f(v) = \rho(g)Tf(v),$$

that is, $Tf \in M^\lambda$. Hence, we have a decomposition of the $\mathbb{C}G \otimes \mathcal{R}^G$-module L (which we shall simply refer to as a $(\mathbf{G}, \mathcal{R}^{\mathbf{G}})$**-module**):

$$L \cong \bigoplus_{\lambda \in \widehat{G}(\rho)} L(\lambda) \otimes M^\lambda. \tag{5.9}$$

We shall need the following variant of the Jacobson density theorem, a proof of which can be found in [**46**, Corollary 4.1.6].

Proposition 5.7. *Let L be a vector space of countable dimension. Assume that a subalgebra $\mathcal{R}$ of $\mathrm{End}(L)$ acts on L irreducibly. Then, for any finite-dimensional subspace X of L, we have $\mathcal{R}|_X = \mathrm{Hom}(X,L)$.*

Theorem 5.8. *Assume that (ρ, L) is a representation of a classical group G of countable dimension, which is a direct sum of finite-dimensional simple G-modules. Further assume that $G, \mathcal{R}, L$ satisfy Conditions (i)-(iii). Then the M^λ's, for $\lambda \in \widehat{G}(\rho)$, are pairwise non-isomorphic irreducible $\mathcal{R}^G$-modules.*

Proof. We shall need the following G-invariant version of Proposition 5.7.

Claim. Let $X \subseteq L$ be a finite-dimensional G-invariant subspace. Then $\mathcal{R}^G|_X = \mathrm{Hom}_G(X,L)$.

Indeed, by G-invariance of X, $\mathrm{Hom}(X,L)$ is a G-module, and the restriction map $\mathcal{R} \xrightarrow{\text{res}} \mathrm{Hom}(X,L)$ given by $r \to r|_X$ is a G-homomorphism. Since $\mathcal{R}$ and $\mathrm{Hom}(X,L)$ are completely reducible G-modules, they contain $\mathcal{R}^G$ and $\mathrm{Hom}_G(X,L)$ as direct summands of trivial submodules. We shall denote by π the respective G-equivariant projection maps. Hence we have the following commutative diagram:

$$
\begin{array}{ccc}
\mathcal{R} & \xrightarrow{\quad\pi\quad} & \mathcal{R}^G \\
{\scriptstyle\text{res}}\downarrow & & \downarrow{\scriptstyle\text{res}} \\
\mathrm{Hom}(X,L) & \xrightarrow{\quad\pi\quad} & \mathrm{Hom}_G(X,L)
\end{array}
\tag{5.10}
$$

Now, suppose that $T \in \mathrm{Hom}_G(X,L)$. Regarding $T \in \mathrm{Hom}(X,L)$, we may apply Proposition 5.7 to find an element $r \in \mathcal{R}$ such that $r|_X = T$. Then we have

$$T = \pi(T) = \pi(r|_X) = \pi(r)|_X,$$

proving the claim, since $\pi(r) \in \mathcal{R}^G$.

Recall $M^\lambda = \mathrm{Hom}_G(L(\lambda), L)$ by definition, for $\lambda \in \widehat{G}(\rho)$. Take two nonzero elements $S, T \in \mathrm{Hom}_G(L(\lambda), L)$. Then $\psi = ST^{-1} : T(L(\lambda)) \to S(L(\lambda))$ is an isomorphism of irreducible G-modules. It follows by the claim above that there exists $u \in \mathcal{R}^G$ that restricts to $\psi : T(L(\lambda)) \to L$. Hence, $uT : L(\lambda) \to S(L(\lambda))$ is an isomorphism of irreducible G-modules. By Schur's Lemma, we must have $S = cuT$ for some scalar c, and this proves the irreducibility of the $\mathcal{R}^G$-module $\mathrm{Hom}_G(L(\lambda), L)$.

It remains to show that, for distinct $\lambda, \mu \in \widehat{G}(\rho)$, an $\mathcal{R}^G$-module homomorphism $\varphi : \mathrm{Hom}_G(L(\lambda), L) \to \mathrm{Hom}_G(L(\mu), L)$ must be zero. To this end, we take an arbitrary element $T \in \mathrm{Hom}_G(L(\lambda), L)$. Set $S := \varphi(T)$, $U := T(L(\lambda)) \oplus S(L(\mu))$, and let $p : U \to S(L(\mu)) \subseteq L$ be the natural projection, which is clearly G-equivariant. By the above claim, there exists $r \in \mathcal{R}^G$ such that $p = r|_U$. Then, it follows from $pT = 0$ that $rT = 0$ and then,

$$0 = \varphi(rT) = r\varphi(T) = rS = pS = S.$$

Hence we conclude that $\varphi = 0$ as desired. □

We will call a decomposition of the $(G, \mathcal{R}^G)$-module L as in (5.9) **strongly multiplicity-free**, in light of Theorem 5.8.

5.1.5. A duality for Weyl-Clifford algebras. Let $U = U_{\bar{0}} \oplus U_{\bar{1}}$ be a vector superspace, and let G be a classical Lie subgroup of $GL(U_{\bar{0}}) \times GL(U_{\bar{1}})$. In all the examples that give rise to concrete Howe dualities in the next sections, $U_{\bar{0}}$ and $U_{\bar{1}}$ are actually direct sums of the natural G-module. The induced G-module $(\rho, \mathcal{S}(U))$ is semisimple. Also, the G-module $\mathfrak{WC}(\boldsymbol{U}) \subseteq \mathrm{End}(\mathcal{S}(U))$, with G-action given by conjugation $g.T = \rho(g)T\rho(g^{-1})$, for $g \in G$ and $T \in \mathfrak{WC}(\boldsymbol{U})$, is semisimple.

Lemma 5.9. *As a module over the Weyl-Clifford algebra $\mathfrak{WC}(\boldsymbol{U})$, $\mathcal{S}(U)$ is irreducible.*

Proof. Identify $\mathcal{S}(U)$ with $\mathbb{C}[x, \eta]$, and identify $\mathfrak{WC}(\boldsymbol{U})$ with the superalgebra of polynomial differential operators on U as in Section 5.1.2. Then, given any nonzero element f in $\mathbb{C}[x, \eta]$, we can find a suitable constant-coefficient differential operator $D \in \mathfrak{WC}(\boldsymbol{U})$ such that $D.f = 1$. Now applying suitable multiplication operators in $\mathfrak{WC}(\boldsymbol{U})$ to 1, we obtain all the monomials $x^\alpha \eta^\beta$ that span $\mathbb{C}[x, \eta]$. The lemma follows. □

Letting $L = \mathcal{S}(U)$ and $\mathcal{R} = \mathfrak{WC}(\boldsymbol{U})$, the triple $(G, L, \mathcal{R})$ satisfies Conditions (i)–(iii) of Section 5.1.4. Hence Theorem 5.8 is applicable in light of Lemma 5.9 and gives us the following.

Theorem 5.10. *Retain the notations above. As a $(G, \mathfrak{WC}(\boldsymbol{U})^G)$-module, $\mathcal{S}(U)$ is strongly multiplicity-free, i.e.,*

$$\mathcal{S}(U) \cong \bigoplus_{\lambda \in \widehat{G}(\rho)} L(\lambda) \otimes M^\lambda.$$

Here the M^λ's are pairwise non-isomorphic irreducible $\mathfrak{WC}(\boldsymbol{U})^G$-modules, while the $L(\lambda)$'s are pairwise non-isomorphic irreducible finite-dimensional G-modules.

Recall by Theorem 4.1 that the algebra $\mathcal{S}(\boldsymbol{U})^G$ is finitely generated. Also recall from (5.3) the symbol map $\sigma : \mathfrak{WC}(\boldsymbol{U}) \to \mathcal{S}(\boldsymbol{U})$.

Proposition 5.11. *Let $\{S_1, \ldots, S_r\}$ be a set of generators of the algebra $\mathcal{S}(\boldsymbol{U})^G$, where G is a classical subgroup of $GL(U_{\bar{0}}) \times GL(U_{\bar{1}})$. Suppose $T_j \in \mathfrak{WC}(\boldsymbol{U})^G$ is chosen so that $\sigma(T_j) = S_j$, for $j = 1, \ldots, r$. Then $\{T_1, \ldots, T_r\}$ generate the algebra $\mathfrak{WC}(\boldsymbol{U})^G$.*

Proof. It follows from Proposition 5.4 that the symbol map σ restricts to an isomorphism from $\mathfrak{WC}(\boldsymbol{U})^G_k$ to $\bigoplus_{j=0}^k \mathcal{S}^j(\boldsymbol{U})^G$, for each $k \geq 0$.

Let $a \in \mathfrak{WC}(\boldsymbol{U})^G_k$. We shall show that a is generated by $T_1, \ldots, T_r$ by induction on k. For $k = 0$, the claim is trivial. Suppose that $k \geq 1$. By assumption $\sigma(a) = f(S_1, S_2, \ldots, S_r)$, for some polynomial f. Now consider the element

$f(T_1,T_2,\ldots,T_r) \in \mathfrak{WC}(\boldsymbol{U})_k^G$. We note that $\sigma(a - f(T_1,\ldots,T_r)) \in \bigoplus_{j=0}^{k-1} \mathcal{S}^j(\boldsymbol{U})$, and hence $a - f(T_1,\ldots,T_r) \in \mathfrak{WC}(\boldsymbol{U})_{k-1}^G$. By induction hypothesis, $a - f(T_1,\ldots,T_r)$ is generated by $T_1,\ldots,T_r$, and hence so is a. □

Proposition 5.12. *Suppose that the generators $T_1,\ldots,T_r$ for the algebra $\mathfrak{WC}(\boldsymbol{U})^G$ can be chosen such that $\mathfrak{g}' := \bigoplus_{i=1}^r \mathbb{C}T_i$ is a Lie subalgebra of $\mathfrak{WC}(\boldsymbol{U})^G$ under the super commutator. Then the $\mathfrak{WC}(\boldsymbol{U})^G$-modules M^λ's (see Theorem 5.10) are pairwise non-isomorphic irreducible $\mathfrak{g}'$-modules, and we have the following $(G,\mathfrak{g}')$-module decomposition:*

$$\mathcal{S}(U) \cong \bigoplus_{\lambda \in \widehat{G}(\rho)} L(\lambda) \otimes M^\lambda.$$

Proof. By assumption, the action of $\mathfrak{g}'$ on $\mathcal{S}(U)$ extends to an action of the universal enveloping superalgebra $U(\mathfrak{g}')$, so we obtain an algebra homomorphism $\rho' : U(\mathfrak{g}') \to \mathrm{End}(\mathcal{S}(U))$. The assumption that $T_1,\ldots,T_r$ generate the algebra $\mathfrak{WC}(\boldsymbol{U})^G$ implies that $\rho'(U(\mathfrak{g}')) = \mathfrak{WC}(\boldsymbol{U})^G$. The proposition now follows from Theorem 5.10. □

We shall refer to $(G,\mathfrak{g}')$ in Proposition 5.12 as a **Howe dual pair**. In the next sections, by specializing to the case when G is a classical group and $U_{\bar{0}}$ and $U_{\bar{1}}$ are direct sums of the natural G-module, we shall be able to make an explicit choice of the generators $T_1,\ldots,T_r$ so that Proposition 5.12 can be applied. Note that $\mathcal{S}(U) = S(U_{\bar{0}})$ when $U_{\bar{1}} = 0$, and that $\mathcal{S}(U) = \wedge(U_{\bar{1}})$ when $U_{\bar{0}} = 0$. In either special case, the $\mathfrak{g}'$ appearing in Proposition 5.12 will be a Lie algebra.

5.2. Howe duality for type A and type Q

In this section, we first formulate Howe duality for the classical group $\mathrm{GL}(V)$ and give a precise multiplicity-free decomposition as predicted by Proposition 5.12. We find explicit formulas for the joint highest weight vectors in the above decomposition, which will also be useful in Section 5.3. In addition, we formulate another Howe duality between a pair of queer Lie superalgebras and provide an explicit multiplicity-free decomposition.

5.2.1. Howe dual pair $(\mathrm{GL}(k),\mathfrak{gl}(m|n))$**.** Take $V = \mathbb{C}^k$, and hence identify $\mathrm{GL}(V)$ with $\mathrm{GL}(k)$ and $\mathfrak{gl}(V)$ with $\mathfrak{gl}(k)$. Given $m,n \geq 0$, we consider the superspace $U = U_{\bar{0}} \oplus U_{\bar{1}}$ with $U_{\bar{0}} = V^m$ and $U_{\bar{1}} = V^n$, and identify naturally $U \equiv V \otimes \mathbb{C}^{m|n}$. We let $e^1,\ldots,e^k$ denote the standard basis for the natural $\mathfrak{gl}(k)$-module $\mathbb{C}^k$, and we let $e_i, i \in I(m|n)$, denote the standard basis for the natural module $\mathbb{C}^{m|n}$ of the general linear Lie superalgebra $\mathfrak{gl}(m|n)$. We denote, for $1 \leq r \leq k$, $1 \leq a \leq m$, $1 \leq b \leq n$,

$$x_{ra} := e^r \otimes e_{\bar{a}}, \qquad \eta_{rb} := e^r \otimes e_b. \tag{5.11}$$

Then the set $\{x_{ra}, \eta_{rb}\}$ is a basis for U. We will denote by $\mathbb{C}[x,\eta]$ the polynomial superalgebra generated by the elements in (5.11), and from now on further identify

$$\mathcal{S}(U) \equiv \mathcal{S}(V \otimes \mathbb{C}^{m|n}) \equiv \mathbb{C}[x,\eta]. \tag{5.12}$$

We introduce the following first-order differential operators

$$\mathcal{E}^{pq} = \sum_{j=1}^{m} x_{pj}\frac{\partial}{\partial x_{qj}} + \sum_{s=1}^{n} \eta_{ps}\frac{\partial}{\partial \eta_{qs}}, \quad 1 \le p,q \le k. \tag{5.13}$$

We also introduce the following first-order differential operators

$$\begin{aligned} \mathcal{E}_{\bar{i}\bar{j}} &= \sum_{p=1}^{k} x_{pi}\frac{\partial}{\partial x_{pj}}, \quad \mathcal{E}_{\bar{i}s} = \sum_{p=1}^{k} x_{pi}\frac{\partial}{\partial \eta_{ps}}, \\ \mathcal{E}_{st} &= \sum_{p=1}^{k} \eta_{ps}\frac{\partial}{\partial \eta_{pt}}, \quad \mathcal{E}_{s\bar{i}} = \sum_{p=1}^{k} \eta_{ps}\frac{\partial}{\partial x_{pi}}, \quad 1 \le i,j \le m, 1 \le s,t \le n. \end{aligned} \tag{5.14}$$

The natural commuting actions of $\mathfrak{gl}(k)$ and $\mathfrak{gl}(m|n)$ on $V \otimes \mathbb{C}^{m|n}$ induce commuting actions on $\mathbb{C}[x,\eta]$. The following lemma is standard and can be verified by a direct computation (see Exercise 5.2(1)).

Lemma 5.13. *The commuting actions of* $\mathfrak{gl}(k)$ *and* $\mathfrak{gl}(m|n)$ *on* $\mathcal{S}(\mathbb{C}^k \otimes \mathbb{C}^{m|n}) = \mathbb{C}[x,\eta]$ *are realized by the formulas* (5.13) *and* (5.14), *respectively.*

We shall denote by $\mathfrak{g}'$ the linear span of the elements in (5.14). Indeed, the elements in (5.14) form a linear basis for $\mathfrak{g}'$. Let $\boldsymbol{U} = U \oplus U^*$ as before. In light of (5.12), we identify $\mathfrak{WC}(\boldsymbol{U})$ with the superalgebra of polynomial differential operators on $\mathbb{C}[x,\eta]$, and so the elements in (5.14) may be regarded as elements in $\mathfrak{WC}(\boldsymbol{U})$.

Theorem 5.14. *The* $\mathfrak{g}'$, *which forms an isomorphic copy of* $\mathfrak{gl}(m|n)$, *generates the algebra of* $\mathrm{GL}(V)$*-invariants in* $\mathfrak{WC}(\boldsymbol{U})$. *Moreover, as a* $(\mathrm{GL}(k), \mathfrak{gl}(m|n))$*-module,* $\mathcal{S}(\mathbb{C}^k \otimes \mathbb{C}^{m|n})$ *is strongly multiplicity-free.*

Proof. Set $G = \mathrm{GL}(k)$. Recall from Proposition 5.4 that the associated graded of the filtered algebra $\mathfrak{WC}(\boldsymbol{U})$ is $\mathcal{S}(\boldsymbol{U})$, and that the action of $\mathfrak{gl}(k)$ on $\mathcal{S}(\boldsymbol{U})$ lifts to an action of G. Let $\{y_{pi}\}$ denote the basis in $U_{\bar{0}}^*$ dual to the basis $\{x_{pi}\}$ for $U_{\bar{0}}$, and let $\{\xi_{pt}\}$ denote the basis in $U_{\bar{1}}^*$ dual to the basis $\{\eta_{pt}\}$ for $U_{\bar{1}}$. By the supersymmetric First Fundamental Theorem for $\mathrm{GL}(V)$ (see Theorem 4.19), the algebra $\mathcal{S}(\boldsymbol{U})^G = \big(S(U_{\bar{0}} \oplus U_{\bar{0}}^*) \otimes \wedge(U_{\bar{1}} \oplus U_{\bar{1}}^*)\big)^G$ is generated by

$$\begin{aligned} z_{\bar{i}\bar{j}} &= \sum_{p=1}^{k} x_{pi}y_{pj}, \quad z_{ts} = \sum_{p=1}^{k} \eta_{pt}\xi_{ps}, \\ z_{\bar{i}t} &= \sum_{p=1}^{k} x_{pi}\xi_{pt}, \quad z_{t\bar{i}} = \sum_{p=1}^{k} \eta_{pt}y_{pi}, \quad 1 \le i,j \le m, 1 \le s,t \le n. \end{aligned} \tag{5.15}$$

Observe that the symbol map σ in Section 5.1.2 sends each $\mathcal{E}_{ab}$ in (5.14) to z_{ab} for $a,b \in I(m|n)$, and hence by Proposition 5.11, the $\mathcal{E}_{ab}$'s form a set of generators for $\mathfrak{WC}(\boldsymbol{U})^G$. Now the theorem follows from Proposition 5.12, since the $\mathcal{E}_{ab}$'s generate the Lie superalgebra $\mathfrak{gl}(m|n)$ by Lemma 5.13. □

Remark 5.15. When $n = 0$, $\mathcal{S}(U) = S(U_{\bar{0}})$ becomes the symmetric algebra on $U_{\bar{0}} = \mathbb{C}^k \otimes \mathbb{C}^m$. When $m = 0$, $\mathcal{S}(U) = \wedge(U_{\bar{1}})$ is the exterior algebra on $U_{\bar{1}} = \mathbb{C}^k \otimes \mathbb{C}^n$. In these two important specializations of Theorem 5.14, $\mathfrak{g}'$ becomes a Lie algebra, namely $\mathfrak{gl}(m)$ and $\mathfrak{gl}(n)$, respectively.

5.2.2. $(\mathrm{GL}(k), \mathfrak{gl}(m|n))$**-Howe duality.** We will first find the multiplicity-free decomposition of the $(GL(V), \mathfrak{gl}(m))$-module $S(\mathbb{C}^k \otimes \mathbb{C}^m)$ explicitly. To do that, we will take advantage of Schur duality which was established in Chapter 3. As the formulation below will involve Lie algebra $\mathfrak{gl}(k)$ for various k, we will add the index k to denote by $L_k(\lambda)$ the irreducible $\mathfrak{gl}(k)$-module of highest weight λ. For dominant integral weight λ, the $\mathfrak{gl}(k)$-module $L_k(\lambda)$ lifts to a $\mathrm{GL}(k)$-module, which will be denoted by the same notation.

Let us consider the natural action of $\mathfrak{gl}(k) \times \mathfrak{gl}(m)$ on $\mathbb{C}^k \otimes \mathbb{C}^m$ and its induced action on the dth symmetric tensor $S^d(\mathbb{C}^k \otimes \mathbb{C}^m)$, for $d \geq 0$. As usual, we identify a polynomial weight for $\mathfrak{gl}(k)$ with a partition of length at most k. Recall $\mathcal{P}_d$ denotes the set of partitions of d.

Theorem 5.16 ($(\mathrm{GL}(k), \mathfrak{gl}(m))$-Howe duality)**.** *As a* $(\mathrm{GL}(k), \mathfrak{gl}(m))$*-module, we have*

$$S^d(\mathbb{C}^k \otimes \mathbb{C}^m) \cong \bigoplus_{\substack{\lambda \in \mathcal{P}_d \\ \ell(\lambda) \leq \min\{k,m\}}} L_k(\lambda) \otimes L_m(\lambda).$$

Proof. We shall show that Schur duality implies Howe duality. By Schur duality (see Theorem 3.11), we have

$$(\mathbb{C}^k)^{\otimes d} \cong \bigoplus_{\substack{\lambda \in \mathcal{P}_d \\ \ell(\lambda) \leq k}} L_k(\lambda) \otimes S^\lambda, \qquad (\mathbb{C}^m)^{\otimes d} \cong \bigoplus_{\substack{\mu \in \mathcal{P}_d \\ \ell(\mu) \leq m}} L_m(\mu) \otimes S^\mu. \tag{5.16}$$

Using (5.16) and denoting by $\Delta\mathfrak{S}_d$ the diagonal subgroup in $\mathfrak{S}_d \times \mathfrak{S}_d$, we have

$$\begin{aligned}
S^d(\mathbb{C}^k \otimes \mathbb{C}^m) &\cong \left((\mathbb{C}^k \otimes \mathbb{C}^m)^{\otimes d}\right)^{\mathfrak{S}_d} \\
&\cong \left((\mathbb{C}^k)^{\otimes d} \otimes (\mathbb{C}^m)^{\otimes d}\right)^{\Delta\mathfrak{S}_d} \\
&\cong \bigoplus_{\substack{\lambda,\mu \in \mathcal{P}_d \\ \ell(\lambda) \leq k, \ell(\mu) \leq m}} L_k(\lambda) \otimes L_m(\mu) \otimes \left(S^\lambda \otimes S^\mu\right)^{\Delta\mathfrak{S}_d} \\
&\cong \bigoplus_{\substack{\lambda \in \mathcal{P}_d \\ \ell(\lambda) \leq \min\{k,m\}}} L_k(\lambda) \otimes L_m(\lambda).
\end{aligned}$$

In the last step, we have used the well-known fact that $S^\lambda \cong (S^\lambda)^*$, and hence by Schur's lemma

$$(S^\lambda \otimes S^\mu)^{\Delta\mathfrak{S}_d} \cong \mathrm{Hom}_{\mathfrak{S}_d}(S^\lambda, S^\mu) \cong \delta_{\lambda,\mu}\mathbb{C}.$$

This completes the proof of the theorem. □

Remark 5.17. It is possible to derive Schur duality from the $(\mathrm{GL}(k), \mathfrak{gl}(m))$-Howe duality by making use of the well-known fact that the zero-weight subspace of the $\mathrm{GL}(d)$-module $L_d(\lambda)$ for a partition λ of d is naturally an $\mathfrak{S}_d$-module, which is isomorphic to the Specht module S^λ. Also, it is possible to derive the FFT for $\mathrm{GL}(k)$ from Howe duality. Hence the FFT for the general linear group, Schur duality, and Howe duality of type A are essentially all equivalent.

Now let us consider the induced action of $\mathfrak{gl}(k) \times \mathfrak{gl}(m)$ on the dth exterior tensor space $\wedge^d(\mathbb{C}^k \otimes \mathbb{C}^m)$, for $0 \le d \le km$. Recall that λ' denotes the conjugate partition of a partition λ. We have the following skew version of Howe duality.

Theorem 5.18 (Skew $(\mathrm{GL}(k), \mathfrak{gl}(m))$-Howe duality). *As a $(\mathrm{GL}(k), \mathfrak{gl}(m))$-module, we have*

$$\wedge^d(\mathbb{C}^k \otimes \mathbb{C}^m) \cong \bigoplus_{\substack{\lambda\in\mathcal{P}_d \\ \ell(\lambda)\le k, \ell(\lambda')\le m}} L_k(\lambda) \otimes L_m(\lambda').$$

Proof. We shall use Schur duality to derive the skew Howe duality. For an $\mathfrak{S}_d$-module M, let us denote by $M^{\mathfrak{S}_d,\mathrm{sgn}}$ the submodule of M that transforms by the sign character. Using (5.16), we have

$$\begin{aligned}
\wedge^d(\mathbb{C}^k \otimes \mathbb{C}^m) &\cong \Big((\mathbb{C}^k \otimes \mathbb{C}^m)^{\otimes d}\Big)^{\mathfrak{S}_d,\mathrm{sgn}} \\
&\cong \Big((\mathbb{C}^k)^{\otimes d} \otimes (\mathbb{C}^m)^{\otimes d}\Big)^{\Delta\mathfrak{S}_d,\mathrm{sgn}} \\
&\cong \bigoplus_{\substack{\lambda,\mu\in\mathcal{P}_d \\ \ell(\lambda)\le k, \ell(\mu)\le m}} L_k(\lambda) \otimes L_m(\mu) \otimes \Big(S^\lambda \otimes S^\mu\Big)^{\Delta\mathfrak{S}_d,\mathrm{sgn}} \\
&\cong \bigoplus_{\substack{\lambda\in\mathcal{P}_d \\ \ell(\lambda)\le k, \ell(\lambda')\le m}} L_k(\lambda) \otimes L_m(\lambda').
\end{aligned}$$

In the last step, we have used the isomorphism $S^\mu \otimes \mathrm{sgn} \cong S^{\mu'}$ from (A.29), and hence

$$(S^\lambda \otimes S^\mu)^{\Delta\mathfrak{S}_d,\mathrm{sgn}} \cong \Big(S^\lambda \otimes S^\mu \otimes \mathrm{sgn}\Big)^{\Delta\mathfrak{S}_d} \cong \mathrm{Hom}_{\mathfrak{S}_d}(S^\lambda, S^\mu \otimes \mathrm{sgn}) \cong \delta_{\lambda,\mu'}\mathbb{C}.$$

The theorem is proved. □

The natural action of $\mathfrak{gl}(k) \times \mathfrak{gl}(m|n)$ on $\mathbb{C}^k \otimes \mathbb{C}^{m|n}$ induces a natural action on the supersymmetric algebra $\mathcal{S}(\mathbb{C}^k \otimes \mathbb{C}^{m|n})$. Recall from (2.4) the weight $\lambda^\natural$ associated to an $(m|n)$-hook partition λ. The following theorem, which is a generalization of both Theorems 5.16 and 5.18, provides an explicit form of the strongly multiplicity-free decomposition in Theorem 5.14.

Theorem 5.19 ($(\mathrm{GL}(k), \mathfrak{gl}(m|n))$-Howe duality)**.** *As a* $(\mathrm{GL}(k), \mathfrak{gl}(m|n))$*-module, we have*

$$\mathcal{S}^d(\mathbb{C}^k \otimes \mathbb{C}^{m|n}) \cong \bigoplus_{\substack{\lambda \in \mathcal{P}_d \\ \ell(\lambda) \leq k, \lambda_{m+1} \leq n}} L_k(\lambda) \otimes L_{m|n}(\lambda^\natural).$$

The proof of Theorem 5.19 is completely analogous to the one for Theorem 5.16, now using the Schur-Sergeev duality (Theorem 3.11) which we recall: as $(\mathfrak{gl}(m|n), \mathfrak{S}_d)$-modules, $(\mathbb{C}^{m|n})^{\otimes d} \cong \oplus_{\lambda \in \mathcal{P}_d, \lambda_{m+1} \leq n} L(\lambda^\natural) \otimes S^\lambda$. We leave this as an exercise (see Exercise 5.3).

Recall that the character of the irreducible $\mathfrak{gl}(k)$-module $L_k(\lambda)$ is given by the Schur function s_λ, while the character of the irreducible $\mathfrak{gl}(m|n)$-module $L_{m|n}(\lambda^\natural)$ is given by the super Schur function $\mathrm{hs}_\lambda(x;y)$ (see Theorem 3.15). Now by computing the characters of both sides of the isomorphism in Theorem 5.19 and summing over all $d \geq 0$, we immediately recover the super Cauchy identity (A.41) in Appendix A:

$$\frac{\prod_{j,k}(1 + y_j z_k)}{\prod_{i,k}(1 - x_i z_k)} = \sum_{\lambda \in \mathcal{P}} \mathrm{hs}_\lambda(x;y) s_\lambda(z).$$

By setting the variables y_j (respectively, x_i) to zero and noting that $\mathrm{hs}_\lambda(x;0) = s_\lambda(x)$ (respectively, $\mathrm{hs}_\lambda(0;y) = s_{\lambda'}(y)$), we further recover the classical Cauchy identities as given in (A.11), which correspond to the Howe duality in Theorem 5.16 and the skew-Howe duality in Theorem 5.18.

Remark 5.20. When we derive the $(\mathrm{GL}(k), \mathfrak{gl}(m|n))$-Howe duality from the Schur-Sergeev duality, we could have replaced $\mathrm{GL}(k)$ by the Lie algebra $\mathfrak{gl}(k)$ without any change. Moreover, using Schur-Sergeev duality in the same fashion, one can derive a more general $(\mathfrak{gl}(k|\ell), \mathfrak{gl}(m|n))$-Howe duality on $\mathcal{S}(\mathbb{C}^{k|\ell} \otimes \mathbb{C}^{m|n})$.

5.2.3. Formulas for highest weight vectors. In this subsection, we shall find explicit formulas for the joint highest weight vectors in the $(\mathrm{GL}(k), \mathfrak{gl}(m|n))$-Howe duality decomposition of Theorem 5.19. These formulas will play a key role in the subsequent section on Howe duality for symplectic and orthogonal groups.

Recalling the notation from (5.13) and (5.14) that the simple root vectors of the standard fundamental systems for $\mathfrak{gl}(k)$ and $\mathfrak{gl}(m|n)$ are respectively ($1 \leq i \leq k-1$, $1 \leq s \leq m-1$, $1 \leq t \leq n-1$):

$$\varepsilon^{i,i+1} = \sum_{j=1}^{m} x_{ij} \frac{\partial}{\partial x_{i+1,j}} + \sum_{s=1}^{n} \eta_{is} \frac{\partial}{\partial \eta_{i+1,s}}, \tag{5.17}$$

(5.18)

$$\mathcal{E}_{\bar{s},\overline{s+1}} = \sum_{p=1}^{k} x_{ps}\frac{\partial}{\partial x_{p,s+1}}, \quad \mathcal{E}_{t,t+1} = \sum_{p=1}^{k} \eta_{pt}\frac{\partial}{\partial \eta_{p,t+1}}, \quad \mathcal{E}_{\overline{m},1} = \sum_{p=1}^{k} x_{pm}\frac{\partial}{\partial \eta_{p1}}.$$

Let λ be an $(m|n)$-hook partition of length at most k. We are looking for the joint highest weight vector with respect to the standard Borel subalgebra of $\mathfrak{gl}(k) \times \mathfrak{gl}(m|n)$, or equivalently the vector annihilated by (5.17) and (5.18), in the summand $L_k(\lambda) \otimes L_{m|n}(\lambda^{\natural})$ in the decomposition of $\mathbb{C}[x,\eta]$. Such a vector is unique up to a scalar multiple, thanks to the multiplicity-free decomposition in Theorem 5.19.

For $1 \leq \ell \leq \min\{k,m\}$, define

$$\diamondsuit_\ell := \det \begin{pmatrix} x_{11} & x_{12} & \cdots & x_{1\ell} \\ x_{21} & x_{22} & \cdots & x_{2\ell} \\ \vdots & \vdots & \vdots & \vdots \\ x_{\ell 1} & x_{\ell 2} & \cdots & x_{\ell\ell} \end{pmatrix}. \tag{5.19}$$

Note that $\diamondsuit_\ell$ is annihilated by all the operators in (5.17) and (5.18), and hence is a joint highest weight vector (see Exercise 5.4(1)).

The **column determinant** of an $\ell \times \ell$ matrix $A = (a_{ij})$ with possibly non-commuting entries is defined to be

$$\mathrm{cdet} A = \sum_{\sigma \in \mathfrak{S}_\ell} (-1)^{\ell(\sigma)} a_{\sigma(1)1} a_{\sigma(2)2} \cdots a_{\sigma(\ell)\ell}.$$

In the case when $m < k$, we introduce the following determinant:

$$\diamondsuit_{t,\ell} := \mathrm{cdet} \begin{pmatrix} x_{11} & x_{12} & \cdots & x_{1m} & \eta_{1t} & \cdots & \eta_{1t} \\ x_{21} & x_{22} & \cdots & x_{2m} & \eta_{2t} & \cdots & \eta_{2t} \\ \vdots & \vdots & \ddots & \vdots & \vdots & \vdots & \vdots \\ x_{\ell 1} & x_{\ell 2} & \cdots & x_{\ell m} & \eta_{\ell t} & \cdots & \eta_{\ell t} \end{pmatrix}, \tag{5.20}$$

for $m < \ell \leq k$ and $1 \leq t \leq n$, where the last $(\ell - m)$ columns are filled with the same vector. For notational convenience, we set $\diamondsuit_{t,\ell} := \diamondsuit_\ell$, for all t and $1 \leq \ell \leq m$.

Remark 5.21. The element $\diamondsuit_{t,\ell}$ is always nonzero (Exercise 5.4). It reduces to (5.19) when $\ell \leq m$, and, up to a scalar multiple, reduces to $\eta_{1t} \cdots \eta_{\ell t}$ when $m = 0$.

For a partition μ with $\mu'_1 \leq k$ and $1 \leq t_1, t_2, \ldots, t_{\mu_1} \leq n$, we define

$$\diamondsuit_{t_1,t_2,\ldots,t_{\mu_1},\mu'} := \diamondsuit_{t_1,\mu'_1} \diamondsuit_{t_2,\mu'_2} \cdots \diamondsuit_{t_{\mu_1},\mu'_{\mu_1}}. \tag{5.21}$$

Now let r be fixed by the conditions $\lambda'_r > m$ and $\lambda'_{r+1} \leq m$ for an $(m|n)$-hook partition λ. Note that we must have $0 \leq r \leq n$. Denote by $\lambda_{\leq r}$ the subdiagram of the Young diagram λ which consists of the first r columns of λ, i.e., the columns of length greater than m.

Lemma 5.22. *The vector* $\diamondsuit_{1,2,\ldots,r,\lambda'_{\leq r}}$ *is a joint highest weight vector for the* $\mathfrak{gl}(k) \times \mathfrak{gl}(m|n)$*-module* $L_k(\lambda_{\leq r}) \otimes L_{m|n}(\lambda^\natural_{\leq r})$ *in the decomposition of* $\mathbb{C}[x,\eta]$.

Proof. Set $\mu = \lambda_{\leq r}$ and $\diamondsuit = \diamondsuit_{1,2,\ldots,r,\mu'}$ in this proof. By Remark 5.21 and a direct computation using (5.13), (5.14), and (5.17), we can verify the following:

(1) $\diamondsuit$ is nonzero.

(2) $\diamondsuit$ has weight $(\mu, \mu^\natural)$ with respect to the action of $\mathfrak{gl}(k) \times \mathfrak{gl}(m|n)$.

(3) Each factor $\diamondsuit_{i,\mu'_i}$ in $\diamondsuit$ is annihilated by the operators in (5.17).

By the strongly multiplicity-free decomposition of Theorem 5.19 (recalling $\mathcal{S}(\mathbb{C}^k \otimes \mathbb{C}^{m|n}) \equiv \mathbb{C}[x,\eta]$), we may identify $L_{m|n}(\mu^\natural) \equiv \mathbb{C}[x,\eta]^{\mathfrak{n}_1}_\mu$, the space of highest weight vectors in the $\mathfrak{gl}(k)$-module $\mathbb{C}[x,\eta]$ of highest weight μ (here $\mathfrak{n}_1$ denotes the nilradical of the standard Borel for $\mathfrak{gl}(k)$). It follows by (3) that $\diamondsuit$ is annihilated by the operators in (5.17), and hence $\diamondsuit \in \mathbb{C}[x,\eta]^{\mathfrak{n}_1}_\mu$ by (1)–(3). Since $\diamondsuit$ has weight $\mu^\natural$ with respect to the action of $\mathfrak{gl}(m|n)$ by (2) again, it must be the unique (up to a scalar multiple) joint highest weight vector in $\mathbb{C}[x,\eta]$ of weight $(\mu,\mu^\natural)$ under the action of $\mathfrak{gl}(k) \times \mathfrak{gl}(m|n)$. □

Theorem 5.23. *Let* λ *be a partition such that* $\ell(\lambda) \leq k$ *and* $\lambda_{m+1} \leq n$*. Then, a joint highest weight vector of weight* $(\lambda, \lambda^\natural)$ *in the* $\mathfrak{gl}(k) \times \mathfrak{gl}(m|n)$*-module* $\mathbb{C}[x,\eta]$ *is given by* $\diamondsuit_{1,2,\ldots,\lambda_1,\lambda'} = \diamondsuit_{1,\lambda'_1}\diamondsuit_{2,\lambda'_2}\cdots\diamondsuit_{\lambda_1,\lambda'_{\lambda_1}}$.

Proof. We have

$$\diamondsuit_{1,2,\ldots,\lambda_1,\lambda'} = \diamondsuit_{1,2,\ldots,r,\lambda'_{\leq r}}\diamondsuit_{\lambda'_{r+1}}\cdots\diamondsuit_{\lambda'_{\lambda_1}}.$$

Thus, it is a product of a nonzero element in $\mathbb{C}[x,\eta]$ and a nonzero element in $\mathbb{C}[x]$, and so it is nonzero. Also we see that it is a product of the joint highest weight vectors $\diamondsuit_{1,2,\ldots,r,\lambda'_{\leq r}}$ (by Lemma 5.22) and $\diamondsuit_\ell$, $\ell = \lambda'_{r+1}, \ldots, \lambda'_{\lambda_1}$, and hence is also a joint highest weight vector. Also it is straightforward to verify by (5.13) and (5.14) that $\diamondsuit_{1,2,\ldots,\lambda_1,\lambda'}$ has the correct weight $(\lambda, \lambda^\natural)$. □

Remark 5.24. When $n = 0$, the highest weight vector in Theorem 5.23 does not involve the odd generators η_{ij}. When $m = 0$, the vector becomes simply a monomial in the anti-commuting variables η_{ij}.

5.2.4. $(\mathfrak{q}(m), \mathfrak{q}(n))$-Howe duality. In this subsection, we obtain a $(\mathfrak{q}(m), \mathfrak{q}(n))$-Howe duality using the $(\mathfrak{q}(n), \mathcal{H}_d)$-Sergeev duality in Theorem 3.49, in a similar fashion as in Section 5.2.2.

Recall that, by definition, $\mathfrak{q}(m)$ consists of $X \in \mathfrak{gl}(m|m)$ **super**-commuting with P given in (1.10), i.e., $\mathfrak{q}(m)$ consists of the $m|m$-block matrices of the form $\begin{pmatrix} A & B \\ B & A \end{pmatrix}$. Recall from Remark 1.11 that we have another nonstandard realization of the Lie superalgebra $\mathfrak{q}(m)$ as $\widetilde{\mathfrak{q}}(m) = \{X \in \mathfrak{gl}(m|m) \mid XP - PX = 0\}$, which consists of the $m|m$-block matrices of the form $\begin{pmatrix} A & B \\ -B & A \end{pmatrix}$. We let $\mathfrak{q}(m)$ act on $\mathbb{C}^{m|m}$ as the nonstandard isomorphic copy $\widetilde{\mathfrak{q}}(m)$, while we let $\mathfrak{q}(n)$ act on $\mathbb{C}^{n|n}$ by linear maps in the standard way, i.e. linear maps super-commuting with P. This induces an action of $\mathfrak{q}(m) \times \mathfrak{q}(n)$ on $\mathbb{C}^{m|m} \otimes \mathbb{C}^{n|n}$.

Define the linear map $P^\Delta : \mathbb{C}^{m|m} \otimes \mathbb{C}^{n|n} \to \mathbb{C}^{m|m} \otimes \mathbb{C}^{n|n}$ by

$$P^\Delta(v \otimes w) := Pv \otimes Pw, \quad \forall v \in \mathbb{C}^{m|m}, w \in \mathbb{C}^{n|n}.$$

Lemma 5.25. *The actions of P^Δ and $\mathfrak{q}(m) \times \mathfrak{q}(n)$ on $\mathbb{C}^{m|m} \otimes \mathbb{C}^{n|n}$ commute with each other.*

Proof. For $\mathbb{Z}_2$-homogeneous $X \in \mathfrak{q}(m)$, $Y \in \mathfrak{q}(n)$, $v \in \mathbb{C}^{m|m}$, and $w \in \mathbb{C}^{n|n}$, we have

$$\begin{aligned} XP^\Delta(v \otimes w) &= X(Pv \otimes Pw) = (XPv) \otimes (Pw) \\ &= PXv \otimes Pw = P^\Delta X(v \otimes w), \\ YP^\Delta(v \otimes w) &= Y(Pv \otimes Pw) = (-1)^{|Y| + |Y||v|} Pv \otimes YPw \\ &= (-1)^{|Y||v|} Pv \otimes PYw = (-1)^{|Y||v|} P^\Delta(v \otimes Yw) \\ &= P^\Delta Y(v \otimes w). \end{aligned}$$

The lemma follows. □

Recall that the finite group Π_d is generated by $a_1, \ldots, a_d, z$ with relations (3.21), and this gives rise to the group $\widetilde{B}_d = \Pi_d \rtimes \mathfrak{S}_d$. The natural action of Π_d (respectively, $\widetilde{B}_d$) on $(\mathbb{C}^{n|n})^{\otimes d}$ factors through an action of the Clifford superalgebra $\mathcal{C}_d$ by Lemma 3.47 (respectively, the superalgebra $\mathcal{H}_d$), where a_i acts as $I^{\otimes i-1} \otimes P \otimes I^{\otimes d-i}$ with P given in (1.10). Hence the diagonal subgroups $\Delta\Pi_d \subseteq \Pi_d \times \Pi_d$ and $\Delta\widetilde{B}_d \subseteq \widetilde{B}_d \times \widetilde{B}_d$ act on the tensor product

$$(\mathbb{C}^{m|m})^{\otimes d} \otimes (\mathbb{C}^{n|n})^{\otimes d} \cong (\mathbb{C}^{m|m} \otimes \mathbb{C}^{n|n})^{\otimes d}.$$

In this subsection, it is more convenient to talk about the groups Π_d and $\widetilde{B}_d$ than about the algebras $\mathcal{C}_d$ and $\mathcal{H}_d$.

Lemma 5.26. *We have a natural identification:*

$$\left((\mathbb{C}^{m|m} \otimes \mathbb{C}^{n|n})^{\otimes d}\right)^{\Delta(\widetilde{B}_d)} \cong \mathcal{S}^d(\mathbb{C}^{mn|mn}),$$

where $(\cdot)^{\Delta(\widetilde{B}_d)}$ denotes the subspace of $\Delta(\widetilde{B}_d)$-invariants.

Proof. Consider first the case when $d = 1$. We observe that $(\mathbb{C}^{m|m} \otimes \mathbb{C}^{n|n})^{P^\Delta}$ consists of elements of the form $v_m \otimes v_n + P(v_m) \otimes P(v_n)$ and $v_m \otimes P(v_n) + P(v_m) \otimes v_n$, where $v_m \in \mathbb{C}^{m|0}$ and $v_n \in \mathbb{C}^{n|0}$. Hence $(\mathbb{C}^{m|m} \otimes \mathbb{C}^{n|n})^{P^\Delta}$ is isomorphic to $\mathbb{C}^{mn|mn}$.

Since $\Delta\Pi_d$ is an even subgroup of $\Delta(\widetilde{B}_d)$ generated by the d copies of P^Δ's, we have

$$\Big((\mathbb{C}^{m|m})^{\otimes d}\otimes(\mathbb{C}^{n|n})^{\otimes d}\Big)^{\Delta(\widetilde{B}_d)}\cong\Big(((\mathbb{C}^{m|m})^{\otimes d}\otimes(\mathbb{C}^{n|n})^{\otimes d})^{\Delta\Pi_d}\Big)^{\Delta\mathfrak{S}_d}$$
$$\cong\Big(((\mathbb{C}^{m|m}\otimes\mathbb{C}^{n|n})^{P^\Delta})^{\otimes d}\Big)^{\Delta\mathfrak{S}_d}$$
$$\cong((\mathbb{C}^{mn|mn})^{\otimes d})^{\mathfrak{S}_d}\cong\mathcal{S}^d(\mathbb{C}^{mn|mn}).$$

This proves the lemma. □

By Sergeev duality in Theorem 3.49, there is a commuting action of $\mathfrak{q}(n)$ and $\widetilde{B}_d$ on $(\mathbb{C}^{n|n})^{\otimes d}$. Hence there is a natural $\mathfrak{q}(m)\times\mathfrak{q}(n)$-action on $((\mathbb{C}^{m|m}\otimes\mathbb{C}^{n|n})^{\otimes d})^{\Delta(\widetilde{B}_d)}$, and by the identification in Lemma 5.26, on $\mathcal{S}^d(\mathbb{C}^{mn|mn})$ as well. We shall write down this action on $\mathcal{S}^d(\mathbb{C}^{mn|mn})$ explicitly in terms of differential operators.

Let x_{pi} and ξ_{pi}, for $1\le p\le m$ and $1\le i\le n$, denote the standard even and odd coordinates of $\mathbb{C}^{mn|mn}$, respectively. We may then identify $\mathcal{S}(\mathbb{C}^{mn|mn})$ with the polynomial superalgebra generated by the x_{pi} and ξ_{pi}, for $1\le p\le m$ and $1\le i\le n$. We introduce the following first-order differential operators:

$$\widetilde{\mathcal{E}}^{pq}=\sum_{j=1}^{n}(x_{pj}\frac{\partial}{\partial x_{qj}}+\xi_{pj}\frac{\partial}{\partial\xi_{qj}}),$$
$$\overline{\mathcal{E}}^{pq}=\sum_{j=1}^{n}(x_{pj}\frac{\partial}{\partial\xi_{qj}}-\xi_{pj}\frac{\partial}{\partial x_{qj}}),\quad 1\le p,q\le m. \tag{5.22}$$

$$\widetilde{\mathcal{E}}_{ij}=\sum_{p=1}^{m}(x_{pi}\frac{\partial}{\partial x_{pj}}+\xi_{pi}\frac{\partial}{\partial\xi_{pj}}),$$
$$\overline{\mathcal{E}}_{ij}=\sum_{p=1}^{m}(x_{pi}\frac{\partial}{\partial\xi_{pj}}+\xi_{pi}\frac{\partial}{\partial x_{pj}}),\quad 1\le i,j\le n. \tag{5.23}$$

The following lemma is proved by a direct computation (see Exercise 5.2(2)).

Lemma 5.27. *The operators $\widetilde{\mathcal{E}}^{pq}$ and $\overline{\mathcal{E}}^{pq}$ in (5.22), for $1\le p,q\le m$, form a copy of $\mathfrak{q}(m)$, while $\widetilde{\mathcal{E}}_{ij}$ and $\overline{\mathcal{E}}_{ij}$ in (5.23), for $1\le i,j\le n$, form a copy of $\mathfrak{q}(n)$. Furthermore, they define commuting actions of $\mathfrak{q}(m)$ and $\mathfrak{q}(n)$ on $\mathcal{S}(\mathbb{C}^{mn|mn})$.*

Recall from (1.52) that, for a partition λ of d with length $\ell(\lambda)$,

$$\delta(\lambda)=\begin{cases}0, & \text{if } \ell(\lambda) \text{ is even},\\ 1, & \text{if } \ell(\lambda) \text{ is odd}.\end{cases}$$

Furthermore, recall that $\mathcal{SP}_d$ denotes the set of strict partitions of d, and $L_n(\lambda)$ stands for the simple $\mathfrak{q}(n)$-module with highest weight λ, for $\lambda\in\mathcal{SP}_d$ of length not exceeding n (see Section 2.1.6).

Theorem 5.28 ($(\mathfrak{q}(m),\mathfrak{q}(n))$-Howe duality)**.** *As a $(\mathfrak{q}(m),\mathfrak{q}(n))$-module, $\mathcal{S}^d(\mathbb{C}^{mn|mn})$ admits the following strongly multiplicity-free decomposition:*

$$\mathcal{S}^d(\mathbb{C}^{mn|mn}) \cong \bigoplus_{\substack{\lambda\in\mathcal{SP}_d \\ \ell(\lambda)\leq\min\{m,n\}}} 2^{-\delta(\lambda)} L_m(\lambda)\otimes L_n(\lambda).$$

Proof. By Sergeev duality (Theorem 3.49), we have as $\mathfrak{q}(m)\times\widetilde{B}_d$ modules

$$(\mathbb{C}^{m|m})^{\otimes d} \cong \bigoplus_{\substack{\lambda\in\mathcal{SP}_d \\ \ell(\lambda)\leq m}} 2^{-\delta(\lambda)} L_m(\lambda)\otimes D^\lambda,$$

where we recall that D^λ denotes an irreducible $\widetilde{B}_d$-module (or equivalently, an irreducible $\mathcal{H}_d$-module). Therefore, combined with Lemma 5.26, this gives us

$$\begin{aligned}
(5.24)\quad \mathcal{S}^d(\mathbb{C}^{mn|mn}) &\cong (((\mathbb{C}^{m|m})^{\otimes d})\otimes((\mathbb{C}^{n|n})^{\otimes d}))^{\Delta(\widetilde{B}_d)} \\
&\cong \bigoplus_{\substack{\lambda,\mu\in\mathcal{SP}_d \\ \ell(\lambda)\leq m,\ell(\mu)\leq n}} 2^{-\delta(\lambda)}2^{-\delta(\mu)}(L_m(\lambda)\otimes D^\lambda\otimes L_n(\mu)\otimes D^\mu)^{\Delta(\widetilde{B}_d)} \\
&\cong \bigoplus_{\substack{\lambda,\mu\in\mathcal{SP}_d \\ \ell(\lambda)\leq m,\ell(\mu)\leq n}} 2^{-\delta(\lambda)}2^{-\delta(\mu)}(L_m(\lambda)\otimes L_n(\mu))\otimes(D^\lambda\otimes D^\mu)^{\Delta(\widetilde{B}_d)}.
\end{aligned}$$

The $\widetilde{B}_d$-module D^λ is self-dual, i.e., $(D^\lambda)^*\cong D^\lambda$, since by Theorem 3.46 and (3.27)–(3.29) all the character values of D^λ are real. Recalling D^λ is of type $\mathtt{M}$ if and only if $\delta(\lambda)=0$, we conclude by super Schur's Lemma 3.4 that

$$\dim(D^\lambda\otimes D^\mu)^{\Delta(\widetilde{B}_d)} = \dim \mathrm{Hom}_{\widetilde{B}_d}(D^\lambda, D^\mu) = 2^{\delta(\lambda)}\delta_{\lambda\mu}.$$

This together with (5.24) completes the proof of the theorem. □

Recall that the character of the irreducible $\mathfrak{q}(n)$-module $L_n(\lambda)$ is given by the Schur Q-function Q_λ up to a 2-power by Theorem 3.51. Now by calculating the characters of both sides of the isomorphism in Theorem 5.28 and summing over all $d\geq 0$, we recover the following Cauchy identity (A.57) in Appendix A:

$$\prod_{i,j}\frac{1+x_iy_j}{1-x_iy_j} = \sum_{\lambda\in\mathcal{SP}} 2^{-\ell(\lambda)} Q_\lambda(x)Q_\lambda(y).$$

5.3. Howe duality for symplectic and orthogonal groups

In this section, we use the First Fundamental Theorem (FFT) for classical Lie groups in Chapter 4 to construct the Howe dual pairs $(\mathrm{Sp}(V),\mathfrak{osp}(2m|2n))$ and $(\mathrm{O}(V),\mathfrak{spo}(2m|2n))$. We also obtain precise multiplicity-free decompositions of these Howe dualities for symplectic and orthogonal groups. The case of orthogonal groups, which are not connected, requires some extra work.

5.3.1. Howe dual pair $(\mathrm{Sp}(V), \mathfrak{osp}(2m|2n))$**.** We take $V = \mathbb{C}^k$, where $k = 2\ell$ is assumed to be even in this subsection. We consider the superspace $U = U_{\bar{0}} \oplus U_{\bar{1}}$ with $U_{\bar{0}} = V^m$ and $U_{\bar{1}} = V^n$, and identify naturally $U \equiv V \otimes \mathbb{C}^{m|n}$ as before. We continue to identify the supersymmetric algebra of U with $\mathbb{C}[x, \eta]$ as in (5.12).

Recall the skew-symmetric matrix (4.7). We let $\mathrm{Sp}(k) = G(V, \omega^-)$ be the symplectic subgroup of $\mathrm{GL}(k)$, consisting of elements preserving the non-degenerate skew-symmetric bilinear form ω^- on $V = \mathbb{C}^k$ corresponding to the matrix (4.7), and let $\mathfrak{sp}(k) \subseteq \mathfrak{gl}(k)$ be its Lie algebra. In order to write down explicitly the Lie algebra $\mathfrak{sp}(k)$, we introduce the following notation. For an $\ell \times \ell$ matrix $A = (a_{ij})$, we denote by $A^\flat = J_\ell A^t J_\ell$ the transpose of A with respect to the "opposite" diagonal:

$$A^\flat = \begin{pmatrix} a_{\ell,\ell} & a_{\ell-1,\ell} & \cdots & a_{2,\ell} & a_{1,\ell} \\ a_{\ell,\ell-1} & a_{\ell-1,\ell-1} & \cdots & a_{2,\ell-1} & a_{1,\ell-1} \\ a_{\ell,\ell-2} & a_{\ell-1,\ell-2} & \cdots & a_{2,\ell-2} & a_{1,\ell-2} \\ \vdots & \vdots & \ddots & \vdots & \vdots \\ a_{\ell,2} & a_{\ell-1,2} & \cdots & a_{2,2} & a_{1,2} \\ a_{\ell,1} & a_{\ell-1,1} & \cdots & a_{2,1} & a_{1,1} \end{pmatrix}. \tag{5.25}$$

It is then straightforward to check that $\mathfrak{sp}(k)$ consists precisely of the following $k \times k$ matrices in $\ell|\ell$-block form:

$$\mathfrak{sp}(k) = \left\{ \begin{pmatrix} A & B \\ C & D \end{pmatrix} \mid D = -A^\flat, B^\flat = B, C^\flat = C \right\}. \tag{5.26}$$

The standard Borel of $\mathfrak{sp}(k)$ consists of the $k \times k$ upper triangular matrices of the form (5.26) and so is compatible with the standard Borel of $\mathfrak{gl}(k)$. Furthermore, the Cartan subalgebra of $\mathfrak{sp}(k)$ is spanned by the basis vectors $E_{ii} - E_{k+1-i,k+1-i}$, for $1 \le i \le k/2$.

The action of the Lie algebra $\mathfrak{sp}(k)$ as the subalgebra of $\mathfrak{gl}(k)$ on $\mathbb{C}[x, \eta]$ given in (5.13) lifts to an action of the Lie group $\mathrm{Sp}(V) = \mathrm{Sp}(k)$. On the other hand, the action ρ of $\mathfrak{gl}(m|n)$ on $\mathbb{C}[x, \eta]$ with $\rho(E_{ab}) = \mathcal{E}_{ab}$, for $a, b \in I(m|n)$, is modified from (5.14) by a shift of scalars on the diagonal matrices:

$$\begin{aligned} \mathcal{E}_{\bar{i}\bar{j}} &= \sum_{p=1}^{k} x_{pi} \frac{\partial}{\partial x_{pj}} + \frac{k}{2}\delta_{ij}, \quad \mathcal{E}_{\bar{i}s} = \sum_{p=1}^{k} x_{pi} \frac{\partial}{\partial \eta_{ps}}, \quad 1 \le i, j \le m, \\ \mathcal{E}_{st} &= \sum_{p=1}^{k} \eta_{ps} \frac{\partial}{\partial \eta_{pt}} - \frac{k}{2}\delta_{st}, \quad \mathcal{E}_{s\bar{i}} = \sum_{p=1}^{k} \eta_{ps} \frac{\partial}{\partial x_{pi}}, \quad 1 \le s, t \le n. \end{aligned} \tag{5.27}$$

Introduce the following additional operators:

$$
\begin{aligned}
\mathcal{J}_{\bar{i}\bar{j}} &= \sum_{p=1}^{\frac{k}{2}} \Big(x_{pi} x_{k+1-p,j} - x_{k+1-p,i} x_{pj} \Big), \\
\mathcal{J}_{\bar{i}s} &= \sum_{p=1}^{\frac{k}{2}} \Big(x_{pi} \eta_{k+1-p,s} - x_{k+1-p,i} \eta_{ps} \Big), \\
\mathcal{J}_{st} &= \sum_{p=1}^{\frac{k}{2}} \Big(\eta_{ps} \eta_{k+1-p,t} - \eta_{k+1-p,s} \eta_{pt} \Big), \\
\Delta_{\bar{i}\bar{j}} &= \sum_{p=1}^{\frac{k}{2}} \Big(\frac{\partial}{\partial x_{pi}} \frac{\partial}{\partial x_{k+1-p,j}} - \frac{\partial}{\partial x_{k+1-p,i}} \frac{\partial}{\partial x_{pj}} \Big), \\
\Delta_{\bar{i}s} &= \sum_{p=1}^{\frac{k}{2}} \Big(\frac{\partial}{\partial x_{pi}} \frac{\partial}{\partial \eta_{k+1-p,s}} - \frac{\partial}{\partial x_{k+1-p,i}} \frac{\partial}{\partial \eta_{ps}} \Big), \quad 1 \le i < j \le m, \\
\Delta_{st} &= \sum_{p=1}^{\frac{k}{2}} \Big(\frac{\partial}{\partial \eta_{ps}} \frac{\partial}{\partial \eta_{k+1-p,t}} - \frac{\partial}{\partial \eta_{k+1-p,s}} \frac{\partial}{\partial \eta_{pt}} \Big), \quad 1 \le s \le t \le n.
\end{aligned} \tag{5.28}
$$

Let us denote by $\mathfrak{g}'$ the linear span of the elements in (5.27) and (5.28). Clearly, the elements in (5.27) and (5.28) form a linear basis for $\mathfrak{g}'$.

Lemma 5.29. (1) *Under the super-commutator, $\mathfrak{g}'$ forms a Lie superalgebra, which is isomorphic to $\mathfrak{osp}(2m|2n)$.*

(2) *The actions of* $\mathrm{Sp}(k)$ *and* $\mathfrak{osp}(2m|2n)$ *on* $\mathcal{S}(\mathbb{C}^k \otimes \mathbb{C}^{m|n})$ *commute.*

Proof. (1) With $U = \mathbb{C}^k \otimes \mathbb{C}^{m|n}$ and $\boldsymbol{U} = U \oplus U^*$, we identify $\mathfrak{WC}(\boldsymbol{U})$ with the superalgebra of differential operators on $\mathcal{S}(U)$. The operators $\mathcal{E}_{\bar{i}\bar{j}}$ and $\mathcal{E}_{st}$ from (5.27) can be written as

$$
\mathcal{E}_{\bar{i}\bar{j}} = \frac{1}{2} \sum_{p=1}^{k} \Big(x_{pi} \frac{\partial}{\partial x_{pj}} + \frac{\partial}{\partial x_{pj}} x_{pi} \Big), \quad \mathcal{E}_{st} = \frac{1}{2} \sum_{p=1}^{k} \Big(\eta_{ps} \frac{\partial}{\partial \eta_{pt}} - \frac{\partial}{\partial \eta_{pt}} \eta_{ps} \Big),
$$

so that $\mathfrak{g}'$ can be regarded as a subspace inside $\mathcal{G} \cong \mathfrak{spo}(\boldsymbol{U})$ as in Proposition 5.5. A direct computation using elements in (5.27) and (5.28) shows that $\mathfrak{g}'$ forms a Lie superalgebra under the super-commutator. Moreover, a further direct computation shows that $\mathfrak{g}' \cong \mathfrak{osp}(2m|2n)$, with identification of root vectors e_α, for a root α, as follows:

$$
\begin{aligned}
e_{\delta_i - \delta_j} &= \mathcal{E}_{\bar{i}\bar{j}}, & e_{\delta_i - \varepsilon_s} &= \mathcal{E}_{\bar{i}s}, \\
e_{\varepsilon_s - \varepsilon_t} &= \mathcal{E}_{st}, & e_{\varepsilon_s - \delta_i} &= \mathcal{E}_{s\bar{i}}, \qquad 1 \le i \neq j \le m, 1 \le s \neq t \le n.
\end{aligned}
$$

$$
\begin{aligned}
e_{\delta_i + \delta_j} &= \mathcal{J}_{\bar{i}\bar{j}}, & e_{\delta_i + \varepsilon_s} &= \mathcal{J}_{\bar{i}s}, \\
e_{\varepsilon_s + \varepsilon_t} &= \mathcal{J}_{st}, & e_{-\delta_i - \delta_j} &= \Delta_{\bar{i}\bar{j}},
\end{aligned}
$$

$$e_{-\varepsilon_s-\varepsilon_t} = \Delta_{st}, \qquad e_{-\delta_i-\varepsilon_s} = \Delta_{\bar{i}s}, \qquad 1 \le i < j \le m, 1 \le s \le t \le n.$$

We omit the details.

(2) Recall that $e^1, \ldots, e^k$ denote the standard basis for $\mathbb{C}^k$. It follows by definition of ω^- that

$$\omega^-(e^p, e^{k+1-p}) = -\omega^-(e^{k+1-p}, e^p) = 1,$$

for $1 \le p \le k/2$, and hence $\sum_{p=1}^{k/2}(e^p \otimes e^{k+1-p} - e^{k+1-p} \otimes e^p) \in \mathbb{C}^k \otimes \mathbb{C}^k$ is $\mathrm{Sp}(k)$-invariant. Thus, $\mathfrak{J}_{\bar{i}\bar{j}}, \mathfrak{J}_{\bar{i}s}, \mathfrak{J}_{st}$ defined above are $\mathrm{Sp}(k)$-invariant in $\mathfrak{WC}(\boldsymbol{U})$, where we recall the definition of the basis (5.11) for U and we have regarded $\frac{\partial}{\partial x_{pi}}$ and $\frac{\partial}{\partial \eta_{ps}}$ as elements in U^*. Similarly, $\Delta_{\bar{i}\bar{j}}, \Delta_{\bar{i}s}, \Delta_{st}$ defined above are $\mathrm{Sp}(k)$-invariant in $\mathfrak{WC}(\boldsymbol{U})$. On the other hand, the elements in (5.27) are $\mathrm{GL}(k)$-invariant (and hence $\mathrm{Sp}(k)$-invariant) in $\mathfrak{WC}(\boldsymbol{U})$.

Hence, for $g \in G$, $x \in \mathfrak{g}'$ and $z \in \mathbb{C}[x,\eta]$, we have $gx.z = xg.z$, whence (2). □

Theorem 5.30 (($\mathrm{Sp}(k), \mathfrak{osp}(2m|2n)$)-Howe duality, I). *$\mathrm{Sp}(k)$ and $\mathfrak{osp}(2m|2n)$ form a Howe dual pair on $\mathcal{S}(\mathbb{C}^k \otimes \mathbb{C}^{m|n})$, and the $(\mathrm{Sp}(k), \mathfrak{osp}(2m|2n))$-module $\mathcal{S}(\mathbb{C}^k \otimes \mathbb{C}^{m|n})$ is strongly multiplicity-free.*

Proof. Set $G = \mathrm{Sp}(k)$. The action of $\mathfrak{sp}(k)$ on $\mathcal{S}(\mathbb{C}^k \otimes \mathbb{C}^{m|n})$ clearly lifts to an action of G. By Lemma 5.29, the actions of G and $\mathfrak{osp}(2m|2n)$ on $\mathcal{S}(U) = \mathcal{S}(\mathbb{C}^k \otimes \mathbb{C}^{m|n})$ commute.

Let $\boldsymbol{U} = U \oplus U^*$. The associated graded for $\mathfrak{WC}(\boldsymbol{U})$ is $\mathcal{S}(\boldsymbol{U})$, and the action of $\mathfrak{sp}(k)$ on $\mathcal{S}(\boldsymbol{U})$ clearly lifts to an action of G. Let $\{y_{pi}\}$ denote the basis in $U_{\bar{0}}^*$ dual to the basis $\{x_{pi}\}$ for $U_{\bar{0}}$, and let $\{\xi_{pt}\}$ denote the basis in $U_{\bar{1}}^*$ dual to the basis $\{\eta_{pt}\}$ for $U_{\bar{1}}$. By the supersymmetric First Fundamental Theorem (FFT) for G (see Theorem 4.19), a generating set $\mathcal{T}$ for the algebra $\mathcal{S}(\boldsymbol{U})^G = \left(S(U_{\bar{0}} \oplus U_{\bar{0}}^*) \otimes \wedge(U_{\bar{1}} \oplus U_{\bar{1}}^*)\right)^G$ consists of $z_{\bar{i}\bar{j}}, z_{st}, z_{\bar{i}s}, z_{s\bar{i}}$ in (5.15), $\mathfrak{J}_{\bar{i}\bar{j}}, \mathfrak{J}_{\bar{i}s}, \mathfrak{J}_{st}$, and

$$\sum_{p=1}^{\frac{k}{2}} \left(y_{pi}y_{k+1-p,j} - y_{k+1-p,i}y_{pj}\right), \quad \sum_{p=1}^{\frac{k}{2}} \left(y_{pi}\xi_{k+1-p,s} - y_{k+1-p,i}\xi_{ps}\right),$$

$$\sum_{p=1}^{\frac{k}{2}} \left(\xi_{ps}\xi_{k+1-p,t} - \xi_{k+1-p,s}\xi_{pt}\right),$$

for $1 \le i,j \le m, 1 \le s,t \le n$. Observe that the symbol map σ sends the basis elements (5.27) and (5.28) for $\mathfrak{g}'$ to the above elements in $\mathcal{T}$ for $\mathcal{S}(\boldsymbol{U})^G$. Now the theorem follows from Proposition 5.12 and Lemma 5.29. □

5.3.2. ($\mathrm{Sp}(V), \mathfrak{osp}(2m|2n)$)**-Howe duality.** Denote by $\mathfrak{u}^+$ (respectively, $\mathfrak{u}^-$) the subalgebra of $\mathfrak{osp}(2m|2n)$ spanned by the Δ (respectively, $\mathfrak{J}$) operators in (5.28). The irreducible highest weight module $L(\mathfrak{osp}(2m|2n), \mu)$ below is understood to be relative to the Borel subalgebra of $\mathfrak{osp}(2m|2n)$ corresponding to the fundamental system specified in the following Dynkin diagram:

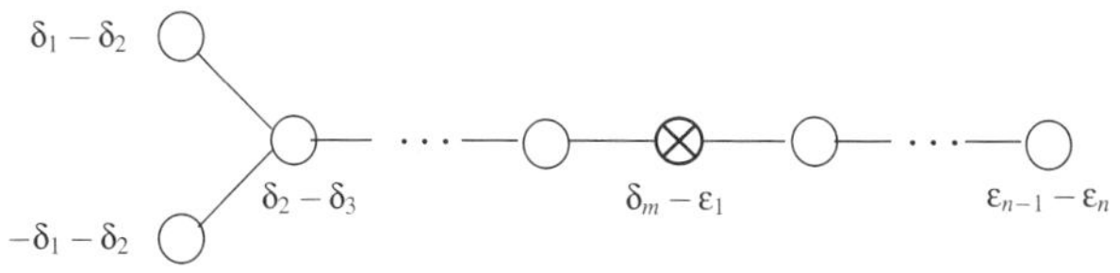

The root vectors associated to the simple roots $-\delta_1 - \delta_2, \delta_i - \delta_{i+1}$ $(1 \le i \le m-1), \delta_m - \varepsilon_1, \varepsilon_t - \varepsilon_{t+1}$ $(1 \le t \le n-1)$, in the notations of (5.27) and (5.28) are, respectively,

$$(5.29) \qquad \Delta_{\bar{1}\bar{2}}, \quad \mathcal{E}_{\bar{i},\overline{i+1}}\ (1 \le i \le m-1), \quad \mathcal{E}_{\bar{m}1}, \quad \mathcal{E}_{t,t+1}\ (1 \le t \le n-1).$$

Then $\mathfrak{gl}(m|n)$ is a Levi subalgebra of $\mathfrak{osp}(2m|2n)$ corresponding to the removal of the simple root $-\delta_1 - \delta_2$. We have a triangular decomposition

$$(5.30) \qquad \mathfrak{osp}(2m|2n) = \mathfrak{u}^- \oplus \mathfrak{gl}(m|n) \oplus \mathfrak{u}^+.$$

In particular, the Lie superalgebras $\mathfrak{gl}(m|n)$ and $\mathfrak{osp}(2m|2n)$ share the same Cartan subalgebra. An element $f \in \mathbb{C}[x,\eta]$ is called **harmonic** if f is annihilated by the subalgebra $\mathfrak{u}^+$. The space of harmonics $\mathbb{C}[x,\eta]^{\mathfrak{u}^+}$ will be denoted by ${}^{\mathrm{Sp}}\mathcal{H}$.

Recall that $\mathcal{P}(m|n)$ denotes the set of all $(m|n)$-hook partitions λ (i.e., $\lambda_{m+1} \le n$), and also recall from (2.4) that, for an $(m|n)$-hook partition λ, $\lambda^\natural$ is a weight for $\mathfrak{gl}(m|n)$. Then $\lambda^\natural + \ell \mathbf{1}_{m|n}$ can be regarded as a weight for the Lie superalgebras $\mathfrak{gl}(m|n)$ and $\mathfrak{osp}(2m|2n)$, where we recall the definition of $\mathbf{1}_{m|n}$ from (1.36).

Theorem 5.31 ((Sp$(k), \mathfrak{osp}(2m|2n)$)-Howe duality, II). *Let $k = 2\ell$. We have the following decomposition as a* (Sp$(k), \mathfrak{osp}(2m|2n)$)*-module:*

$$\mathcal{S}(\mathbb{C}^k \otimes \mathbb{C}^{m|n}) = \bigoplus_{\lambda \in \mathcal{P}(m|n), \ell(\lambda) \le \ell} L(\mathrm{Sp}(k), \lambda) \otimes L\big(\mathfrak{osp}(2m|2n), \lambda^\natural + \ell \mathbf{1}_{m|n}\big).$$

Proof. Observing that $M^{\mathfrak{u}^+}$ is a $\mathfrak{gl}(m|n)$-module for any $\mathfrak{osp}(2m|2n)$-module M by the triangular decomposition (5.30), we may regard the space of harmonics ${}^{\mathrm{Sp}}\mathcal{H}$ as an $(\mathrm{Sp}(k), \mathfrak{gl}(m|n))$-module. Furthermore, since each graded subspace of $\mathcal{S}(\mathbb{C}^k \otimes \mathbb{C}^{m|n})$ as a $\mathfrak{gl}(m|n)$-module is a submodule of a tensor power of $\mathbb{C}^{m|n}$ and hence is completely reducible by Theorem 3.11, $L(\mathfrak{osp}(2m|2n), \mu)^{\mathfrak{u}^+}$ is also a completely reducible $\mathfrak{gl}(m|n)$-module for any irreducible $\mathfrak{osp}(2m|2n)$-module $L(\mathfrak{osp}(2m|2n), \mu)$ appearing in $\mathcal{S}(\mathbb{C}^k \otimes \mathbb{C}^{m|n})$. It follows by the irreducibility of $L(\mathfrak{osp}(2m|2n), \mu)$ that $L(\mathfrak{osp}(2m|2n), \mu)^{\mathfrak{u}^+}$ is an irreducible $\mathfrak{gl}(m|n)$-module, and a highest weight vector of highest weight μ in $L(\mathfrak{osp}(2m|2n), \mu)$ remains a highest weight vector in $L(\mathfrak{osp}(2m|2n), \mu)^{\mathfrak{u}^+}$. Therefore, we must have

$$(5.31) \qquad L(\mathfrak{osp}(2m|2n), \mu)^{\mathfrak{u}^+} \cong L_{m|n}(\mu).$$

By Theorem 5.30, the $(\mathrm{Sp}(k), \mathfrak{osp}(2m|2n))$-module $\mathcal{S}(\mathbb{C}^k \otimes \mathbb{C}^{m|n})$ is strongly multiplicity-free, and hence by (5.31), the $(\mathrm{Sp}(k), \mathfrak{gl}(m|n))$-module ${}^{\mathrm{Sp}}\mathcal{H}$ is also strongly multiplicity-free. To establish the explicit form of decomposition as stated in the theorem, it suffices (and is indeed equivalent) to establish the following explicit decomposition of ${}^{\mathrm{Sp}}\mathcal{H}$ as an $(\mathrm{Sp}(k), \mathfrak{gl}(m|n))$-module:

$$
{}^{\mathrm{Sp}}\mathcal{H} \cong \bigoplus_{\lambda \in \mathcal{P}(m|n), \ell(\lambda) \leq \ell} L(\mathrm{Sp}(k), \lambda) \otimes L_{m|n}(\lambda^{\natural} + \ell \mathbf{1}_{m|n}). \tag{5.32}
$$

We first consider the limit case $n = \infty$ with the space of harmonics denoted by ${}^{\mathrm{Sp}}\mathcal{H}^{\infty}$, where the condition $\lambda \in \mathcal{P}(m|n)$ reduces to $\lambda \in \mathcal{P}$. Set $\diamondsuit = \diamondsuit_{1,2,\ldots,\lambda_1,\lambda'}$. Observe by Theorem 5.23 that the vector $\diamondsuit$ associated to any partition λ with $\ell(\lambda) \leq \ell$ is a joint $\mathfrak{sp}(k) \times \mathfrak{gl}(m|n)$-highest weight vector and $\diamondsuit$ is a polynomial independent of the variables x_{pi} and η_{pt} for $p \geq \ell + 1$. Each Δ-operator in (5.28) is a second-order differential operator whose summands always involve differentiation with respect to one of these variables for $p \geq \ell + 1$. Thus, $\diamondsuit$ is annihilated by $\mathfrak{u}^{+}$, and hence we have $\diamondsuit \in {}^{\mathrm{Sp}}\mathcal{H}^{\infty}$. One also easily checks that the vector $\diamondsuit$ has weight $(\lambda, \lambda^{\natural} + \ell \mathbf{1}_{m|n})$ under the transformations of $\mathfrak{sp}(k) \times \mathfrak{osp}(2m|2n)$. Therefore, all the summands on the right-hand side of (5.32) indeed occur in ${}^{\mathrm{Sp}}\mathcal{H}^{\infty}$ (with multiplicity one), and all irreducible representations of $\mathrm{Sp}(k)$ occur. We conclude that (5.32) must hold (when $n = \infty$).

Now consider the finite n case. We may regard $\mathcal{S}(\mathbb{C}^d \otimes \mathbb{C}^{m|n}) \subseteq \mathcal{S}(\mathbb{C}^d \otimes \mathbb{C}^{m|\infty})$ with compatible actions of $\mathfrak{osp}(2m|2n) \subseteq \mathfrak{osp}(2m|2\infty)$. We claim the following equality for spaces of joint highest weight vectors holds:

$$
({}^{\mathrm{Sp}}\mathcal{H})^{\mathfrak{n}_1 \times \mathfrak{n}(n)} = ({}^{\mathrm{Sp}}\mathcal{H}^{\infty})^{\mathfrak{n}_1 \times \mathfrak{n}(\infty)} \cap \mathbb{C}[x, \eta].
$$

Here $\mathfrak{n}_1$ (respectively, $\mathfrak{n}(n)$) denotes the nilradical of the standard Borel for $\mathfrak{sp}(k)$ (respectively, the nilradical generated by (5.29) for $\mathfrak{osp}(2m|2n)$). Note by definition that $({}^{\mathrm{Sp}}\mathcal{H})^{\mathfrak{n}_1 \times \mathfrak{n}(n)} \supseteq ({}^{\mathrm{Sp}}\mathcal{H}^{\infty})^{\mathfrak{n}_1 \times \mathfrak{n}(\infty)} \cap \mathbb{C}[x, \eta]$. Since each summand of the operators in (5.27) and the Δ-operators in (5.28) that lie in $\mathfrak{osp}(2m|2\infty) \backslash \mathfrak{osp}(2m|2n)$ must involve $\frac{\partial}{\partial \eta_{ps}}$ for $s > n$, we conclude that $({}^{\mathrm{Sp}}\mathcal{H})^{\mathfrak{n}_1 \times \mathfrak{n}(n)} \subseteq ({}^{\mathrm{Sp}}\mathcal{H}^{\infty})^{\mathfrak{n}_1 \times \mathfrak{n}(\infty)} \cap \mathbb{C}[x, \eta]$, and hence the claim is proved. Observe from the explicit formulas of the joint highest vectors in ${}^{\mathrm{Sp}}\mathcal{H}^{\infty}$ (see Theorem 5.23) that precisely those vectors corresponding to $(m|n)$-hook partitions will not involve the variables η_{ps} for $s > n$, and hence they lie in $({}^{\mathrm{Sp}}\mathcal{H}^{\infty})^{\mathfrak{n}_1 \times \mathfrak{n}(\infty)} \cap \mathbb{C}[x, \eta]$. This together with the claim above completes the proof of the theorem. □

The irreducible $\mathfrak{osp}(2m|2n)$-modules appearing in the $(\mathrm{Sp}(k), \mathfrak{osp}(2m|2n))$-Howe duality decomposition above (for varied k) are called the **oscillator modules** of $\mathfrak{osp}(2m|2n)$. In general the $\mathfrak{osp}(2m|2n)$-oscillator modules are infinite dimensional, as their highest weights evaluated at the simple root $-\delta_1 - \delta_2$ is $-\lambda_1 - \lambda_2 - k < 0$.

Specializing $n = 0$ in Theorem 5.31 gives us the following.

Corollary 5.32 ($(\mathrm{Sp}(k), \mathfrak{so}(2m))$-Howe duality). *Let $k = 2\ell$. We have the following decomposition of $(\mathrm{Sp}(k), \mathfrak{so}(2m))$-modules:*

$$S(\mathbb{C}^k \otimes \mathbb{C}^m) \cong \bigoplus_{\lambda \in \mathcal{P}, \ell(\lambda) \leq \min\{\ell, m\}} L(\mathrm{Sp}(k), \lambda) \otimes L\Big(\mathfrak{so}(2m), \sum_{i=1}^{m} (\lambda_i + \ell)\delta_i\Big).$$

We now specialize $m = 0$ in Theorem 5.31. The irreducible highest weight $\mathfrak{sp}(2n)$-modules in Corollary 5.33 below are relative to the Borel subalgebra of $\mathfrak{sp}(2n)$ corresponding to the fundamental system specified in the following Dynkin diagram:

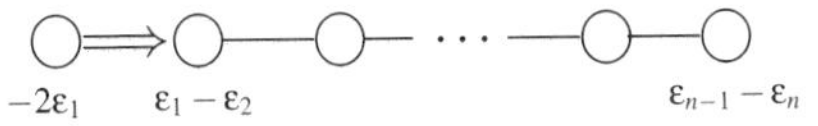

Corollary 5.33 ($(\mathrm{Sp}(k), \mathfrak{sp}(2n))$-Howe duality). *Let $k = 2\ell$. We have the following decomposition of $(\mathrm{Sp}(k), \mathfrak{sp}(2n))$-modules:*

$$\wedge(\mathbb{C}^k \otimes \mathbb{C}^n) \cong \bigoplus_{\substack{\lambda \in \mathcal{P} \\ \ell(\lambda) \leq \ell, \ell(\lambda') \leq n}} L(\mathrm{Sp}(k), \lambda) \otimes L\Big(\mathfrak{sp}(2n), \sum_{i=1}^{n} (\lambda_i' - \ell)\varepsilon_i\Big).$$

5.3.3. Irreducible modules of $\mathrm{O}(V)$. Take $V = \mathbb{C}^k$. Let $\mathrm{O}(k) = G(V, \omega^+)$ be the orthogonal subgroup of $\mathrm{GL}(k)$ consisting of elements preserving the non-degenerate symmetric bilinear form ω^+ on $V = \mathbb{C}^k$ associated to the $k \times k$ matrix J_k in (4.6), and let $\mathfrak{so}(k) \subseteq \mathfrak{gl}(k)$ be its Lie algebra. Let $\mathrm{SO}(V) = \mathrm{O}(V) \cap \mathrm{SL}(V)$ denote the special orthogonal group. The Lie algebras $\mathfrak{so}(k)$, for $k = 2\ell$ and $k = 2\ell + 1$ with $\ell \in \mathbb{Z}_+$, consist precisely of the following $k \times k$ matrices in $\ell|\ell$- and $\ell|1|\ell$-block forms, respectively (recall $A^\flat$ from (5.25)):

$$\mathfrak{so}(2\ell) = \left\{ \begin{pmatrix} A & B \\ C & D \end{pmatrix} \mid D = -A^\flat, B^\flat = -B, C^\flat = -C \right\}, \tag{5.33}$$

$$\mathfrak{so}(2\ell+1) = \left\{ \begin{pmatrix} A & x & B \\ y & 0 & z \\ C & w & D \end{pmatrix} \mid D = -A^\flat, B^\flat = -B, C^\flat = -C \right\}, \tag{5.34}$$

where, in addition, the $\ell \times 1$ matrices $x = (x_{i1})$ and $w = (w_{i1})$ and the $1 \times \ell$ matrices $y = (y_{1i})$ and $z = (z_{1i})$ are related by $z_{1i} = -x_{\ell+1-i,1}$ and $y_{1i} = -w_{\ell+1-i,1}$, for all $1 \leq i \leq \ell$. The standard Borel of $\mathfrak{so}(k)$, for k even and odd, is the subalgebra of upper triangular matrices in $\mathfrak{so}(k)$, and hence is compatible with the standard Borel of $\mathfrak{gl}(k)$. The Cartan subalgebra of $\mathfrak{so}(k)$ is spanned by the basis vectors $E_{ii} - E_{k+1-i,k+1-i}$, for $1 \leq i \leq \ell$, where $k = 2\ell$ or $2\ell + 1$. The following fact is well known (cf. [**46**]).

Lemma 5.34. *A finite-dimensional simple $\mathfrak{so}(k)$-module $L(\mathfrak{so}(k), \lambda)$ of highest weight $\lambda = \sum_{i=1}^{\ell} \lambda_i \varepsilon_i$ lifts to an $\mathrm{SO}(k)$-module if and only if*

(1) $(\lambda_1, \ldots, \lambda_\ell)$ *is a partition, for odd* $k = 2\ell + 1$,

(2) $(\lambda_1, \ldots, \lambda_{\ell-1}, |\lambda_\ell|)$ *is a partition, for even* $k = 2\ell$.

Next we will describe the classification of simple $O(k)$-modules and give a parametrization in terms of partitions. Let det denote the one-dimensional determinant module of $O(k)$.

First consider the case when $k = 2\ell + 1$ is odd. In this case, $-I \in O(k) \backslash SO(k)$ and $O(k)$ is the direct product $SO(k) \times \mathbb{Z}_2$. Associated to a partition λ with $\ell(\lambda) \le \ell$, we let $L(O(k), \lambda)$ stand for the irreducible $O(k)$-module that restricts to the irreducible $SO(k)$-module $L(SO(k), \lambda)$ and on which the element $-I$ acts trivially. We shall denote the irreducible $O(k)$-module $L(O(k), \lambda) \otimes \det$ by $L(O(k), \widetilde{\lambda})$, where the Young diagram of the partition $\widetilde{\lambda}$ is obtained from λ by replacing the first column λ'_1 by $k - \lambda'_1$. This way we have obtained a complete parametrization of the simple $O(k)$-modules $L(O(k), \mu)$ in terms of partitions μ with $\mu'_1 + \mu'_2 \le k = 2\ell + 1$.

Now consider the case when $k = 2\ell$ is even.

Proposition 5.35. *Let* $k = 2\ell$ *be even. A simple* $O(k)$*-module* L *is exactly one of the following (where all* $\lambda_i \in \mathbb{Z}_+$*):*

(1) *L is a direct sum of two irreducible* $SO(k)$*-modules of highest weights* $(\lambda_1, \cdots, \lambda_{\ell-1}, \lambda_\ell)$ *and* $(\lambda_1, \cdots, \lambda_{\ell-1}, -\lambda_\ell)$*, where* $\lambda_\ell > 0$*. In this case,* $L \cong L \otimes \det$.

(2) *For* $\lambda = (\lambda_1, \cdots, \lambda_{\ell-1}, 0)$*, there are exactly two inequivalent irreducible* $O(k)$*-modules that restrict to the irreducible* $SO(k)$*-module of highest weight* λ*. If one of these modules is* L*, then the other one is* $L \otimes \det$.

We shall denote the simple $O(k)$-module in Proposition 5.35(1) by $L(O(k), \lambda)$. Let $\tau \in O(k) \setminus SO(k)$ denote the element that switches the basis vector e^ℓ with $e^{\ell+1}$ and fixes all other standard basis vectors e^i's of $\mathbb{C}^k$. We declare $L(O(k), \lambda)$ to be the $O(k)$-module in Proposition 5.35(2) on which τ acts on an $SO(k)$-highest weight vector trivially. Note that τ transforms an $SO(k)$-highest weight vector in the $O(k)$-module $L(O(k), \lambda) \otimes \det$ by -1. We shall denote the simple $O(k)$-module $L(O(k), \lambda) \otimes \det$ by $L(O(k), \widetilde{\lambda})$, where as before the Young diagram of the partition $\widetilde{\lambda}$ is obtained from λ by replacing the first column λ'_1 by $k - \lambda'_1$. In this way, we have obtained a complete parametrization of the simple $O(k)$-modules $L(O(k), \mu)$ in terms of partitions μ with $\mu'_1 + \mu'_2 \le k = 2\ell$.

Summarizing the above discussion for k odd and even, we have obtained the following uniform description.

Proposition 5.36. *A complete set of pairwise non-isomorphic simple* $O(k)$*-modules consists of* $L(O(k), \mu)$*, where* μ *runs over all partitions of length no greater than* k *with* $\mu'_1 + \mu'_2 \le k$.

5.3.4. Howe dual pair $(\mathrm{O}(k), \mathfrak{spo}(2m|2n))$**.** We consider the superspace $U = U_{\bar{0}} \oplus U_{\bar{1}}$ with $U_{\bar{0}} = V^m$ and $U_{\bar{1}} = V^n$, and identify naturally $U \equiv V \otimes \mathbb{C}^{m|n}$ as before. We continue the identification $\mathcal{S}(U) \equiv \mathbb{C}[x,\eta]$ as in (5.12).

The action of the Lie algebra $\mathfrak{so}(k)$ as a subalgebra of $\mathfrak{gl}(k)$ on $\mathbb{C}[x,\eta]$ given in (5.13) lifts to an action of Lie group $\mathrm{O}(V) = \mathrm{O}(k)$. On the other hand, recall the action of $\mathfrak{gl}(m|n)$ on $\mathbb{C}[x,\eta]$ given by (5.27). Introduce the following additional operators on $\mathbb{C}[x,\eta]$:

$$
\begin{aligned}
&\mathfrak{Z}_{\bar{i}\bar{j}} = \sum_{p=1}^{k} x_{pi} x_{k+1-p,j}, && \nabla_{\bar{i}\bar{j}} = \sum_{p=1}^{k} \frac{\partial}{\partial x_{pi}} \frac{\partial}{\partial x_{k+1-p,j}}, && \\
&\mathfrak{Z}_{\bar{i}s} = \sum_{p=1}^{k} x_{pi} \eta_{k+1-p,s}, && \nabla_{\bar{i}s} = \sum_{p=1}^{k} \frac{\partial}{\partial x_{pi}} \frac{\partial}{\partial \eta_{k+1-p,s}}, && 1 \le i \le j \le m, \\
&\mathfrak{Z}_{st} = \sum_{p=1}^{k} \eta_{ps} \eta_{k+1-p,t}, && \nabla_{st} = \sum_{p=1}^{k} \frac{\partial}{\partial \eta_{ps}} \frac{\partial}{\partial \eta_{k+1-p,t}}, && 1 \le s < t \le n.
\end{aligned} \tag{5.35}
$$

Let us denote by $\mathfrak{g}'$ the linear span of the elements in (5.27) and (5.35). Note that the elements in (5.27) and (5.35) form a linear basis for $\mathfrak{g}'$.

Lemma 5.37. (1) *Under the super-commutator, $\mathfrak{g}'$ forms a Lie superalgebra, which is isomorphic to $\mathfrak{spo}(2m|2n)$.*

(2) *The actions of $\mathrm{O}(k)$ and $\mathfrak{spo}(2m|2n)$ on $\mathcal{S}(\mathbb{C}^k \otimes \mathbb{C}^{m|n})$ commute.*

Proof. (1) Letting $U = \mathbb{C}^k \otimes \mathbb{C}^{m|n}$ and $\boldsymbol{U} = U \oplus U^*$, we identify $\mathfrak{WC}(\boldsymbol{U})$ with the superalgebra of differential operators on $\mathcal{S}(U)$. We can then regard $\mathfrak{g}'$ as a subspace inside $\mathcal{G} \cong \mathfrak{spo}(\boldsymbol{U})$ as in Proposition 5.5. In a completely analogous fashion as in the proof of Lemma 5.29, we can show that $\mathfrak{g}' \cong \mathfrak{spo}(2m|2n)$ with an explicit identification of root vectors with elements in (5.27) and (5.35). We leave it to the reader to fill in the details (Exercise 5.9).

(2) Recall the standard basis $e^1, \ldots, e^k$ for $\mathbb{C}^k$. By definition of ω^+, we have

$$\omega^+(e^p, e^{k+1-p}) = 1,$$

for $1 \le p \le k$, and hence $\sum_{p=1}^{k} e^p \otimes e^{k+1-p} \in \mathbb{C}^k \otimes \mathbb{C}^k$ is $\mathrm{O}(k)$-invariant. It follows that $\mathfrak{Z}_{\bar{i}\bar{j}}, \mathfrak{Z}_{\bar{i}s}, \mathfrak{Z}_{st}$ defined above are $\mathrm{O}(k)$-invariant in $\mathfrak{WC}(\boldsymbol{U})$. Dually, the elements $\nabla_{\bar{i}\bar{j}}, \nabla_{\bar{i}s}, \nabla_{st}$ are also $\mathrm{O}(k)$-invariant in $\mathfrak{WC}(\boldsymbol{U})$. On the other hand, the elements in (5.27) are clearly $\mathrm{O}(k)$-invariant in $\mathfrak{WC}(\boldsymbol{U})$. Hence (2) follows. □

Theorem 5.38 ($(\mathrm{O}(k), \mathfrak{spo}(2m|2n))$-Howe duality, I)**.** *$\mathrm{O}(k)$ and $\mathfrak{spo}(2m|2n)$ form a Howe dual pair on $\mathcal{S}(\mathbb{C}^k \otimes \mathbb{C}^{m|n})$, and the $(\mathrm{O}(k), \mathfrak{spo}(2m|2n))$-module $\mathcal{S}(\mathbb{C}^k \otimes \mathbb{C}^{m|n})$ is strongly multiplicity-free.*

Proof. Set $G = \mathrm{O}(k)$. The action of $\mathfrak{so}(k)$ on $\mathcal{S}(\mathbb{C}^k \otimes \mathbb{C}^{m|n})$ lifts to an action of G. By Lemma 5.37, the actions of G and $\mathfrak{spo}(2m|2n)$ on $\mathcal{S}(U) = \mathcal{S}(\mathbb{C}^k \otimes \mathbb{C}^{m|n})$ commute.

Let $\boldsymbol{U} = U \oplus U^*$. The associated graded for $\mathfrak{WC}(\boldsymbol{U})$ is $\mathcal{S}(\boldsymbol{U})$. Thus, the action of $\mathfrak{so}(k)$ on $\mathfrak{WC}(\boldsymbol{U})$ lifts to an action of G. Let $\{y_{pi}\}$ denote the basis in $U_{\bar{0}}^*$ dual to the basis $\{x_{pi}\}$ for $U_{\bar{0}}$, and let $\{\xi_{pt}\}$ denote the basis in $U_{\bar{1}}^*$ dual to the basis $\{\eta_{pt}\}$ for $U_{\bar{1}}$. By the supersymmetric First Fundamental Theorem for G (Theorem 4.19), a generating set $\mathcal{T}$ for the algebra $\mathcal{S}(\boldsymbol{U})^G = \big(S(U_{\bar{0}} \oplus U_{\bar{0}}^*) \otimes \wedge(U_{\bar{1}} \oplus U_{\bar{1}}^*)\big)^G$ consists of $z_{\bar{i}\bar{j}}, z_{st}, z_{\bar{i}s}, z_{s\bar{i}}$ in (5.15), $\mathfrak{I}_{\bar{i}\bar{j}}, \mathfrak{I}_{\bar{i}s}, \mathfrak{I}_{st}$, and

$$\sum_{p=1}^{k} y_{pi} y_{k+1-p,j}, \quad \sum_{p=1}^{k} y_{pi} \xi_{k+1-p,s}, \quad \sum_{p=1}^{k} \xi_{ps} \xi_{k+1-p,t},$$

for $1 \le i,j \le m, 1 \le s,t \le n$. The symbol map σ sends the basis elements (5.27) and (5.35) for $\mathfrak{g}'$ to the above elements in $\mathcal{T}$, and hence the theorem follows from Proposition 5.12 and Lemma 5.37. □

5.3.5. $(\mathrm{O}(V), \mathfrak{spo}(2m|2n))$**-Howe duality.** Denote by $\mathfrak{u}^+$ (respectively, $\mathfrak{u}^-$) the subalgebra of $\mathfrak{spo}(2m|2n)$ spanned by the ∇ (respectively, $\mathfrak{I}$) operators in (5.35). The irreducible highest weight module $L(\mathfrak{spo}(2m|2n), \mu)$ below is understood to be relative to the Borel subalgebra of $\mathfrak{spo}(2m|2n)$ corresponding to the fundamental system specified in the following Dynkin diagram:

The root vectors associated to the simple roots $-2\delta_1, \delta_i - \delta_{i+1}$ $(1 \le i \le m-1), \delta_m - \varepsilon_1, \varepsilon_t - \varepsilon_{t+1}$ $(1 \le t \le n-1)$, in the notations of (5.27) and (5.35), are respectively

$$\nabla_{\bar{1}\bar{1}}, \quad \mathcal{E}_{\bar{i},\overline{i+1}}\ (1 \le i \le m-1), \quad \mathcal{E}_{\bar{m}1}, \quad \mathcal{E}_{t,t+1}\ (1 \le t \le n-1). \tag{5.36}$$

Then $\mathfrak{gl}(m|n)$ is a Levi subalgebra of $\mathfrak{spo}(2m|2n)$ corresponding to the removal of the simple root $-2\delta_1$. We have a triangular decomposition

$$\mathfrak{spo}(2m|2n) = \mathfrak{u}^- \oplus \mathfrak{gl}(m|n) \oplus \mathfrak{u}^+. \tag{5.37}$$

In particular, the Lie superalgebras $\mathfrak{gl}(m|n)$ and $\mathfrak{spo}(2m|2n)$ share the same Cartan subalgebra. For $\lambda \in \mathcal{P}(m|n)$, $\lambda^\natural + \frac{k}{2}\mathbf{1}_{m|n}$ is a weight for $\mathfrak{gl}(m|n)$, and hence can be regarded as a weight for $\mathfrak{spo}(2m|2n)$ as well.

Theorem 5.39 ($(\mathrm{O}(k), \mathfrak{spo}(2m|2n))$-Howe duality, II)**.** *We have the following decomposition of* $(\mathrm{O}(k), \mathfrak{spo}(2m|2n))$*-modules:*

$$\mathcal{S}(\mathbb{C}^k \otimes \mathbb{C}^{m|n}) = \bigoplus_{\substack{\lambda \in \mathcal{P}(m|n) \\ \lambda_1' + \lambda_2' \le k}} L(\mathrm{O}(k), \lambda) \otimes L\Big(\mathfrak{spo}(2m|2n), \lambda^\natural + \frac{k}{2}\mathbf{1}_{m|n}\Big).$$

Proof. Denote the space of harmonics $\mathbb{C}[x,\eta]^{\mathfrak{u}^+}$ by ${}^{\mathrm{O}}\mathcal{H}$. Then $M^{\mathfrak{u}^+}$ is a $\mathfrak{gl}(m|n)$-module for any $\mathfrak{spo}(2m|2n)$-module M by the triangular decomposition (5.37), so that ${}^{\mathrm{O}}\mathcal{H}$ is an $(\mathrm{O}(k), \mathfrak{gl}(m|n))$-module. Similar to the proof of Theorem 5.31, it is enough to establish the following decomposition of ${}^{\mathrm{O}}\mathcal{H}$ as an $(\mathrm{O}(k), \mathfrak{gl}(m|n))$-module:

$$
{}^{\mathrm{O}}\mathcal{H} \cong \bigoplus_{\substack{\lambda\in\mathcal{P}(m|n)\\ \lambda'_1+\lambda'_2\leq k}} L(\mathrm{O}(k),\lambda) \otimes L_{m|n}\Big(\lambda^{\natural} + \frac{k}{2}\mathbf{1}_{m|n}\Big). \tag{5.38}
$$

The validity of (5.38) for finite n follows from the case when $n = \infty$ in a completely parallel way as in the proof of Theorem 5.31. Hence it remains to establish (5.38) in the limit case $n = \infty$, where the condition $\lambda \in \mathcal{P}(m|n)$ reduces to $\lambda \in \mathcal{P}$.

Assume now $n = \infty$ in the remainder of the proof. Set $\diamondsuit = \diamondsuit_{1,2,\ldots,\lambda_1,\lambda'}$, for $\lambda \in \mathcal{P}$ with $\lambda'_1 + \lambda'_2 \leq k$. We claim that $\diamondsuit \in {}^{\mathrm{O}}\mathcal{H}$, i.e, $\diamondsuit$ is annihilated by $\mathfrak{u}^+$. By Theorem 5.23 $\diamondsuit$ is a joint $\mathfrak{so}(k) \times \mathfrak{gl}(m|n)$-highest weight vector. To establish the claim, it suffices to show that $\diamondsuit$ is annihilated by the simple root vector $\nabla_{\bar{1}\bar{1}}$, since $\diamondsuit$ is annihilated by the remaining simple root vectors in (5.36) that lie in $\mathfrak{gl}(m|n)$.

To show that $\nabla_{\bar{1}\bar{1}}(\diamondsuit) = 0$, we recall that $\nabla_{\bar{1}\bar{1}} = \sum_{p=1}^{k} \frac{\partial}{\partial x_{p1}} \frac{\partial}{\partial x_{k+1-p,1}}$ and recall from (5.21) that $\diamondsuit = \diamondsuit_{1,\lambda'_1} \cdots \diamondsuit_{\lambda_1,\lambda'_{\lambda_1}}$. We observe that each summand in the expansion of the double differentiation $\nabla_{\bar{1}\bar{1}}(\diamondsuit)$ is either of the form (i) $\frac{\partial \diamondsuit_{i,\lambda'_i}}{\partial x_{p1}} \frac{\partial \diamondsuit_{j,\lambda'_j}}{\partial x_{k+1-p,1}} \cdots$ for $i \neq j$, or of the form (ii) $\frac{\partial^2 \diamondsuit_{i,\lambda'_i}}{\partial x_{p1} \partial x_{k+1-p,1}} \cdots$. Since $\lambda'_1 + \lambda'_2 \leq k$ by assumption, we have $\lambda'_i + \lambda'_j \leq k$ for $i \neq j$. Note by definitions (5.19) and (5.20) that the x_{pq}'s appearing in $\diamondsuit_{i,\lambda'_i}$ satisfy the constraints $p \leq \lambda'_i$ and the x_{rs}'s appearing in $\diamondsuit_{j,\lambda'_j}$ satisfy $r \leq \lambda'_j$, and hence $p + r \leq \lambda'_i + \lambda'_j \leq k$. Thus, the derivative in (i) must be zero. Expanding the determinant $\diamondsuit_{i,\lambda'_i}$ along the first column, we obtain $\diamondsuit_{i,\lambda'_i} = \sum_p x_{p1} A_p$, where A_p is an expression not containing any x_{q1}. Therefore, the derivative in (ii) must also be zero, and hence $\nabla_{\bar{1}\bar{1}}(\diamondsuit) = 0$.

When $k = 2\ell$, recall the element $\tau \in \mathrm{O}(k) \setminus \mathrm{SO}(k)$ defined in Section 5.3.3. When $k = 2\ell + 1$, we let τ be the linear map on $\mathbb{C}^k$ that fixes e^p, for all $p \neq \ell + 1$, and that sends $e^{\ell+1}$ to $-e^{\ell+1}$. By examining how τ transforms the $\mathfrak{so}(k)$-highest weight vector $\diamondsuit$, for $\lambda'_1 + \lambda'_2 \leq k$, one concludes that the $\mathrm{O}(k)$-module generated by $\diamondsuit$ is $L(\mathrm{O}(k), \lambda)$. Another direct computation implies that $\diamondsuit$ has $\mathfrak{spo}(2m|2n)$-weight $\lambda^{\natural} + \frac{k}{2}\mathbf{1}_{m|n}$. Thus, all irreducible representations of $\mathrm{O}(k)$ occur in ${}^{\mathrm{O}}\mathcal{H}$, and all the summands on the right-hand side of (5.38) occur in the space of harmonics ${}^{\mathrm{O}}\mathcal{H}$. Therefore, (5.38) holds in the case when $n = \infty$, and the theorem is proved. □

The simple $\mathfrak{spo}(2m|2n)$-modules appearing in the above $(\mathrm{O}(k), \mathfrak{spo}(2m|2n))$-Howe duality decomposition (for varied k) will be called the **oscillator modules** of $\mathfrak{spo}(2m|2n)$. The $\mathfrak{spo}(2m|2n)$-oscillator modules are, in general, infinite dimensional, as their highest weights evaluated at the simple root $-2\delta_1$ is $-2\lambda_1 - k < 0$.

Specializing $n = 0$ in Theorem 5.39 gives us the following.

Corollary 5.40 ($(\mathrm{O}(k),\mathfrak{sp}(2m))$-Howe duality). *We have the following decomposition of $(\mathrm{O}(k),\mathfrak{sp}(2m))$-modules:*

$$S(\mathbb{C}^k\otimes\mathbb{C}^m)\cong\bigoplus_{\ell(\lambda)\le m,\lambda'_1+\lambda'_2\le k}L(\mathrm{O}(k),\lambda)\otimes L\Big(\mathfrak{sp}(2m),\sum_{i=1}^m(\lambda_i+\frac{k}{2})\delta_i\Big).$$

We now specialize $m=0$ in Theorem 5.39. The irreducible highest weight $\mathfrak{so}(2n)$-modules in Corollary 5.41 below are relative to the Borel subalgebra of $\mathfrak{so}(2n)$ corresponding to the fundamental system specified in the following Dynkin diagram:

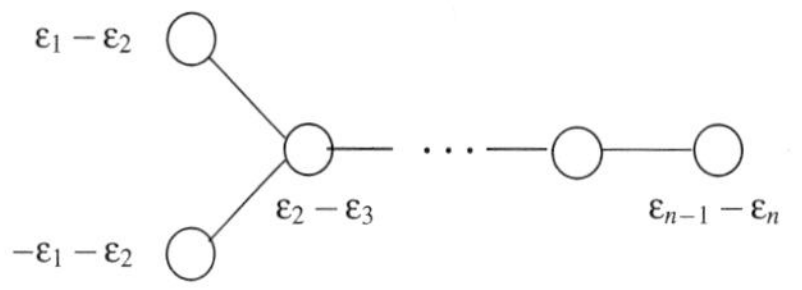

Corollary 5.41 ($(\mathrm{O}(k),\mathfrak{so}(2n))$-Howe duality). *We have the following decomposition of $(\mathrm{O}(k),\mathfrak{so}(2n))$-modules:*

$$\wedge(\mathbb{C}^k\otimes\mathbb{C}^n)\cong\bigoplus_{\ell(\lambda')\le n,\lambda'_1+\lambda'_2\le k}L(\mathrm{O}(k),\lambda)\otimes L\Big(\mathfrak{so}(2n),\sum_{i=1}^n(\lambda'_i-\frac{k}{2})\varepsilon_i\Big).$$

5.4. Howe duality for infinite-dimensional Lie algebras

In this section, we formulate Howe dualities between classical Lie groups and infinite-dimensional Lie algebras in fermionic Fock spaces. The results of this section will be applied in Section 5.5 to derive character formulas for the irreducible oscillator modules of the Lie superalgebras arising in the Howe duality decompositions in Section 5.3.

5.4.1. Lie algebras $\mathfrak{a}_\infty$, $\mathfrak{c}_\infty$, and $\mathfrak{d}_\infty$. We define the infinite-dimensional Lie algebras $\mathfrak{a}_\infty\equiv\widehat{\mathfrak{gl}}_\infty$ and its Lie subalgebras $\mathfrak{c}_\infty$ and $\mathfrak{d}_\infty$ of type C and D.

The Lie algebra $\mathfrak{a}_\infty\equiv\widehat{\mathfrak{gl}}_\infty$. Denote by $\mathfrak{gl}_\infty$ the Lie algebra of all matrices $(a_{ij})_{i,j\in\mathbb{Z}}$ with $a_{ij}=0$ for all but finitely many i's and j's. Let the degree of the elementary matrix E_{ij} be $j-i$. This defines the $\mathbb{Z}$-principal gradation on $\mathfrak{gl}_\infty=\bigoplus_{k\in\mathbb{Z}}\mathfrak{gl}_{\infty,k}$. Denote by $\mathfrak{a}_\infty\equiv\widehat{\mathfrak{gl}}_\infty=\mathfrak{gl}_\infty\oplus\mathbb{C}K$ the central extension associated to the following 2-cocycle τ:

$$\tau(A,B)=\mathrm{Tr}\,([J,A]B),\tag{5.39}$$

where $J=\sum_{j\le 0}E_{ii}$. Denoting by $[\cdot,\cdot]'$ the Lie bracket on $\mathfrak{gl}_\infty$, we introduce a bracket $[\cdot,\cdot]$ on the central extension $\mathfrak{a}_\infty=\mathfrak{gl}_\infty\oplus\mathbb{C}K$ by

$$[A+aK,B+bK]:=[A,B]'+\tau(A,B)K,\quad A,B\in\mathfrak{gl}_\infty,\ a,b\in\mathbb{C}.$$

By definition, a **2-cocycle** τ satisfies the following conditions: for $A,B,C \in \mathfrak{gl}_\infty$,

$$\text{(i) } \tau(A,B) = -\tau(B,A); \quad \text{(ii) } \tau(A,[B,C]) + \tau(B,[C,A]) + \tau(C,[A,B]) = 0.$$

The 2-cocyle conditions on τ ensure that the bracket $[\cdot,\cdot]$ is skew-symmetric and satisfies the Jacobi identity. More explicitly, we have the following commutation relation on $\mathfrak{a}_\infty$: for $i,j,m,n \in \mathbb{Z}$,

$$[E_{ij}, E_{mn}] = \delta_{jm}E_{in} - \delta_{in}E_{mj} + \delta_{jm}\delta_{in}\big(\theta\{i \le 0\} - \theta\{j \le 0\}\big)K,$$

where the **Boolean characteristic function** is defined to be

$$\theta\{P\} = \begin{cases} 1, & \text{if the statement } P \text{ is true,} \\ 0, & \text{if the statement } P \text{ is false.} \end{cases} \tag{5.40}$$

Remark 5.42. The Lie algebra $\mathfrak{a}_\infty$ can be "completed" in the following sense (see [**37**]). Denote by $\widetilde{\mathfrak{gl}}_\infty$ the completed Lie algebra of matrices $(a_{ij})_{i,j\in\mathbb{Z}}$ with $a_{ij} = 0$ for $|i-j| \gg 0$. The formula for τ in (5.39) also makes sense for $\widetilde{\mathfrak{gl}}_\infty$ and thus we obtain a central extension of $\widetilde{\mathfrak{gl}}_\infty$ (cf. [**65, 87**]). While τ is a 2-coboundary on $\mathfrak{gl}_\infty$, leading to a trivial central extension $\mathfrak{a}_\infty$ as we have defined in this book, τ on $\widetilde{\mathfrak{gl}}_\infty$ is no longer a 2-coboundary, and it gives a nontrivial central extension of $\widetilde{\mathfrak{gl}}_\infty$. Similar remarks apply to the classical subalgebras of $\mathfrak{gl}_\infty$ and $\widetilde{\mathfrak{gl}}_\infty$ below in this section. The completed Lie algebras and their central extensions are important because of their relations to various subalgebras such as Heisenberg, Virasoro, and affine Lie algebras, but they play no particular role in this book. Actually in Chapter 6, the "uncompleted" variants are preferred as we need to compare these with finite rank Lie algebras. The main results in this section make sense when formulated for these completed Lie algebras.

We extend the $\mathbb{Z}$-gradation of the Lie algebra $\mathfrak{gl}_\infty$ to $\mathfrak{a}_\infty$ by putting the degree of K to be 0. In particular, we have a triangular decomposition

$$\mathfrak{a}_\infty = \mathfrak{a}_{\infty+} \oplus \mathfrak{a}_{\infty 0} \oplus \mathfrak{a}_{\infty -},$$

where

$$\mathfrak{a}_{\infty\pm} = \bigoplus_{j\in\mathbb{N}} \mathfrak{gl}_{\infty,\pm j}, \qquad \mathfrak{a}_{\infty 0} = \mathfrak{gl}_{\infty,0} \oplus \mathbb{C}K.$$

Denote by ε_i for $i \in \mathbb{Z}$ the element in $(\mathfrak{a}_{\infty 0})^*$ determined by $\varepsilon_i(E_{jj}) = \delta_{ij}$ and $\varepsilon_i(K) = 0$, for $j \in \mathbb{Z}$. Then the root system of $\mathfrak{a}_\infty$ is $\{\varepsilon_i - \varepsilon_j \mid i,j \in \mathbb{Z}, i \neq j\}$ with a fundamental system $\{\varepsilon_i - \varepsilon_{i+1} \mid i \in \mathbb{Z}\}$. The corresponding set of simple coroots is given by $\{H_i^{\mathfrak{a}} := E_{ii} - E_{i+1,i+1} + \delta_{i,0}K \mid i \in \mathbb{Z}\}$. Let $\Lambda_j^{\mathfrak{a}}$, for $j \in \mathbb{Z}$, be the jth fundamental weight for $\mathfrak{a}_\infty$, i.e., $\Lambda_j^{\mathfrak{a}}(H_i^{\mathfrak{a}}) = \delta_{ij}$, for all $i \in \mathbb{Z}$, and $\Lambda_j^{\mathfrak{a}}(K) = 1$. A direct computation shows that

$$\Lambda_j^{\mathfrak{a}} = \begin{cases} \Lambda_0^{\mathfrak{a}} - \sum_{k=j+1}^{0} \varepsilon_k, & \text{for } j < 0, \\ \Lambda_0^{\mathfrak{a}} + \sum_{k=1}^{j} \varepsilon_k, & \text{for } j \ge 1. \end{cases} \tag{5.41}$$

Let $L(\mathfrak{a}_\infty, \Lambda)$ denote the irreducible highest weight $\mathfrak{a}_\infty$-module of highest weight $\Lambda \in (\mathfrak{a}_{\infty 0})^*$ with respect to the standard Borel subalgebra $\mathfrak{a}_{\infty +} \oplus \mathfrak{a}_{\infty 0}$. The **level** of Λ is given by the scalar $\Lambda(K) = \sum_{i\in\mathbb{Z}} \Lambda(H_i^{\mathfrak{a}})$.

The Lie algebra $\mathfrak{c}_\infty$. Now consider the natural $\mathfrak{gl}_\infty$-module V_0 with basis $\{v_i \mid i \in \mathbb{Z}\}$ such that $E_{ij}v_k = \delta_{jk}v_i$. Let C be the following skew-symmetric bilinear form on V_0:

$$C(v_i, v_j) = (-1)^i \delta_{i,1-j}, \quad \forall i, j \in \mathbb{Z}.$$

Denote by $\overline{\mathfrak{c}}_\infty$ the Lie subalgebra of $\mathfrak{gl}_\infty$ that preserves the bilinear form C:

$$\begin{aligned}\overline{\mathfrak{c}}_\infty &= \{a \in \mathfrak{gl}_\infty \mid C(a(u), v) + C(u, a(v)) = 0, \forall u, v \in V_0\} \\ &= \{(a_{ij})_{i,j\in\mathbb{Z}} \in \mathfrak{gl}_\infty \mid a_{ij} = -(-1)^{i+j} a_{1-j,1-i}\}.\end{aligned}$$

Denote by $\mathfrak{c}_\infty = \overline{\mathfrak{c}}_\infty \oplus \mathbb{C}K$ the central extension of $\overline{\mathfrak{c}}_\infty$ associated to the 2-cocycle (5.39) restricted to $\overline{\mathfrak{c}}_\infty$. Then $\mathfrak{c}_\infty$ inherits from $\mathfrak{a}_\infty$ a natural triangular decomposition:

$$\mathfrak{c}_\infty = \mathfrak{c}_{\infty +} \oplus \mathfrak{c}_{\infty 0} \oplus \mathfrak{c}_{\infty -},$$

where $\mathfrak{c}_{\infty \pm} = \mathfrak{c}_\infty \cap \mathfrak{a}_{\infty \pm}$, $\mathfrak{c}_{\infty 0} = \mathfrak{c}_\infty \cap \mathfrak{a}_{\infty 0}$. Let $\{\varepsilon_i \mid i \in \mathbb{N}\}$ be the basis in $(\mathfrak{c}_{\infty 0})^*$ dual to the basis $\{E_{i,i} - E_{-i+1,-i+1} \mid i \in \mathbb{N}\}$ in $\mathfrak{c}_{\infty 0}$. The Dynkin diagram for $\mathfrak{c}_\infty$ with a standard simple system is given as follows:

(5.42)

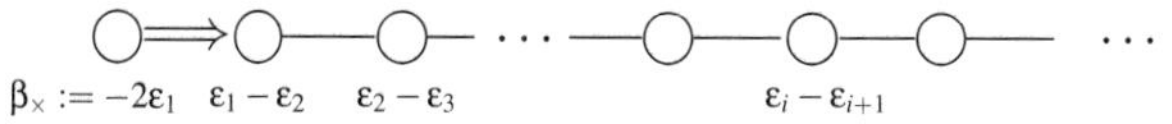

A set of simple coroots $\{H_i^{\mathfrak{c}}, i \geq 0\}$ for $\mathfrak{c}_\infty$ is

$$\begin{aligned} H_i^{\mathfrak{c}} &= E_{ii} + E_{-i,-i} - E_{i+1,i+1} - E_{1-i,1-i}, \quad i \in \mathbb{N}, \\ H_0^{\mathfrak{c}} &= E_{0,0} - E_{1,1} + K. \end{aligned}$$

Then we denote by $\Lambda_j^{\mathfrak{c}} \in (\mathfrak{c}_{\infty 0})^*$, for $j \in \mathbb{Z}_+$, the jth fundamental weight of $\mathfrak{c}_\infty$, i.e., $\Lambda_j^{\mathfrak{c}}(H_i^{\mathfrak{c}}) = \delta_{ij}$ for all $i \in \mathbb{Z}_+$. Note that $\Lambda_j^{\mathfrak{c}}(K) = 1$ for all j. A direct computation shows that

$$\Lambda_j^{\mathfrak{c}} = \sum_{k=1}^{j} \varepsilon_k + \Lambda_0^{\mathfrak{c}}, \quad \text{for } j \geq 1. \tag{5.43}$$

Denote by $L(\mathfrak{c}_\infty, \Lambda)$ the irreducible highest weight module of $\mathfrak{c}_\infty$ of highest weight $\Lambda \in (\mathfrak{c}_{\infty 0})^*$, whose level is given by $\Lambda(K) = \sum_{i \geq 0} \Lambda(H_i^{\mathfrak{c}})$.

The Lie algebra $\mathfrak{d}_\infty$. We denote by $\overline{\mathfrak{d}}_\infty$ the Lie subalgebra of $\mathfrak{gl}_\infty$ preserving the following symmetric bilinear form D on V_0:

$$D(v_i, v_j) = \delta_{i,1-j}, \quad \forall i, j \in \mathbb{Z}.$$

Namely, we have

$$\begin{aligned}\overline{\mathfrak{d}}_\infty &= \{a \in \mathfrak{gl}_\infty \mid D(a(u), v) + D(u, a(v)) = 0, \forall u, v \in V_0\} \\ &= \{(a_{ij})_{i,j\in\mathbb{Z}} \in \mathfrak{gl}_\infty \mid a_{ij} = -a_{1-j,1-i}\}.\end{aligned}$$

Denote by $\mathfrak{d}_\infty = \overline{\mathfrak{d}}_\infty \oplus \mathbb{C}K$ the central extension associated to the 2-cocycle (5.39) restricted to $\overline{\mathfrak{d}}_\infty$. Then $\mathfrak{d}_\infty$ has a triangular decomposition induced from $\widehat{\mathfrak{gl}}_\infty$:

$$\mathfrak{d}_\infty = \mathfrak{d}_{\infty +} \oplus \mathfrak{d}_{\infty 0} \oplus \mathfrak{d}_{\infty -},$$

where $\mathfrak{d}_{\infty \pm} = \mathfrak{d}_\infty \cap \mathfrak{a}_{\infty \pm}$ and $\mathfrak{d}_{\infty 0} = \mathfrak{d}_\infty \cap \mathfrak{a}_{\infty 0}$.

The Cartan subalgebra $\mathfrak{d}_{\infty 0}$ of $\mathfrak{d}_\infty$ is the same as $\mathfrak{c}_{\infty 0}$ (both as subalgebras of $\mathfrak{a}_\infty$), and we let $\varepsilon_j \in (\mathfrak{d}_{\infty 0})^*$ be defined in the same way as in the case of $\mathfrak{c}_\infty$. The Dynkin diagram with a standard fundamental system for $\mathfrak{d}_\infty$ is given as follows:

(5.44)
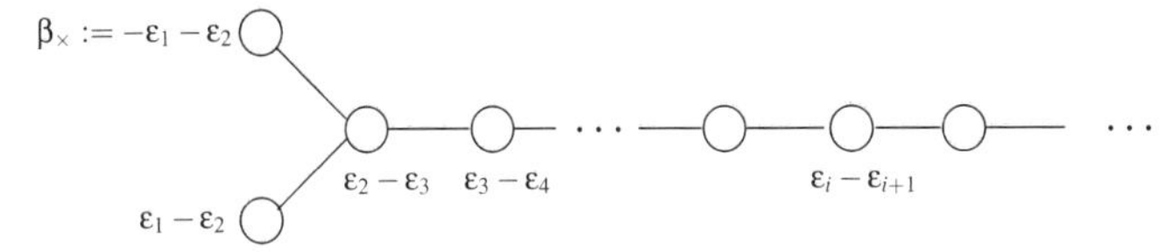

A set of simple coroots $\{H_i^{\mathfrak{d}} \mid i \geq 0\}$ for $\mathfrak{d}_\infty$ is given by

$$\begin{aligned} H_i^{\mathfrak{d}} &= E_{ii} + E_{-i,-i} - E_{i+1,i+1} - E_{1-i,1-i} \quad (i \in \mathbb{N}), \\ H_0^{\mathfrak{d}} &= E_{0,0} + E_{-1,-1} - E_{2,2} - E_{1,1} + 2K. \end{aligned}$$

Denote by $\Lambda_j^{\mathfrak{d}} \in (\mathfrak{d}_{\infty 0})^*$, for $j \in \mathbb{Z}_+$, the jth fundamental weight of $\mathfrak{d}_\infty$, i.e, $\Lambda_j^{\mathfrak{d}}(H_i^{\mathfrak{d}}) = \delta_{ij}$ for $i \in \mathbb{Z}_+$. Denote by $L(\mathfrak{d}_\infty, \Lambda)$ the irreducible highest weight $\mathfrak{d}_\infty$-module of highest weight Λ, whose level is $\Lambda(K) = \frac{1}{2}\Lambda(H_0^{\mathfrak{d}}) + \frac{1}{2}\Lambda(H_1^{\mathfrak{d}}) + \sum_{i \geq 2} \Lambda(H_i^{\mathfrak{d}})$.

5.4.2. The fermionic Fock space. Recall from (A.68) that $\mathcal{F}$ denotes the fermionic Fock space generated by a pair of fermions $\psi^\pm(z)$, whose components $\psi_r^\pm$, $r \in \frac{1}{2} + \mathbb{Z}$, satisfy the Clifford commutation relation (A.66).

We shall denote by $\mathcal{F}^\ell$ the fermionic Fock space of ℓ pairs of fermions

$$\psi^{\pm,p}(z) = \sum_{r \in \frac{1}{2} + \mathbb{Z}} \psi_r^{\pm,p} z^{-r-\frac{1}{2}}, \qquad 1 \leq p \leq \ell.$$

More precisely, we let $\widehat{\mathcal{C}}^\ell$ be the Clifford algebra generated by $\psi_r^{\pm,p}$, where $1 \leq p \leq \ell$ and $r \in \frac{1}{2} + \mathbb{Z}$, with anti-commutation relations given by

$$[\psi_r^{+,p}, \psi_s^{-,q}]_+ = \delta_{p,q}\delta_{r,-s} I, \qquad [\psi_r^{+,p}, \psi_s^{+,q}]_+ = [\psi_r^{-,p}, \psi_s^{-,q}]_+ = 0,$$

for all $r, s \in \frac{1}{2} + \mathbb{Z}$, $1 \leq p, q \leq \ell$. Then $\mathcal{F}^\ell$ is the simple $\widehat{\mathcal{C}}^\ell$-module generated by the vacuum vector $|0\rangle$, which satisfies the condition $\psi_r^{\pm,p}|0\rangle = 0$ for all $r > 0$ and $1 \leq p \leq \ell$. Here I is identified with the identity operator on $\mathcal{F}^\ell$.

Introduce a neutral fermionic field

$$\phi(z) = \sum_{r \in \frac{1}{2} + \mathbb{Z}} \phi_r z^{-r-\frac{1}{2}}$$

whose components satisfy the following anti-commutation relation:

$$[\phi_r, \phi_s]_+ = \delta_{r,-s} I, \quad r, s \in \frac{1}{2} + \mathbb{Z}.$$

We denote by $\widehat{\mathcal{C}}^{\ell+\frac{1}{2}}$ the Clifford algebra generated by ϕ_r and $\psi_r^{\pm,p}$ subject to the additional anti-commutation relation $[\phi_r, \psi_s^{\pm,p}]_+ = 0$, where $1 \le p \le \ell$. By the Fock space $\mathcal{F}^{\ell+\frac{1}{2}}$ we mean the simple $\widehat{\mathcal{C}}^{\ell+\frac{1}{2}}$-module generated by the vacuum vector $|0\rangle$, which satisfies the condition $\phi_r|0\rangle = \psi_r^{\pm,p}|0\rangle = 0$ for all $r > 0$ and $1 \le p \le \ell$.

The algebras $\widehat{\mathcal{C}}^{\ell}$ and $\widehat{\mathcal{C}}^{\ell+\frac{1}{2}}$ are naturally filtered algebras by letting the degree of each $\psi_r^{\pm,p}$ and ϕ_r be 1 and the degree of I be 0. The associated graded algebras are exterior algebras. We introduce natural $\frac{1}{2}\mathbb{Z}_+$-gradations (called the **principal gradation**) on $\mathcal{F}^\ell$ and $\mathcal{F}^{\ell+\frac{1}{2}}$ by the eigenvalues of the degree operator d on $\mathcal{F}^\ell$ and $\mathcal{F}^{\ell+\frac{1}{2}}$ defined by

$$d|0\rangle = 0, \quad \left[d, \psi_{-r}^{\pm,p}\right] = r\psi_{-r}^{\pm,p}, \quad [d, \phi_{-r}] = r\phi_{-r}, \quad \forall r, p.$$

Every graded subspace of $\mathcal{F}^\ell$ and $\mathcal{F}^{\ell+\frac{1}{2}}$ with respect to the principal gradation is finite dimensional.

Introduce the **normal ordered product** (denoted by : :)

(5.45)
$$:\psi_r^{+,p}\psi_s^{-,q}: = \begin{cases} -\psi_s^{-,q}\psi_r^{+,p}, & \text{if } s = -r < 0, \\ \psi_r^{+,p}\psi_s^{-,q}, & \text{otherwise,} \end{cases}$$
$$:\psi_r^{-,p}\psi_s^{+,q}: = \begin{cases} -\psi_s^{+,q}\psi_r^{-,p}, & \text{if } s = -r < 0, \\ \psi_r^{-,p}\psi_s^{+,q}, & \text{otherwise.} \end{cases}$$

Also let $:\psi_r^{+,p}\psi_s^{+,q}: = \psi_r^{+,p}\psi_s^{+,q}$, $:\psi_r^{-,p}\psi_s^{-,q}: = \psi_r^{-,p}\psi_s^{-,q}$, $:\psi_r^{\pm,p}\phi_s^q: = \psi_r^{\pm,p}\phi_s^q$, for all p, q, r, s.

5.4.3. $(\mathrm{GL}(\ell), \mathfrak{a}_\infty)$**-Howe duality.** Let

$$\sum_{i,j\in\mathbb{Z}} \mathcal{E}_{ij}^* z^{i-1} w^{-j} = \sum_{p=1}^{\ell} :\psi^{+,p}(z)\psi^{-,p}(w):.$$

Equivalently, we have

(5.46)
$$\mathcal{E}_{ij}^* = \sum_{p=1}^{\ell} :\psi_{\frac{1}{2}-i}^{+,p}\psi_{j-\frac{1}{2}}^{-,p}:.$$

Consider the following generating functions

$$\sum_{n\in\mathbb{Z}} e_*^{pq}(n) z^{-n-1} = :\psi^{+,p}(z)\psi^{-,q}(z):, \quad p, q = 1, \dots, \ell.$$

Such generating functions are "vertex operators" in the theory of vertex algebras, but we will avoid such terminology in this book. We will only need $e_*^{pq}(0)$ below, and hence introduce a short-hand notation $\mathcal{E}_*^{pq} \equiv e_*^{pq}(0)$. It follows that

$$\mathcal{E}_*^{pq} = \sum_{r \in \frac{1}{2}+\mathbb{Z}} :\psi_{-r}^{+,p}\psi_r^{-,q}:, \quad p,q = 1,\ldots,\ell. \tag{5.47}$$

Lemma 5.43. (1) *Sending E_{ij} to $\mathcal{E}_{ij}^*$, for $i,j \in \mathbb{Z}$, defines a representation of the Lie algebra $\mathfrak{a}_\infty$ on $\mathcal{F}^\ell$ of level ℓ.*

(2) *Sending E^{pq} to $\mathcal{E}_*^{pq}$ in (5.47), for $1 \le p,q \le \ell$, defines an action of the Lie algebra $\mathfrak{gl}(\ell)$, which lifts to an action of* $\mathrm{GL}(\ell)$ *on $\mathcal{F}^\ell$.*

(3) $\mathrm{GL}(\ell)$ *and $\mathfrak{a}_\infty$ form a Howe dual pair on $\mathcal{F}^\ell$.*

Proof. (1) We need to check that

$$[\mathcal{E}_{ij}^*, \mathcal{E}_{mn}^*] = \delta_{jm}\mathcal{E}_{in}^* - \delta_{in}\mathcal{E}_{mj}^* + \delta_{jm}\delta_{in}\big(\theta\{i \le 0\} - \theta\{j \le 0\}\big)\ell I.$$

Let us check in detail the case when $m = j$ and $n = i$. Since clearly $[\mathcal{E}_{ii}^*, \mathcal{E}_{ii}^*] = 0$, we can assume that $i \neq j$. It follows from (5.46) that

$$\begin{aligned}
[\mathcal{E}_{ij}^*, \mathcal{E}_{ji}^*] &= \sum_{p=1}^{\ell} \left[\psi_{\frac{1}{2}-i}^{+,p}\psi_{j-\frac{1}{2}}^{-,p}, \psi_{\frac{1}{2}-j}^{+,p}\psi_{i-\frac{1}{2}}^{-,p}\right] \\
&= \sum_{p=1}^{\ell} \left(\psi_{\frac{1}{2}-i}^{+,p}\psi_{j-\frac{1}{2}}^{-,p}\psi_{\frac{1}{2}-j}^{+,p}\psi_{i-\frac{1}{2}}^{-,p} - \psi_{\frac{1}{2}-j}^{+,p}\psi_{i-\frac{1}{2}}^{-,p}\psi_{\frac{1}{2}-i}^{+,p}\psi_{j-\frac{1}{2}}^{-,p}\right) \\
&= \sum_{p=1}^{\ell} \left(\psi_{\frac{1}{2}-i}^{+,p}\psi_{i-\frac{1}{2}}^{-,p} - \psi_{\frac{1}{2}-j}^{+,p}\psi_{j-\frac{1}{2}}^{-,p}\right) \\
&= \sum_{p=1}^{\ell} \left(:\psi_{\frac{1}{2}-i}^{+,p}\psi_{i-\frac{1}{2}}^{-,p}: - :\psi_{\frac{1}{2}-j}^{+,p}\psi_{j-\frac{1}{2}}^{-,p}:\right) + \big(\theta\{i \le 0\} - \theta\{j \le 0\}\big)\ell I \\
&= \mathcal{E}_{ii}^* - \mathcal{E}_{jj}^* + \big(\theta\{i \le 0\} - \theta\{j \le 0\}\big)\ell I.
\end{aligned}$$

The remaining cases are similar.

(2) We need to check that $[\mathcal{E}_*^{pq}, \mathcal{E}_*^{uv}] = \delta_{qu}\mathcal{E}_*^{pv} - \delta_{pv}\mathcal{E}_*^{uq}$ for $1 \le p,q,u,v \le \ell$. Indeed, it follows by (5.47) that

$$\begin{aligned}
[\mathcal{E}_*^{pq}, \mathcal{E}_*^{uv}] &= \sum_{r \in \frac{1}{2}+\mathbb{Z}} [:\psi_{-r}^{+,p}\psi_r^{-,q}:, :\psi_{-r}^{+,u}\psi_r^{-,v}:] \\
&= \sum_{r \in \frac{1}{2}+\mathbb{Z}} \left(\delta_{qu}:\psi_{-r}^{+,p}\psi_r^{-,v}: - \delta_{vp}:\psi_{-r}^{+,u}\psi_r^{-,q}:\right) = \delta_{qu}\mathcal{E}_*^{pv} - \delta_{pv}\mathcal{E}_*^{uq}.
\end{aligned}$$

As the operators $\mathcal{E}_*^{pq}$ have degree zero in the principal gradation, the action of $\mathfrak{gl}(\ell)$ preserves each (principally) graded subspace of $\mathcal{F}^\ell$, which is finite dimensional and of integral weight. Hence the action of $\mathfrak{gl}(\ell)$ lifts to an action of the Lie group $\mathrm{GL}(\ell)$ on $\mathcal{F}^\ell$.

(3) It follows by (5.46) and (5.47) that

$$\begin{aligned}
[\mathcal{E}^{pq}_*, \mathcal{E}^*_{ij}] &= \Big[\sum_{r\in\frac{1}{2}+\mathbb{Z}} :\psi^{+,p}_{-r}\psi^{-,q}_{r}:, \sum_{u=1}^{\ell} :\psi^{+,u}_{\frac{1}{2}-i}\psi^{-,u}_{j-\frac{1}{2}}: \Big] \\
&= \Big[:\psi^{+,p}_{\frac{1}{2}-i}\psi^{-,q}_{i-\frac{1}{2}}:, :\psi^{+,q}_{\frac{1}{2}-i}\psi^{-,q}_{j-\frac{1}{2}}: \Big] + \Big[:\psi^{+,p}_{\frac{1}{2}-j}\psi^{-,q}_{j-\frac{1}{2}}:, :\psi^{+,p}_{\frac{1}{2}-i}\psi^{-,p}_{j-\frac{1}{2}}: \Big] \\
&= :\psi^{+,p}_{\frac{1}{2}-i}\psi^{-,q}_{j-\frac{1}{2}}: - :\psi^{+,p}_{\frac{1}{2}-i}\psi^{-,q}_{j-\frac{1}{2}}: = 0.
\end{aligned}$$

Hence the actions of $\mathrm{GL}(\ell)$ and $\mathfrak{a}_\infty$ on $\mathcal{F}^\ell$ commute.

The statement on the Howe dual pair can be proved by a similar argument as for Theorem 5.14, and we rephrase briefly the argument as follows. Set $G = \mathrm{GL}(\ell)$, $U = \mathbb{C}^\ell \otimes \mathbb{C}^\infty \oplus \mathbb{C}^{\ell *} \otimes \mathbb{C}^\infty$, where $\mathbb{C}^\infty$ is a vector space with a basis $\{w_r \mid r \in -\frac{1}{2} - \mathbb{Z}_+\}$. The dual basis in $\mathbb{C}^{\infty *}$ is denoted by $\{w_{-r} \mid r \in -\frac{1}{2} - \mathbb{Z}_+\}$. We take a basis $\{v^{+,p} \mid 1 \le p \le \ell\}$ for $\mathbb{C}^\ell$, and a dual basis $\{v^{-,p} \mid 1 \le p \le \ell\}$ for $\mathbb{C}^{\ell *}$. The Clifford algebra $\widehat{\mathfrak{C}}^\ell$ can be naturally identified with $\mathfrak{WC}(\boldsymbol{U})$ for the purely odd space $\boldsymbol{U} = U \oplus U^*$ by setting $\psi^{\pm,p}_r \equiv v^{\pm,p} \otimes w_r$, so that its associated graded can be identified with $\wedge(\boldsymbol{U})$. Observe that the images under the symbol map σ (defined in 5.1.2) of $\mathcal{E}^*_{ij}$ in (5.46) form a set of generators for $\wedge(\boldsymbol{U})^G$ by the supersymmetric First Fundamental Theorem of invariant theory for $\mathrm{GL}(V)$ (see Theorem 4.19). Now the claim on the Howe dual pair follows from Proposition 5.12. □

Recall the well-known fact that a simple $\mathfrak{gl}(\ell)$-module $L(\mathfrak{gl}(\ell),\lambda)$ of highest weight $\lambda = \sum_{i=1}^{\ell} \lambda_i \delta_i$ (with respect to the standard Borel subalgebra spanned by E^{pq} for $p \le q$) lifts to a simple $\mathrm{GL}(\ell)$-module $L(\mathrm{GL}(\ell),\lambda)$ if and only if $(\lambda_1,\ldots,\lambda_\ell)$ is a **generalized partition** in the sense that all $\lambda_i \in \mathbb{Z}$ and $\lambda_1 \ge \ldots \ge \lambda_\ell$. By abuse of notation, we shall identify a weight λ for $\mathfrak{gl}(\ell)$ with a generalized partition of the form $\lambda = (\lambda_1,\ldots,\lambda_\ell)$.

For a generalized partition $\lambda = (\lambda_1,\ldots,\lambda_\ell)$ we define, for $i \in \mathbb{Z}$,

$$\lambda'_i := \begin{cases} |\{j \mid \lambda_j \ge i\}|, & \text{for } i \ge 1, \\ -|\{j \mid \lambda_j < i\}|, & \text{for } i \le 0. \end{cases}$$

Theorem 5.44 ($(\mathrm{GL}(\ell),\mathfrak{a}_\infty)$-Howe duality). *As a $(\mathrm{GL}(\ell),\mathfrak{a}_\infty)$-module, we have*

$$\mathcal{F}^\ell \cong \bigoplus_\lambda L(\mathrm{GL}(\ell),\lambda) \otimes L(\mathfrak{a}_\infty, \Lambda^{\mathfrak{a}}(\lambda)), \tag{5.48}$$

where the summation is over all generalized partitions of the form $\lambda = (\lambda_1,\ldots,\lambda_\ell)$, and $\Lambda^{\mathfrak{a}}(\lambda) := \Lambda^{\mathfrak{a}}_{\lambda_1} + \ldots + \Lambda^{\mathfrak{a}}_{\lambda_\ell} = \ell\Lambda^{\mathfrak{a}}_0 + \sum_{i\in\mathbb{Z}} \lambda'_i \varepsilon_i$.

Proof. By Lemma 5.43 and Proposition 5.12, we have a strongly multiplicity-free decomposition of the $(\mathrm{GL}(\ell),\mathfrak{a}_\infty)$-module $\mathcal{F}^\ell$. To obtain the explicit decomposition as given in the theorem, we will exhibit an explicit formula for a joint highest weight vector associated to each λ; see Proposition 5.45 below. The theorem follows as we observe by Proposition 5.45 that every simple $\mathrm{GL}(\ell)$-module appears in the decomposition (5.48) of $\mathcal{F}^\ell$. □

For $1 \le p \le \ell$ and $m \ge 1$, we denote

$$\Xi_m^{+,p} := \psi^{+,p}_{-m+\frac{1}{2}} \cdots \psi^{+,p}_{-\frac{3}{2}} \psi^{+,p}_{-\frac{1}{2}},$$
$$\Xi_m^{-,p} := \psi^{-,p}_{-m+\frac{1}{2}} \cdots \psi^{-,p}_{-\frac{3}{2}} \psi^{-,p}_{-\frac{1}{2}}.$$

We make the convention that $\Xi_0^{\pm,p} = 1$.

Proposition 5.45. *Given a generalized partition of the form $\lambda = (\lambda_1, \dots, \lambda_\ell)$, let i, j be such that*

$$\lambda_1 \ge \cdots \ge \lambda_i > \lambda_{i+1} = \cdots = \lambda_{j-1} = 0 > \lambda_j \ge \cdots \ge \lambda_\ell.$$

Then the joint highest weight vector in $\mathcal{F}^\ell$ associated to λ with respect to the standard Borel for $\mathfrak{gl}(\ell) \times \mathfrak{a}_\infty$ is

$$v_\lambda^{\mathfrak{a}} = \Xi^{+,1}_{\lambda_1} \Xi^{+,2}_{\lambda_2} \cdots \Xi^{+,i}_{\lambda_i} \cdot \Xi^{-,j}_{-\lambda_j} \Xi^{-,j+1}_{-\lambda_{j+1}} \cdots \Xi^{-,\ell}_{-\lambda_\ell} |0\rangle, \tag{5.49}$$

whose weights with respect to $\mathfrak{gl}(\ell)$ and $\mathfrak{a}_\infty$ are λ and $\Lambda^{\mathfrak{a}}(\lambda)$, respectively.

Proof. The vector $v_\lambda^{\mathfrak{a}}$ in (5.49) is indeed a highest weight vector for $\mathfrak{gl}(\ell)$ and $\mathfrak{a}_\infty$, respectively, since applying any positive root vector in either $\mathfrak{gl}(\ell)$ or $\mathfrak{a}_\infty$ to $v_\lambda^{\mathfrak{a}}$ is either manifestly zero or gives rise to two identical ψ^{**}_* in the resulting monomial, whence also zero (see Exercise 5.10). Another direct calculation shows that the highest weight of $v_\lambda^{\mathfrak{a}}$ for $\mathfrak{gl}(\ell)$ is $(\lambda_1, \dots, \lambda_\ell)$.

We recall that $\mathcal{E}^*_{mm} = \sum_{p=1}^\ell :\psi^{+,p}_{\frac{1}{2}-m} \psi^{-,p}_{m-\frac{1}{2}}:$. When $m \ge 1$ and $n \ge 1$, one easily computes

$$[\mathcal{E}^*_{mm}, \psi^{+,q}_{-n+\frac{1}{2}}] = \delta_{m,n} \psi^{+,q}_{-n+\frac{1}{2}}, \quad [\mathcal{E}^*_{mm}, \psi^{-,q}_{-n+\frac{1}{2}}] = 0.$$

For $m \le 0$ and $n \ge 1$ we have

$$[\mathcal{E}^*_{mm}, \psi^{-,q}_{-n+\frac{1}{2}}] = -\delta_{-m+1,n} \psi^{-,q}_{-n+\frac{1}{2}}, \quad [\mathcal{E}^*_{mm}, \psi^{+,q}_{-n+\frac{1}{2}}] = 0.$$

This implies that the weight of the vector $v_\lambda^{\mathfrak{a}}$ with respect to the action of $\mathfrak{a}_\infty$ equals $\ell\Lambda_0^{\mathfrak{a}} + \sum_i \lambda_i' \varepsilon_i$, which is also equal to $\Lambda^{\mathfrak{a}}_{\lambda_1} + \ldots + \Lambda^{\mathfrak{a}}_{\lambda_\ell}$ by (5.41). □

5.4.4. $(\mathrm{Sp}(k), \mathfrak{c}_\infty)$**-Howe duality.** Let $k = 2\ell$. It follows by (5.46) that

$$\begin{aligned} &\sum_{i,j\in\mathbb{Z}} (\mathcal{E}^*_{ij} - (-1)^{i+j} \mathcal{E}^*_{1-j,1-i}) z^{i-1} w^{-j} \\ &= \sum_{p=1}^\ell \left(:\psi^{+,p}(z)\psi^{-,p}(w): + :\psi^{+,p}(-w)\psi^{-,p}(-z): \right). \end{aligned}$$

Equivalently, we have

$$\mathcal{E}^*_{ij} - (-1)^{i+j} \mathcal{E}^*_{1-j,1-i} = \sum_{p=1}^\ell \left(:\psi^{+,p}_{\frac{1}{2}-i} \psi^{-,p}_{j-\frac{1}{2}}: - (-1)^{i+j} :\psi^{+,p}_{j-\frac{1}{2}} \psi^{-,p}_{\frac{1}{2}-i}: \right). \tag{5.50}$$

Let $p,q = 1,\dots,\ell$. Consider the following generating functions

$$\sum_{n\in\mathbb{Z}} \tilde{e}^{pq}(n)z^{-n-1} = :\psi^{-,p}(z)\psi^{-,q}(-z):,$$
$$\sum_{n\in\mathbb{Z}} \tilde{e}^{pq}_{**}(n)z^{-n-1} = :\psi^{+,p}(z)\psi^{+,q}(-z):.$$

We introduce the following short-hand notation:

$$\tilde{\mathcal{E}}^{pq} \equiv \tilde{e}^{pq}(0), \quad \tilde{\mathcal{E}}^{pq}_{**} \equiv \tilde{e}^{pq}_{**}(0).$$

It follows that

$$\tilde{\mathcal{E}}^{pq} = \sum_{r\in\frac{1}{2}+\mathbb{Z}} (-1)^{-r-\frac{1}{2}} :\psi^{-,p}_{-r}\psi^{-,q}_{r}:,$$
$$\tilde{\mathcal{E}}^{pq}_{**} = \sum_{r\in\frac{1}{2}+\mathbb{Z}} (-1)^{-r-\frac{1}{2}} :\psi^{+,p}_{-r}\psi^{+,q}_{r}:. \tag{5.51}$$

Note that $\tilde{\mathcal{E}}^{pq} = \tilde{\mathcal{E}}^{qp}$ and $\tilde{\mathcal{E}}^{pq}_{**} = \tilde{\mathcal{E}}^{qp}_{**}$.

Lemma 5.46. *Let $k = 2\ell$.*

(1) *The formula* (5.50) *defines an action of Lie algebra $\mathfrak{c}_\infty$ on $\mathcal{F}^\ell$ of level ℓ.*

(2) *The operators $\tilde{\mathcal{E}}^{pq}$ and $\tilde{\mathcal{E}}^{pq}_{**}$ in* (5.51) *together with $\mathcal{E}^{pq}_{*}$ in* (5.47)*, for $1 \le p,q \le \ell$, define an action of the Lie algebra $\mathfrak{sp}(k)$ on $\mathcal{F}^\ell$, which lifts to an action of* $\mathrm{Sp}(k)$.

(3) $\mathrm{Sp}(k)$ *and $\mathfrak{c}_\infty$ form a Howe dual pair on $\mathcal{F}^\ell$.*

Proof. (1) This follows from Lemma 5.43(1), (5.50), and the definition of $\mathfrak{c}_\infty$.

(2) Let us denote a standard basis for $V = \mathbb{C}^\ell \oplus \mathbb{C}^{\ell *}$ by $v^{\pm,p}$, for $p = 1,\dots,\ell$, and denote a standard basis for $\mathbb{C}^\infty$ by w_r with $r \in (-\frac{1}{2} - \mathbb{Z}_+)$. Letting $U = V \otimes \mathbb{C}^\infty$, we naturally identify $\mathcal{F}^\ell$ with $\wedge(U)$ by setting $\psi^{\pm,p}_r \equiv v^{\pm,p} \otimes w_r$. Note that V is a symplectic space with a natural pairing, and so is U. Hence, we have a natural action of $\mathfrak{sp}(V) = \mathfrak{sp}(k)$ on $\mathcal{F}^\ell$. The formula (5.51) is simply a precise way of writing down this action in terms of coordinates. Indeed, comparing these formulas with (5.26) we can see that the matrices there with $B = C = 0$ correspond to the $\mathcal{E}^{pq}_*$'s, while the ones with $A = C = 0$ correspond to the $\tilde{\mathcal{E}}^{pq}_{**}$'s, and the ones with $A = B = 0$ correspond to the $\tilde{\mathcal{E}}^{pq}$'s. Alternatively, it can be verified by a direct computation that the operators in (5.51) generate $\mathfrak{sp}(k)$. Note that the action of $\mathfrak{sp}(k)$ preserves the principal gradation of $\mathcal{F}^\ell$ and every graded subspace of $\mathcal{F}^\ell$ is finite dimensional. Thus, the action of $\mathfrak{sp}(k)$ lifts to that of $\mathrm{Sp}(k)$.

(3) As part of the proof of Lemma 5.43, we see that $\mathcal{E}^{pq}_*$ commutes with the action of $\mathfrak{c}_\infty$. We will now check that $[\tilde{\mathcal{E}}^{uv}, \mathfrak{c}_\infty] = 0$, and leave the similar verification that $[\tilde{\mathcal{E}}^{uv}_{**}, \mathfrak{c}_\infty] = 0$ to the reader. Indeed, we have that

$$[\tilde{\mathcal{E}}^{uv}, \mathcal{E}^*_{ij} - (-1)^{i+j}\mathcal{E}^*_{1-j,1-i}]$$

$$=\Big[\sum_{r\in\frac12+\mathbb{Z}}(-1)^{-r-\frac12}{:}\psi^{-,u}_{-r}\psi^{-,v}_{r}{:},\sum_{p=1}^{\ell}({:}\psi^{+,p}_{\frac12-i}\psi^{-,p}_{j-\frac12}{:}-(-1)^{i+j}{:}\psi^{+,p}_{j-\frac12}\psi^{-,p}_{\frac12-i}{:})\Big]$$

$$=\Big[(-1)^i\psi^{-,v}_{\frac12-i}\psi^{-,u}_{i-\frac12}+(-1)^i\psi^{-,u}_{\frac12-i}\psi^{-,v}_{i-\frac12}+(-1)^j\psi^{-,u}_{\frac12-j}\psi^{-,v}_{j-\frac12}+(-1)^j\psi^{-,v}_{\frac12-j}\psi^{-,u}_{j-\frac12},$$
$$\psi^{+,u}_{\frac12-i}\psi^{-,u}_{j-\frac12}+\psi^{+,v}_{\frac12-i}\psi^{-,v}_{j-\frac12}-(-1)^{i+j}\psi^{+,u}_{j-\frac12}\psi^{-,u}_{\frac12-i}-(-1)^{i+j}\psi^{+,v}_{j-\frac12}\psi^{-,v}_{\frac12-i}\Big]$$

$$=(-1)^i\psi^{-,v}_{\frac12-i}\psi^{-,u}_{j-\frac12}+(-1)^i\psi^{-,u}_{\frac12-i}\psi^{-,v}_{j-\frac12}-(-1)^i\psi^{-,u}_{\frac12-i}\psi^{-,v}_{j-\frac12}-(-1)^i\psi^{-,v}_{\frac12-i}\psi^{-,u}_{j-\frac12}=0.$$

Hence the actions of $\mathrm{Sp}(k)$ and $\mathfrak{c}_\infty$ on $\mathcal{F}^\ell$ commute.

The Howe dual pair claim follows by the same type of argument as the one given in the proof of Lemma 5.43(3) using now the First Fundamental Theorem of invariant theory for $\mathrm{Sp}(k)$ (see Theorem 4.19). □

Theorem 5.47 (($\mathrm{Sp}(k),\mathfrak{c}_\infty$)-Howe duality). *As an $(\mathrm{Sp}(k),\mathfrak{c}_\infty)$-module, we have*

$$\mathcal{F}^\ell\cong\bigoplus_{\lambda\in\mathcal{P}(\ell)}L(\mathrm{Sp}(k),\lambda)\otimes L(\mathfrak{c}_\infty,\Lambda^{\mathfrak{c}}(\lambda)),$$

where $\Lambda^{\mathfrak{c}}(\lambda):=\ell\Lambda^{\mathfrak{c}}_0+\sum_{i\geq1}\lambda'_i\epsilon_i=\Lambda^{\mathfrak{c}}_{\lambda_1}+\ldots+\Lambda^{\mathfrak{c}}_{\lambda_\ell}$. The joint highest weight vector in $\mathcal{F}^\ell$ associated to λ with respect to the standard Borel for $\mathfrak{sp}(k)\times\mathfrak{c}_\infty$ is $v^{\mathfrak{c}}_\lambda:=\Xi^{+,1}_{\lambda_1}\Xi^{+,2}_{\lambda_2}\cdots\Xi^{+,\ell}_{\lambda_\ell}|0\rangle$.

Proof. We observe that $v^{\mathfrak{c}}_\lambda$ has weights λ and $\ell\Lambda^{\mathfrak{c}}_0+\sum_{i\geq1}\lambda'_i\epsilon_i$ with respect to the actions of $\mathfrak{sp}(k)$ and $\mathfrak{c}_\infty$, respectively, the latter of which equals $\Lambda^{\mathfrak{c}}_{\lambda_1}+\ldots+\Lambda^{\mathfrak{c}}_{\lambda_\ell}$ by (5.43). Also $\tilde{\mathcal{E}}^{pq}_{**}$ annihilates $v^{\mathfrak{c}}_\lambda$, since in both expressions only $\psi^{+,p}_*$'s are involved and $\tilde{\mathcal{E}}^{pq}_{**}|0\rangle=0$. Since $v^{\mathfrak{c}}_\lambda$ is known to be a joint $(\mathfrak{gl}(\ell),\mathfrak{a}_\infty)$-highest weight vector by Proposition 5.45, we conclude that $v^{\mathfrak{c}}_\lambda$ is a joint $(\mathfrak{sp}(k),\mathfrak{c}_\infty)$-highest weight vector.

By Lemma 5.46 and Proposition 5.12, $\mathcal{F}^\ell$ is strongly multiplicity-free as an $(\mathrm{Sp}(k),\mathfrak{c}_\infty)$-module. Since $v^{\mathfrak{c}}_\lambda$ is a nonzero vector in $\mathfrak{F}^\ell$ of weight λ under the action of $\mathfrak{sp}(k)$, for every $\lambda\in\mathcal{P}(\ell)$, we also observe that all finite-dimensional $\mathrm{Sp}(k)$-modules appear in the decomposition of $\mathcal{F}^\ell$. So the multiplicity-free $(\mathrm{Sp}(k),\mathfrak{c}_\infty)$-module decomposition of $\mathcal{F}^\ell$ follows. □

Let $z_1,\ldots,z_\ell,y_1,y_2,\ldots$ be indeterminates. The character of the $\mathrm{Sp}(k)$-module $\mathcal{F}^\ell$ is the trace of the operator $\prod_{p=1}^{\ell}z_p^{\mathcal{E}^{pp}_*}$ on $\mathcal{F}^\ell$, while the character of the $\mathfrak{c}_\infty$-module $\mathcal{F}^\ell$ is the trace of the operator $\prod_{i\in\mathbb{N}}y_i^{\mathcal{E}^*_{ii}-\mathcal{E}^*_{1-i,1-i}}$ on $\mathcal{F}^\ell$. Computing the trace of $\prod_{i\in\mathbb{N}}y_i^{\mathcal{E}^*_{ii}-\mathcal{E}^*_{1-i,1-i}}\prod_{p=1}^{\ell}z_p^{\mathcal{E}^{pp}_*}$ on both sides of the isomorphism in Theorem 5.47, we obtain the following character identity:

$$\prod_{i=1}^{\infty}\prod_{p=1}^{\ell}(1+y_iz_p)(1+y_iz_p^{-1})=\sum_{\lambda\in\mathcal{P}(\ell)}\mathrm{ch}L(\mathrm{Sp}(k),\lambda)\,\mathrm{ch}L(\mathfrak{c}_\infty,\Lambda^{\mathfrak{c}}(\lambda)),\tag{5.52}$$

where by definition $\mathrm{ch}L(\mathfrak{c}_\infty,\Lambda^{\mathfrak{c}}(\lambda))=\mathrm{Tr}|_{L(\mathfrak{c}_\infty,\Lambda^{\mathfrak{c}}(\lambda))}\prod_{i\in\mathbb{N}}y_i^{\mathcal{E}^*_{ii}-\mathcal{E}^*_{1-i,1-i}}$.

5.4.5. $(\mathrm{O}(k),\mathfrak{d}_\infty)$-Howe duality. Let $k=2\ell$. It follows by (5.46) that

$$\sum_{i,j\in\mathbb{Z}}\left(\mathcal{E}^*_{i,j}-\mathcal{E}^*_{1-j,1-i}\right)z^{i-1}w^{-j}=\sum_{p=1}^{\ell}\left(:\psi^{+,p}(z)\psi^{-,p}(w):-:\psi^{+,p}(w)\psi^{-,p}(z):\right).$$

Equivalently, we have the following operators on $\mathcal{F}^\ell$:

$$\mathcal{E}^*_{i,j}-\mathcal{E}^*_{1-j,1-i}=\sum_{p=1}^{\ell}\left(:\psi^{+,p}_{\frac12-i}\psi^{-,p}_{j-\frac12}:-:\psi^{+,p}_{j-\frac12}\psi^{-,p}_{\frac12-i}:\right). \tag{5.53}$$

Let $p,q=1,\dots,\ell$. We introduce the following generating functions

$$\textstyle\sum_{n\in\mathbb{Z}}e^{pq}(n)z^{-n-1}=:\psi^{-,p}(z)\psi^{-,q}(z):,$$
$$\textstyle\sum_{n\in\mathbb{Z}}e^{pq}_{**}(n)z^{-n-1}=:\psi^{+,p}(z)\psi^{+,q}(z):.$$

We will adopt the following short-hand notation:

$$\check{\mathcal{E}}^{pq}\equiv e^{pq}(0),\quad \check{\mathcal{E}}^{pq}_{**}\equiv e^{pq}_{**}(0).$$

It follows that

$$\check{\mathcal{E}}^{pq}=\sum_{r\in\frac12+\mathbb{Z}}:\psi^{-,p}_{-r}\psi^{-,q}_{r}:,\qquad \check{\mathcal{E}}^{pq}_{**}=\sum_{r\in\frac12+\mathbb{Z}}:\psi^{+,p}_{-r}\psi^{+,q}_{r}:. \tag{5.54}$$

Note that $\check{\mathcal{E}}^{pq}=-\check{\mathcal{E}}^{qp}$ and $\check{\mathcal{E}}^{pq}_{**}=-\check{\mathcal{E}}^{qp}_{**}$.

Lemma 5.48. (1) *The formula* (5.53) *defines an action of the Lie algebra $\mathfrak{d}_\infty$ on $\mathcal{F}^\ell$ of level ℓ.*

(2) *The operators $\check{\mathcal{E}}^{pq},\check{\mathcal{E}}^{pq}_{**}$ in* (5.54) *together with $\mathcal{E}^{pq}_*$ in* (5.47)*, for $1\le p,q\le\ell$, define an action of $\mathfrak{so}(2\ell)$ on $\mathcal{F}^\ell$, which lifts to an action of* $\mathrm{O}(2\ell)$.

(3) $\mathrm{O}(2\ell)$ *and $\mathfrak{d}_\infty$ form a Howe dual pair on $\mathcal{F}^\ell$.*

Proof. The proof is completely analogous to that of Lemma 5.46 for type C. For example, (2) can be easily obtained by comparing the formulas in (5.54) and (5.47) with the one in (5.33), similar to the type C case. We leave the details to the reader (Exercise 5.13). □

Let $k=2\ell+1$. We introduce the following operators on $\mathcal{F}^{\ell+\frac12}$, for $i,j\in\mathbb{Z}$:

$$\mathcal{E}^*_{ij}-\mathcal{E}^*_{1-j,1-i}=\sum_{p=1}^{\ell}\left(:\psi^{+,p}_{\frac12-i}\psi^{-,p}_{j-\frac12}:-:\psi^{+,p}_{j-\frac12}\psi^{-,p}_{\frac12-i}:\right)+:\phi^{p}_{\frac12-i}\phi^{p}_{j-\frac12}:. \tag{5.55}$$

Equivalently, we have

$$\begin{aligned}&\sum_{i,j\in\mathbb{Z}}\left(\mathcal{E}^*_{ij}-\mathcal{E}^*_{1-j,1-i}\right)z^{i-1}w^{-j}\\ =\;&\textstyle\sum_{p=1}^{\ell}\left(:\psi^{+,p}(z)\psi^{-,p}(w):-:\psi^{+,p}(w)\psi^{-,p}(z):\right)+:\phi(z)\phi(w):.\end{aligned}$$

We introduce the following additional generating functions:

$$\sum_{n\in\mathbb{Z}} e^p(n)z^{-n-1} = :\psi^{-,p}(z)\phi(z):,$$
$$\sum_{n\in\mathbb{Z}} e^p_*(n)z^{-n-1} = :\psi^{+,p}(z)\phi(z):, \quad p=1,\dots,\ell.$$

Introducing the short-hand notation $\check{\mathcal{E}}^p_* = e^p_*(0)$ and $\check{\mathcal{E}}^p = e^p(0)$, we have the following formulas:

$$\check{\mathcal{E}}^p = \sum_{r\in\frac{1}{2}+\mathbb{Z}} :\psi^{-,p}_{-r}\phi_r:, \qquad \check{\mathcal{E}}^p_* = \sum_{r\in\frac{1}{2}+\mathbb{Z}} :\psi^{+,p}_{-r}\phi_r:. \tag{5.56}$$

The following is a counterpart for $k=2\ell+1$ of Lemma 5.48.

Lemma 5.49. (1) *The formula* (5.55) *defines an action of the Lie algebra* $\mathfrak{d}_\infty$ *on* $\mathcal{F}^{\ell+\frac{1}{2}}$.

(2) *The operators* $\check{\mathcal{E}}^{pq}_{**}, \check{\mathcal{E}}^{pq}$ *in* (5.54), $\check{\mathcal{E}}^p_*, \check{\mathcal{E}}^p$ *in* (5.56), *together with* $\mathcal{E}^{pq}_*$ *in* (5.47), *for* $1\le p,q\le \ell$, *define an action of the Lie algebra* $\mathfrak{so}(2\ell+1)$ *on* $\mathcal{F}^{\ell+\frac{1}{2}}$, *which lifts to an action of* $\mathrm{O}(2\ell+1)$.

(3) $\mathrm{O}(2\ell+1)$ *and* $\mathfrak{d}_\infty$ *form a Howe dual pair when acting on* $\mathcal{F}^{\ell+\frac{1}{2}}$.

Proof. The proof is analogous to that of Lemma 5.46 in type C and will be left to the reader (Exercise 5.13). □

Recall the partition parametrization of simple $O(k)$-modules from Proposition 5.36. The next theorem treats even and odd k uniformly.

Theorem 5.50 (($\mathrm{O}(k),\mathfrak{d}_\infty$)-Howe duality)**.** *As an* $(\mathrm{O}(k),\mathfrak{d}_\infty)$*-module we have*

$$\mathcal{F}^{\frac{k}{2}} \cong \bigoplus_{\lambda\in\mathcal{P},\lambda'_1+\lambda'_2\le k} L(\mathrm{O}(k),\lambda)\otimes L\left(\mathfrak{d}_\infty,\Lambda^{\mathfrak{d}}(\lambda)\right),$$

where $\Lambda^{\mathfrak{d}}(\lambda) := k\Lambda^{\mathfrak{d}}_0 + \sum_{i\ge1}\lambda'_i\varepsilon_i$.

Proof. By Lemmas 5.48 and 5.49, we have a strongly multiplicity-free decomposition of the $(\mathrm{O}(k),\mathfrak{d}_\infty)$-module $\mathcal{F}^{\frac{k}{2}}$. The proof is completed by finding explicit joint highest weight vectors with prescribed weights and noting that every simple $\mathrm{O}(k)$-module appears in the decomposition of $\mathcal{F}^{\frac{k}{2}}$, analogous to the proofs of Theorems 5.44 and 5.47. We will refer the reader to [**129**, Theorem 3.2(2)] and [**129**, Theorem 4.1(2)] for the precise forms of these joint highest weight vectors for k even and odd, respectively. □

Computing the trace of $\prod_{i\in\mathbb{N}} y_i^{\mathcal{E}^*_{ii}-\mathcal{E}^*_{1-i,1-i}} \prod_{p=1}^{\ell} z_p^{\mathcal{E}^{pp}_*}$ on both sides of the identity in Theorem 5.50, we obtain the following character identity (recall the Boolean characteristic function (5.40)):

$$\prod_{i=1}^{\infty}(1+y_i)^{\theta\{k \text{ is odd}\}} \cdot \prod_{i=1}^{\infty}\prod_{p=1}^{\ell}(1+y_iz_p)(1+y_iz_p^{-1}) \tag{5.57}$$
$$= \sum_{\lambda\in\mathcal{P},\lambda_1'+\lambda_2'\le k} \mathrm{ch}\, L(\mathrm{O}(k),\lambda)\,\mathrm{ch}\, L(\mathfrak{d}_\infty,\Lambda^{\mathfrak{d}}(\lambda)),$$

where $\mathrm{ch}\, L(\mathfrak{d}_\infty,\Lambda^{\mathfrak{d}}(\lambda)) = \mathrm{Tr}|_{L(\mathfrak{d}_\infty,\Lambda^{\mathfrak{d}}(\lambda))}\prod_{i\in\mathbb{N}} y_i^{\mathcal{E}_{ii}^*-\mathcal{E}_{1-i,1-i}^*}$.

5.5. Character formula for Lie superalgebras

In this section, we obtain character formulas for the irreducible oscillator modules of the Lie superalgebras of type $\mathfrak{osp}$ constructed in Section 5.3 as a simple application of the Howe dualities established in Section 5.4.

5.5.1. Characters for modules of Lie algebras $\mathfrak{c}_\infty$ and $\mathfrak{d}_\infty$. For $\mathfrak{x}\in\{\mathfrak{c},\mathfrak{d}\}$, let $\mathfrak{x}_{\infty 0}$ be the Cartan subalgebra, and let W be the Weyl group of the Lie algebras $\mathfrak{x}_\infty$ defined in Section 5.4. Let $\mathfrak{l}$ be the Levi subalgebra of $\mathfrak{c}_\infty$ (respectively of $\mathfrak{d}_\infty$) corresponding to the removal of the simple root $\beta_\times$ in the Dynkin diagram (5.42) (respectively in the Dynkin diagram (5.44)), and let W_0 denote the Weyl group of $\mathfrak{l}$. Also, let Φ^+ and $\Phi_{\mathfrak{l}}^+$ denote the sets of positive roots of $\mathfrak{x}_\infty$ and $\mathfrak{l}$, respectively, corresponding to the respective Dynkin diagrams. Let $\mathfrak{u}^+$ and $\mathfrak{u}^-$ be, respectively, the nilradical and opposite nilradical associated to $\mathfrak{l}$ so that we have $\mathfrak{x}_\infty = \mathfrak{u}^-\oplus\mathfrak{l}\oplus\mathfrak{u}^+$. Let W_r^0 be the set of the minimal length representatives of the right cosets $W_0\backslash W$ of length r for $\mathfrak{x}_\infty$. It is well known that the Weyl group W can be written as $W=W_0W^0$ with $W^0=\bigsqcup_{r\ge 0}W_r^0$, and we have

$$W^0 = \{w\in W \mid w(-\Phi^+)\cap\Phi^+\subseteq\Phi^+\setminus\Phi_{\mathfrak{l}}^+\}. \tag{5.58}$$

For $\mu\in(\mathfrak{x}_{\infty 0})^*$ and $w\in W$, we set

$$w\circ\mu := w(\mu+\rho_{\mathfrak{x}})-\rho_{\mathfrak{x}},$$

where $\rho_{\mathfrak{x}}\in(\mathfrak{x}_{\infty 0})^*$ is determined by $\langle\rho_{\mathfrak{x}},H_i^{\mathfrak{x}}\rangle=1$, for every simple coroot $H_i^{\mathfrak{x}}$.

Let λ be as in Theorem 5.47 in the case of $\mathfrak{c}_\infty$ and as in Theorem 5.50 in the case of $\mathfrak{d}_\infty$. Since $\langle\Lambda^{\mathfrak{x}}(\lambda),H_j^{\mathfrak{x}}\rangle\in\mathbb{Z}_+$, for every j, it follows by (5.58) that

$$\langle w\circ\Lambda^{\mathfrak{x}}(\lambda),H_i^{\mathfrak{x}}\rangle\in\mathbb{Z}_+, \text{ for } w\in W^0 \text{ and } H_i^{\mathfrak{x}}\in\mathfrak{l}.$$

In our setting, noting that W_0 is the group of permutations of $\mathbb{N}$ that fix all but finitely many numbers, we may find explicitly a partition

$$\lambda_w^{\mathfrak{x}} = \lambda_w = ((\lambda_w)_1,(\lambda_w)_2,\ldots)$$

such that $w\circ\Lambda^{\mathfrak{x}}(\lambda)$ can be written as

$$w\circ\Lambda^{\mathfrak{x}}(\lambda) = \begin{cases} k\Lambda_0^{\mathfrak{x}}+\sum_{j>0}(\lambda_w)_j\varepsilon_j, & \text{if } \mathfrak{x}=\mathfrak{d},\\ \frac{k}{2}\Lambda_0^{\mathfrak{x}}+\sum_{j>0}(\lambda_w)_j\varepsilon_j, & \text{if } \mathfrak{x}=\mathfrak{c}.\end{cases}$$

Recall that s_μ denotes the Schur function associated to a partition μ (see, e.g., A.1). Let

$$\mathrm{D}^{\mathfrak{x}} := \begin{cases} \prod_{1\le i\le j}(1-y_iy_j), & \text{for } \mathfrak{x}=\mathfrak{c}, \\ \prod_{1\le i<j}(1-y_iy_j), & \text{for } \mathfrak{x}=\mathfrak{d}. \end{cases}$$

Note that $\mathrm{D}^{\mathfrak{x}} = \prod_{\alpha\in\Phi^+\setminus\Phi_{\mathfrak{l}}^+}(1-e^{-\alpha})$, where we recall that $\Phi_{\mathfrak{l}}^+$ is the set of the positive roots in $\mathfrak{l}$.

Proposition 5.51. *Let λ be as in Theorem 5.47 in the case of $\mathfrak{c}_\infty$ and as in Theorem 5.50 in the case of $\mathfrak{d}_\infty$. We have the following character formula:*

$$\mathrm{ch}\,L(\mathfrak{x}_\infty,\Lambda^{\mathfrak{x}}(\lambda)) = \frac{1}{\mathrm{D}^{\mathfrak{x}}}\sum_{r=0}^{\infty}(-1)^r\sum_{w\in W_r^0} s_{\lambda_w^{\mathfrak{x}}}(y_1,y_2,\ldots), \qquad \textit{for } \mathfrak{x}=\mathfrak{c},\mathfrak{d}.$$

Proof. For any such λ, set $\Lambda=\Lambda^{\mathfrak{x}}(\lambda)$. The module $L(\mathfrak{x}_\infty,\Lambda)$ is integrable and hence affords the Weyl-Kac character formula (see e.g. [**64**])

$$\mathrm{ch}\,L(\mathfrak{x}_\infty,\Lambda) = \sum_{\sigma\in W}(-1)^{\ell(\sigma)}\frac{e^{\sigma(\Lambda+\rho_{\mathfrak{x}})-\rho_{\mathfrak{x}}}}{\prod_{\alpha\in\Phi^+}(1-e^{-\alpha})}, \tag{5.59}$$

where we recall that Φ^+ denotes the set of positive roots. Let $\rho_{\mathfrak{l}}\in(\mathfrak{x}_{\infty 0})^*$ be the element determined by $\rho_{\mathfrak{l}}(H_j)=1$, for all $j\in\mathbb{N}$, and $\rho_{\mathfrak{l}}(H_0^{\mathfrak{x}})=0$. Then $\tau(\rho_{\mathfrak{x}}-\rho_{\mathfrak{l}}) = \rho_{\mathfrak{x}}-\rho_{\mathfrak{l}}$, for $\tau\in W_0$. We now rewrite (5.59) as follows:

$$\begin{aligned}
\mathrm{ch}\,L(\mathfrak{x}_\infty,\Lambda) &= \sum_{w\in W^0}\sum_{\tau\in W_0}(-1)^{\ell(\tau)}(-1)^{\ell(w)}\frac{e^{\tau w(\Lambda+\rho_{\mathfrak{x}})-\rho_{\mathfrak{x}}}}{\prod_{\alpha\in\Phi^+}(1-e^{-\alpha})} \\
&= \sum_{w\in W^0}\sum_{\tau\in W_0}\frac{(-1)^{\ell(w)}}{\mathrm{D}^{\mathfrak{x}}}(-1)^{\ell(\tau)}\frac{e^{\tau w(\Lambda+\rho_{\mathfrak{x}})-\tau(\rho_{\mathfrak{x}})+\tau(\rho_{\mathfrak{x}})-\rho_{\mathfrak{x}}}}{\prod_{\alpha\in\Phi_{\mathfrak{l}}^+}(1-e^{-\alpha})} \\
&= \sum_{w\in W^0}\sum_{\tau\in W_0}\frac{(-1)^{\ell(w)}}{\mathrm{D}^{\mathfrak{x}}}(-1)^{\ell(\tau)}\frac{e^{\tau\left(w(\Lambda+\rho_{\mathfrak{x}})-\rho_{\mathfrak{x}}\right)+\tau(\rho_{\mathfrak{l}})-\rho_{\mathfrak{l}}}}{\prod_{\alpha\in\Phi_{\mathfrak{l}}^+}(1-e^{-\alpha})} \\
&= \sum_{w\in W^0}\frac{(-1)^{\ell(w)}}{\mathrm{D}^{\mathfrak{x}}}\sum_{\tau\in W_0}(-1)^{\ell(\tau)}\frac{e^{\tau(w\circ\Lambda+\rho_{\mathfrak{l}})-\rho_{\mathfrak{l}}}}{\prod_{\alpha\in\Phi_{\mathfrak{l}}^+}(1-e^{-\alpha})} \\
&= \frac{1}{\mathrm{D}^{\mathfrak{x}}}\sum_{r=0}^{\infty}(-1)^r\sum_{w\in W_r^0}s_{\lambda_w}(y_1,y_2,\ldots).
\end{aligned}$$

In the last identity we have used the fact that $\mathfrak{l}$ is of type $A_{+\infty}$, and so the irreducible $\mathfrak{l}$-character of integrable highest weight $w\circ\Lambda$ is simply the Schur function s_{λ_w}. □

5.5.2. Characters of oscillator $\mathfrak{osp}(2m|2n)$-modules. Recall that hs_λ denotes the super Schur function defined in Appendix (A.37). By the character of a module of $\mathrm{Sp}(k)$ or of $\mathfrak{osp}(2m|2n)$, we mean the trace of the operator $\prod_{1\le p\le\ell} z_p^{\mathcal{E}_{pp}-\mathcal{E}_{k+1-p,k+1-p}}$ or $\prod_{1\le i\le m}\prod_{1\le j\le n} x_i^{\mathcal{E}_{\bar{i}\bar{i}}} y_j^{\mathcal{E}_{jj}}$, respectively. We have the following character formula for the irreducible oscillator $\mathfrak{osp}(2m|2n)$-modules appearing in the Howe duality decomposition in Theorem 5.31.

Theorem 5.52. *Let $x = \{x_1, \ldots, x_m\}$ and $y = \{y_1, \ldots, y_n\}$. For $\lambda \in \mathcal{P}(\ell)$ with $\lambda_{m+1} \leq n$, we have the following character formula:*

$$\begin{aligned}&\mathrm{ch}L(\mathfrak{osp}(2m|2n), \lambda^\natural + \ell\mathbf{1}_{m|n})\\ &= \left(\frac{x_1\cdots x_m}{y_1\cdots y_n}\right)^\ell \cdot \frac{\prod_{\substack{1\leq i\leq m\\1\leq s\leq n}}(1+x_iy_s)\cdot\sum_{r=0}^\infty(-1)^r\sum_{w\in W_r^0} hs_{\lambda_w^c}(y;x)}{\prod_{1\leq i<j\leq m}\prod_{1\leq s\leq t\leq n}(1-x_ix_j)(1-y_sy_t)}.\end{aligned}$$

Proof. Computing the trace of the operator $\prod_{p=1}^\ell z_p^{\mathcal{E}_{pp}-\mathcal{E}_{k+1-p,k+1-p}}\prod_{i=1}^m\prod_{j=1}^n x_i^{\mathcal{E}_{\bar{i}\bar{i}}}y_j^{\mathcal{E}_{jj}}$ on both sides of the isomorphism in Theorem 5.31, we obtain that

$$\begin{aligned}&\left(\frac{x_1\cdots x_m}{y_1\cdots y_n}\right)^\ell\prod_{p=1}^\ell\prod_{i=1}^m\prod_{j=1}^n\frac{(1+y_jz_p^{-1})(1+y_jz_p)}{(1-x_iz_p^{-1})(1-x_iz_p)}\\ &\qquad= \sum_{\substack{\lambda\in\mathcal{P}(\ell)\\ \lambda_{m+1}\leq n}} \mathrm{ch}L(\mathrm{Sp}(2\ell),\lambda)\,\mathrm{ch}L(\mathfrak{osp}(2m|2n),\lambda^\natural+\ell\mathbf{1}_{m|n}).\end{aligned} \tag{5.60}$$

Next, by formal algebraic manipulations starting from (5.52), we shall obtain a new identity with the same left-hand side as (5.60). Replacing $\mathrm{ch}L(\mathfrak{c}_\infty, \Lambda^\mathfrak{c}(\lambda))$ in (5.52) by the expression in Proposition 5.51, we obtain an identity of symmetric functions in variables $y_1, y_2, \ldots$. We replace y_{n+i} by x_i, for all $i \geq 1$, and get the following identity:

$$\begin{aligned}&\prod_{j=1}^\infty\prod_{i=1}^n\prod_{p=1}^\ell(1+y_iz_p)(1+y_iz_p^{-1})(1+x_jz_p)(1+x_jz_p^{-1}) = \sum_{\lambda\in\mathcal{P}(\ell)}\mathrm{ch}L(\mathrm{Sp}(k),\lambda)\\ &\times\prod_{1\leq s\leq t\leq n}\prod_{1\leq i\leq j}\frac{1}{(1-x_iy_s)(1-x_ix_j)(1-y_sy_t)}\sum_{r=0}^\infty(-1)^r\sum_{w\in W_r^0}s_{\lambda_w^c}(y_1,\ldots,y_n;x_1,\ldots).\end{aligned}$$

Now we apply the standard involution ω_x (see (A.7)) to both sides of this new identity on the ring of symmetric functions in the variables $x_1, x_2, \ldots$. In the process we use the identities (A.7), (A.24), and (A.38):

$$\omega_x\Big(\prod_{i\geq1}(1-y_sx_i)^{-1}\Big) = \prod_{i\geq1}(1+y_sx_i),$$

$$\omega_x\Big(\prod_{1\leq i\leq j}(1-x_ix_j)^{-1}\Big) = \prod_{1\leq i<j}(1-x_ix_j)^{-1},$$

$$\omega_x\big(s_\mu(y,x)\big) = \mathrm{hs}_\mu(y;x).$$

Finally, in the resulting identity we set $x_j = 0$ for $j \geq m+1$ and multiply both sides by $\left(\frac{x_1\cdots x_m}{y_1\cdots y_n}\right)^\ell$. In this way we obtain the following identity which shares the same left-hand side as (5.60):

$$\left(\frac{x_1\cdots x_m}{y_1\cdots y_n}\right)^{\ell}\prod_{p=1}^{\ell}\prod_{i=1}^{m}\prod_{j=1}^{n}\frac{(1+y_jz_p^{-1})(1+y_jz_p)}{(1-x_iz_p^{-1})(1-x_iz_p)}=\sum_{\substack{\lambda\in\mathcal{P}(\ell)\\ \lambda_{m+1}\le n}}\operatorname{ch}L(\mathrm{Sp}(2\ell),\lambda)$$

$$\times\left(\frac{x_1\cdots x_m}{y_1\cdots y_n}\right)^{\ell}\frac{\prod_{\substack{1\le i\le m\\ 1\le s\le n}}(1+x_iy_s)\sum_{r=0}^{\infty}(-1)^r\sum_{w\in W_r^0}hs_{\lambda_w^c}(y;x)}{\prod_{1\le i<j\le m}\prod_{1\le s\le t\le n}(1-x_ix_j)(1-y_sy_t)}.$$

The theorem follows by comparing this identity with (5.60) and noting the linear independence of the characters $\operatorname{ch}L(\mathrm{Sp}(2\ell),\lambda)$. □

5.5.3. Characters for oscillator $\mathfrak{spo}(2m|2n)$-modules. The goal of this subsection is to find a character formula for the irreducible oscillator $\mathfrak{spo}(2m|2n)$-modules, which appear in the $(\mathrm{O}(k),\mathfrak{spo}(2m|2n))$-Howe duality (Theorem 5.39).

First, let $k=2\ell+1$ be odd and let λ be a partition with $\lambda_1'+\lambda_2'\le k$. Following the notations of Section 5.3.3 we denote by $\widetilde{\lambda}$ the partition obtained from λ by replacing the first column by $k-\lambda_1'$. Since the restrictions to $\mathrm{SO}(k)$ of the modules $L(\mathrm{O}(k),\lambda)$ and $L(\mathrm{O}(k),\widetilde{\lambda})$ are isomorphic, the usual character will not distinguish them. Recall $-I\in\mathrm{O}(k)\backslash\mathrm{SO}(k)$. So we define the (enhanced) character $\operatorname{ch}M$ of an $\mathrm{O}(k)$-module M to be the trace of the operator $\varepsilon^{-I}\prod_{p=1}^{\ell}z_p^{\mathcal{E}_{pp}-\mathcal{E}_{k+1-p,k+1-p}}$, where ε is an additional formal variable such that $\varepsilon^2=1$. The character of an $\mathfrak{spo}(2m|2n)$-module is defined as usual to be the trace of the operator $\prod_{i=1}^{m}\prod_{j=1}^{n}x_i^{\mathcal{E}_{\bar{i}\bar{i}}}y_j^{\mathcal{E}_{jj}}$.

Theorem 5.53. *Let $k=2\ell+1$. Let $x=\{x_1,\ldots,x_m\}$ and $y=\{y_1,\ldots,y_n\}$. For $\lambda\in\mathcal{P}(m|n)$ with $\lambda_1'+\lambda_2'\le k$, we have the following character formula:*

$$\operatorname{ch}L\Big(\mathfrak{spo}(2m|2n),\lambda^{\natural}+\frac{k}{2}\mathbf{1}_{m|n}\Big)$$
$$=\left(\frac{x_1\cdots x_m}{y_1\cdots y_n}\right)^{\frac{k}{2}}\cdot\frac{\prod_{\substack{1\le i\le m\\ 1\le s\le n}}(1+x_iy_s)\cdot\sum_{r=0}^{\infty}(-1)^r\sum_{w\in W_r^0}hs_{\lambda_w^0}(y;x)}{\prod_{1\le i\le j\le m}\prod_{1\le s<t\le n}(1-x_ix_j)(1-y_sy_t)}.$$

Proof. From the formula of the highest weight vector of $L(\mathrm{O}(k),\lambda)$ in the proof of Theorem 5.39, we observe that

$$\operatorname{ch}L(\mathrm{O}(k),\lambda)=\begin{cases}\varepsilon^{|\lambda|}\operatorname{ch}L(\mathfrak{so}(k),\lambda), & \text{if } \ell(\lambda)\le\ell,\\ \varepsilon^{|\lambda|}\operatorname{ch}L(\mathfrak{so}(k),\widetilde{\lambda}), & \text{if } \ell(\lambda)>\ell.\end{cases}$$

Since $\varepsilon^{|\lambda|}=\varepsilon^{|\widetilde{\lambda}|+1}$, we conclude that the (enhanced) characters $\operatorname{ch}L(\mathrm{O}(k),\lambda)$, where $\lambda_1'+\lambda_2'\le k$, are linearly independent.

Computing the trace of the operator $\varepsilon^{-I}\prod_{p=1}^{\ell}z_p^{\mathcal{E}_{pp}-\mathcal{E}_{k+1-p,k+1-p}}\prod_{i,j}x_i^{\mathcal{E}_{\bar{i}\bar{i}}}y_j^{\mathcal{E}_{jj}}$ on both sides of the isomorphism in Theorem 5.39, we obtain the following character identity:

$$\left(\frac{x_1\cdots x_m}{y_1\cdots y_n}\right)^{\frac{k}{2}} \prod_{p=1}^{\ell}\prod_{i=1}^{m}\prod_{j=1}^{n} \frac{(1+\varepsilon y_j z_p^{-1})(1+\varepsilon y_j z_p)(1+\varepsilon y_j)}{(1-\varepsilon x_i z_p^{-1})(1-\varepsilon x_i z_p)(1-\varepsilon x_i)} \tag{5.61}$$
$$= \sum_{\substack{\lambda'_1+\lambda'_2\le k \\ \lambda_{m+1}\le n}} \mathrm{ch} L(\mathrm{O}(k),\lambda)\, \mathrm{ch} L\Big(\mathfrak{spo}(2m|2n),\lambda^\natural+\frac{k}{2}\mathbf{1}_{m|n}\Big).$$

Analogous to the proof of Theorem 5.52, by means of algebraic manipulations starting from (5.57), we shall obtain a new identity with the same left-hand side as (5.61). Replacing $\mathrm{ch} L(\mathfrak{d}_\infty,\Lambda^{\mathfrak{d}}(\lambda))$ in (5.57) by the expression in Proposition 5.51 and taking into account the eigenvalue of the element $-I$, we obtain an identity of symmetric functions in the variables $y_1,y_2,\ldots$. We replace y_{n+i} by x_i, for all $i\ge 1$, and then apply the involution ω_x on the ring of symmetric functions in the variables $x_1,x_2,\ldots$. Finally, we set $x_j=0$ for $j\ge m+1$ and multiply both sides of the resulting identity by $\left(\frac{x_1\cdots x_m}{y_1\cdots y_n}\right)^{\frac{k}{2}}$. In this way we obtain

$$\left(\frac{x_1\cdots x_m}{y_1\cdots y_n}\right)^{\frac{k}{2}} \prod_{p=1}^{\ell}\prod_{i,j} \frac{(1+\varepsilon y_j z_p^{-1})(1+\varepsilon y_j z_p)(1+\varepsilon y_j)}{(1-\varepsilon x_i z_p^{-1})(1-\varepsilon x_i z_p)(1-\varepsilon x_i)}$$
$$= \sum_{\substack{\lambda'_1+\lambda'_2\le k \\ \lambda_{m+1}\le n}} \mathrm{ch} L(\mathrm{O}(k),\lambda)$$
$$\times \left(\frac{x_1\cdots x_m}{y_1\cdots y_n}\right)^{\frac{k}{2}} \frac{\prod_{\substack{1\le i\le m\\1\le s\le n}}(1+x_iy_s)\sum_{r=0}^{\infty}(-1)^r\sum_{w\in W_r^{\mathfrak{d}}} hs_{\lambda_w^{\mathfrak{d}}}(y;x)}{\prod_{1\le i\le j\le m}\prod_{1\le s<t\le n}(1-x_ix_j)(1-y_sy_t)}.$$

The theorem follows by comparing this identity with (5.61) and noting that the (enhanced) characters $\mathrm{ch} L(\mathrm{O}(k),\lambda)$ are linearly independent. □

We now turn to the case when $k=2\ell$ is even. In this case, the trick of introducing the extra variable ε does not work, and we obtain the following.

Theorem 5.54. *Let $k=2\ell$. Let $x=\{x_1,\ldots,x_m\}$ and $y=\{y_1,\ldots,y_n\}$. For $\lambda\in\mathcal{P}(m|n)$ with $\lambda'_1+\lambda'_2\le k$, we have the following character identity:*

$$\mathrm{ch} L(\mathfrak{spo}(2m|2n),\lambda^\natural+\ell\mathbf{1}_{m|n})+\mathrm{ch} L(\mathfrak{spo}(2m|2n),\widetilde{\lambda}^\natural+\ell\mathbf{1}_{m|n})$$
$$=\left(\frac{x_1\cdots x_m}{y_1\cdots y_n}\right)^{\ell}\cdot\frac{\prod_{\substack{1\le i\le m\\1\le s\le n}}(1+x_iy_s)\cdot\sum_{r=0}^{\infty}(-1)^r\sum_{w\in W_r^{\mathfrak{d}}}\big(hs_{\lambda_w^{\mathfrak{d}}}(y;x)+hs_{\widetilde{\lambda}_w^{\mathfrak{d}}}(y;x)\big)}{\prod_{1\le i\le j\le m}\prod_{1\le s<t\le n}(1-x_ix_j)(1-y_sy_t)}.$$

Under the additional assumption that $\lambda_\ell=0$, $\widetilde{\lambda}=\lambda$, the theorem becomes

$$\mathrm{ch} L(\mathfrak{spo}(2m|2n),\lambda^\natural+\ell\mathbf{1}_{m|n})$$

$$= \left(\frac{x_1\cdots x_m}{y_1\cdots y_n}\right)^{\ell} \cdot \frac{\prod_{\substack{1\le i\le m\\ 1\le s\le n}}(1+x_iy_s)\cdot\sum_{r=0}^{\infty}(-1)^r\sum_{w\in W_r^0} hs_{\lambda_w^0}(y;x)}{\prod_{1\le i\le j\le m}\prod_{1\le s<t\le n}(1-x_ix_j)(1-y_sy_t)}.$$

Proof. Computing the trace of the operator $z_p^{\mathcal{E}_{pp}-\mathcal{E}^{k+1-p,k+1-p}}\prod_{i,j}x_i^{\mathcal{E}_{\bar{i}\bar{i}}}y_j^{\mathcal{E}_{jj}}$ on both sides of the isomorphism in Theorem 5.39, we obtain

$$\begin{aligned}&\left(\frac{x_1\cdots x_m}{y_1\cdots y_n}\right)^{\ell}\prod_{p=1}^{\ell}\prod_{i=1}^{m}\prod_{j=1}^{n}\frac{(1+y_jz_p^{-1})(1+y_jz_p)}{(1-x_iz_p^{-1})(1-x_iz_p)}\\ &\quad=\sum_{\substack{\lambda_1'+\lambda_2'\le k\\ \lambda_{m+1}\le n}}\mathrm{ch}L(\mathrm{O}(k),\lambda)\,\mathrm{ch}L(\mathfrak{spo}(2m|2n),\lambda^{\natural}+\ell\mathbf{1}_{m|n}).\end{aligned}\tag{5.62}$$

In the same way as in the proof of Theorem 5.53, using a formal algebraic manipulation based on (5.57), we obtain the following identity that shares the same left-hand side as (5.62):

$$\begin{aligned}\left(\frac{x_1\cdots x_m}{y_1\cdots y_n}\right)^{\ell}\prod_{p=1}^{\ell}\prod_{i=1}^{m}\prod_{j=1}^{n}\frac{(1+y_jz_p^{-1})(1+y_jz_p)}{(1-x_iz_p^{-1})(1-x_iz_p)} &= \sum_{\substack{\lambda_1'+\lambda_2'\le k\\ \lambda_{m+1}\le n}}\mathrm{ch}L(\mathrm{O}(k),\lambda)\\ \times\left(\frac{x_1\cdots x_m}{y_1\cdots y_n}\right)^{\ell}&\cdot\frac{\prod_{\substack{1\le i\le m\\ 1\le s\le n}}(1+x_iy_s)\cdot\sum_{r=0}^{\infty}(-1)^r\sum_{w\in W_r^0} hs_{\lambda_w^0}(y;x)}{\prod_{1\le i\le j\le m}\prod_{1\le s<t\le n}(1-x_ix_j)(1-y_sy_t)}.\end{aligned}$$

The theorem follows by comparing this identity with (5.62) and noting that the characters of $L(\mathrm{O}(k),\lambda)$ and $L(\mathrm{O}(k),\widetilde{\lambda})$ coincide. □

5.6. Exercises

Exercise 5.1. (1) Verify (5.7) and (5.8).

(2) Let $a,a_1\in\boldsymbol{U}_{\bar{0}}$, $c,c_1\in\boldsymbol{U}_{\bar{1}}$. Prove the following identity in $\mathfrak{WC}(\boldsymbol{U})$:

$$\langle[ac,a_1],c_1\rangle'+\langle a_1,[ac,c_1]_+\rangle'=0.$$

Exercise 5.2. Prove:

(1) The formulas in (5.13) and (5.14) define commuting actions of $\mathfrak{gl}(k)$ and $\mathfrak{gl}(m|n)$ on $\mathbb{C}[x,\eta]$.

(2) The formulas in (5.22) and (5.23) define commuting actions of $\mathfrak{q}(m)$ and $\mathfrak{q}(n)$ on $\mathbb{C}[x,\xi]$.

Exercise 5.3. Prove Theorem 5.19 using the Schur-Sergeev duality. In particular, the $\mathfrak{gl}(m|n)$-modules $\mathcal{S}^k(\mathbb{C}^{m|n})$ and $\Lambda^k(\mathbb{C}^{m|n})$ are irreducible, for all $k\in\mathbb{N}$.

Exercise 5.4. Recall $\diamondsuit_\ell$ and $\diamondsuit_{t,\ell}$ from (5.19) and (5.20), respectively. Prove:

(1) $\diamondsuit_\ell$ is annihilated by the operators in (5.17) and (5.18).

(2) $\diamondsuit_{t,\ell}$ is a nonzero in $\mathbb{C}[x,\eta]$. (Hint: Apply a sequence of differential operators to it to obtain $\diamondsuit_\ell$.)

Exercise 5.5. Verify (1)–(3) in the proof of Lemma 5.22.

Exercise 5.6. Verify directly that the operators in (5.18) annihilate the vector $\diamondsuit$ in the proof of Lemma 5.22.

Exercise 5.7. Let $\mathfrak{g} = \mathfrak{q}(m)$ and let $\mathbb{C}^{m|m}$ be its natural module. Prove that the $\mathfrak{g}$-module $\mathcal{S}^k(\mathbb{C}^{m|m})$ is irreducible, for all $k \in \mathbb{N}$.

Exercise 5.8. Suppose that $k = 2\ell$ and $m,n,p,q \in \mathbb{Z}_+$. We have a natural embedding $\mathfrak{osp}(2m|2n) \oplus \mathfrak{osp}(2p|2q) \subseteq \mathfrak{osp}(2m+2p|2n+2q)$. Prove:

(1) For $\gamma \in \mathcal{P}(m+p,n+q) \cap \mathcal{P}(\ell)$, we have the following isomorphism of $\mathfrak{osp}(2m|2n) \oplus \mathfrak{osp}(2p|2q)$-modules:

$$L(\mathfrak{osp}(2m+2p|2n+2q),\gamma^\natural + \ell\mathbf{1}_{m+p|n+q})$$
$$\cong \bigoplus_{\lambda\in\mathcal{P}(m|n),\mu\in\mathcal{P}(p|q)} \left(L(\mathfrak{osp}(2m|2n),\lambda^\natural + \ell\mathbf{1}_{m|n}) \otimes L(\mathfrak{osp}(2p|2q),\mu^\natural + \ell\mathbf{1}_{p|q})\right)^{a^\gamma_{\lambda\mu}},$$

where $L(\mathrm{Sp}(k),\lambda) \otimes L(\mathrm{Sp}(k),\mu) \cong \bigoplus_\gamma L(\mathrm{Sp}(k),\gamma)^{a^\gamma_{\lambda\mu}}$, for $\lambda,\mu,\gamma \in \mathcal{P}(\ell)$.

(2) We have a tensor product decomposition of $\mathfrak{osp}(2m|2n)$-modules:

$$L(\mathfrak{osp}(2m|2n),\lambda^\natural + \ell\mathbf{1}_{m|n}) \otimes L(\mathfrak{osp}(2m|2n),\mu^\natural + \ell\mathbf{1}_{p|q})$$
$$\cong \bigoplus_{\gamma\in\mathcal{P}(m|n),\ell(\gamma)\le k} L(\mathfrak{osp}(2m|2n),\gamma^\natural + k\mathbf{1}_{m|n})^{b^\gamma_{\lambda\mu}},$$

where $\lambda,\mu \in \mathcal{P}(m|n) \cap \mathcal{P}(\ell)$, and $b^\gamma_{\lambda\mu}$ are determined by the $\mathrm{Sp}(k)$-module isomorphism $L(\mathrm{Sp}(2k),\gamma) \cong \bigoplus_{\lambda,\mu} \left(L(\mathrm{Sp}(k),\lambda) \otimes L(\mathrm{Sp}(k),\mu)\right)^{b^\gamma_{\lambda\mu}}$.

(Hint: Use Theorem 5.31.)

Exercise 5.9. Prove Lemma 5.37(1).

Exercise 5.10. Let λ be a generalized partition and let $v^{\mathfrak{a}}_\lambda$ be the vector defined in (5.49). Recall $\mathcal{E}^*_{ij}$ from (5.46) and $\mathcal{E}^{pq}_*$ from (5.47). Prove:

(1) $\mathcal{E}^*_{i,i+1} v^{\mathfrak{a}}_\lambda = 0$, for all $i \in \mathbb{Z}$.

(2) $\mathcal{E}^{p,p+1}_* v^{\mathfrak{a}}_\lambda = 0$, for all $1 \le p \le \ell - 1$.

Exercise 5.11. Consider the $\mathfrak{gl}(k) \times \mathfrak{gl}(1|1)$-Howe duality on the space $\mathbb{C}[x,\eta]$. We write $x_{1i} = x_i$ and $\eta_{1i} = dx_i$, for $i = 1,\dots,k$ so that $\mathbb{C}[x,\eta]$ gets identified with the polynomial differential forms on $\mathbb{C}^k$, and the operator $\mathcal{E}_{1\bar{1}}$ with the total differential $d = \sum_{i=1}^k dx_i \frac{\partial}{\partial x_i}$. Let Ω^i denote the space of i-forms, for $i = 1,\dots,k$, and set $\Omega^{-1} = 0$. Put $d_i = d|_{\Omega^i}$. Prove:

(1) d is a differential, i.e., $d^2 = 0$.

(2) Every representation of $\mathfrak{gl}(1|1)$ that appears in the Howe duality decomposition is typical, except for the trivial representation.

(3) Poincaré's lemma, i.e., $\ker d_i/\operatorname{im} d_{i-1} = \begin{cases} \mathbb{C}, & \text{if } i=0, \\ 0, & \text{otherwise.} \end{cases}$

Exercise 5.12. The natural action of $\mathrm{O}(k)$ on $\mathbb{C}^k$ given by (5.33) or (5.34) induces an action of $\mathrm{O}(k)$ on $\mathbb{C}[x] = \mathbb{C}[x_1,\ldots,x_k]$. Let $J = \sum_{p=1}^k x_p x_{p-k+1}$, $\Delta = \sum_{p=1}^k \frac{\partial}{\partial x_p}\frac{\partial}{\partial x_{k-p+1}}$, and let $\mathcal{H} = \ker\Delta$ be the space of **spherical harmonics**. Prove:

(1) $\mathcal{H}$ is equal to the $\mathrm{O}(k)$-module generated by the polynomials x_1^m, for all $m \in \mathbb{Z}_+$.

(2) Every polynomial in $\mathbb{C}[x]$ can be written in the form $\sum_i f_i(J)h_i$, where $f_i(J)$ is a polynomial in J and $h_i \in \mathcal{H}$.

(Hint: Use Theorem 5.39.)

Exercise 5.13. Prove Lemma 5.48 and Lemma 5.49.

Exercise 5.14. We identify the action of $\mathfrak{gl}(k)$ on the symmetric square $S^2(\mathbb{C}^k)$ of its natural module $\mathbb{C}^k$ via that of the differential operators $\mathcal{E}^{p,q}$ given in (5.13) with $m=1$ and $n=0$. Set $y_{ij} := x_{i1}x_{j1}$, for $1 \le i,j \le k$, so that $y_{ij} = y_{ji}$, and identify $S(S^2(\mathbb{C}^k))$ with the polynomial algebra $\mathbb{C}[y]$. Let Γ be the symmetric matrix $(y_{ij})_{1\le i,j\le k}$, and let Γ_r be the determinant of the rth principal minor of Γ, for $1 \le r \le k$. Prove:

(1) The polynomial $\Gamma_r \in \mathbb{C}[y]$ is annihilated by the operators in (5.17) and has $\mathfrak{gl}(k)$-weight corresponding to the partition (2^r).

(2) We have the following isomorphism of $\mathfrak{gl}(k)$-modules:

$$S(S^2(\mathbb{C}^k)) \cong \bigoplus_{\ell(\lambda)\le k;\lambda \text{ even}} L_k(\lambda).$$

Here λ **even** means that every part of λ is even. (Hint: Compute their characters and use the first identity in (A.23).)

Exercise 5.15. We identify the action of $\mathfrak{gl}(k)$ on the exterior square $\wedge^2(\mathbb{C}^k)$ of its natural module via that of the differential operators $\mathcal{E}^{p,q}$ as in (5.13) with $m=0$ and $n=1$. Set $z_{ij} := \eta_{i1}\eta_{j1}$, for $1 \le i,j \le k$, so that $z_{ij} = -z_{ji}$, and identify $S(\wedge^2(\mathbb{C}^k))$ with the polynomial algebra $\mathbb{C}[z]$. Let Υ be the skew-symmetric matrix $(z_{ij})_{1\le i,j\le k}$, and let $\Upsilon_{2\ell}$ be the Pfaffian (see (A.49)) of the $2\ell\times 2\ell$ matrix obtained by taking the first 2ℓ rows and columns of Υ, for $1 \le 2\ell \le k$. Prove:

(1) The polynomial $\Upsilon_{2\ell} \in \mathbb{C}[z]$ is annihilated by the operators in (5.17) and has $\mathfrak{gl}(k)$-weight corresponding to the partition $(1^{2\ell})$.

(2) We have the following $\mathfrak{gl}(k)$-module isomorphism:

$$S(\wedge^2(\mathbb{C}^k)) \cong \bigoplus_{\lambda_1\le k;\lambda \text{ even}} L_k(\lambda').$$

(Hint: Use the second identity in (A.23).)

Exercise 5.16. Recall the Littlewood-Richardson coefficients $c^{\mu}_{\lambda\nu}$, for $\lambda,\mu,\nu \in \mathcal{P}$, from (A.17). Let $\lambda,\mu \in \mathcal{P}(m|n)$, and let $x = (x_1,\dots,x_m)$ and $y = (y_1,\dots,y_n)$ be indeterminates. Prove:

$$\mathrm{hs}_\nu(x;y)\cdot \mathrm{hs}_\lambda(x;y) = \sum_{\mu\in\mathcal{P}(m|n)} c^{\mu}_{\lambda\nu}\mathrm{hs}_\mu(x;y).$$

Exercise 5.17. For $\lambda \in \mathcal{P}(\ell)$ recall the oscillator module $L(\mathfrak{g},\lambda+\ell\mathbf{1}_\infty)$ of $\mathfrak{g} = \mathfrak{so}(2\infty)$ (see Corollary 5.32 with $m=\infty$). For $\lambda\in\mathcal{P}(\ell)$ and $\mu\in\mathcal{P}(r)$, prove:

(1) We have a tensor product decomposition of the form

$$L(\mathfrak{g},\lambda+\ell\mathbf{1}_\infty)\otimes L(\mathfrak{g},\mu+r\mathbf{1}_\infty) = \bigoplus_{\nu\in\mathcal{P}(\ell+r)} L(\mathfrak{g},\nu+(\ell+r)\mathbf{1}_\infty)^{a^{\nu}_{\lambda\mu}},$$

where $a^{\nu}_{\lambda\mu}\in\mathbb{Z}_+$ denote the respective multiplicities.

(2) If, in addition, $\lambda,\mu\in\mathcal{P}(m)$, then we have

$$L(\mathfrak{so}(2m),\lambda+\ell\mathbf{1}_m)\otimes L(\mathfrak{so}(2m),\mu+r\mathbf{1}_m) = \bigoplus_{\nu\in\mathcal{P}(m)} L(\mathfrak{so}(2m),\nu+(\ell+r)\mathbf{1}_m)^{a^{\nu}_{\lambda\mu}}.$$

(3) If, in addition, $\lambda_1,\mu_1\le n$, then we have

$$L(\mathfrak{sp}(2n),\lambda'-\ell\mathbf{1}_n)\otimes L(\mathfrak{sp}(2n),\mu'-r\mathbf{1}_n) = \bigoplus_{\nu_1\le n} L(\mathfrak{sp}(2n),\nu'-(\ell+r)\mathbf{1}_n)^{a^{\nu}_{\lambda\mu}}.$$

(4) If, in addition, $\lambda,\mu\in\mathcal{P}(m|n)$, then we have

$$L(\mathfrak{osp}(2m|2n),\lambda^\natural+\ell\mathbf{1}_{m|n})\otimes L(\mathfrak{osp}(2m|2n),\mu^\natural+r\mathbf{1}_{m|n}) = \bigoplus_{\nu\in\mathcal{P}(m|n)} L(\mathfrak{osp}(2m|2n),\nu^\natural+(\ell+r)\mathbf{1}_{m|n})^{a^{\nu}_{\lambda\mu}}.$$

Exercise 5.18. For $\lambda\in\mathcal{P}$ with $\lambda'_1+\lambda'_2\le k$, recall the oscillator module $L(\mathfrak{g},\lambda+\frac{k}{2}\mathbf{1}_\infty)$ of $\mathfrak{g}=\mathfrak{sp}(2\infty)$ (see Corollary 5.40 with $m=\infty$). For $\lambda,\mu\in\mathcal{P}$ with $\lambda'_1+\lambda'_2\le k$ and $\mu'_1+\mu'_2\le r$, prove:

(1) We have a tensor product decomposition of the form

$$L(\mathfrak{g},\lambda+\frac{k}{2}\mathbf{1}_\infty)\otimes L(\mathfrak{g},\mu+\frac{r}{2}\mathbf{1}_\infty) = \bigoplus_{\nu'_1+\nu'_2\le k+r} L(\mathfrak{g},\nu+\frac{k+r}{2}\mathbf{1}_\infty)^{b^{\nu}_{\lambda\mu}},$$

where $b^{\nu}_{\lambda\mu}\in\mathbb{Z}_+$ denote the respective multiplicities.

(2) If, in addition, $\lambda,\mu\in\mathcal{P}(m)$, then we have

$$L(\mathfrak{sp}(2m),\lambda+\frac{k}{2}\mathbf{1}_m)\otimes L(\mathfrak{sp}(2m),\mu+\frac{r}{2}\mathbf{1}_m)$$

$$= \bigoplus_{\nu \in \mathcal{P}(m)} L(\mathfrak{sp}(2m), \nu + \frac{k+r}{2}\mathbf{1}_m)^{b^{\nu}_{\lambda\mu}}.$$

(3) If, in addition, $\lambda_1, \mu_1 \leq n$, then we have

$$L(\mathfrak{so}(2n), \lambda' - \frac{k}{2}\mathbf{1}_n) \otimes L(\mathfrak{so}(2n), \mu' - \frac{r}{2}\mathbf{1}_n)$$
$$= \bigoplus_{\nu_1 \leq n} L(\mathfrak{so}(2n), \nu' - \frac{k+r}{2}\mathbf{1}_n)^{b^{\nu}_{\lambda\mu}}.$$

(4) If, in addition, $\lambda, \mu \in \mathcal{P}(m|n)$, then we have

$$L(\mathfrak{spo}(2m|2n), \lambda^{\natural} + \frac{k}{2}\mathbf{1}_{m|n}) \otimes L(\mathfrak{spo}(2m|2n), \mu^{\natural} + \frac{r}{2}\mathbf{1}_{m|n})$$
$$= \bigoplus_{\nu \in \mathcal{P}(m|n)} L(\mathfrak{spo}(2m|2n), \nu^{\natural} + \frac{k+r}{2}\mathbf{1}_{m|n})^{b^{\nu}_{\lambda\mu}}.$$

Notes

Section 5.1. The materials on Weyl-Clifford algebra and connections to classical Lie algebras/superalgebras are standard. The general duality theorem, Theorem 5.8, is taken from Goodman-Wallach [**46**, 4.2.1], and it can be regarded as an abstract generalization of the original formulation (see Theorem 5.10 and Proposition 5.12) of Howe duality [**51**] (Howe's paper as a preprint dates back to 1976).

Section 5.2. The $(\mathrm{GL}(k), \mathfrak{gl}(m))$-Howe duality of Theorem 5.16 (respectively, its skew version of Theorem 5.18), due to Howe [**51, 52**], offers a representation theoretical interpretation of the classical Cauchy identity (respectively, its dual version). It is equivalent to Schur duality as well as the First Fundamental Theorem of invariant theory for general linear groups.

The Howe dual pair $(\mathrm{GL}(k), \mathfrak{gl}(m|n))$ (Theorem 5.14) already appeared in Howe's classical paper [**51**], and the precise multiplicity-free $(\mathrm{GL}(k), \mathfrak{gl}(m|n))$-Howe duality decomposition (Theorem 5.19) was obtained independently by Brini-Palareti-Teolis [**10**], Sergeev [**113**], and Cheng-Wang [**29**]. The formula for the joint highest weight vectors in the $(\mathrm{GL}(k), \mathfrak{gl}(m|n))$-Howe duality (Theorem 5.23) was due to Cheng-Wang [**29**], and it unifies the formulas of Howe [**52**] in two (non-super) special cases of $m = 0$ and $n = 0$. We note that a formula of highest weight vectors for the more general $(\mathfrak{gl}(k|\ell), \mathfrak{gl}(m|n))$-Howe duality was also obtained in *loc. cit.*. The $(\mathfrak{q}(m), \mathfrak{q}(n))$-Howe duality was due to Sergeev [**113**] and Cheng-Wang [**28**] independently, and it gives a representation-theoretic interpretation of the Cauchy identity for Schur Q-functions.

Section 5.3. The Howe dual pairs $(\mathrm{Sp}(k), \mathfrak{osp}(2m|2n))$ and $(\mathrm{O}(k), \mathfrak{spo}(2m|2n))$ on $\mathcal{S}(\mathbb{C}^k \otimes \mathbb{C}^{m|n})$ (Theorems 5.30 and 5.38) were formulated in Howe [**51**]; they were based on, and in turn can be regarded as a Lie theoretical reformulation of, the

First Fundamental Theorem of invariant theory for classical groups. The orthogonal group $\mathrm{O}(k)$ is disconnected. We follow Howe [**52**] to present a parametrization of the simple $\mathrm{O}(k)$-modules in terms of partitions. The strongly multiplicity-free decompositions in $(\mathrm{Sp}(k), \mathfrak{osp}(2m|2n))$- and $(\mathrm{O}(k), \mathfrak{spo}(2m|2n))$-Howe dualities (Theorems 5.31 and 5.39) were obtained in Cheng-Zhang [**36**], where the highest weight vector formula in Section 5.2 plays a key role in the proofs. The shorter proofs presented in this book follow Cheng-Kwon-Wang [**21**]. We easily recover the Howe duality decompositions in the non-super setting, which were due to Howe [**52**] with somewhat different arguments, by setting either m or n to zero.

There are additional Howe dualities involving the spin groups (see Howe [**52**]) as well as $\mathfrak{spo}(2m|2n+1)$ (see Cheng-Kwon-Wang [**21**, Appendix A]). There is also a type A Howe duality involving infinite-dimensional irreducible modules of $\mathfrak{gl}(n)$ or more generally $\mathfrak{gl}(m|n)$ (see Kashiwara-Vergne [**69**] and Cheng-Lam-Zhang [**27**]). We do not treat these cases in the book and refer the reader to the original papers for details.

Section 5.4. Howe duality between classical groups and infinite-dimensional Lie algebras was systematically developed by Wang [**129**], and it has been used in Kac-Wang-Yan [**67**] for the study of representation theory of classical Lie subalgebras of $W_{1+\infty}$. The $(\mathrm{GL}(\ell), \mathfrak{a}_\infty)$-Howe duality (Theorem 5.44) can also be recovered as a limit case of a duality for affine Lie algebras of type $\mathfrak{gl}$ given earlier by Igor Frenkel [**44**], and the proof here follows [**129**]. The $(\mathrm{Sp}(k), \mathfrak{c}_\infty)$-Howe duality and $(\mathrm{O}(k), \mathfrak{d}_\infty)$-Howe duality (Theorems 5.47 and 5.50) are due to Wang [**129**], where one can find several more Howe dualities in bosonic and fermionic Fock spaces not covered in the book.

Section 5.5. Irreducible characters for the oscillator modules of Lie superalgebras were computed in Cheng-Zhang [**36**] following the approach of Cheng-Lam [**22**]. A character formula of Enright [**41**] played an essential role there. The simpler and more elementary approach presented in this book follows Cheng-Kwon-Wang [**21**] and bypasses Enright's formula via a comparison with Howe duality involving infinite-dimensional Lie algebras in Section 5.4.

The reader is referred to Cheng-Kwon-Wang [**21**], where the calculation of the characters via a comparison of two Howe dualities has been refined to compute the corresponding $\mathfrak{u}^-$-homology groups with coefficients in the oscillator modules. The approach there follows the strategy of Aribaud [**2**] and Cheng-Kwon [**19**].

Exercise 5.8 is an example of Kudla's seesaw pair [**75, 52**], while Exercises 5.11 and 5.12 were taken from [**51**]. The formulas for the highest weight vectors in Exercises 5.14 and 5.15 already appeared in [**52**]. More general formulas for the highest weight vectors in the supersymmetric tensor of the supersymmetric square of the natural representation of the general linear Lie superalgebra can be found in [**29**]. We remark that Exercises 5.17 and 5.18 make direct connections between the tensor product decompositions of infinite-dimensional oscillator representations of type D and C Lie algebras and those of finite-dimensional representations of type C and D Lie algebras, respectively. This observation was made in [**35**], and indeed they also follow from super duality in the next chapter.

Chapter 6

Super duality

In this chapter, we develop a *super duality* approach to obtain a complete and conceptual solution of the irreducible character problem in certain parabolic Bernstein-Gelfand-Gelfand (BGG) categories for general linear and ortho-symplectic Lie superalgebras. These parabolic categories contain all the finite-dimensional irreducible modules of these Lie superalgebras.

Super duality is an equivalence of categories between parabolic BGG categories for Lie superalgebras $\overline{\mathcal{O}}_{\overline{\mathfrak{g}}}$ and their Lie algebra counterparts $\mathcal{O}_{\mathfrak{g}}$ at an infinite-rank limit, which can be concisely summarized in the following diagram:

$$
(6.1)\qquad
\begin{array}{ccccccccc}
\widetilde{\mathcal{O}}_{\widetilde{\mathfrak{g}}} \xrightarrow{\;T\;} \mathcal{O}_{\mathfrak{g}} : & \cdots \longrightarrow & \mathcal{O}_{\mathfrak{g}_{n+1}} & \xrightarrow{\mathfrak{tr}_n} & \mathcal{O}_{\mathfrak{g}_n} & \xrightarrow{\mathfrak{tr}_{n-1}} & \mathcal{O}_{\mathfrak{g}_{n-1}} & \longrightarrow \cdots \\
{}_{\overline{T}}\searrow \quad \downarrow\cong \quad & & & & & & & \\
\overline{\mathcal{O}}_{\overline{\mathfrak{g}}} : & \cdots \longrightarrow & \overline{\mathcal{O}}_{\overline{\mathfrak{g}}_{n+1}} & \xrightarrow{\mathfrak{tr}_n} & \overline{\mathcal{O}}_{\overline{\mathfrak{g}}_n} & \xrightarrow{\mathfrak{tr}_{n-1}} & \overline{\mathcal{O}}_{\overline{\mathfrak{g}}_{n-1}} & \longrightarrow \cdots
\end{array}
$$

A weak version on the Grothendieck group level of the equivalence $\overline{\mathcal{O}}_{\overline{\mathfrak{g}}} \cong \mathcal{O}_{\mathfrak{g}}$, which is established first in an elementary way, already implies a solution of the irreducible character problem of these Lie superalgebras. It is further shown that the corresponding Kostant $\mathfrak{u}$-homology groups, or equivalently the Kazhdan-Lusztig-Vogan polynomials, are matched under super duality. These results are established by introducing an auxiliary Lie superalgebra $\widetilde{\mathfrak{g}}$, its parabolic BGG category $\widetilde{\mathcal{O}}_{\widetilde{\mathfrak{g}}}$, and two functors T and $\overline{T}$. The equivalence $\overline{\mathcal{O}}_{\overline{\mathfrak{g}}} \cong \mathcal{O}_{\mathfrak{g}}$ follows by establishing the two category equivalences T and $\overline{T}$.

The category $\mathcal{O}_{\mathfrak{g}}$ is the inverse limit of a sequence of categories $\{\mathcal{O}_{\mathfrak{g}_n}\}_{n\geq 0}$, where $\mathfrak{tr}_n$ are the truncation functors. The Kazhdan-Lusztig solutions for the categories $\mathcal{O}_{\mathfrak{g}_n}$ for all $n \geq 0$ of the finite-rank Lie algebras $\mathfrak{g}_n$ are shown to amount to the Kazhdan-Lusztig solution for the category $\mathcal{O}_{\mathfrak{g}}$. Similarly, the Kazhdan-Lusztig solutions for the categories $\overline{\mathcal{O}}_{\bar{\mathfrak{g}}_n}$ for all $n \geq 0$ of the finite-rank Lie superalgebras $\bar{\mathfrak{g}}_n$ amount to the Kazhdan-Lusztig solution for the category $\overline{\mathcal{O}}_{\bar{\mathfrak{g}}}$. By chasing the diagram (6.1), we solve the irreducible character problem in the parabolic BGG categories $\overline{\mathcal{O}}_{\bar{\mathfrak{g}}_n}$ for the finite-rank basic Lie superalgebras $\bar{\mathfrak{g}}_n$ of type $\mathfrak{gl}$ or $\mathfrak{osp}$ via the classical parabolic Kazhdan-Lusztig polynomials.

6.1. Lie superalgebras of classical types

This section starts with a discussion of the infinite-rank Lie superalgebras $\mathfrak{g}$, $\bar{\mathfrak{g}}$, and $\widetilde{\mathfrak{g}}$, whose representation theories we will investigate in later sections. Here, $\mathfrak{g}$ is a classical Lie algebra, $\bar{\mathfrak{g}}$ is the corresponding Lie superalgebra of classical type, and $\widetilde{\mathfrak{g}}$ is an auxiliary Lie superalgebra that contains both $\mathfrak{g}$ and $\bar{\mathfrak{g}}$ as subalgebras and that plays the role of an intermediary between them. We present explicit matrix realizations of these Lie superalgebras together with their finite-dimensional counterparts, $\widetilde{\mathfrak{g}}_n^{\mathfrak{x}}$, $\bar{\mathfrak{g}}_n^{\mathfrak{x}}$, and $\mathfrak{g}_n^{\mathfrak{x}}$, for $n \in \mathbb{N}$. All of these superalgebras implicitly depend on a fixed type $\mathfrak{x} = \mathfrak{a}, \mathfrak{b}, \mathfrak{b}^\bullet, \mathfrak{c}, \mathfrak{d}$.

6.1.1. Head, tail, and master diagrams. For $m \in \mathbb{Z}_+$, consider a vector space with basis $\{\varepsilon_{-m}, \ldots, \varepsilon_{-1}\} \cup \{\varepsilon_r | r \in \frac{1}{2}\mathbb{N}\}$ and a symmetric bilinear form $(\cdot|\cdot)$ given by

$$(\varepsilon_r | \varepsilon_s) = (-1)^{2r} \delta_{rs}, \qquad r, s \in \{-m, \ldots, -1\} \cup \frac{1}{2}\mathbb{N}.$$

We introduce the following notation for roots which we shall need shortly:

(6.2)
$$\alpha_\times := \varepsilon_{-1} - \varepsilon_{1/2}, \quad \alpha_j := \varepsilon_j - \varepsilon_{j+1}, \quad -m \leq j \leq -2,$$
$$\beta_\times := \varepsilon_{-1} - \varepsilon_1, \quad \alpha_r := \varepsilon_r - \varepsilon_{r+1/2}, \quad \beta_r := \varepsilon_r - \varepsilon_{r+1}, \quad r \in \frac{1}{2}\mathbb{N}.$$

We define the **tail diagrams** to be the following three Dynkin diagrams $\mathfrak{T}_n$, $\overline{\mathfrak{T}}_n$, and $\widetilde{\mathfrak{T}}_n$ with prescribed fundamental systems denoted by $\Pi(\mathfrak{T}_n)$, $\Pi(\overline{\mathfrak{T}}_n)$, and $\Pi(\widetilde{\mathfrak{T}}_n)$, for $n \in \mathbb{N}$:

$\mathfrak{T}_n$: $\bigcirc\!\!-\!\!\bigcirc\!\!-\!\!\bigcirc\!\!-\cdots-\!\!\bigcirc\!\!-\!\!\bigcirc$
$\phantom{\mathfrak{T}_n:}\ \ \beta_\times\ \ \beta_1\ \ \beta_2\ \ \ \ \ \ \ \beta_{n-2}\ \beta_{n-1}$

$\overline{\mathfrak{T}}_n$: $\otimes\!\!-\!\!\bigcirc\!\!-\!\!\bigcirc\!\!-\cdots-\!\!\bigcirc\!\!-\!\!\bigcirc$
$\phantom{\overline{\mathfrak{T}}_n:}\ \ \alpha_\times\ \ \beta_{1/2}\ \beta_{3/2}\ \ \ \ \ \beta_{n-5/2}\ \beta_{n-3/2}$

$\widetilde{\mathfrak{T}}_n$: $\otimes\!\!-\!\!\otimes\!\!-\!\!\otimes\!\!-\cdots-\!\!\otimes\!\!-\!\!\otimes$
$\phantom{\widetilde{\mathfrak{T}}_n:}\ \ \alpha_\times\ \ \alpha_{1/2}\ \ \alpha_1\ \ \ \ \ \ \alpha_{n-1}\ \alpha_{n-1/2}$

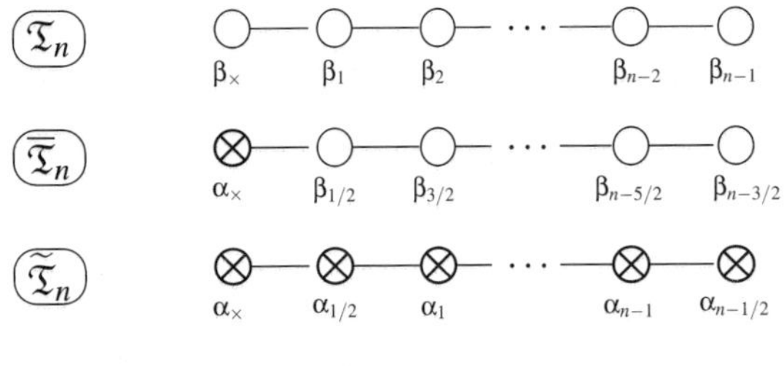

Corresponding to these three diagrams we have the Lie superalgebras $\mathfrak{gl}(1+n)$, $\mathfrak{gl}(1|n)$, and $\mathfrak{gl}(1+n|n)$. For $n = \infty$, we have analogous diagrams that are the Dynkin diagrams of the corresponding infinite-rank Lie superalgebras.

We shall choose another Dynkin diagram $\boxed{\mathfrak{k}}$, with fundamental system $\Pi(\mathfrak{k})$, called a **head diagram**, to connect with one of the three tail diagrams to produce the following three new Dynkin diagrams, which will be called the **master diagrams** ($n \in \mathbb{N} \cup \{\infty\}$):

$$(6.3) \qquad \boxed{\mathfrak{k}}\!-\!\!-\!\boxed{\mathfrak{T}_n} \qquad\qquad \boxed{\mathfrak{k}}\!-\!\!-\!\boxed{\overline{\mathfrak{T}}_n} \qquad\qquad \boxed{\mathfrak{k}}\!-\!\!-\!\boxed{\widetilde{\mathfrak{T}}_n}$$

The three master diagrams will be denoted by $\boxed{\mathfrak{g}_n}$, $\boxed{\overline{\mathfrak{g}}_n}$, and $\boxed{\widetilde{\mathfrak{g}}_n}$ respectively, and the associated Lie superalgebras $\mathfrak{g}_n, \overline{\mathfrak{g}}_n$, and $\widetilde{\mathfrak{g}}_n$ will be introduced subsequently. We denote the fundamental systems corresponding to the three master diagrams by

$$(6.4) \qquad \Pi_n := \Pi(\mathfrak{k}) \sqcup \Pi(\mathfrak{T}_n), \quad \overline{\Pi}_n := \Pi(\mathfrak{k}) \sqcup \Pi(\overline{\mathfrak{T}}_n), \quad \widetilde{\Pi}_n := \Pi(\mathfrak{k}) \sqcup \Pi(\widetilde{\mathfrak{T}}_n).$$

When $n = \infty$, we shall make it a convention to drop the subscript n, and denote these master diagrams, fundamental systems, and Lie superalgebras by $\boxed{\mathfrak{g}}$, $\boxed{\overline{\mathfrak{g}}}$, and $\boxed{\widetilde{\mathfrak{g}}}$, Π, $\overline{\Pi}$, and $\widetilde{\Pi}$, $\mathfrak{g}$, $\overline{\mathfrak{g}}$, and $\widetilde{\mathfrak{g}}$, respectively.

The head diagram $\boxed{\mathfrak{k}}$ used in this book is always chosen to be one of the Dynkin diagrams $\boxed{\mathfrak{k}^{\mathfrak{x}}}$ defined below. Accordingly, we will add the superscript $\mathfrak{x}$ to the general notations in Section 6.1.1 to write $\Pi_n^{\mathfrak{x}}$, $\boxed{\mathfrak{g}_n^{\mathfrak{x}}}$, $\mathfrak{g}_n^{\mathfrak{x}}$, and so on, when it is needed to specify the type $\mathfrak{x}$. For $\mathfrak{x} = \mathfrak{a}, \mathfrak{b}, \mathfrak{b}^\bullet, \mathfrak{c}, \mathfrak{d}$ and $m \geq 1$, introduce the Lie (super)algebras $\mathfrak{k}^{\mathfrak{x}}$ with Dynkin diagrams $\boxed{\mathfrak{k}^{\mathfrak{x}}}$ and prescribed fundamental systems $\Pi(\mathfrak{k}^{\mathfrak{x}})$ as follows (see (6.2) for notation of α_j):

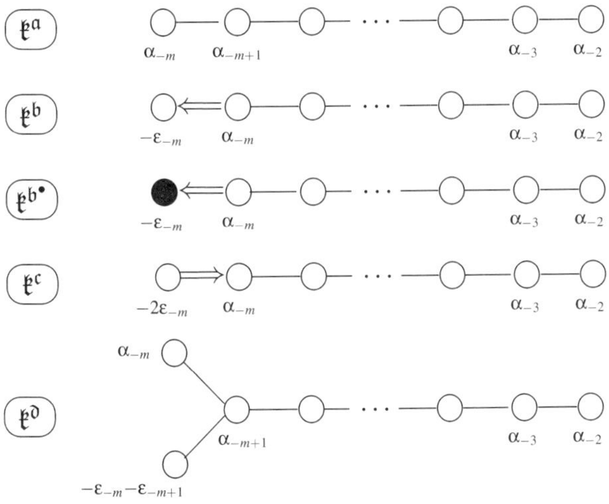

We have the following identifications of Lie algebras: $\mathfrak{k}^{\mathfrak{a}} = \mathfrak{gl}(m)$, $\mathfrak{k}^{\mathfrak{b}} = \mathfrak{so}(2m+1)$, $\mathfrak{k}^{\mathfrak{b}^\bullet} = \mathfrak{osp}(1|2m)$, $\mathfrak{k}^{\mathfrak{c}} = \mathfrak{sp}(2m)$, and $\mathfrak{k}^{\mathfrak{d}} = \mathfrak{so}(2m)$.

Remark 6.1. The Lie superalgebra $\mathfrak{osp}(1|2m)$ behaves like a finite-dimensional simple Lie algebra in the sense that every finite-dimensional $\mathfrak{osp}(1|2m)$-module is completely reducible by Corollary 2.33. Furthermore, according to Section 2.2.6, the linkage is controlled entirely by the Weyl group, just like for finite-dimensional simple Lie algebras. Thus, we shall slightly abuse the terminology and refer to $\mathfrak{osp}(1|2m)$ also as a classical Lie algebra in this chapter.

6.1.2. The index sets. Let us introduce some notation and conventions for index sets, which will be needed for defining the Lie superalgebras $\mathfrak{g}^{\mathfrak{x}}$, $\overline{\mathfrak{g}}^{\mathfrak{x}}$, and $\widetilde{\mathfrak{g}}^{\mathfrak{x}}$ associated to the master diagrams (6.3).

For $m \in \mathbb{Z}_+$, we introduce the following totally ordered set $\widetilde{\mathbb{I}}_m$:

$$\cdots < \overline{\frac{3}{2}} < \overline{1} < \overline{\frac{1}{2}} < \underbrace{\overline{-1} < \cdots < \overline{-m}}_{m} < \overline{0} < \underbrace{-m < \cdots < -1}_{m} < \frac{1}{2} < 1 < \frac{3}{2} < \cdots$$

We further introduce the following subsets of $\widetilde{\mathbb{I}}_m$:

$$\mathbb{I}_m := \Big\{\underbrace{\overline{-1},\ldots,\overline{-m}}_{m},\overline{0},\underbrace{-m,\ldots,-1}_{m}\Big\} \cup \{\overline{1},\overline{2},\overline{3},\ldots\} \cup \{1,2,3,\ldots\},$$

$$\overline{\mathbb{I}}_m := \Big\{\underbrace{\overline{-1},\ldots,\overline{-m}}_{m},\overline{0},\underbrace{-m,\ldots,-1}_{m}\Big\} \cup \Big\{\overline{\frac{1}{2}},\overline{\frac{3}{2}},\overline{\frac{5}{2}},\ldots\Big\} \cup \Big\{\frac{1}{2},\frac{3}{2},\frac{5}{2},\ldots\Big\},$$

$$\widetilde{\mathbb{I}}_m^+ := \Big\{-m,\ldots,-1,\frac{1}{2},1,\frac{3}{2},2,\ldots\Big\}.$$

For $\mathbb{X} = \mathbb{I}_m, \overline{\mathbb{I}}_m$, or $\widetilde{\mathbb{I}}_m$, define

$$\mathbb{X}^\times := \mathbb{X} \setminus \{\overline{0}\}, \quad \mathbb{X}^+ := \mathbb{X} \cap \widetilde{\mathbb{I}}_m^+.$$

6.1.3. Infinite-rank Lie superalgebras. We shall provide explicit matrix realizations of infinite-rank Lie superalgebras $\mathfrak{g}^{\mathfrak{x}}$, $\overline{\mathfrak{g}}^{\mathfrak{x}}$, and $\widetilde{\mathfrak{g}}^{\mathfrak{x}}$, for $\mathfrak{x} = \mathfrak{a}, \mathfrak{b}, \mathfrak{b}^\bullet, \mathfrak{c}, \mathfrak{d}$. Actually $\mathfrak{g}^{\mathfrak{x}}$, for $\mathfrak{x} = \mathfrak{a}, \mathfrak{b}, \mathfrak{c}, \mathfrak{d}$, is a Lie algebra.

Lie superalgebras of type $\mathfrak{a}$. For $m \in \mathbb{Z}_+$, let $\widetilde{V}_m$ be the infinite-dimensional superspace over $\mathbb{C}$ with ordered basis $\{v_i | i \in \widetilde{\mathbb{I}}_m\}$, whose $\mathbb{Z}_2$-grading is specified as follows:

$$|v_r| = |v_{\overline{r}}| = \overline{0} \quad (r \in \mathbb{Z} \setminus \{0\}), \qquad |v_s| = |v_{\overline{s}}| = \overline{1} \quad (s \in \frac{1}{2} + \mathbb{Z}_+).$$

The parity of the vector $v_{\overline{0}}$ is to be specified. With respect to this basis, a linear map on $\widetilde{V}_m$ may be identified with a complex matrix $(a_{rs})_{r,s \in \widetilde{\mathbb{I}}_m}$. Let $\mathfrak{gl}(\widetilde{V}_m)$ denote the Lie superalgebra consisting of $(a_{rs})_{r,s \in \widetilde{\mathbb{I}}_m}$ with $a_{rs} = 0$ for all but finitely many a_{rs}'s. Denote as usual by $E_{rs} \in \mathfrak{gl}(\widetilde{V}_m)$ the elementary matrix with 1 at the rth row and sth column and zero elsewhere.

The superspaces V_m and $\overline{V}_m$ are defined to be the subspaces of $\widetilde{V}_m$ with ordered basis $\{v_i\}$ indexed by $\mathbb{I}_m$ and $\overline{\mathbb{I}}_m$, respectively. The subspaces of V_m, $\overline{V}_m$, and $\widetilde{V}_m$ with basis vectors v_i, with i indexed by $\mathbb{I}_m^\times$, $\overline{\mathbb{I}}_m^\times$, and $\widetilde{\mathbb{I}}_m^\times$, are denoted by $V_m^\times$, $\overline{V}_m^\times$ and $\widetilde{V}_m^\times$, respectively. Similarly, the subspaces with basis vectors v_i, for i indexed by

$\mathbb{I}_m^+$, $\overline{\mathbb{I}}_m^+$ and $\widetilde{\mathbb{I}}_m^+$, are denoted by V_m^+, $\overline{V}_m^+$ and $\widetilde{V}_m^+$, respectively. We summarize these vector superspaces together with the index sets for their bases in Table 1 below:

Table 1

Superspaces	V_m	$\overline{V}_m$	$\widetilde{V}_m$	$V_m^\times$	$\overline{V}_m^\times$	$\widetilde{V}_m^\times$	V_m^+	$\overline{V}_m^+$	$\widetilde{V}_m^+$
Index sets	$\mathbb{I}_m$	$\overline{\mathbb{I}}_m$	$\widetilde{\mathbb{I}}_m$	$\mathbb{I}_m^\times$	$\overline{\mathbb{I}}_m^\times$	$\widetilde{\mathbb{I}}_m^\times$	$\mathbb{I}_m^+$	$\overline{\mathbb{I}}_m^+$	$\widetilde{\mathbb{I}}_m^+$

Let W be one of the superspaces $\widetilde{V}_m, \widetilde{V}_m^\times, \widetilde{V}_m^+, V_m, V_m^\times, V_m^+, \overline{V}_m, \overline{V}_m^\times$ or $\overline{V}_m^+$, regarded as a subspace of $\widetilde{V}_m$. Then W gives rise to the Lie superalgebra $\mathfrak{gl}(W)$ as a subalgebra of $\mathfrak{gl}(\widetilde{V}_m)$. The **standard Cartan subalgebra** of $\mathfrak{gl}(W)$ is spanned by the basis $\{E_{rr}\}$, with corresponding dual basis $\{\varepsilon_r\}$, where r runs over the index set corresponding to W. The **standard Borel subalgebra** of $\mathfrak{gl}(W)$ is spanned by E_{rs}, with $r \leq s$.

The fundamental systems corresponding to the standard Borel subalgebras of the Lie superalgebras $\mathfrak{gl}(\widetilde{V}_m^+)$, $\mathfrak{gl}(V_m^+)$, and $\mathfrak{gl}(\overline{V}_m^+)$ are precisely $\widetilde{\Pi}^{\mathfrak{a}}$, $\Pi^{\mathfrak{a}}$, and $\overline{\Pi}^{\mathfrak{a}}$, respectively, and the corresponding Dynkin diagrams are the master diagrams $\boxed{\widetilde{\mathfrak{g}}^{\mathfrak{a}}}$, $\boxed{\mathfrak{g}^{\mathfrak{a}}}$ and $\boxed{\overline{\mathfrak{g}}^{\mathfrak{a}}}$, respectively. Therefore, $\mathfrak{gl}(\widetilde{V}_m^+)$, $\mathfrak{gl}(V_m^+)$, and $\mathfrak{gl}(\overline{V}_m^+)$ are matrix realizations of the Lie superalgebras $\widetilde{\mathfrak{g}}^{\mathfrak{a}}$, $\mathfrak{g}^{\mathfrak{a}}$, and $\overline{\mathfrak{g}}^{\mathfrak{a}}$, respectively. We summarize this in Table 2 below.

Table 2. Matrix forms for Lie superalgebras of type $\mathfrak{a}$

Lie superalgebras	$\widetilde{\mathfrak{g}}^{\mathfrak{a}}$	$\mathfrak{g}^{\mathfrak{a}}$	$\overline{\mathfrak{g}}^{\mathfrak{a}}$
Matrix forms	$\mathfrak{gl}(\widetilde{V}_m^+)$	$\mathfrak{gl}(V_m^+)$	$\mathfrak{gl}(\overline{V}_m^+)$

Associated to the fundamental systems $\Pi^{\mathfrak{a}}$, $\overline{\Pi}^{\mathfrak{a}}$, and $\widetilde{\Pi}^{\mathfrak{a}}$, we have the following positive systems:

$$
\begin{aligned}
{}^{\mathfrak{a}}\widetilde{\Phi}^+ &= \{\varepsilon_r - \varepsilon_s | r < s \ (r, s \in \widetilde{\mathbb{I}}_m^+)\}, \\
{}^{\mathfrak{a}}\Phi^+ &= \{\varepsilon_i - \varepsilon_j | i < j \ (i, j \in \mathbb{I}_m^+)\}, \\
{}^{\mathfrak{a}}\overline{\Phi}^+ &= \{\varepsilon_r - \varepsilon_s | r < s \ (r, s \in \overline{\mathbb{I}}_m^+)\}.
\end{aligned}
$$

Lie superalgebras of types $\mathfrak{b}^\bullet, \mathfrak{c}$. We set $|v_{\bar{0}}| = \bar{1}$. For $m \in \mathbb{Z}_+$, define a non-degenerate skew-supersymmetric bilinear form $(\cdot|\cdot)$ on the superspace $\widetilde{V}_m$ by

$$
\begin{aligned}
&(v_r|v_s) = (v_{\bar{r}}|v_{\bar{s}}) = 0, \quad (v_r|v_{\bar{s}}) = \delta_{rs} = -(-1)^{|v_r|\cdot|v_s|}(v_{\bar{s}}|v_r), \quad r, s \in \widetilde{\mathbb{I}}_m^+, \\
&(v_{\bar{0}}|v_{\bar{0}}) = 1, \quad (v_{\bar{0}}|v_r) = (v_{\bar{0}}|v_{\bar{r}}) = 0, \quad r \in \widetilde{\mathbb{I}}_m^+.
\end{aligned}
\tag{6.5}
$$

We obtain non-degenerate skew-supersymmetric bilinear forms by restriction, which are again denoted by $(\cdot|\cdot)$, to the subspaces $\widetilde{V}_m^{\times}$, V_m, $V_m^{\times}$, $\overline{V}_m$ and $\overline{V}_m^{\times}$.

Let W be one of the superspaces $\widetilde{V}_m$, $\widetilde{V}_m^{\times}$, V_m, $V_m^{\times}$, $\overline{V}_m$ or $\overline{V}_m^{\times}$, respectively. Recall from Section 1.1.3 that the Lie superalgebra $\mathfrak{spo}(W)$ is the subalgebra of $\mathfrak{gl}(W)$ preserving the form defined in (6.5). That is, for $\varepsilon \in \mathbb{Z}_2$,

$$\mathfrak{spo}(W)_\varepsilon = \{T \in \mathfrak{gl}(W)_\varepsilon \mid (Tv|w) = -(-1)^{\varepsilon|v|}(v|Tw), \forall v, w \in W\}.$$

The **standard Cartan subalgebra** of $\mathfrak{spo}(W)$ has a basis given by

$$E_r := E_{rr} - E_{\overline{r},\overline{r}},$$

where r runs over the index sets $\widetilde{\mathbb{I}}_m^+$, $\widetilde{\mathbb{I}}_m^+$, $\mathbb{I}_m^+$, $\mathbb{I}_m^+$, $\overline{\mathbb{I}}_m^+$, $\overline{\mathbb{I}}_m^+$, respectively. We denote the corresponding dual basis by $\{\varepsilon_r\}$. The **standard Borel subalgebra** of $\mathfrak{spo}(W)$ is obtained by taking the intersection of $\mathfrak{spo}(W)$ with the standard Borel subalgebra of $\mathfrak{gl}(W)$. One checks in a straightforward fashion that the fundamental systems associated to these standard Borel subalgebras are precisely $\widetilde{\Pi}^{\mathfrak{b}^\bullet}$, $\widetilde{\Pi}^{\mathfrak{c}}$, $\Pi^{\mathfrak{b}^\bullet}$, $\Pi^{\mathfrak{c}}$, $\overline{\Pi}^{\mathfrak{b}^\bullet}$, $\overline{\Pi}^{\mathfrak{c}}$, respectively. Thus, we have obtained matrix realizations of the Lie superalgebras, for $m \geq 1$, in Table 3 below.

Table 3. Matrix forms for Lie superalgebras of types $\mathfrak{b}^\bullet, \mathfrak{c}$

Superalgebras	$\widetilde{\mathfrak{g}}^{\mathfrak{b}^\bullet}$	$\mathfrak{g}^{\mathfrak{b}^\bullet}$	$\overline{\mathfrak{g}}^{\mathfrak{b}^\bullet}$	$\widetilde{\mathfrak{g}}^{\mathfrak{c}}$	$\mathfrak{g}^{\mathfrak{c}}$	$\overline{\mathfrak{g}}^{\mathfrak{c}}$
Matrix forms	$\mathfrak{spo}(\widetilde{V}_m)$	$\mathfrak{spo}(V_m)$	$\mathfrak{spo}(\overline{V}_m)$	$\mathfrak{spo}(\widetilde{V}_m^{\times})$	$\mathfrak{sp}(V_m^{\times})$	$\mathfrak{spo}(\overline{V}_m^{\times})$

Associated to the fundamental systems $\Pi^{\mathfrak{x}}$, $\overline{\Pi}^{\mathfrak{x}}$, and $\widetilde{\Pi}^{\mathfrak{x}}$, for $\mathfrak{x} = \mathfrak{b}^\bullet, \mathfrak{c}$, we have the following positive systems:

$$^{\mathfrak{b}^\bullet}\widetilde{\Phi}^+ = \{\pm\varepsilon_r - \varepsilon_s | r < s \ (r,s \in \widetilde{\mathbb{I}}_m^+)\} \cup \{-2\varepsilon_i \ (i \in \mathbb{I}_m^+)\} \cup \{-\varepsilon_r \ (r \in \widetilde{\mathbb{I}}_m^+)\},$$

$$^{\mathfrak{c}}\widetilde{\Phi}^+ = \{\pm\varepsilon_r - \varepsilon_s | r < s \ (r,s \in \widetilde{\mathbb{I}}_m^+)\} \cup \{-2\varepsilon_i \ (i \in \mathbb{I}_m^+)\},$$

$$^{\mathfrak{b}^\bullet}\Phi^+ = \{\pm\varepsilon_i - \varepsilon_j | i < j \ (i,j \in \mathbb{I}_m^+)\} \cup \{-\varepsilon_i, -2\varepsilon_i \ (i \in \mathbb{I}_m^+)\},$$

$$^{\mathfrak{c}}\Phi^+ = \{\pm\varepsilon_i - \varepsilon_j | i < j \ (i,j \in \mathbb{I}_m^+)\} \cup \{-2\varepsilon_i \ (i \in \mathbb{I}_m^+)\},$$

$$^{\mathfrak{b}^\bullet}\overline{\Phi}^+ = \{\pm\varepsilon_r - \varepsilon_s | r < s \ (r,s \in \overline{\mathbb{I}}_m^+)\} \cup \{-2\varepsilon_i \ (-m \leq i \leq -1)\} \cup \{-\varepsilon_r \ (r \in \overline{\mathbb{I}}_m^+)\},$$

$$^{\mathfrak{c}}\overline{\Phi}^+ = \{\pm\varepsilon_r - \varepsilon_s | r < s \ (r,s \in \overline{\mathbb{I}}_m^+)\} \cup \{-2\varepsilon_i \ (-m \leq i \leq -1)\}.$$

Lie superalgebras of types $\mathfrak{b}, \mathfrak{d}$. Now we set $|v_{\overline{0}}| = \bar{0}$. Define a supersymmetric bilinear form $(\cdot|\cdot)$ on the superspace $\widetilde{V}_m$ by

$$(v_r|v_s) = (v_{\overline{r}}|v_{\overline{s}}) = 0, \quad (v_r|v_{\overline{s}}) = \delta_{rs} = (-1)^{|v_r|\cdot|v_s|}(v_{\overline{s}}|v_r), \quad r,s \in \widetilde{\mathbb{I}}_m^+,$$
$$(v_{\overline{0}}|v_{\overline{0}}) = 1, \quad (v_{\overline{0}}|v_r) = (v_{\overline{0}}|v_{\overline{r}}) = 0, \quad r \in \widetilde{\mathbb{I}}_m^+. \tag{6.6}$$

By restriction, we obtain non-degenerate supersymmetric bilinear forms, which will also be denoted by $(\cdot|\cdot)$ on the subspaces $\widetilde{V}_m^{\times}$, V_m, $V_m^{\times}$, $\overline{V}_m$ and $\overline{V}_m^{\times}$.

Let W be one of the spaces $\widetilde{V}_m$, $\widetilde{V}_m^{\times}$, V_m, $V_m^{\times}$, $\overline{V}_m$ and $\overline{V}_m^{\times}$, respectively. Recall from Section 1.1.3 that the Lie superalgebra $\mathfrak{osp}(W)$ is the subalgebra of $\mathfrak{gl}(W)$ preserving the form given by (6.6). The **standard Cartan subalgebra** of $\mathfrak{osp}(W)$ has the basis $\{E_r\}$, with corresponding dual basis $\{\varepsilon_r\}$, where r runs over the index sets $\widetilde{\mathbb{I}}_m^+$, $\widetilde{\mathbb{I}}_m^+$, $\mathbb{I}_m^+$, $\mathbb{I}_m^+$, $\overline{\mathbb{I}}_m^+$, $\overline{\mathbb{I}}_m^+$, respectively. As before, the **standard Borel subalgebra** of $\mathfrak{osp}(W)$ is obtained by intersecting $\mathfrak{osp}(W)$ with the standard Borel subalgebra of $\mathfrak{gl}(W)$. One computes that the associated fundamental systems are precisely $\widetilde{\Pi}^{\mathfrak{b}}$, $\widetilde{\Pi}^{\mathfrak{d}}$, $\Pi^{\mathfrak{b}}$, $\Pi^{\mathfrak{d}}$, $\overline{\Pi}^{\mathfrak{b}}$, $\overline{\Pi}^{\mathfrak{d}}$, respectively. Thus, we have obtained the following matrix realizations of Lie superalgebras, for $m \geq 1$, in Table 4.

Table 4. Matrix forms for Lie superalgebras of types $\mathfrak{b}, \mathfrak{d}$

Superalgebras	$\widetilde{\mathfrak{g}}^{\mathfrak{b}}$	$\mathfrak{g}^{\mathfrak{b}}$	$\overline{\mathfrak{g}}^{\mathfrak{b}}$	$\widetilde{\mathfrak{g}}^{\mathfrak{d}}$	$\mathfrak{g}^{\mathfrak{d}}$	$\overline{\mathfrak{g}}^{\mathfrak{d}}$
Matrix forms	$\mathfrak{osp}(\widetilde{V}_m)$	$\mathfrak{so}(V_m)$	$\mathfrak{osp}(\overline{V}_m)$	$\mathfrak{osp}(\widetilde{V}_m^{\times})$	$\mathfrak{so}(V_m^{\times})$	$\mathfrak{osp}(\overline{V}_m^{\times})$

Associated to the fundamental systems $\widetilde{\Pi}^{\mathfrak{x}}$, $\Pi^{\mathfrak{x}}$, and $\overline{\Pi}^{\mathfrak{x}}$, for $\mathfrak{x} = \mathfrak{b}, \mathfrak{d}$, we have the following positive systems:

$$
\begin{aligned}
{}^{\mathfrak{b}}\widetilde{\Phi}^+ &= \{\pm\varepsilon_r - \varepsilon_s | r < s \; (r, s \in \widetilde{\mathbb{I}}_m^+)\} \cup \{-2\varepsilon_s \; (s \in \overline{\mathbb{I}}_0^+)\} \cup \{-\varepsilon_r \; (r \in \widetilde{\mathbb{I}}_m^+)\}, \\
{}^{\mathfrak{d}}\widetilde{\Phi}^+ &= \{\pm\varepsilon_r - \varepsilon_s | r < s \; (r, s \in \widetilde{\mathbb{I}}_m^+)\} \cup \{-2\varepsilon_s \; (s \in \overline{\mathbb{I}}_0^+)\}, \\
{}^{\mathfrak{b}}\Phi^+ &= \{\pm\varepsilon_i - \varepsilon_j | i < j \; (i, j \in \mathbb{I}_m^+)\} \cup \{-\varepsilon_i \; (i \in \mathbb{I}_m^+)\}, \\
{}^{\mathfrak{d}}\Phi^+ &= \{\pm\varepsilon_i - \varepsilon_j | i < j \; (i, j \in \mathbb{I}_m^+)\}, \\
{}^{\mathfrak{b}}\overline{\Phi}^+ &= \{\pm\varepsilon_r - \varepsilon_s | r < s \; (r, s \in \overline{\mathbb{I}}_m^+)\} \cup \{-2\varepsilon_s \; (s \in \overline{\mathbb{I}}_0^+)\} \cup \{-\varepsilon_r \; (r \in \overline{\mathbb{I}}_m^+)\}, \\
{}^{\mathfrak{d}}\overline{\Phi}^+ &= \{\pm\varepsilon_r - \varepsilon_s | r < s \; (r, s \in \overline{\mathbb{I}}_m^+)\} \cup \{-2\varepsilon_s \; (s \in \overline{\mathbb{I}}_0^+)\}.
\end{aligned}
$$

6.1.4. The case of $m = 0$. Let $\mathfrak{x} = \mathfrak{a}, \mathfrak{b}, \mathfrak{b}^{\bullet}, \mathfrak{c}, \mathfrak{d}$. The sets ${}^{\mathfrak{x}}\widetilde{\Phi}^+$, ${}^{\mathfrak{x}}\Phi^+$, and ${}^{\mathfrak{x}}\overline{\Phi}^+$ for $m = 0$ still make sense, and they are the positive systems for the Lie superalgebras $\widetilde{\mathfrak{g}}^{\mathfrak{x}}$, $\mathfrak{g}^{\mathfrak{x}}$, and $\overline{\mathfrak{g}}^{\mathfrak{x}}$, whose Dynkin diagrams and fundamental systems are as follows. Some of the Dynkin diagrams in the case of $m = 0$ differ somewhat from the counterparts in the case when $m \geq 1$, and this is why we treat the two cases separately. For $m = 0$, the corresponding Lie superalgebras are realized similarly as in Section 6.1.3, and we shall use the same notation as before.

$\widetilde{\mathfrak{g}}^{\mathfrak{a}}$

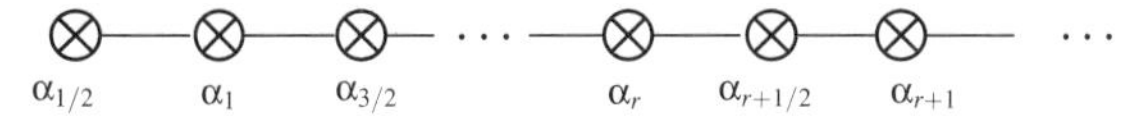

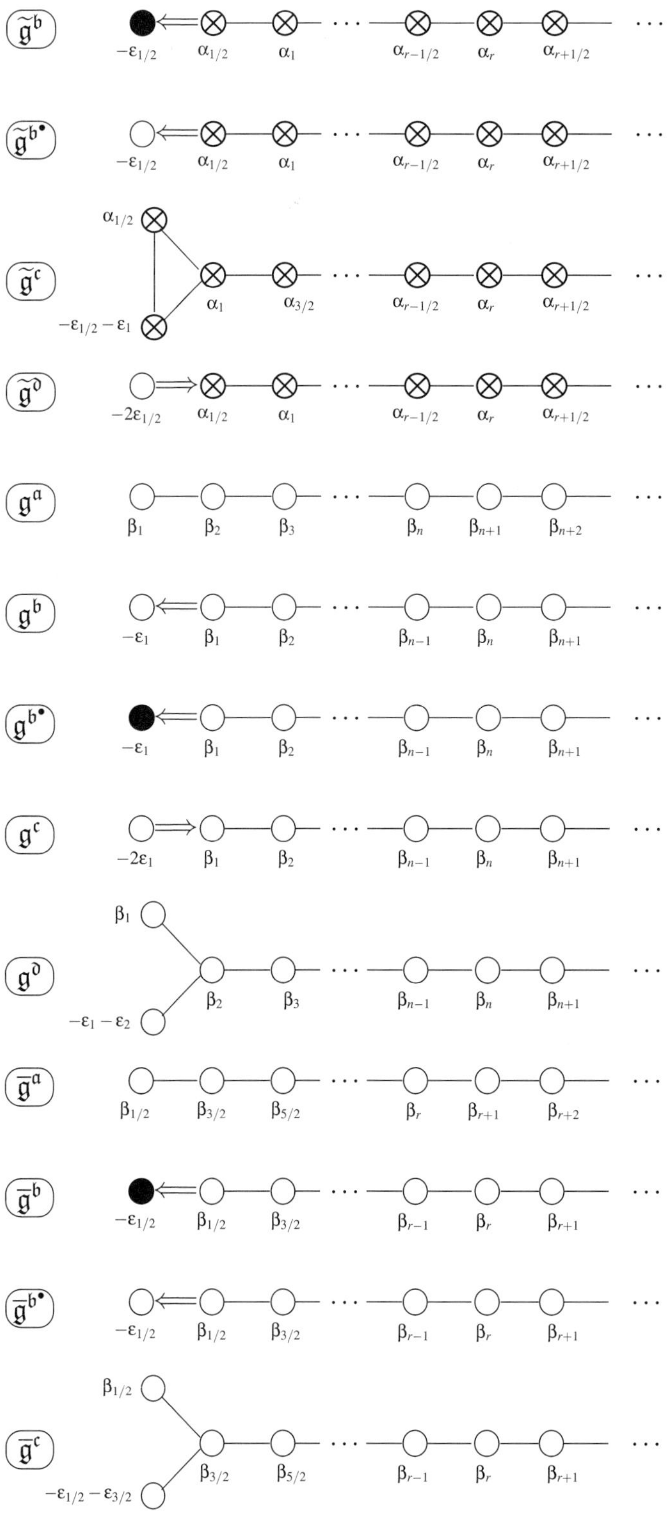
$\widetilde{\mathfrak{g}}^{\mathfrak{b}}$
$-\epsilon_{1/2}$ $\alpha_{1/2}$ α_1 $\alpha_{r-1/2}$ α_r $\alpha_{r+1/2}$
$\widetilde{\mathfrak{g}}^{\mathfrak{b}^\bullet}$
$-\epsilon_{1/2}$ $\alpha_{1/2}$ α_1 $\alpha_{r-1/2}$ α_r $\alpha_{r+1/2}$
$\widetilde{\mathfrak{g}}^{\mathfrak{c}}$
$\alpha_{1/2}$ $-\epsilon_{1/2}-\epsilon_1$ α_1 $\alpha_{3/2}$ $\alpha_{r-1/2}$ α_r $\alpha_{r+1/2}$
$\widetilde{\mathfrak{g}}^{\mathfrak{d}}$
$-2\epsilon_{1/2}$ $\alpha_{1/2}$ α_1 $\alpha_{r-1/2}$ α_r $\alpha_{r+1/2}$
$\mathfrak{g}^{\mathfrak{a}}$
β_1 β_2 β_3 β_n β_{n+1} β_{n+2}
$\mathfrak{g}^{\mathfrak{b}}$
$-\epsilon_1$ β_1 β_2 β_{n-1} β_n β_{n+1}
$\mathfrak{g}^{\mathfrak{b}^\bullet}$
$-\epsilon_1$ β_1 β_2 β_{n-1} β_n β_{n+1}
$\mathfrak{g}^{\mathfrak{c}}$
$-2\epsilon_1$ β_1 β_2 β_{n-1} β_n β_{n+1}
$\mathfrak{g}^{\mathfrak{d}}$
β_1 $-\epsilon_1-\epsilon_2$ β_2 β_3 β_{n-1} β_n β_{n+1}
$\overline{\mathfrak{g}}^{\mathfrak{a}}$
$\beta_{1/2}$ $\beta_{3/2}$ $\beta_{5/2}$ β_r β_{r+1} β_{r+2}
$\overline{\mathfrak{g}}^{\mathfrak{b}}$
$-\epsilon_{1/2}$ $\beta_{1/2}$ $\beta_{3/2}$ β_{r-1} β_r β_{r+1}
$\overline{\mathfrak{g}}^{\mathfrak{b}^\bullet}$
$-\epsilon_{1/2}$ $\beta_{1/2}$ $\beta_{3/2}$ β_{r-1} β_r β_{r+1}
$\overline{\mathfrak{g}}^{\mathfrak{c}}$
$\beta_{1/2}$ $-\epsilon_{1/2}-\epsilon_{3/2}$ $\beta_{3/2}$ $\beta_{5/2}$ β_{r-1} β_r β_{r+1}

$\boxed{\overline{\mathfrak{g}}^{\mathfrak{d}}}$ 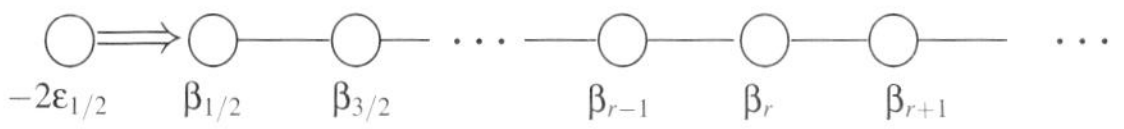

We have the following identifications of Dynkin diagrams when $m=0$:

$$\boxed{\mathfrak{g}^{\mathfrak{a}}} = \boxed{\overline{\mathfrak{g}}^{\mathfrak{a}}}, \boxed{\mathfrak{g}^{\mathfrak{b}}} = \boxed{\overline{\mathfrak{g}}^{\mathfrak{b}^\bullet}}, \boxed{\mathfrak{g}^{\mathfrak{b}^\bullet}} = \boxed{\overline{\mathfrak{g}}^{\mathfrak{b}}}, \boxed{\mathfrak{g}^{\mathfrak{c}}} = \boxed{\overline{\mathfrak{g}}^{\mathfrak{d}}}, \boxed{\mathfrak{g}^{\mathfrak{d}}} = \boxed{\overline{\mathfrak{g}}^{\mathfrak{c}}}.$$

6.1.5. Finite-dimensional Lie superalgebras. Fix $m \in \mathbb{Z}_+$. Let W be one of the superspaces $\widetilde{V}_m$, $\overline{V}_m$, V_m, $\widetilde{V}^+_m$, $\overline{V}^+_m$, V^+_m, $\widetilde{V}^\times_m$, $\overline{V}^\times_m$, or $V^\times_m$. For $n \in \mathbb{N}$, let W_n stand for the subspace of W spanned by the vectors $v_r \in W$ with $\overline{n} \le r \le n$. We consider the Lie superalgebras $\mathfrak{gl}(W_n)$, for $W = \widetilde{V}^+_m, \overline{V}^+_m, V^+_m$, and the Lie superalgebras $\mathfrak{spo}(W_n)$ and $\mathfrak{osp}(W_n)$, for $W = \widetilde{V}_m, \widetilde{V}^\times_m, V_m, V^\times_m, \overline{V}_m, \overline{V}^\times_m$. As in the case of $n=\infty$, these provide matrix realizations of Lie superalgebras $\mathfrak{g}^{\mathfrak{x}}_n$, $\overline{\mathfrak{g}}^{\mathfrak{x}}_n$, and $\widetilde{\mathfrak{g}}^{\mathfrak{x}}_n$, whose associated Dynkin diagrams $\boxed{\mathfrak{g}^{\mathfrak{x}}_n}$, $\boxed{\overline{\mathfrak{g}}^{\mathfrak{x}}_n}$, and $\boxed{\widetilde{\mathfrak{g}}^{\mathfrak{x}}_n}$ are given in (6.3) (see Table 5 below). Note that $\mathfrak{g}^{\mathfrak{x}}_n$ is actually a Lie algebra, for $\mathfrak{x} = \mathfrak{a}, \mathfrak{b}, \mathfrak{c}, \mathfrak{d}$.

Table 5. Identifications for finite-dimensional Lie (super)algebras

Type $\mathfrak{x}$	$\mathfrak{g}^{\mathfrak{x}}_n$	$\overline{\mathfrak{g}}^{\mathfrak{x}}_n$	$\widetilde{\mathfrak{g}}^{\mathfrak{x}}_n$
$\mathfrak{a}$	$\mathfrak{gl}(m+n)$	$\mathfrak{gl}(m\|n)$	$\mathfrak{gl}(m+n\|n)$
$\mathfrak{b}$	$\mathfrak{so}(2(m+n)+1)$	$\mathfrak{osp}(2m+1\|2n)$	$\mathfrak{osp}(2(m+n)+1\|2n)$
$\mathfrak{b}^\bullet$	$\mathfrak{osp}(1\|2(m+n))$	$\mathfrak{spo}(2m\|2n+1)$	$\mathfrak{spo}(2(m+n)\|2n+1)$
$\mathfrak{c}$	$\mathfrak{sp}(2(m+n))$	$\mathfrak{spo}(2m\|2n)$	$\mathfrak{spo}(2(m+n)\|2n)$
$\mathfrak{d}$	$\mathfrak{so}(2(m+n))$	$\mathfrak{osp}(2m\|2n)$	$\mathfrak{osp}(2(m+n)\|2n)$

6.1.6. Central extensions. We assume that $\mathfrak{x} = \mathfrak{b}, \mathfrak{b}^\bullet, \mathfrak{c}, \mathfrak{d}$ in this subsection. We shall replace the Lie superalgebras $\mathfrak{g}^{\mathfrak{x}}$, $\overline{\mathfrak{g}}^{\mathfrak{x}}$, and $\widetilde{\mathfrak{g}}^{\mathfrak{x}}$ and their finite-dimensional analogues by their central extensions, and study the representations of their central extensions instead. As these central extensions are trivial, we can easily recover the representations of $\mathfrak{g}^{\mathfrak{x}}$, $\overline{\mathfrak{g}}^{\mathfrak{x}}$, and $\widetilde{\mathfrak{g}}^{\mathfrak{x}}$ from those of their central extensions. The use of central extensions will be more conceptual and convenient for formulations of truncation functors and super duality in later sections.

Let $m \in \mathbb{Z}_+$. The supertrace Str defined in Section 1.1 makes sense for $\mathfrak{gl}(\widetilde{V}_m)$. For $X, Y \in \mathfrak{gl}(\widetilde{V}_m)$, we have

$$\mathrm{Str}(XY) = (-1)^{|X||Y|}\mathrm{Str}(YX). \tag{6.7}$$

Let $\mathfrak{J} := E_{\overline{0}\overline{0}} + \sum_{r \le \overline{\frac{1}{2}}} E_{rr}$. This allows us to define $\tau : \mathfrak{gl}(\widetilde{V}_m) \times \mathfrak{gl}(\widetilde{V}_m) \to \mathbb{C}$ by

$$\tau(A,B) := \mathrm{Str}([\mathfrak{J},A]B) = \mathrm{Str}(\mathfrak{J}AB - A\mathfrak{J}B), \quad A, B \in \mathfrak{gl}(\widetilde{V}_m).$$

It is easy to show that τ is a (trivial) 2-cocycle (see Section 5.4.1), and hence defines a central extension $\widehat{\mathfrak{gl}}(\widetilde{V}_m)$ of $\mathfrak{gl}(\widetilde{V}_m)$ by the one-dimensional center $\mathbb{C}K$. That is, we have $\widehat{\mathfrak{gl}}(\widetilde{V}_m) = \mathfrak{gl}(\widetilde{V}_m) \oplus \mathbb{C}K$ as a vector superspace with Lie bracket

$$[\widehat{X}, \widehat{Y}] = \widehat{[X,Y]} + \tau(X,Y)K,$$

where we have used $\widehat{X}$ to denote the element in $\widehat{\mathfrak{gl}}(\widetilde{V}_m)$ that corresponds to the element $X \in \mathfrak{gl}(\widetilde{V}_m)$. This central extension is trivial, since it is straightforward to check that an isomorphism φ from the Lie superalgebra $\widehat{\mathfrak{gl}}(\widetilde{V}_m)$ to the direct sum of Lie superalgebras $\mathfrak{gl}(\widetilde{V}_m) \oplus \mathbb{C}K$ is given by

$$\varphi(\widehat{X}) = X - \mathrm{Str}(\mathfrak{J}X)K, \qquad \varphi(K) = K. \tag{6.8}$$

Now for W being one of the spaces $\widetilde{V}_m^{\times}, V_m, V_m^{\times}, \overline{V}_m, \overline{V}_m^{\times}$, the restriction of τ to the subalgebras $\mathfrak{gl}(W)$ gives rise to respective central extensions, which in turn induce central extensions of the ortho-symplectic subalgebras. The central extension of $\mathfrak{g}^{\mathfrak{x}}$, $\overline{\mathfrak{g}}^{\mathfrak{x}}$, and $\widetilde{\mathfrak{g}}^{\mathfrak{x}}$ arising this way will also be denoted by $\mathfrak{g}^{\mathfrak{x}}$, $\overline{\mathfrak{g}}^{\mathfrak{x}}$, and $\widetilde{\mathfrak{g}}^{\mathfrak{x}}$, respectively, by abuse of notation. We trust that this will not cause confusion, as only these central extensions will be used in the remainder of this chapter. For $\mathfrak{x} = \mathfrak{a}$, the Lie superalgebras $\mathfrak{g}^{\mathfrak{a}}$, $\overline{\mathfrak{g}}^{\mathfrak{a}}$, and $\widetilde{\mathfrak{g}}^{\mathfrak{a}}$ are already suitable for super duality later on, and so their central extensions will not be needed.

We note that $\mathfrak{g}^{\mathfrak{x}}$ and $\overline{\mathfrak{g}}^{\mathfrak{x}}$ are naturally subalgebras of $\widetilde{\mathfrak{g}}^{\mathfrak{x}}$. The standard Cartan subalgebras of $\mathfrak{g}^{\mathfrak{x}}$, $\overline{\mathfrak{g}}^{\mathfrak{x}}$, and $\widetilde{\mathfrak{g}}^{\mathfrak{x}}$ will be denoted by $\mathfrak{h}^{\mathfrak{x}}$, $\overline{\mathfrak{h}}^{\mathfrak{x}}$, and $\widetilde{\mathfrak{h}}^{\mathfrak{x}}$, respectively. Then $\mathfrak{h}^{\mathfrak{x}}$, $\overline{\mathfrak{h}}^{\mathfrak{x}}$, or $\widetilde{\mathfrak{h}}^{\mathfrak{x}}$ has a basis $\{K, \widehat{E}_r\}$ with dual basis $\{\Lambda_0, \varepsilon_r\}$, where r runs over the index sets $\mathbb{I}_m^+$, $\overline{\mathbb{I}}_m^+$, or $\widetilde{\mathbb{I}}_m^+$, respectively. Here Λ_0 is determined by

$$\Lambda_0(K) = 1, \quad \Lambda_0(\widehat{E}_r) = 0,$$

for all admissible r in each case. The $\mathbb{C}$-span of Λ_0 and the ε_r's in the dual of $\mathfrak{h}^{\mathfrak{x}}$ (respectively, $\overline{\mathfrak{h}}^{\mathfrak{x}}$ and $\widetilde{\mathfrak{h}}^{\mathfrak{x}}$) will be denoted by $\mathfrak{h}^{\mathfrak{x}*}$ (respectively, $\overline{\mathfrak{h}}^{\mathfrak{x}*}$ and $\widetilde{\mathfrak{h}}^{\mathfrak{x}*}$), where r lies in $\mathbb{I}_m^+$ (respectively, $\overline{\mathbb{I}}_m^+$ and $\widetilde{\mathbb{I}}_m^+$).

In the case when $\mathfrak{x} = \mathfrak{a}$ it will also be advantageous to set $\widehat{E}_r \equiv E_{rr}$, for $r > \overline{0}$. For notational convenience later on, we shall declare $\Lambda_0^{\mathfrak{a}}$ to be 0.

From now on, we shall adopt a convention of dropping the superscript $\mathfrak{x}$. So for example, we shall write $\mathfrak{g}$, $\overline{\mathfrak{g}}$, and $\widetilde{\mathfrak{g}}$ for $\mathfrak{g}^{\mathfrak{x}}$, $\overline{\mathfrak{g}}^{\mathfrak{x}}$, and $\widetilde{\mathfrak{g}}^{\mathfrak{x}}$, with associated Dynkin diagrams ($\mathfrak{g}$), ($\overline{\mathfrak{g}}$), and ($\widetilde{\mathfrak{g}}$), respectively, where $\mathfrak{x}$ denotes a fixed type among $\mathfrak{a}, \mathfrak{b}, \mathfrak{b}^{\bullet}, \mathfrak{c}, \mathfrak{d}$.

6.2. The module categories

In this section, we define the parabolic Bernstein-Gelfand-Gelfand (BGG) categories of modules for the Lie superalgebras $\widetilde{\mathfrak{g}}$, $\overline{\mathfrak{g}}$, and $\mathfrak{g}$, and also their finite-rank counterparts. We introduce the truncation functors to relate the module categories of the infinite-rank Lie superalgebras to their finite-rank counterparts. It is shown that the truncation functors send irreducible and parabolic Verma modules to irreducible and parabolic Verma modules, respectively, or to zero.

6.2.1. Category of polynomial modules revisited. Recall from Section 6.1.5 that $\widetilde{V}^+_{0,k}$ is the $(k|k)$-dimensional superspace spanned by v_r, for $r = 1/2, 1, 3/2, \ldots, k-1/2, k$, for $k \in \mathbb{N}$, and $\widetilde{V}^+_0 = \widetilde{V}^+_{0,\infty}$. We introduce the short-hand notation

$$\widetilde{\mathfrak{l}}^+ = \mathfrak{gl}(\widetilde{V}^+_0), \qquad \widetilde{\mathfrak{l}}^+_k = \mathfrak{gl}(\widetilde{V}^+_{0,k}), \;\; k \in \mathbb{N}.$$

In this subsection, which is an infinite-rank counterpart of Section 3.2.6, we shall show that the category of polynomial $\widetilde{\mathfrak{l}}^+$-modules is semisimple. This will be needed in Section 6.2.2, where $\widetilde{\mathfrak{l}}^+$ appears as a direct summand of a Levi subalgebra for $\widetilde{\mathfrak{g}}$.

Let $\lambda = (\lambda_1, \lambda_2, \ldots)$ be a partition. For $j \in \mathbb{N}$, we denote

$$(6.9) \qquad \theta(\lambda)_j = \max\{\lambda_j - j, 0\}, \quad \theta(\lambda)_{j-1/2} = \max\{\lambda'_j - j + 1, 0\}.$$

One recognizes $(\theta(\lambda)_{1/2}, \theta(\lambda)_{3/2}, \ldots \mid \theta(\lambda)_1, \theta(\lambda)_2, \ldots)$ as the modified Frobenius coordinates of λ' of Example 2.56(3). Define

$$\lambda^\theta := \sum_{r \in \frac{1}{2}\mathbb{N}} \theta(\lambda)_r \varepsilon_r.$$

(For a definition of λ^θ in a general setting, see (6.16) below.)

The **standard Borel subalgebra** of $\widetilde{\mathfrak{l}}^+_k$ corresponds to the fundamental system consisting of (all odd) simple roots $\varepsilon_r - \varepsilon_{r+1/2}$, for $r = 1/2, 1, 3/2, \ldots, k-1/2$. By Theorem 2.55, the irreducible polynomial representations of $\widetilde{\mathfrak{l}}^+_k \cong \mathfrak{gl}(k|k)$ are parameterized by the set $\mathcal{P}(k|k)$ of $(k|k)$-hook partitions, and that the set of highest weights of these representations is precisely $\{\lambda^\theta \mid \lambda \in \mathcal{P}(k|k)\}$. We have natural inclusions of Lie superalgebras:

$$\widetilde{\mathfrak{l}}^+_1 \subset \widetilde{\mathfrak{l}}^+_2 \subset \cdots \subset \widetilde{\mathfrak{l}}^+_{k-1} \subset \widetilde{\mathfrak{l}}^+_k \subset \cdots,$$

compatible with their respective standard Borel subalgebras.

Let $k \in \mathbb{N} \cup \{\infty\}$. Suppose that $V = \bigoplus_\mu V_\mu$ is an $\widetilde{\mathfrak{l}}^+_k$-module that is semisimple with respect to the action of its standard Cartan subalgebra such that $V_\mu = 0$ unless μ satisfies the following polynomial weight conditions: $(-1)^{2r}(\mu, \varepsilon_r) \in \mathbb{Z}_+$, for $r \in \frac{1}{2}\mathbb{N}$, and $(\mu, \varepsilon_r) = 0$, for $r \gg 0$. Let $n \in \mathbb{N}$ with $n < k$. We form the following subspace of V:

$$\mathrm{tr}^k_n V := \bigoplus_\nu V_\nu,$$

where the summation is over ν satisfying $(\nu, \varepsilon_r) = 0$, for $r > n$. (tr^k_n as a functor will be defined and studied systematically in a general setting in Section 6.2.5.)

Let $\lambda \in \mathcal{P}(k|k)$. The character of the irreducible $\widetilde{\mathfrak{l}}^+_k$-module $L(\widetilde{\mathfrak{l}}^+_k, \lambda^\theta)$ is given by $hs_\lambda(x_1, \ldots, x_k; y_{1/2}, y_{3/2}, \ldots, y_{k-1/2})$ according to Theorem 3.15, where $x_j = e^{\varepsilon_j}$

and $y_{j-1/2} = e^{\varepsilon_{j-1/2}}$ for $j \in \mathbb{N}$. Hence the character of the $\widetilde{\mathfrak{l}}_n^+$-module $\mathrm{tr}_n^k\big(L(\widetilde{\mathfrak{l}}_k^+, \lambda^\theta)\big)$ is equal to $\mathrm{hs}_\lambda(x_1, \ldots, x_n; y_{1/2}, y_{3/2}, \ldots, y_{n-1/2})$ or zero, depending on whether λ is an $(n|n)$-hook partition. Hence, for $n < k$, we have

$$\mathrm{tr}_n^k\big(L(\widetilde{\mathfrak{l}}_k^+, \lambda^\theta)\big) \cong \begin{cases} L(\widetilde{\mathfrak{l}}_n^+, \lambda^\theta), & \text{if } \lambda \text{ is an } (n|n)\text{-hook partition}, \\ 0, & \text{otherwise}. \end{cases} \tag{6.10}$$

From now on, the natural inclusions $L(\widetilde{\mathfrak{l}}_n^+, \lambda^\theta) \subseteq L(\widetilde{\mathfrak{l}}_k^+, \lambda^\theta)$ for any $\lambda \in \mathcal{P}(n|n)$ will be understood in the sense of the above isomorphism, for $n < k$. One checks that $\bigcup_{k \geq n} L(\widetilde{\mathfrak{l}}_k^+, \lambda^\theta)$ is an irreducible highest weight $\widetilde{\mathfrak{l}}^+$-module of highest weight λ^θ. Hence,

$$L(\widetilde{\mathfrak{l}}^+, \lambda^\theta) = \bigcup_{k \geq n} L(\widetilde{\mathfrak{l}}_k^+, \lambda^\theta),$$

and $L(\widetilde{\mathfrak{l}}^+, \lambda^\theta)$, for $\lambda \in \mathcal{P}$, has character given by the super Schur function hs_λ.

Lemma 6.2. *Let* $\lambda, \mu \in \mathcal{P}$. *Then* $\mathrm{Ext}^1_{\widetilde{\mathfrak{l}}^+}\big(L(\widetilde{\mathfrak{l}}^+, \lambda^\theta), L(\widetilde{\mathfrak{l}}^+, \mu^\theta)\big) = 0$.

Proof. Consider a short exact sequence of $\widetilde{\mathfrak{l}}^+$-modules

$$0 \longrightarrow L(\widetilde{\mathfrak{l}}^+, \lambda^\theta) \longrightarrow E \longrightarrow L(\widetilde{\mathfrak{l}}^+, \mu^\theta) \longrightarrow 0. \tag{6.11}$$

First suppose that $\lambda = \mu$. Then by weight consideration, the two-dimensional μ^θ-weight subspace of E is a highest weight space, and thus E contains two proper submodules and the short exact sequence must split.

Now suppose that $\lambda \neq \mu$. Choose $n > \max\{|\lambda|, |\mu|\}$. Applying tr_n^∞ to every term in (6.11), we obtain by (6.10) the following short exact sequence of polynomial $\widetilde{\mathfrak{l}}_n^+$-modules:

$$0 \longrightarrow L(\widetilde{\mathfrak{l}}_n^+, \lambda^\theta) \longrightarrow \mathrm{tr}_n^\infty E \longrightarrow L(\widetilde{\mathfrak{l}}_n^+, \mu^\theta) \longrightarrow 0. \tag{6.12}$$

By Theorem 3.27, (6.12) splits. Hence, we can find a singular vector w in the $\widetilde{\mathfrak{l}}_n^+$-module $\mathrm{tr}_n^\infty E$ so that $U(\widetilde{\mathfrak{l}}_n^+)w = L(\widetilde{\mathfrak{l}}_n^+, \mu^\theta)$. Since there are no weight vectors in E of weight $\mu^\theta + \varepsilon_i - \varepsilon_{i+1/2}$, for $i \geq n$, w is a singular vector in E with respect to the standard Borel subalgebra of $\widetilde{\mathfrak{l}}^+$, too.

Now we consider the $\widetilde{\mathfrak{l}}^+$-submodule L generated by w in E. We claim that $L \cong L(\widetilde{\mathfrak{l}}^+, \mu^\theta)$, and so (6.11) splits. For otherwise L contains $L(\widetilde{\mathfrak{l}}^+, \lambda^\theta)$, and in particular L contains a highest weight vector v of weight λ^θ. This implies that $\mu^\theta - \lambda^\theta$ is a positive integral combination of positive roots of $\widetilde{\mathfrak{l}}^+$. By the choice of n, $\mu^\theta - \lambda^\theta$ has to be a positive integral combination of positive roots of $\widetilde{\mathfrak{l}}_n^+$, and hence $v \in U(\widetilde{\mathfrak{l}}_n^+)w$ and $U(\widetilde{\mathfrak{l}}_n^+)w \supseteq L(\widetilde{\mathfrak{l}}_n^+, \lambda^\theta)$. This contradicts $U(\widetilde{\mathfrak{l}}_n^+)w = L(\widetilde{\mathfrak{l}}_n^+, \mu^\theta)$. So (6.11) splits, and the lemma is proved. □

Lemma 6.3. *As an* $\widetilde{\mathfrak{l}}^+$*-module* $(\widetilde{V}_0^+)^{\otimes d}$ *is completely reducible and we have*

$$(\widetilde{V}_0^+)^{\otimes d} \cong \bigoplus_{\lambda \in \mathcal{P}_d} L(\widetilde{\mathfrak{l}}^+, \lambda^\theta)^{d_\lambda},$$

where d_λ denotes the multiplicity of $L(\widetilde{\mathfrak{l}}^+, \lambda^\theta)$ in $(\widetilde{V}_0^+)^{\otimes d}$. Furthermore, d_λ equals the dimension of the Specht module of $\mathfrak{S}_d$ corresponding to λ.

Proof. Set $W = (\widetilde{V}_0^+)^{\otimes d}$. Fix an integer $n \geq d$. By Theorem 3.11, we have an isomorphism of $\widetilde{\mathfrak{l}}_n^+$-modules

$$\mathfrak{tr}_n^\infty W = (\widetilde{V}_{0,n}^+)^{\otimes d} \cong \bigoplus_{\lambda \in \mathcal{P}_d} L(\widetilde{\mathfrak{l}}_n^+, \lambda^\theta)^{d_\lambda}.$$

Now let w be a singular vector in the $\widetilde{\mathfrak{l}}_n^+$-module $\mathfrak{tr}_n^\infty W$ of weight λ^θ, for a fixed $\lambda \in \mathcal{P}_d$. Observe, as in the proof of Lemma 6.2, that w is also a singular vector with respect to the standard Borel subalgebra of $\widetilde{\mathfrak{l}}^+$ in W. This implies that for each $\lambda \in \mathcal{P}_d$, the module $L(\widetilde{\mathfrak{l}}^+, \lambda^\theta)$ appears in W as a composition factor with multiplicity at least d_λ. A comparison of $\widetilde{\mathfrak{l}}^+$-characters shows that each $L(\widetilde{\mathfrak{l}}^+, \lambda^\theta)$ appears in W with multiplicity exactly d_λ, and they are all the composition factors of W. Now the lemma follows from Lemma 6.2. □

The notions of polynomial weights and polynomial modules for $\widetilde{\mathfrak{l}}^+$ can be defined just as in Definition 3.25. As in Proposition 3.26, the irreducible polynomial $\widetilde{\mathfrak{l}}^+$-modules are precisely $L(\widetilde{\mathfrak{l}}^+, \lambda^\theta)$, for $\lambda \in \mathcal{P}$. The following theorem is an infinite-rank analogue of Theorem 3.27, and it follows easily from Lemmas 6.2 and 6.3.

Theorem 6.4. *The category of polynomial modules of $\widetilde{\mathfrak{l}}^+ = \mathfrak{gl}(\widetilde{V}_0^+)$ is a semisimple tensor category.*

In the extreme case when $\mathfrak{k} = 0$, we have $\widetilde{\mathfrak{g}} = \mathfrak{gl}(\widetilde{V}_0^+)$, and the category of polynomial modules of $\mathfrak{gl}(\widetilde{V}_0^+)$ is simply the category $\widetilde{\mathcal{O}}$ defined below in Section 6.2.3.

6.2.2. Parabolic subalgebras and dominant weights. Recall the Lie superalgebras $\mathfrak{g}$, $\overline{\mathfrak{g}}$, and $\widetilde{\mathfrak{g}}$ from Section 6.1.6, which implicitly depend on a fixed $m \in \mathbb{Z}_+$. We shall in addition fix an arbitrary subset Y_0 of $\Pi(\mathfrak{k})$. Let Y, $\overline{Y}$, and $\widetilde{Y}$ be the following subsets of Π, $\overline{\Pi}$, and $\widetilde{\Pi}$, respectively:

$$(6.13)\qquad \begin{gathered} Y = Y_0 \sqcup \Pi(\mathfrak{T}) \setminus \{\beta_\times\}, \\ \overline{Y} = Y_0 \sqcup \Pi(\overline{\mathfrak{T}}) \setminus \{\alpha_\times\}, \quad \widetilde{Y} = Y_0 \sqcup \Pi(\widetilde{\mathfrak{T}}) \setminus \{\alpha_\times\}. \end{gathered}$$

As fixing Y_0 also fixes the sets Y, $\overline{Y}$, and $\widetilde{Y}$, we will make the convention of suppressing them from notation below.

Let $\mathfrak{l}$, $\overline{\mathfrak{l}}$, and $\widetilde{\mathfrak{l}}$ be the **standard Levi subalgebras** of $\mathfrak{g}$, $\overline{\mathfrak{g}}$, and $\widetilde{\mathfrak{g}}$ corresponding to the subsets Y, $\overline{Y}$, and $\widetilde{Y}$, respectively. The standard Borel subalgebras of $\mathfrak{g}$, $\overline{\mathfrak{g}}$, and $\widetilde{\mathfrak{g}}$, spanned by the central element K and upper triangular matrices, are denoted by $\mathfrak{b}$, $\overline{\mathfrak{b}}$, and $\widetilde{\mathfrak{b}}$, respectively. Let $\mathfrak{p} = \mathfrak{l} + \mathfrak{b}$, $\overline{\mathfrak{p}} = \overline{\mathfrak{l}} + \overline{\mathfrak{b}}$, and $\widetilde{\mathfrak{p}} = \widetilde{\mathfrak{l}} + \widetilde{\mathfrak{b}}$ be the corresponding **parabolic subalgebras** with **nilradicals** $\mathfrak{u}$, $\overline{\mathfrak{u}}$, and $\widetilde{\mathfrak{u}}$ and **opposite nilradicals** $\mathfrak{u}^-$, $\overline{\mathfrak{u}}^-$, and $\widetilde{\mathfrak{u}}^-$, respectively.

Let $^{-}\lambda = (\lambda_{-m}, \ldots, \lambda_{-1}) \in \mathbb{C}^m$ and let $^{+}\lambda$ be a partition. Recall also that h_α denotes the coroot of a root α. Associated to a given $d \in \mathbb{C}$ and a tuple $(^{-}\lambda; ^{+}\lambda)$ such that $\langle \sum_{i=-m}^{-1} \lambda_i \varepsilon_i, h_\alpha \rangle \in \mathbb{Z}_+$ for all $\alpha \in Y_0 \subseteq \Pi(\mathfrak{k})$, we define the weights

$$\lambda := d\Lambda_0 + \sum_{i=-m}^{-1} \lambda_i \varepsilon_i + \sum_{j \in \mathbb{N}} {}^{+}\lambda_j \varepsilon_j \in \mathfrak{h}^*, \tag{6.14}$$

$$\lambda^\natural := d\Lambda_0 + \sum_{i=-m}^{-1} \lambda_i \varepsilon_i + \sum_{s \in \frac{1}{2}+\mathbb{Z}_+} {}^{+}\lambda'_{s+\frac{1}{2}} \varepsilon_s \in \overline{\mathfrak{h}}^*, \tag{6.15}$$

$$\lambda^\theta := d\Lambda_0 + \sum_{i=-m}^{-1} \lambda_i \varepsilon_i + \sum_{r \in \frac{1}{2}\mathbb{N}} \theta(^{+}\lambda)_r \varepsilon_r \in \widetilde{\mathfrak{h}}^*. \tag{6.16}$$

These weights $\lambda, \lambda^\natural, \lambda^\theta$ will be referred to as **dominant weights**. We denote by $P^+ \subseteq \mathfrak{h}^*$, $\overline{P}^+ \subseteq \overline{\mathfrak{h}}^*$, and $\widetilde{P}^+ \subseteq \widetilde{\mathfrak{h}}^*$ the sets of dominant weights of the forms (6.14), (6.15), and (6.16), for all $d \in \mathbb{C}$, respectively. We will also identify an element $\lambda \in P^+$ of the form (6.14) with the tuple $(d\Lambda_0, {}^{-}\lambda, {}^{+}\lambda)$. The next lemma follows by definition.

Lemma 6.5. *We have the following bijective maps:*

$$\natural : P^+ \longrightarrow \overline{P}^+, \qquad \lambda \mapsto \lambda^\natural,$$
$$\theta : P^+ \longrightarrow \widetilde{P}^+, \qquad \lambda \mapsto \lambda^\theta.$$

For $\lambda \in P^+$, let $L(\mathfrak{l}, \lambda)$ denote the irreducible highest weight $\mathfrak{l}$-module of highest weight λ. We extend $L(\mathfrak{l}, \lambda)$ to a $\mathfrak{p}$-module by letting $\mathfrak{u}$ act trivially. Define the **parabolic Verma $\mathfrak{g}$-module** $\Delta(\lambda)$ as

$$\Delta(\lambda) := \mathrm{Ind}_{\mathfrak{p}}^{\mathfrak{g}} L(\mathfrak{l}, \lambda) = U(\mathfrak{g}) \otimes_{U(\mathfrak{p})} L(\mathfrak{l}, \lambda),$$

and denote its unique irreducible quotient $\mathfrak{g}$-module by $L(\lambda)$.

Similarly, for $\lambda \in P^+$, we define the irreducible $\bar{\mathfrak{l}}$-module $L(\bar{\mathfrak{l}}, \lambda^\natural)$, the parabolic Verma $\bar{\mathfrak{g}}$-module $\overline{\Delta}(\lambda^\natural)$, and its unique irreducible quotient $\bar{\mathfrak{g}}$-module $\overline{L}(\lambda^\natural)$. We also similarly define the irreducible $\widetilde{\mathfrak{l}}$-module $L(\widetilde{\mathfrak{l}}, \lambda^\theta)$, the parabolic Verma $\widetilde{\mathfrak{g}}$-module $\widetilde{\Delta}(\lambda^\theta)$, and its unique irreducible quotient $\widetilde{\mathfrak{g}}$-module $\widetilde{L}(\lambda^\theta)$.

6.2.3. The categories $\mathcal{O}$, $\overline{\mathcal{O}}$, and $\widetilde{\mathcal{O}}$.

We now introduce a version of parabolic BGG category $\mathcal{O}$ of $\mathfrak{g}$-modules.

Definition 6.6. Let $\mathcal{O}$ be the category of $\mathfrak{g}$-modules M such that M is a semisimple $\mathfrak{h}$-module with finite-dimensional weight subspaces M_γ, $\gamma \in \mathfrak{h}^*$, satisfying

(i) M decomposes over $\mathfrak{l}$ as a direct sum of $L(\mathfrak{l}, \mu)$ for $\mu \in P^+$.

(ii) There exist finitely many weights $\lambda^1, \lambda^2, \ldots, \lambda^k \in P^+$ (depending on M) such that if γ is a weight in M, then $\gamma \in \lambda^i - \sum_{\alpha \in \Pi} \mathbb{Z}_+ \alpha$, for some i. (Recall that Π is the fundamental system for $\mathfrak{g}$.)

Analogously we define the category $\overline{\mathcal{O}}$ of $\overline{\mathfrak{g}}$-modules using $\overline{\mathfrak{h}}^*, \overline{P}^+, \overline{\mathfrak{l}}, \overline{\Pi}$, and the category $\widetilde{\mathcal{O}}$ of $\widetilde{\mathfrak{g}}$-modules using $\widetilde{\mathfrak{h}}^*, \widetilde{P}^+, \widetilde{\mathfrak{l}}, \widetilde{\Pi}$. The morphisms in $\mathcal{O}$, $\overline{\mathcal{O}}$, and $\widetilde{\mathcal{O}}$ are all (not necessarily even) $\mathfrak{g}$-, $\overline{\mathfrak{g}}$-, and $\widetilde{\mathfrak{g}}$-homomorphisms, respectively.

Proposition 6.7. *Let $\mu \in P^+$. The following statements hold.*

(1) *The restrictions to $\mathfrak{l}$ of the $\mathfrak{g}$-modules $\Delta(\mu)$ and $L(\mu)$ decompose as direct sums of $L(\mathfrak{l}, \nu)$, for $\nu \in P^+$.*

(2) *The restrictions to $\overline{\mathfrak{l}}$ of the $\overline{\mathfrak{g}}$-modules $\overline{\Delta}(\mu^\natural)$ and $\overline{L}(\mu^\natural)$ decompose as direct sums of $L(\overline{\mathfrak{l}}, \nu^\natural)$, for $\nu \in P^+$.*

(3) *The restrictions to $\widetilde{\mathfrak{l}}$ of the $\widetilde{\mathfrak{g}}$-modules $\widetilde{\Delta}(\mu^\theta)$ and $\widetilde{L}(\mu^\theta)$ decompose as direct sums of $L(\widetilde{\mathfrak{l}}, \nu^\theta)$, for $\nu \in P^+$.*

Proof. (1) As a Lie (super)algebra we have $\mathfrak{l} \cong \mathfrak{k}_0 \oplus \mathfrak{gl}(V_0^+)$, where $\mathfrak{k}_0$ is a Levi subalgebra of $\mathfrak{k}$ corresponding to Y_0 (our convention is that $K \in \mathfrak{k}_0$), and V_0^+ was defined in Section 6.1.3. Hence an irreducible $\mathfrak{l}$-module of highest weight $\nu \in P^+$ is isomorphic to a tensor product of a finite-dimensional irreducible $\mathfrak{k}_0$-module and a polynomial $\mathfrak{gl}(V_0^+)$-module. Note that the opposite nilradical $\mathfrak{u}^-$ as an $\mathfrak{l}$-module is a direct sum of irreducible modules with highest weights lying in P^+. More explicitly, we have the following isomorphisms of $\mathfrak{l}$-modules:

$$\mathfrak{u}^- \cong \begin{cases} \mathbb{C}^{m*} \otimes V_0^+ \oplus \mathfrak{u}_0^-, & \text{for } \mathfrak{x} = \mathfrak{a} \\ \mathbb{C}^{2m+1} \otimes V_0^+ \oplus \wedge^2(V_0^+) \oplus \mathfrak{u}_0^-, & \text{for } \mathfrak{x} = \mathfrak{b} \\ \mathbb{C}^{2m|1} \otimes V_0^+ \oplus S^2(V_0^+) \oplus \mathfrak{u}_0^-, & \text{for } \mathfrak{x} = \mathfrak{b}^\bullet \\ \mathbb{C}^{2m} \otimes V_0^+ \oplus S^2(V_0^+) \oplus \mathfrak{u}_0^-, & \text{for } \mathfrak{x} = \mathfrak{c} \\ \mathbb{C}^{2m} \otimes V_0^+ \oplus \wedge^2(V_0^+) \oplus \mathfrak{u}_0^-, & \text{for } \mathfrak{x} = \mathfrak{d}. \end{cases}$$

Here $V_0^+ \cong \mathbb{C}^\infty$ is a purely even space and it is isomorphic to the natural module of $\mathfrak{gl}(V_0^+) \cong \mathfrak{gl}(\infty)$, $\mathfrak{u}_0^-$ is the opposite nilradical of $\mathfrak{k}$ corresponding to the Levi subalgebra $\mathfrak{k}_0$, and furthermore $\mathbb{C}^m$, $\mathbb{C}^{2m}$, $\mathbb{C}^{2m+1}$, and $\mathbb{C}^{2m|1}$ are the natural $\mathfrak{k}$-modules on which $\mathfrak{k}_0$ acts semisimply by restriction. It is evident that the tensor product of two irreducible $\mathfrak{l}$-modules with highest weights in P^+ decomposes into a direct sum of irreducibles with highest weights in P^+. Recall that $\mathcal{S}(U)$ stands for the supersymmetric algebra of a superspace U. Since, as $\mathfrak{l}$-modules, we have

$$\Delta(\mu) \cong \mathcal{S}(\mathfrak{u}^-) \otimes L(\mathfrak{l}, \mu),$$

it follows that $\Delta(\mu)$ is a direct sum of irreducible $\mathfrak{l}$-modules with highest weights in P^+. Since $L(\mu)$ is a quotient of $\Delta(\mu)$, the same holds for $L(\mu)$. This proves (1).

(2) The Levi subalgebra of $\overline{\mathfrak{g}}$ is $\overline{\mathfrak{l}} = \mathfrak{k}_0 \oplus \mathfrak{gl}(\overline{V}_0^+)$, which is isomorphic to $\mathfrak{l}$. The opposite nilradical $\overline{\mathfrak{u}}^-$ is a direct sum of irreducible $\overline{\mathfrak{l}}$-modules with highest weights

of the form $\nu^\natural$, where $\nu \in P^+$. Explicitly, we have

$$\overline{\mathfrak{u}}^- \cong \begin{cases} \mathbb{C}^{m*} \otimes \overline{V}_0^+ \oplus \mathfrak{u}_0^-, & \text{for } \mathfrak{x} = \mathfrak{a} \\ \mathbb{C}^{2m+1} \otimes \overline{V}_0^+ \oplus S^2(\overline{V}_0^+) \oplus \mathfrak{u}_0^-, & \text{for } \mathfrak{x} = \mathfrak{b} \\ \mathbb{C}^{2m|1} \otimes \overline{V}_0^+ \oplus \wedge^2(\overline{V}_0^+) \oplus \mathfrak{u}_0^-, & \text{for } \mathfrak{x} = \mathfrak{b}^\bullet \\ \mathbb{C}^{2m} \otimes \overline{V}_0^+ \oplus \wedge^2(\overline{V}_0^+) \oplus \mathfrak{u}_0^-, & \text{for } \mathfrak{x} = \mathfrak{c} \\ \mathbb{C}^{2m} \otimes \overline{V}_0^+ \oplus S^2(\overline{V}_0^+) \oplus \mathfrak{u}_0^-, & \text{for } \mathfrak{x} = \mathfrak{d}. \end{cases}$$

A main difference from (1) is that $\overline{V}_0^+ \cong \mathbb{C}^\infty$ here is a purely odd space, yet the notation for the exterior and the symmetric squares of $\overline{V}_0^+$ here are understood in the non-super sense. Now we have $\overline{\Delta}(\lambda^\natural) = \mathcal{S}(\bar{\mathfrak{u}}^-) \otimes L(\bar{\mathfrak{l}}, \lambda^\natural)$, and so a verbatim argument as in (1) establishes (2).

(3) The Levi subalgebra $\widetilde{\mathfrak{l}}$ of $\widetilde{\mathfrak{g}}$ is isomorphic to $\mathfrak{k}_0 \oplus \mathfrak{gl}(\widetilde{V}_0^+)$. As an $\widetilde{\mathfrak{l}}$-module, $\widetilde{\mathfrak{u}}^-$ is isomorphic to a direct sum of irreducible modules, each of which is a tensor product of a finite-dimensional irreducible $\mathfrak{k}_0$-module and an irreducible polynomial module of $\mathfrak{gl}(\widetilde{V}_0^+)$. Recall the polynomial modules of $\mathfrak{gl}(\widetilde{V}_0^+)$ form a semisimple tensor category by Theorem 6.4. We have

$$\widetilde{\mathfrak{u}}^- \cong \begin{cases} \mathbb{C}^{m*} \otimes \widetilde{V}_0^+ \oplus \mathfrak{u}_0^-, & \text{for } \mathfrak{x} = \mathfrak{a} \\ \mathbb{C}^{2m+1} \otimes \widetilde{V}_0^+ \oplus \wedge^2(\widetilde{V}_0^+) \oplus \mathfrak{u}_0^-, & \text{for } \mathfrak{x} = \mathfrak{b} \\ \mathbb{C}^{2m|1} \otimes \widetilde{V}_0^+ \oplus \mathcal{S}^2(\widetilde{V}_0^+) \oplus \mathfrak{u}_0^-, & \text{for } \mathfrak{x} = \mathfrak{b}^\bullet \\ \mathbb{C}^{2m} \otimes \widetilde{V}_0^+ \oplus \mathcal{S}^2(\widetilde{V}_0^+) \oplus \mathfrak{u}_0^-, & \text{for } \mathfrak{x} = \mathfrak{c} \\ \cong \mathbb{C}^{2m} \otimes \widetilde{V}_0^+ \oplus \wedge^2(\widetilde{V}_0^+) \oplus \mathfrak{u}_0^-, & \text{for } \mathfrak{x} = \mathfrak{d}. \end{cases}$$

Here the exterior squares are understood in the super sense for the superspace $\widetilde{V}_0^+$. So each irreducible $\widetilde{\mathfrak{l}}$-submodule of $\widetilde{\mathfrak{u}}^-$ has highest weight lying in $\widetilde{P}^+$. Thus, by Theorem 6.4, $\widetilde{\Delta}(\lambda^\theta)$, for $\lambda \in P^+$, is also a direct sum of irreducible $\widetilde{\mathfrak{l}}$-modules with highest weights in $\widetilde{P}^+$, and so is its quotient $\widetilde{L}(\lambda^\theta)$. This proves (3). □

As an immediate consequence of Proposition 6.7, we have the following.

Corollary 6.8. *Let* $\lambda \in P^+$.

(1) *The modules* $\Delta(\lambda)$ *and* $L(\lambda)$ *lie in* $\mathcal{O}$.

(2) *The modules* $\overline{\Delta}(\lambda^\natural)$ *and* $\overline{L}(\lambda^\natural)$ *lie in* $\overline{\mathcal{O}}$.

(3) *The modules* $\widetilde{\Delta}(\lambda^\theta)$ *and* $\widetilde{L}(\lambda^\theta)$ *lie in* $\widetilde{\mathcal{O}}$.

6.2.4. The categories $\mathcal{O}_n$, $\overline{\mathcal{O}}_n$, and $\widetilde{\mathcal{O}}_n$. For $n \in \mathbb{N}$, recall the sets $\Pi_n, \overline{\Pi}_n, \widetilde{\Pi}_n$ of simple roots for the Dynkin diagrams (6.3) and the associated finite-dimensional Lie superalgebras $\mathfrak{g}_n$, $\overline{\mathfrak{g}}_n$, and $\widetilde{\mathfrak{g}}_n$ from Section 6.1.5. These Lie superalgebras $\mathfrak{g}_n$, $\overline{\mathfrak{g}}_n$, and $\widetilde{\mathfrak{g}}_n$ can be identified naturally with the subalgebras of $\mathfrak{g}$, $\overline{\mathfrak{g}}$, and $\widetilde{\mathfrak{g}}$ generated by the central element K and the root vectors associated to the fundamental systems in (6.3). Moreover, we have natural inclusions $\mathfrak{g}_n \subset \mathfrak{g}_{n+1}, \overline{\mathfrak{g}}_n \subset \overline{\mathfrak{g}}_{n+1}$, and $\widetilde{\mathfrak{g}}_n \subseteq \widetilde{\mathfrak{g}}_{n+1}$, with $\mathfrak{g} = \cup_n \mathfrak{g}_n$, $\overline{\mathfrak{g}} = \cup_n \overline{\mathfrak{g}}_n$, and $\widetilde{\mathfrak{g}} = \cup_n \widetilde{\mathfrak{g}}_n$. The standard Cartan and Borel subalgebras of $\mathfrak{g}_n$ are

$$\mathfrak{h}_n = \mathfrak{h} \cap \mathfrak{g}_n, \qquad \mathfrak{b}_n = \mathfrak{b} \cap \mathfrak{g}_n,$$

respectively. Similarly, we write $\overline{\mathfrak{h}}_n$ and $\widetilde{\mathfrak{h}}_n$ ($\overline{\mathfrak{b}}_n$ and $\widetilde{\mathfrak{b}}_n$) for the standard Cartan (Borel) subalgebras of $\mathfrak{g}_n$ and $\widetilde{\mathfrak{g}}_n$, respectively.

Recall the notation $\lambda \in P^+$, $\lambda^\natural$, and λ^θ from (6.14), (6.15), and (6.16). Given $\lambda \in P^+$ with ${}^+\lambda_j = 0$ for $j > n$, we may regard λ as a weight in $\mathfrak{h}_n^*$ in a natural way. Similarly, for $\lambda \in P^+$ with ${}^+\lambda_j' = 0$ for $j > n$, we regard $\lambda^\natural$ as a weight in $\overline{\mathfrak{h}}_n^*$. Finally, for $\lambda \in P^+$ with $\theta({}^+\lambda)_j = 0$ for $j \in \frac{1}{2}\mathbb{N}$ with $j > n$, we regard λ^θ as a weight in $\widetilde{\mathfrak{h}}_n^*$. These weights $\lambda, \lambda^\natural, \lambda^\theta$ in $\mathfrak{h}_n^*$, $\overline{\mathfrak{h}}_n^*$, and $\widetilde{\mathfrak{h}}_n^*$, respectively, with the above constraints, will be called **dominant weights**. The subsets of dominant weights in $\mathfrak{h}_n^*$, $\overline{\mathfrak{h}}_n^*$, and $\widetilde{\mathfrak{h}}_n^*$ will be denoted by P_n^+, $\overline{P}_n^+$, and $\widetilde{P}_n^+$, respectively.

Corresponding to a fixed $Y_0 \subseteq \Pi(\mathfrak{k})$, the Levi and parabolic subalgebras of the finite-rank Lie superalgebra $\mathfrak{g}_n$ are

$$\mathfrak{l}_n = \mathfrak{l} \cap \mathfrak{g}_n, \qquad \mathfrak{p}_n = \mathfrak{p} \cap \mathfrak{g}_n,$$

respectively. This allows us to define the corresponding parabolic Verma and irreducible $\mathfrak{g}_n$-modules $\Delta_n(\mu)$ and $L_n(\mu)$ with highest weight $\mu \in P_n^+$. The corresponding category of $\mathfrak{g}_n$-modules is denoted by $\mathcal{O}_n$, which is defined as in Definition 6.6, now with $\mathfrak{h}$, $\mathfrak{l}$, P^+, and Π therein replaced by $\mathfrak{h}_n$, $\mathfrak{l}_n$, P_n^+, and Π_n, respectively.

The statements in the previous paragraph admit obvious counterparts for the Lie superalgebras $\overline{\mathfrak{g}}_n$ and $\widetilde{\mathfrak{g}}_n$ as well. We introduce the self-explanatory notation $\overline{\Delta}_n(\upsilon)$, $\overline{L}_n(\upsilon)$, $\overline{\mathcal{O}}_n$, $\overline{\mathfrak{l}}$, $\overline{\mathfrak{p}}$ for $\overline{\mathfrak{g}}_n$, and $\widetilde{\Delta}_n(\nu)$, $\widetilde{L}_n(\nu)$, $\widetilde{\mathcal{O}}_n$ $\widetilde{\mathfrak{l}}$, $\widetilde{\mathfrak{p}}$ for $\widetilde{\mathfrak{g}}_n$, where $\upsilon \in \overline{P}_n^+$ and $\nu \in \widetilde{P}_n^+$.

6.2.5. Truncation functors. Let $n < k \leq \infty$. For $M \in \mathcal{O}_k$, we write $M = \bigoplus_\gamma M_\gamma$, where $\gamma \in \sum_{i=-m}^{-1} \mathbb{C}\varepsilon_i + \sum_{0<j\leq k} \mathbb{Z}_+\varepsilon_j + \mathbb{C}\Lambda_0$, according to its weight space decomposition. The **truncation functor** $\mathfrak{tr}_n^k : \mathcal{O}_k \to \mathcal{O}_n$ is defined by

$$\mathfrak{tr}_n^k(M) = \bigoplus_\nu M_\nu,$$

where the summation is over $\nu \in \sum_{i=-m}^{-1} \mathbb{C}\varepsilon_i + \sum_{0<j\leq n} \mathbb{Z}_+\varepsilon_j + \mathbb{C}\Lambda_0$. When it is clear from the context we shall also write $\mathfrak{tr}_n$ instead of $\mathfrak{tr}_n^k$. Analogously, truncation functors $\mathfrak{tr}_n^k : \overline{\mathcal{O}}_k \to \overline{\mathcal{O}}_n$ and $\mathfrak{tr}_n^k : \widetilde{\mathcal{O}}_k \to \widetilde{\mathcal{O}}_n$ are defined. These functors are obviously exact. (The notation $\mathfrak{tr}_n^k$ used earlier in Section 6.2.1 corresponds to the extreme case here when $\mathfrak{k} = 0$.)

Recall from Sections 6.1.3 and 6.1.6 the notation E_j and $\widehat{E}_j$ for basis elements of the Cartan subalgebras.

Proposition 6.9. *Let $n < k \leq \infty$ and $X = L, \Delta$.*

(1) *For* $\mu \in P_k^+$ *we have* $\mathfrak{tr}_n(X_k(\mu)) = \begin{cases} X_n(\mu), & \textit{if } \langle \mu, \widehat{E}_j \rangle = 0, \forall j > n, \\ 0, & \textit{otherwise}. \end{cases}$

(2) *For* $\mu \in \overline{P}_k^+$ *we have* $\mathfrak{tr}_n(\overline{X}_k(\mu)) = \begin{cases} \overline{X}_n(\mu), & \textit{if } \langle \mu, \widehat{E}_j \rangle = 0, \forall j > n, \\ 0, & \textit{otherwise}. \end{cases}$

(3) *For* $\mu \in \widetilde{P}_k^+$ *we have* $\mathfrak{tr}_n(\widetilde{X}_k(\mu)) = \begin{cases} \widetilde{X}_n(\mu), & \textit{if } \langle \mu, \widehat{E}_j \rangle = 0, \forall j > n, \\ 0, & \textit{otherwise}. \end{cases}$

Proof. We shall only show (1), as similar arguments prove (2) and (3). Since $\mathfrak{tr}_n^k \circ \mathfrak{tr}_k^l = \mathfrak{tr}_n^l$, it suffices to show (1) for $k = \infty$.

First suppose that $\langle \mu, \widehat{E}_{n+1} \rangle > 0$. Then every weight ν of $L(\mathfrak{l}, \mu)$ also satisfies $\langle \nu, \widehat{E}_{n+1} \rangle > 0$ by the choice of fundamental systems of $\mathfrak{g}$ and hence of $\mathfrak{l}$. Recall from the proof of Proposition 6.7 that as an $\mathfrak{l}$-module $\mathfrak{u}^-$ decomposes into a direct sum of irreducibles with highest weights in P^+. Thus, every weight υ in $\Delta(\mu) = S(\mathfrak{u}^-) \otimes L(\mathfrak{l}, \mu)$ must also satisfy $\langle \upsilon, \widehat{E}_{n+1} \rangle > 0$. Therefore, $\mathfrak{tr}_n(\Delta(\mu)) = 0$, and hence $\mathfrak{tr}_n(L(\mu)) = 0$.

Now suppose that $\langle \mu, \widehat{E}_{n+1} \rangle = 0$. Since $\mu \in P_k^+$, this implies that $\langle \mu, \widehat{E}_j \rangle = 0$ for all $j > n$. Let $\mathfrak{l}'$ and $\mathfrak{p}'$ denote the standard Levi and parabolic subalgebras of $\mathfrak{g}$ corresponding to the removal of the vertex β_n of the Dynkin diagram of $\mathfrak{g}$. Then $\mathfrak{l}' \cong \mathfrak{g}_n \oplus \mathfrak{gl}(W)$, where W is the subspace of V_0^+ spanned by the vectors v_i, $i > n$. Consider the parabolic Verma module $\mathrm{Ind}_{\mathfrak{p}'}^{\mathfrak{g}} L_n(\mu)$, where $L_n(\mu)$ extends to an $\mathfrak{l}'$-module by the trivial action of $\mathfrak{gl}(W)$ and is then extended to a $\mathfrak{p}'$-module in a trivial way. Clearly, $L(\mu)$ is the unique irreducible quotient of $\mathrm{Ind}_{\mathfrak{p}'}^{\mathfrak{g}} L_n(\mu)$. Since $\mathfrak{tr}_n(\mathrm{Ind}_{\mathfrak{p}'}^{\mathfrak{g}} L_n(\mu)) = L_n(\mu)$ and $\mathfrak{tr}_n(L(\mu))$ is a nonzero $\mathfrak{g}_n$-module (as it contains a nonzero vector of weight μ), we conclude that $\mathfrak{tr}_n(L(\mu)) = L_n(\mu)$.

To complete the proof, we observe that $\Delta(\mu) \cong S(\mathfrak{u}^-) \otimes L(\mathfrak{l}, \mu)$, and that the $\mathfrak{g}_n$-module $\mathfrak{tr}_n(\Delta(\mu))$ has highest weight μ with character equal to $\mathrm{ch}\, \Delta_n(\mu)$. From these we conclude that $\mathfrak{tr}_n(\Delta(\mu)) = \Delta_n(\mu)$. □

6.3. The irreducible character formulas

In this section, two functors $T : \widetilde{\mathcal{O}} \to \mathcal{O}$ and $\overline{T} : \widetilde{\mathcal{O}} \to \overline{\mathcal{O}}$ are introduced, and they are shown to send irreducible and parabolic Verma modules to irreducible and parabolic Verma modules, respectively. The main ingredients for studying these functors are two sequences of odd reflections on the standard Borel subalgebra of $\widetilde{\mathfrak{g}}$, leading to new Borel subalgebras which are "approximately compatible" with the standard Borel subalgebras of $\mathfrak{g}$ and $\overline{\mathfrak{g}}$. As a consequence, the solution of the irreducible character problem in $\mathcal{O}$ via the classical Kazhdan-Lusztig theory also provides a complete solution to the irreducible character problem for $\overline{\mathcal{O}}$ and $\widetilde{\mathcal{O}}$.

6.3.1. Two sequences of Borel subalgebras of $\widetilde{\mathfrak{g}}$. Let us recall briefly the basics of odd reflections from Chapter 1, Sections 1.3 and 1.5, which are applied now to the Lie superalgebra $\widetilde{\mathfrak{g}}$. Under the odd reflection with respect to an isotropic odd simple root α in a fundamental system $\widetilde{\Pi}'$, the resulting new fundamental system $\widetilde{\Pi}^\alpha$ is given by (1.44) as follows:

$$\widetilde{\Pi}^\alpha = \{\beta \in \widetilde{\Pi}' \mid (\beta,\alpha) = 0, \beta \neq \alpha\} \cup \{\beta + \alpha \mid \beta \in \widetilde{\Pi}', (\beta,\alpha) \neq 0\} \cup \{-\alpha\}.$$

Let us denote the Borel subalgebra corresponding to $\widetilde{\Pi}'$ by $\mathfrak{b}'$ and the resulting new Borel subalgebra by $\mathfrak{b}^\alpha$. According to Lemma 1.40, an irreducible $\widetilde{\mathfrak{g}}$-module of $\mathfrak{b}'$-highest weight λ is also a $\mathfrak{b}^\alpha$-highest weight module with $\mathfrak{b}^\alpha$-highest weight equal to either λ or $\lambda - \alpha$, depending on whether $(\lambda, \alpha) = 0$. It is instructive to review Examples 1.35 and 1.41 to see how the highest weights change under a sequence of odd reflections, as they offer in a simpler setting a similar pattern to what we shall see below.

Recall from (6.2) the odd roots α_r and the even roots β_r, for $r \in \frac{1}{2}\mathbb{N}$. Recall from Section 6.1 that the standard Dynkin diagram associated to $\widetilde{\mathfrak{g}}$ is given by

(6.17) 

Fix $n \in \mathbb{N}$. We shall specify a sequence of $\frac{n(n+1)}{2}$ odd reflections, which will transform the standard diagram (6.17) into a new diagram with a new fundamental system of the form (6.19) below. The ordered sequence of $\frac{n(n+1)}{2}$ odd roots we use are:

$$\begin{array}{c} \varepsilon_{1/2} - \varepsilon_1, \\ \varepsilon_{3/2} - \varepsilon_2, \quad \varepsilon_{1/2} - \varepsilon_2, \\ \varepsilon_{5/2} - \varepsilon_3, \quad \varepsilon_{3/2} - \varepsilon_3, \quad \varepsilon_{1/2} - \varepsilon_3, \\ \dots, \dots, \dots, \dots, \\ \varepsilon_{n-1/2} - \varepsilon_n, \quad \varepsilon_{n-3/2} - \varepsilon_n, \quad \dots, \quad \varepsilon_{3/2} - \varepsilon_n, \quad \varepsilon_{1/2} - \varepsilon_n. \end{array} \tag{6.18}$$

Let us go into the details. First, applying the odd reflection corresponding to $\alpha_{1/2} = \varepsilon_{1/2} - \varepsilon_1$, we obtain the following diagram:

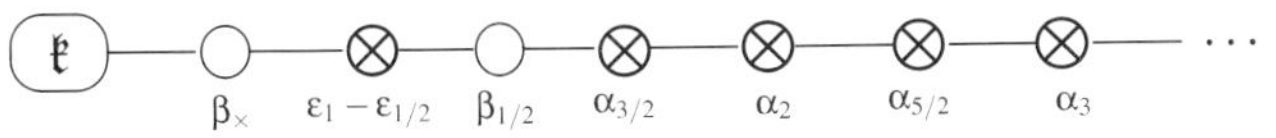

Next, applying to it consecutively the two odd reflections corresponding to $\alpha_{3/2} = \varepsilon_{3/2} - \varepsilon_2$ and $\alpha_{1/2} + \alpha_1 + \alpha_{3/2} = \varepsilon_{1/2} - \varepsilon_2$, we obtain the following two diagrams:

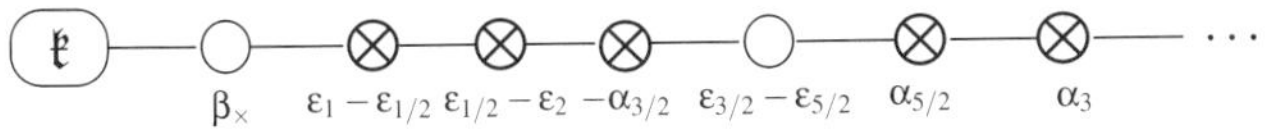

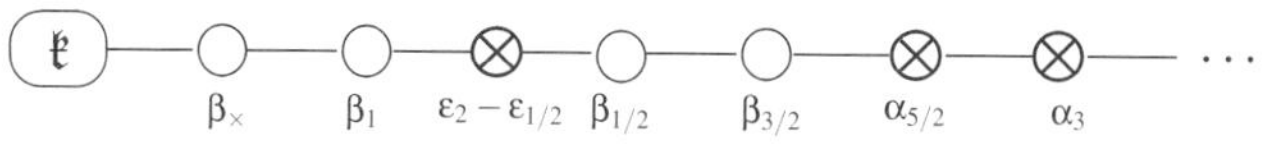

Now, to the last diagram above we apply consecutively the three odd reflections corresponding to $\alpha_{5/2} = \varepsilon_{5/2} - \varepsilon_3$, $\alpha_{3/2} + \alpha_2 + \alpha_{5/2} = \varepsilon_{3/2} - \varepsilon_3$, and $\alpha_{1/2} + \alpha_1 + \alpha_{3/2} + \alpha_2 + \alpha_{5/2} = \varepsilon_{1/2} - \varepsilon_3$ and obtain the following three diagrams:

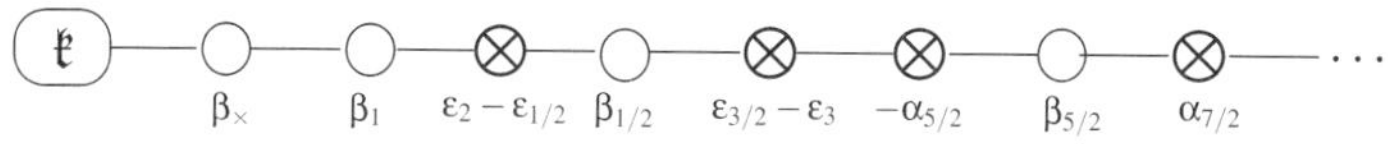

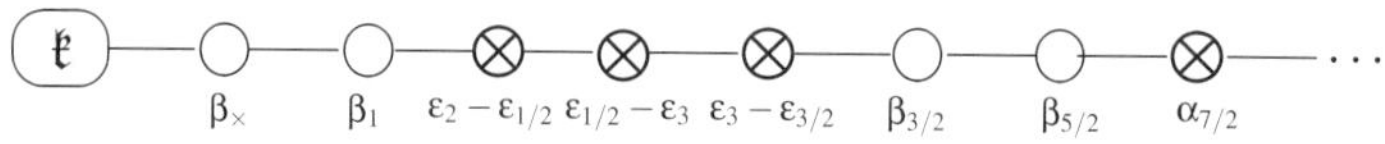

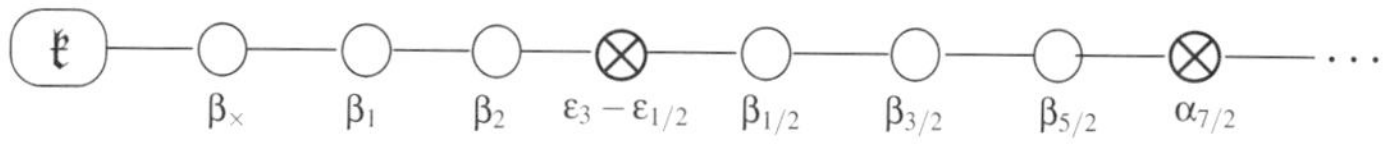

We continue this way until we finally apply the n odd reflections corresponding to $\alpha_{n-1/2}, \alpha_{n-3/2} + \alpha_{n-1} + \alpha_{n-1/2}, \ldots, \sum_{i=1}^{2n-1} \alpha_{i/2}$, which equal $\varepsilon_{n-1/2} - \varepsilon_n, \varepsilon_{n-3/2} - \varepsilon_n, \ldots, \varepsilon_{1/2} - \varepsilon_n$, respectively. The resulting new Borel subalgebra for $\widetilde{\mathfrak{g}}$ in the end will be denoted by $\widetilde{\mathfrak{b}}^c(n)$, and the corresponding positive system will be denoted by $\widetilde{\Phi}^c(n)$. The corresponding fundamental system, denoted by $\widetilde{\Pi}^c(n)$, is listed in the following Dynkin diagram (recall the tail diagram $\boxed{\mathfrak{T}_n}$ from Section 6.1.1):

(6.19) $\quad \boxed{\mathfrak{k}} \!-\!\boxed{\mathfrak{T}_n}\!-\!\otimes\!-\!\bigcirc\!-\cdots-\!\bigcirc\!-\!\otimes\!-\!\otimes\!-\!\cdots$
$\qquad\qquad\qquad\quad \varepsilon_n - \varepsilon_{1/2} \quad \beta_{1/2} \qquad\qquad \beta_{n-1/2}\ \alpha_{n+1/2}\ \alpha_{n+1}$

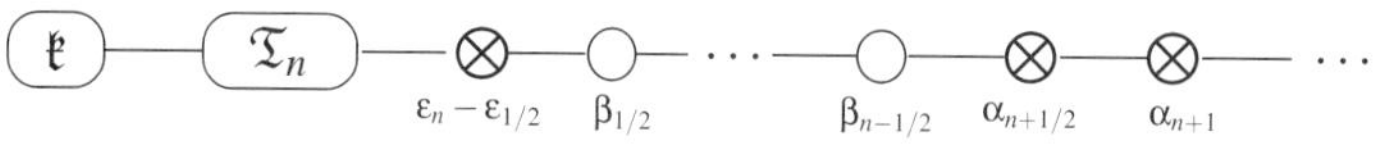

The crucial point here is that the subdiagram to the left of the first $\otimes$ in (6.19) is the Dynkin diagram of $\mathfrak{g}_n$, and the precise detail of the remaining part of (6.19) is not needed.

On the other hand, starting with the standard Dynkin diagram (6.17) of $\widetilde{\mathfrak{g}}$, we may apply a different sequence of $\frac{n(n+1)}{2}$ odd reflections associated to the following ordered sequence of $\frac{n(n+1)}{2}$ odd roots:

(6.20)
$$\begin{aligned}
&\varepsilon_1 - \varepsilon_{3/2}, \\
&\varepsilon_2 - \varepsilon_{5/2}, \quad \varepsilon_1 - \varepsilon_{5/2}, \\
&\varepsilon_3 - \varepsilon_{7/2}, \quad \varepsilon_2 - \varepsilon_{7/2}, \quad \varepsilon_1 - \varepsilon_{7/2}, \\
&\cdots, \cdots, \cdots, \cdots, \\
&\varepsilon_n - \varepsilon_{n+1/2}, \quad \varepsilon_{n-1} - \varepsilon_{n+1/2}, \quad \cdots, \quad \varepsilon_2 - \varepsilon_{n+1/2}, \quad \varepsilon_1 - \varepsilon_{n+1/2}.
\end{aligned}$$

Since the procedure here is completely parallel to the previous one for the sequence (6.18), we skip the details. The resulting new Borel subalgebra for $\widetilde{\mathfrak{g}}$ in the end will be denoted by $\widetilde{\mathfrak{b}}^s(n)$, and the corresponding positive system will be denoted by

$\widetilde{\Phi}^s(n)$. The corresponding fundamental system, denoted by $\widetilde{\Pi}^s(n)$, is listed in the following Dynkin diagram (recall the tail diagram $\overline{\mathfrak{T}_n}$ from Section 6.1.1):

(6.21) $\quad \mathfrak{k} - \overline{\mathfrak{T}_{n+1}} \underset{\varepsilon_{n+1/2}-\varepsilon_1}{\otimes} \underset{\beta_1}{\bigcirc} \cdots \underset{\beta_n}{\bigcirc} \underset{\alpha_{n+1}}{\otimes} \underset{\alpha_{n+3/2}}{\otimes} \cdots$

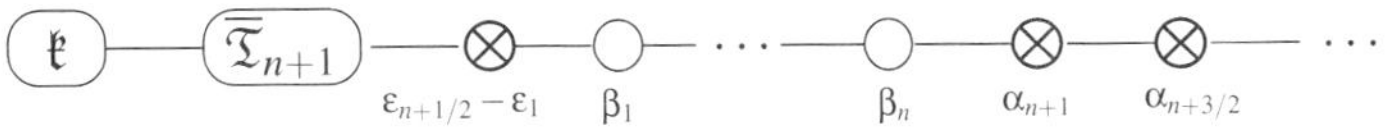

Again the crucial fact here is that the subdiagram to the left of the odd simple root $\varepsilon_{n+1/2} - \varepsilon_1$ in (6.21) is the Dynkin diagram of $\overline{\mathfrak{g}}_{n+1}$.

Recall that Φ^+ and $\overline{\Phi}^+$ denote the standard positive systems of $\mathfrak{g}$ and $\overline{\mathfrak{g}}$, respectively. Also, $\mathfrak{b}$ and $\overline{\mathfrak{b}}$ denote the corresponding standard Borel subalgebras of $\mathfrak{g}$ and $\overline{\mathfrak{g}}$, respectively. By definition, $\mathfrak{g}$ and $\overline{\mathfrak{g}}$ are naturally subalgebras of $\widetilde{\mathfrak{g}}$.

Proposition 6.10. *We have* $\Phi^+ \subseteq \widetilde{\Phi}^c(n)$, *and* $\mathfrak{b} = \widetilde{\mathfrak{b}}^c(n) \cap \mathfrak{g}$. *Also,* $\overline{\Phi}^+ \subseteq \widetilde{\Phi}^s(n)$, *and* $\overline{\mathfrak{b}} = \widetilde{\mathfrak{b}}^s(n) \cap \overline{\mathfrak{g}}$.

Proof. It suffices to check that $\Pi \subseteq \widetilde{\Phi}^c(n)$ and $\overline{\Pi} \subseteq \widetilde{\Phi}^s(n)$, where Π and $\overline{\Pi}$ are the standard fundamental systems of $\mathfrak{g}$ and $\overline{\mathfrak{g}}$, respectively. These inclusions can then be observed directly from (6.19) and (6.21). $\square$

6.3.2. Odd reflections and highest weight modules.

Recall the standard Levi subalgebra $\widetilde{\mathfrak{l}}$ of $\widetilde{\mathfrak{g}}$ with nilradical $\widetilde{\mathfrak{u}}$ and opposite nilradical $\widetilde{\mathfrak{u}}^-$ from Section 6.2.2.

Lemma 6.11. *The sequences of odd reflections (6.18) and (6.20) leave the set of roots of $\widetilde{\mathfrak{u}}$ and the set of roots of $\widetilde{\mathfrak{u}}^-$ invariant.*

Proof. The isotropic odd roots used in the sequences of odd reflections (6.18) and (6.20) are of the form $\alpha_i + \alpha_{i+1/2} + \ldots + \alpha_j$ for $0 < i < j$; hence, they are all roots of $\widetilde{\mathfrak{l}}$. Since $\widetilde{\mathfrak{u}}$ and $\widetilde{\mathfrak{u}}^-$ are both invariant under the action of $\widetilde{\mathfrak{l}}$, the lemma follows by applying Lemma 1.30. $\square$

We denote by $\widetilde{\mathfrak{b}}^c_{\widetilde{\mathfrak{l}}}(n)$ and $\widetilde{\mathfrak{b}}^s_{\widetilde{\mathfrak{l}}}(n)$ the Borel subalgebras of the Levi subalgebra $\widetilde{\mathfrak{l}}$ corresponding to the fundamental systems $\widetilde{\Pi}^c(n) \cap \sum_{\alpha \in \widetilde{Y}} \mathbb{Z}\alpha$ and $\widetilde{\Pi}^s(n) \cap \sum_{\alpha \in \widetilde{Y}} \mathbb{Z}\alpha$, respectively (here we recall $\widetilde{Y}$ from (6.13)). By Lemma 6.11 we have, for all $n \in \mathbb{N}$,

(6.22) $\qquad \widetilde{\mathfrak{b}}^c(n) = \widetilde{\mathfrak{b}}^c_{\widetilde{\mathfrak{l}}}(n) + \widetilde{\mathfrak{u}}, \qquad \widetilde{\mathfrak{b}}^s(n) = \widetilde{\mathfrak{b}}^s_{\widetilde{\mathfrak{l}}}(n) + \widetilde{\mathfrak{u}}.$

Below we will regard $P^+ \subseteq \widetilde{\mathfrak{h}}^*$ and $\overline{P}^+ \subseteq \widetilde{\mathfrak{h}}^*$, via $\mathfrak{h}^* \subseteq \widetilde{\mathfrak{h}}^*$ and $\overline{\mathfrak{h}}^* \subseteq \widetilde{\mathfrak{h}}^*$.

Proposition 6.12. *Let* $\lambda = (d\Lambda_0, {}^-\lambda, {}^+\lambda) \in P^+$, *where* ${}^+\lambda = ({}^+\lambda_1, {}^+\lambda_2, \ldots)$ *is a partition. Let $n \in \mathbb{N}$.*

    (1) *Suppose that $\ell({}^+\lambda) \leq n$. Then the highest weight of $L(\widetilde{\mathfrak{l}}, \lambda^\theta)$ with respect to the Borel subalgebra $\widetilde{\mathfrak{b}}^c_{\widetilde{\mathfrak{l}}}(n)$ is λ.*

(2) *Suppose that* $^+\lambda_1 \le n$. *Then the highest weight of* $L(\widetilde{\mathfrak{l}}, \lambda^\theta)$ *with respect to the Borel subalgebra* $\widetilde{\mathfrak{b}}^s_{\mathfrak{l}}(n)$ *is* $\lambda^\natural$.

Proof. The proof of (2) using the sequence (6.20) is completely parallel to the proof of (1) which uses (6.18). We shall only prove (1).

We observe that the odd reflections (6.18) only affect the tail diagram $(\widetilde{\mathfrak{T}})$ and leave the head diagram $(\mathfrak{k})$ unchanged, and the summand

(6.23) $$\lambda|_{\mathfrak{k}} := d\Lambda_0 + \sum_{i=-m}^{-1} \lambda_i \varepsilon_i$$

of λ^θ in (6.16) is unchanged by these odd reflections.

We will show more generally by induction on k that, after applying the first $k(k+1)/2$ odd reflections of the sequence (6.18), the highest weight of $L(\widetilde{\mathfrak{l}}, \lambda^\theta)$ with respect to the Borel subalgebra $\widetilde{\mathfrak{b}}^c_{\mathfrak{l}}(k)$ becomes

(6.24) $$\begin{aligned}\lambda_{[k]} =& \lambda|_{\mathfrak{k}} + \sum_{i=1}^{k} {}^+\lambda_i \varepsilon_i + \sum_{i=1}^{k} \langle {}^+\lambda'_i - k \rangle \varepsilon_{i-1/2} \\ &+ \sum_{j \ge k+1} \langle {}^+\lambda'_j - j + 1 \rangle \varepsilon_{j-1/2} + \sum_{j \ge k+1} \langle {}^+\lambda_j - j \rangle \varepsilon_j,\end{aligned}$$

where $\langle q \rangle := \max\{q, 0\}$ for $q \in \mathbb{Z}$. Part (1) follows as a special case of (6.24) for $k = n$. Note that $\lambda_{[0]} = \lambda^\theta = \lambda|_{\mathfrak{k}} + \sum_{j \ge 1} \langle \lambda'_j - j + 1 \rangle \varepsilon_{j-1/2} + \sum_{j \ge 1} \langle \lambda_j - j \rangle \varepsilon_j$.

Suppose that $k = 1$. If $\ell({}^+\lambda) < 1$, then ${}^+\lambda = 0$ and $\lambda^\theta = \lambda|_{\mathfrak{k}}$. So in particular $\langle \lambda^\theta, \widehat{E}_{1/2} + \widehat{E}_1 \rangle = 0$, and thus by Lemma 1.40, the $\widetilde{\mathfrak{b}}^c_{\mathfrak{l}}(1)$-highest weight is $\lambda_{[1]} = \lambda^\theta$. If $\ell({}^+\lambda) \ge 1$, then $\langle \lambda^\theta, \widehat{E}_{1/2} + \widehat{E}_1 \rangle > 0$, as we recall that

$$\lambda^\theta = \lambda|_{\mathfrak{k}} + {}^+\lambda'_1 \varepsilon_{1/2} + ({}^+\lambda_1 - 1)\varepsilon_1 + \cdots .$$

Hence, by Lemma 1.40, the $\widetilde{\mathfrak{b}}^c_{\mathfrak{l}}(1)$-highest weight after the odd reflection with respect to $\varepsilon_{1/2} - \varepsilon_1$ is

$$\lambda_{[1]} = \lambda|_{\mathfrak{k}} + {}^+\lambda_1 \varepsilon_1 + ({}^+\lambda'_1 - 1)\varepsilon_{1/2} + \cdots ,$$

proving (6.24) in the case $k = 1$.

Now by the induction hypothesis, suppose that $\lambda_{[k]}$ is the new highest weight after applying the first $k(k+1)/2$ odd reflections of the sequence (6.18).

If $\ell({}^+\lambda) \le k$, then $\lambda_{[k]} = \lambda|_{\mathfrak{k}} + \sum_{i=1}^{k} {}^+\lambda_i \varepsilon_i$. Therefore, for $1 \le i \le k+1$, we have $\langle \lambda_{[k]}, \widehat{E}_{i-1/2} + \widehat{E}_{k+1} \rangle = 0$, and by Lemma 1.40, the odd reflections with respect to $\varepsilon_{i-1/2} - \varepsilon_{k+1}$ leave $\lambda_{[k]}$ unchanged. Thus we have $\lambda_{[k+1]} = \lambda_{[k]}$, and in this case we are done.

Now assume that $\ell({}^+\lambda) \ge k+1$. Let $s = {}^+\lambda_{k+1}$. We further separate this into two cases (i) and (ii) below.

(i) Suppose that ${}^+\lambda_{k+1} \geq k+1$. Then ${}^+\lambda'_{k+1} \geq k+1$, and hence we can rewrite

$$\lambda_{[k]} = \lambda|_{\mathfrak{t}} + \sum_{i=1}^{k} {}^+\lambda_i \varepsilon_i + \sum_{i=1}^{k} ({}^+\lambda'_i - k)\varepsilon_{i-1/2} + ({}^+\lambda'_{k+1} - k)\varepsilon_{k+1/2}$$
$$+ ({}^+\lambda_{k+1} - k - 1)\varepsilon_{k+1} + \sum_{j \geq k+2} \langle {}^+\lambda'_j - j + 1\rangle \varepsilon_{j-1/2} + \sum_{j \geq k+2} \langle {}^+\lambda_j - j\rangle \varepsilon_j.$$

Now we apply the next $(k+1)$ odd reflections in the sequence (6.18) consecutively. As $\langle \lambda_{[k]}, \widehat{E}_{k+1/2} + \widehat{E}_{k+1}\rangle > 0$, by Lemma 1.40 we calculate the new weight after the odd reflection with respect to $\varepsilon_{k+1/2} - \varepsilon_{k+1}$ to be

$$\lambda_{[k,1]} = \lambda|_{\mathfrak{t}} + \sum_{i=1}^{k} {}^+\lambda_i \varepsilon_i + \sum_{i=1}^{k} ({}^+\lambda'_i - k)\varepsilon_{i-1/2} + ({}^+\lambda'_{k+1} - k - 1)\varepsilon_{k+1/2}$$
$$+ ({}^+\lambda_{k+1} - k)\varepsilon_{k+1} + \sum_{j \geq k+2} \langle {}^+\lambda'_j - j + 1\rangle \varepsilon_{j-1/2} + \sum_{j \geq k+2} \langle {}^+\lambda_j - j\rangle \varepsilon_j.$$

Now $\langle \lambda_{[k,1]}, \widehat{E}_{k-1/2} + \widehat{E}_{k+1}\rangle > 0$, so by Lemma 1.40 we calculate the new weight after the odd reflection with respect to $\varepsilon_{k-1/2} - \varepsilon_{k+1}$ to be

$$\begin{aligned}\lambda_{[k,2]} =& \lambda|_{\mathfrak{t}} + \sum_{i=1}^{k} {}^+\lambda_i \varepsilon_i + \sum_{i=1}^{k-1} ({}^+\lambda'_i - k)\varepsilon_{i-1/2} + ({}^+\lambda'_k - k - 1)\varepsilon_{k-1/2} \\ &+ ({}^+\lambda'_{k+1} - k - 1)\varepsilon_{k+1/2} + ({}^+\lambda_{k+1} - k + 1)\varepsilon_{k+1} \\ &+ \sum_{j \geq k+2} \langle {}^+\lambda'_j - j + 1\rangle \varepsilon_{j-1/2} + \sum_{j \geq k+2} \langle {}^+\lambda_j - j\rangle \varepsilon_j.\end{aligned}$$

Continuing this way, after a total of $(k+1)$ odd reflections we end up with the weight

$$\begin{aligned}\lambda_{[k,k+1]} =& \lambda|_{\mathfrak{t}} + \sum_{i=1}^{k+1} {}^+\lambda_i \varepsilon_i + \sum_{i=1}^{k+1} ({}^+\lambda'_i - k - 1)\varepsilon_{i-1/2} \\ &+ \sum_{j \geq k+2} \langle {}^+\lambda'_j - j + 1\rangle \varepsilon_{j-1/2} + \sum_{j \geq k+2} \langle {}^+\lambda_j - j\rangle \varepsilon_j,\end{aligned}$$

which is exactly $\lambda_{[k+1]}$. The induction step is completed in this case.

(ii) Now suppose that ${}^+\lambda_{k+1} = s < k+1$. Since ${}^+\lambda'$ is a partition, the weight (6.24) becomes

$$\lambda_{[k]} = \lambda|_{\mathfrak{t}} + \sum_{i=1}^{k} {}^+\lambda_i \varepsilon_i + \sum_{i=1}^{s} ({}^+\lambda'_i - k)\varepsilon_{i-1/2},$$

where ${}^+\lambda'_i - k > 0$, for $i \leq s$. It follows by Lemma 1.40 that odd reflections with respect to $\varepsilon_{k+1/2} - \varepsilon_{k+1}, \cdots, \varepsilon_{s+1/2} - \varepsilon_{k+1}$ leave $\lambda_{[k]}$ unchanged, while odd reflections with respect to $\varepsilon_{s-1/2} - \varepsilon_{k+1}, \cdots, \varepsilon_{1/2} - \varepsilon_{k+1}$ do affect $\lambda_{[k]}$. In a similar way, the new weight after these $(k+1)$ odd reflections is calculated to be

$$\lambda_{[k,k+1]} = \lambda|_{\mathfrak{k}} + \sum_{i=1}^{k+1} {}^{+}\lambda_i \varepsilon_i + \sum_{i=1}^{s} ({}^{+}\lambda_i' - k - 1)\varepsilon_{i-1/2},$$

which is again equal to $\lambda_{[k+1]}$. The induction step is completed. □

Proposition 6.13. *Let $n \in \mathbb{N}$ and $\lambda = (d\Lambda_0, {}^{-}\lambda, {}^{+}\lambda) \in P^+$. Let $\widetilde{V}(\lambda^\theta) \in \widetilde{\mathcal{O}}$ be a highest weight $\widetilde{\mathfrak{g}}$-module of highest weight λ^θ with respect to the standard Borel subalgebra $\widetilde{\mathfrak{b}}$. The following statements hold.*

(1) *Suppose that $\ell({}^{+}\lambda) \leq n$. Then $\widetilde{V}(\lambda^\theta)$ is a highest weight module of highest weight λ with respect to the Borel subalgebra $\widetilde{\mathfrak{b}}^c(n)$.*

(2) *Suppose that ${}^{+}\lambda_1 \leq n$. Then $\widetilde{V}(\lambda^\theta)$ is a highest weight module of highest weight $\lambda^\natural$ with respect to the Borel subalgebra $\widetilde{\mathfrak{b}}^s(n)$.*

Proof. By definition, $\widetilde{\Delta}(\lambda^\theta) = U(\widetilde{\mathfrak{g}}) \otimes_{U(\widetilde{\mathfrak{p}})} L(\widetilde{\mathfrak{l}}, \lambda^\theta)$, and hence its $\widetilde{\mathfrak{g}}$-quotient $\widetilde{V}(\lambda^\theta)$ contains a unique copy of $L(\widetilde{\mathfrak{l}}, \lambda^\theta)$ that is annihilated by $\widetilde{\mathfrak{u}}$. By Proposition 6.12(1), $L(\widetilde{\mathfrak{l}}, \lambda^\theta)$ has $\widetilde{\mathfrak{b}}^c_{\mathfrak{l}}(n)$-highest weight λ, and let us denote by v a corresponding $\widetilde{\mathfrak{b}}^c_{\mathfrak{l}}(n)$-highest weight vector. Thus, by (6.22), v is also a $\widetilde{\mathfrak{b}}^c(n)$-singular vector in $\widetilde{V}(\lambda^\theta)$ of weight λ. The vector v clearly generates the $\widetilde{\mathfrak{l}}$-module $L(\widetilde{\mathfrak{l}}, \lambda^\theta)$ and hence the $\widetilde{\mathfrak{g}}$-module $\widetilde{V}(\lambda^\theta)$, proving (1).

The proof of (2) based on Proposition 6.12(2) is similar and hence omitted. □

6.3.3. The functors T and $\overline{T}$. By definition, $\mathfrak{g}$ and $\overline{\mathfrak{g}}$ are naturally subalgebras of $\widetilde{\mathfrak{g}}$, $\mathfrak{l}$ and $\overline{\mathfrak{l}}$ are subalgebras of $\widetilde{\mathfrak{l}}$, while $\mathfrak{h}$ and $\overline{\mathfrak{h}}$ are subalgebras of $\widetilde{\mathfrak{h}}$. Also, we have inclusions of the restricted duals $\mathfrak{h}^* \subseteq \widetilde{\mathfrak{h}}^*$ and $\overline{\mathfrak{h}}^* \subseteq \widetilde{\mathfrak{h}}^*$.

Given a semisimple $\widetilde{\mathfrak{h}}$-module $\widetilde{M} = \bigoplus_{\gamma \in \widetilde{\mathfrak{h}}^*} \widetilde{M}_\gamma$, we form the following subspaces of $\widetilde{M}$:

$$T(\widetilde{M}) := \bigoplus_{\gamma \in \mathfrak{h}^*} \widetilde{M}_\gamma, \quad \text{and} \quad \overline{T}(\widetilde{M}) := \bigoplus_{\gamma \in \overline{\mathfrak{h}}^*} \widetilde{M}_\gamma. \tag{6.25}$$

Note that $T(\widetilde{M})$ is an $\mathfrak{h}$-submodule of $\widetilde{M}$, and $\overline{T}(\widetilde{M})$ is an $\overline{\mathfrak{h}}$-submodule of $\widetilde{M}$. One checks that if $\widetilde{M}$ is also an $\widetilde{\mathfrak{l}}$-module, then $T(\widetilde{M})$ is an $\mathfrak{l}$-submodule of $\widetilde{M}$ and $\overline{T}(\widetilde{M})$ is an $\overline{\mathfrak{l}}$-submodule of $\widetilde{M}$. Furthermore, if $\widetilde{M}$ is a $\widetilde{\mathfrak{g}}$-module, then $T(\widetilde{M})$ is a $\mathfrak{g}$-submodule of $\widetilde{M}$ and $\overline{T}(\widetilde{M})$ is a $\overline{\mathfrak{g}}$-submodule of $\widetilde{M}$.

The direct sum decompositions in $\widetilde{M}$ give rise to the natural projections

$$T_{\widetilde{M}} : \widetilde{M} \longrightarrow T(\widetilde{M}) \quad \text{and} \quad \overline{T}_{\widetilde{M}} : \widetilde{M} \longrightarrow \overline{T}(\widetilde{M})$$

that are $\mathfrak{h}$- and $\overline{\mathfrak{h}}$-module homomorphisms, respectively. If $\widetilde{f} : \widetilde{M} \to \widetilde{N}$ is an $\widetilde{\mathfrak{h}}$-homomorphism, then the following maps induced by restrictions of $\widetilde{f}$,

$$T[\widetilde{f}] : T(\widetilde{M}) \longrightarrow T(\widetilde{N}) \quad \text{and} \quad \overline{T}[\widetilde{f}] : \overline{T}(\widetilde{M}) \longrightarrow \overline{T}(\widetilde{N}),$$

are also $\mathfrak{h}$- and $\overline{\mathfrak{h}}$-module homomorphisms, respectively. Also if $\widetilde{f}:\widetilde{M}\to\widetilde{N}$ is a $\widetilde{\mathfrak{g}}$-module homomorphism, then $T_{\widetilde{M}}$ and $T[\widetilde{f}]$ (respectively, $\overline{T}_{\widetilde{M}}$ and $\overline{T}[\widetilde{f}]$) are $\mathfrak{g}$-module (respectively, $\overline{\mathfrak{g}}$-module) homomorphisms. It follows that T and $\overline{T}$ define exact functors from the category of $\widetilde{\mathfrak{h}}$-semisimple $\widetilde{\mathfrak{g}}$-modules to the category of $\mathfrak{h}$-semisimple $\mathfrak{g}$-modules and the category of $\overline{\mathfrak{h}}$-semisimple $\overline{\mathfrak{g}}$-modules, respectively. Also, we have the following commutative diagrams:

$$
\begin{array}{ccc}
\widetilde{M} & \xrightarrow{\;\widetilde{f}\;} & \widetilde{N} \\
\big\downarrow{\scriptstyle T_{\widetilde{M}}} & & \big\downarrow{\scriptstyle T_{\widetilde{N}}} \\
T(\widetilde{M}) & \xrightarrow{\;T[\widetilde{f}]\;} & T(\widetilde{N})
\end{array}
\qquad\qquad
\begin{array}{ccc}
\widetilde{M} & \xrightarrow{\;\widetilde{f}\;} & \widetilde{N} \\
\big\downarrow{\scriptstyle \overline{T}_{\widetilde{M}}} & & \big\downarrow{\scriptstyle \overline{T}_{\widetilde{N}}} \\
\overline{T}(\widetilde{M}) & \xrightarrow{\;\overline{T}[\widetilde{f}]\;} & \overline{T}(\widetilde{N})
\end{array}
\tag{6.26}
$$

Lemma 6.14. *For $\lambda\in P^+$, we have*

$$T\big(L(\widetilde{\mathfrak{l}},\lambda^\theta)\big)=L(\mathfrak{l},\lambda),\qquad \overline{T}\big(L(\widetilde{\mathfrak{l}},\lambda^\theta)\big)=L(\overline{\mathfrak{l}},\lambda^\natural).$$

Proof. We prove the first formula using a comparison of the characters. The proof of the second formula is similar and will be omitted.

Let $\lambda=(d\Lambda_0,{}^-\lambda,{}^+\lambda)\in P^+$. To be consistent with the notation of $\lambda|_{\mathfrak{k}}$ in (6.23), it is convenient to make the convention that $\mathfrak{k}$ contains the central element K, and let $L(\widetilde{\mathfrak{l}}\cap\mathfrak{k},\lambda|_{\mathfrak{k}})$ denote the irreducible $\widetilde{\mathfrak{l}}\cap\mathfrak{k}$-module of highest weight $\lambda|_{\mathfrak{k}}$. It follows from the discussion in Section 6.2.1 that

$$\mathrm{ch}\,L(\widetilde{\mathfrak{l}},\lambda^\theta)=\mathrm{ch}\,L(\widetilde{\mathfrak{l}}\cap\mathfrak{k},\lambda|_{\mathfrak{k}})\,hs_{+\lambda'}(x_{1/2},x_{3/2},\ldots;x_1,x_2,\ldots),\tag{6.27}$$

where $x_r:=e^{\varepsilon_r}$ for $r\in\frac{1}{2}\mathbb{N}$.

Note that $\widetilde{\mathfrak{l}}\cap\mathfrak{k}=\mathfrak{l}\cap\mathfrak{k}$. Now $L(\widetilde{\mathfrak{l}},\lambda^\theta)$ is completely reducible as an $\mathfrak{l}$-module, since a polynomial module of $\mathfrak{gl}(\widetilde{V}_0^+)=\mathfrak{gl}(\infty|\infty)$ is completely reducible when restricted to $\mathfrak{gl}(V_0^+)=\mathfrak{gl}(\infty)$. On the character level, applying T to $L(\widetilde{\mathfrak{l}},\lambda^\theta)$ corresponds to setting $x_{1/2},x_{3/2},x_{5/2},\ldots$ in the character formula (6.27) to zero. Thus, by (A.37), $T\big(L(\widetilde{\mathfrak{l}},\lambda^\theta)\big)$ is an $\mathfrak{l}$-module with character $\mathrm{ch}\,L(\widetilde{\mathfrak{l}}\cap\mathfrak{k},\lambda|_{\mathfrak{k}})\,s_{+\lambda}(x_1,x_2,\ldots)$, which is precisely the character of $L(\mathfrak{l},\lambda)$. This proves the first formula. □

Proposition 6.15. *T is an exact functor from $\widetilde{\mathcal{O}}$ to $\mathcal{O}$, and $\overline{T}$ is an exact functor from $\widetilde{\mathcal{O}}$ to $\overline{\mathcal{O}}$.*

Proof. In light of Proposition 6.14, it suffices to show that if $\widetilde{M}\in\widetilde{\mathcal{O}}$, then $T(\widetilde{M})\in\mathcal{O}$ and $\overline{T}(\widetilde{M})\in\overline{\mathcal{O}}$. We shall only show that $T(\widetilde{M})\in\mathcal{O}$, as the argument for $\overline{T}(\widetilde{M})\in\overline{\mathcal{O}}$ is analogous.

First consider the case when the head diagram is non-degenerate, i.e., $m>0$ (see Section 6.1.1). Let $\widetilde{M}\in\widetilde{\mathcal{O}}$. Then there exists $\zeta_1^\theta,\zeta_2^\theta,\ldots,\zeta_r^\theta$, with $\zeta_1,\ldots,\zeta_r\in P^+$, such that any weight of $\widetilde{M}$ is bounded by some ζ_i^θ. Ignoring $d\Lambda_0$ recall that we have $\zeta_j = ({}^-\zeta_j, {}^+\zeta_j)$, where ${}^+\zeta_j$ is a partition with $|{}^+\zeta_j| = k_j$. For each ζ_j, let P_j be the following finite subset of $\mathfrak{h}^*$:

$$P_j := \{({}^-\zeta_j, \mu) \mid \mu \in \mathcal{P} \text{ with } |\mu| = k_j\}.$$

Set $M := T(\widetilde{M})$ and $P(M) := \bigcup_{j=1}^{r} P_j$.

Claim. Given any weight ν of M, there exists $\gamma \in P(M)$ such that $\gamma - \nu \in \mathbb{Z}_+\Pi$.

It suffices to prove the claim for $\nu \in P^+$. Since ν is also a weight of $\widetilde{M}$, we have $\zeta_i^\theta - \nu \in \mathbb{Z}_+\widetilde{\Pi}$, for some i. Thus

$$\zeta_i^\theta - \nu = p\alpha_\times + {}^-\kappa + {}^+\kappa,$$

where $p \in \mathbb{Z}_+$, ${}^-\kappa \in \sum_{\alpha \in \Pi(\mathfrak{k})} \mathbb{Z}_+\alpha$, and ${}^+\kappa \in \sum_{\beta \in \Pi(\widetilde{\mathfrak{T}}) \setminus \alpha_\times} \mathbb{Z}_+\beta$. This implies that ${}^+\nu$ is a partition of size $k_i + p$, and hence there exists $\gamma \in P(M)$ such that ${}^+\nu$ is obtained from the partition ${}^+\gamma$ by adding p boxes to it. For every such a box, we record the row number in which it was added in the multiset J with $|J| = p$. Then we have

$$\nu = \gamma - {}^-\kappa - \sum_{j \in J} (\varepsilon_{-1} - \varepsilon_j),$$

and hence $\gamma - \nu \in \mathbb{Z}_+\Pi$. Thus, we conclude that $M \in \mathcal{O}$.

The limit case $m = 0$ now can be proved case-by-case by slight modification of the argument above (see Exercise 6.1). □

Proposition 6.16. *Let $\lambda \in P^+$. If $\widetilde{V}(\lambda^\theta) \in \widetilde{\mathcal{O}}$ is a highest weight $\widetilde{\mathfrak{g}}$-module of highest weight λ^θ, then $T(\widetilde{V}(\lambda^\theta))$ and $\overline{T}(\widetilde{V}(\lambda^\theta))$ are highest weight $\mathfrak{g}$- and $\overline{\mathfrak{g}}$-modules of highest weights λ and $\lambda^\natural$, respectively.*

Proof. We will only prove the statement for $T(\widetilde{V}(\lambda^\theta))$, as the case of $\overline{T}(\widetilde{V}(\lambda^\theta))$ is analogous.

By Proposition 6.13, for $n > \ell({}^+\lambda)$, $\widetilde{V}(\lambda^\theta)$ is a $\widetilde{\mathfrak{b}}^c(n)$-highest weight module of highest weight λ, and thus a nonzero vector v in $\widetilde{V}(\lambda^\theta)$ of weight λ (unique up to a scalar multiple and independent of n) is a $\widetilde{\mathfrak{b}}^c(n)$-highest weight vector of $\widetilde{V}(\lambda^\theta)$. Evidently we have $v \in T(\widetilde{V}(\lambda^\theta))$, and by Proposition 6.10, v is a $\mathfrak{b}$-singular vector in $T(\widetilde{V}(\lambda^\theta))$.

The $\mathfrak{g}$-module $T(\widetilde{V}(\lambda^\theta))$, regarded as an $\mathfrak{l}$-module, is completely reducible by Proposition 6.7 and Lemma 6.14. To complete the proof, it suffices to show that every vector $w \in T(\widetilde{V}(\lambda^\theta))$ of weight $\mu = (d\Lambda_0, {}^-\mu, {}^+\mu) \in P^+$ lies in $U(\mathfrak{n}^-)v$, where $\mathfrak{n}^-$ denotes the opposite nilradical of $\mathfrak{b}$.

To that end, we choose n such that $n > \max\{\ell({}^+\lambda), \ell({}^+\mu)\}$. Then we have $w \in U(\widetilde{\mathfrak{n}}^c(n)^-)v$, where $\widetilde{\mathfrak{n}}^c(n)^-$ denotes the opposite nilradical of $\widetilde{\mathfrak{b}}^c(n)$. Now the condition $n > \max\{\ell({}^+\lambda), \ell({}^+\mu)\}$ implies that

$$\lambda - \mu = \sum_{i=-m}^{-1} a_i \varepsilon_i + \sum_{j=1}^{n-1} b_j \varepsilon_j, \quad a_i, b_j \in \mathbb{Z}. \tag{6.28}$$

(We emphasize that only ε_j with *integer* indices j appear in (6.28).) But $\lambda - \mu$ is also a $\mathbb{Z}_+$-linear combination of simple roots from $\widetilde{\Pi}^c(n)$, i.e.,

$$\lambda - \mu = \sum_{\alpha \in \widetilde{\Pi}^c(n)} a_\alpha \alpha,$$

with all $a_\alpha \in \mathbb{Z}_+$ and finitely many $a_\alpha > 0$. We have the following.

Claim. $\lambda - \mu$ is a $\mathbb{Z}_+$-linear combination of $\Pi_n = \Pi(\mathfrak{k}) \sqcup \{\beta_{-1}, \beta_1, \ldots, \beta_{n-1}\}$.

The claim then implies that $w \in U(\mathfrak{n}^-)v$, and the proposition follows.

It remains to prove the claim. Assume on the contrary that there were some $\alpha \in \widetilde{\Pi}^c(n) \setminus \Pi_n = \{\varepsilon_n - \varepsilon_{1/2}, \beta_{1/2}, \ldots, \beta_{n-1/2}, \alpha_{n+1/2}, \alpha_{n+1}, \ldots\}$ with $a_\alpha \neq 0$. If $a_{\varepsilon_n - \varepsilon_{\frac{1}{2}}} \neq 0$, then we must also have $a_\alpha \neq 0$, for $\alpha = \beta_{1/2}, \ldots, \beta_{n-1/2}, \alpha_{n+1/2}$ by the constraint (6.28). Similarly, if $a_{\beta_{i-1/2}} \neq 0$ for some $1 \leq i \leq n$, then we must also have $a_\alpha \neq 0$, for $\alpha = \beta_{i+1/2}, \ldots, \beta_{n-1/2}, \alpha_{n+1/2}$ as well. In all cases, $\langle \lambda - \mu, \widehat{E}_r \rangle \neq 0$, for some $r \geq n$, contradicting (6.28). □

6.3.4. Character formulas. The following theorem is the first main result of this chapter.

Theorem 6.17. *Let* $\lambda \in P^+$. *We have*

$$T\big(\widetilde{\Delta}(\lambda^\theta)\big) = \Delta(\lambda), \quad T\big(\widetilde{L}(\lambda^\theta)\big) = L(\lambda);$$
$$\overline{T}\big(\widetilde{\Delta}(\lambda^\theta)\big) = \overline{\Delta}(\lambda^\natural), \quad \overline{T}\big(\widetilde{L}(\lambda^\theta)\big) = \overline{L}(\lambda^\natural).$$

Proof. We will prove only the statements involving T. The statements involving $\overline{T}$ can be proved in an analogous way.

Let us write $\Delta(\lambda) = U(\mathfrak{u}^-) \otimes_{\mathbb{C}} L(\mathfrak{l}, \lambda)$ and $\widetilde{\Delta}(\lambda^\theta) = U(\widetilde{\mathfrak{u}}^-) \otimes_{\mathbb{C}} L(\widetilde{\mathfrak{l}}, \lambda^\theta)$. We observe that all the weights in $U(\mathfrak{u}^-)$, $L(\mathfrak{l}, \lambda)$, $U(\widetilde{\mathfrak{u}}^-)$, and $L(\widetilde{\mathfrak{l}}, \lambda^\theta)$ are of the form $\sum_{j<0} a_j \varepsilon_j + \sum_{r>0} b_r \varepsilon_r$ with $b_r \in \mathbb{Z}_+$ (possibly modulo $d\Lambda_0$). Since $T(U(\widetilde{\mathfrak{u}}^-)) = U(\mathfrak{u}^-)$, it follows by Lemma 6.14 that $\operatorname{ch} T\big(\widetilde{\Delta}(\lambda^\theta)\big) = \operatorname{ch} \Delta(\lambda)$. Since $T\big(\widetilde{\Delta}(\lambda^\theta)\big)$ is a highest weight module of highest weight λ by Proposition 6.16, we have $T\big(\widetilde{\Delta}(\lambda^\theta)\big) = \Delta(\lambda)$.

Set $\widetilde{M} = \widetilde{L}(\lambda^\theta)$. Suppose that $M := T(\widetilde{M})$ is reducible. Since by Proposition 6.16 the module M is a highest weight $\mathfrak{g}$-module, it must have a $\mathfrak{b}$-singular vector, say w of weight $\mu \in P^+$, inside M that is not a highest weight vector. We choose $n > \max\{\ell(^+\lambda), \ell(^+\mu)\}$. By Proposition 6.13, the $\widetilde{\mathfrak{g}}$-module $\widetilde{M}$ is a $\widetilde{\mathfrak{b}}^c(n)$-highest weight module of highest weight λ, say with a highest weight vector v_λ. By Proposition 6.16, v_λ is a $\mathfrak{b}$-highest weight vector of M, and hence $w \in U(\mathfrak{a})v_\lambda$, where $\mathfrak{a}$ is the subalgebra of $\mathfrak{n}^-$ generated by the root vectors corresponding to the roots in $-\Pi = -\Pi(\mathfrak{k}) \sqcup \{-\beta_{-1}, -\beta_1, \ldots, -\beta_k\}$, for some k.

Now choose $q > \max\{n, k+1\}$. Note that v_λ is also a $\widetilde{\mathfrak{b}}^c(q)$-highest weight vector of the $\widetilde{\mathfrak{g}}$-module $\widetilde{M}$ of weight λ. Since w is $\mathfrak{b}$-singular, it is annihilated by the root vectors corresponding to the roots in $\Pi(\mathfrak{k}) \sqcup \{\beta_{-1}, \beta_1, \beta_2, \ldots\}$. But w is also annihilated by the root vectors corresponding to the roots in $\widetilde{\Pi}^c(q)$ complementary to the subset $\Pi = \Pi(\mathfrak{k}) \sqcup \{\beta_{-1}, \beta_1, \beta_2, \ldots, \beta_{q-1}\}$, since these root vectors commute with $\mathfrak{a}$ and $w \in U(\mathfrak{a})v_\lambda$. It follows that w is a $\widetilde{\mathfrak{b}}^c(q)$-singular vector in $\widetilde{M}$, contradicting the irreducibility of $\widetilde{M}$. □

It is standard to write

$$\mathrm{ch} L(\lambda) = \sum_{\mu \in P^+} a_{\mu\lambda} \mathrm{ch} \Delta(\mu), \tag{6.29}$$

for $\lambda \in P^+$ and $a_{\mu\lambda} \in \mathbb{Z}$. For $\mathfrak{x} = \mathfrak{a}, \mathfrak{b}, \mathfrak{c}, \mathfrak{d}$, the triangular transition matrix $(a_{\mu\lambda})$ in Theorem 6.18 or its inverse is known in the Kazhdan-Lusztig theory, since the Kazhdan-Lusztig polynomials determine the composition multiplicities of parabolic Verma modules in the parabolic BGG category $\mathcal{O}$ of $\mathfrak{g}$-modules; see, e.g., Soergel [**116**, p. 455]. The infinite rank of $\mathfrak{g}$ here does not cause any difficulty in light of Proposition 6.9. We have the following character formulas for $\overline{\mathfrak{g}}$ and $\widetilde{\mathfrak{g}}$.

Theorem 6.18. *Let $\lambda \in P^+$. Then*

(1) $\mathrm{ch} \overline{L}(\lambda^\natural) = \sum_{\mu \in P^+} a_{\mu\lambda} \mathrm{ch} \overline{\Delta}(\mu^\natural)$,

(2) $\mathrm{ch} \widetilde{L}(\lambda^\theta) = \sum_{\mu \in P^+} a_{\mu\lambda} \mathrm{ch} \widetilde{\Delta}(\mu^\theta)$.

Proof. This follows immediately from Theorem 6.17 and (6.29). □

Remark 6.19. Theorem 6.18 together with Proposition 6.9 provide a complete solution à la Kazhdan-Lusztig to the irreducible character problem in the category $\overline{\mathcal{O}}_n$ for the finite-rank basic Lie superalgebras of type $\mathfrak{gl}$ and $\mathfrak{osp}$.

6.4. Kostant homology and KLV polynomials

In this section, we formulate and study the Kostant homology groups of the Lie superalgebras $\widetilde{\mathfrak{u}}^-$, $\mathfrak{u}^-$, and $\bar{\mathfrak{u}}^-$, with coefficients in modules belonging to the respective categories $\mathcal{O}$, $\overline{\mathcal{O}}$, and $\widetilde{\mathcal{O}}$. We show that the functors $T : \widetilde{\mathcal{O}} \to \mathcal{O}$ and $\overline{T} : \widetilde{\mathcal{O}} \to \overline{\mathcal{O}}$ match perfectly the corresponding Kostant homology groups (and also Kostant cohomology groups). Such matchings are then interpreted as equalities of the Kazhdan-Lusztig-Vogan polynomials for the categories $\mathcal{O}$, $\overline{\mathcal{O}}$, and $\widetilde{\mathcal{O}}$.

Some basic facts on Lie algebra homology and cohomology needed in this section can be found in the book of Kumar [**76**].

6.4.1. Homology and cohomology of Lie superalgebras. Let $L = L_{\bar{0}} \oplus L_{\bar{1}}$ be a vector superspace, and let $\mathcal{T}(L)$ be the tensor superalgebra of L. Then $\mathcal{T}(L) = \bigoplus_{n=0}^{\infty} L^{\otimes n}$ is a $\mathbb{Z}$-graded associative superalgebra. Recall that the exterior superalgebra of L is the quotient superalgebra $\Lambda(L) := \mathcal{T}(L)/J$, where J is the homogeneous two-sided ideal of $\mathcal{T}(L)$ generated by elements of the form $x \otimes y +$

$(-1)^{|x||y|}y\otimes x$, where x and y are $\mathbb{Z}_2$-homogeneous elements of L. Then $\wedge(L) = \bigoplus_{n=0}^{\infty}\wedge^n L$ is also a $\mathbb{Z}$-graded associative superalgebra. For elements $x_1, x_2, \ldots, x_k$ in L, the image of the element $x_1\otimes x_2\otimes\cdots\otimes x_k$ under the canonical quotient map from $L^{\otimes k}$ to $\wedge^k(L)$ will be denoted by $x_1x_2\cdots x_k$. We emphasize that the exterior algebra defined and used in this subsection is always understood in the super sense.

Now let $L = L_{\bar 0}\oplus L_{\bar 1}$ be a Lie superalgebra of countable dimension. For an L-module V we define

$$C_n(L,V) = \wedge^n(L)\otimes V, \qquad C_\bullet(L,V) = \bigoplus_{n\geq 0} C_n(L,V).$$

Define the **boundary operator** $d := \bigoplus_n d_n : C_\bullet(L,V)\to C_\bullet(L,V)$ by

$$\begin{aligned} & d_n(x_1x_2\cdots x_n\otimes v) \\ & := \sum_{1\leq s<t\leq n} (-1)^{s+t+|x_s|\sum_{i=1}^{s-1}|x_i|+|x_t|\sum_{j=1}^{t-1}|x_j|+|x_s||x_t|}[x_s,x_t]x_1\cdots\widehat{x_s}\cdots\widehat{x_t}\cdots x_n\otimes v \\ & + \sum_{s=1}^{n}(-1)^{s+|x_s|\sum_{i=s+1}^{n}|x_i|}x_1\cdots\widehat{x_s}\cdots x_n\otimes x_s v, \end{aligned} \tag{6.30}$$

for $x_i\in L$ homogeneous and $v\in V$. Here $[x_s,x_t]\in L$ denotes the supercommutator and $\widehat{y}$ indicates that the term y is omitted as usual. It is standard to check that $d^2=0$ (for Lie algebras see, e.g., Kumar [**76**] and for Lie superalgebras see, e.g., [**68**] for a proof).

The kth **Lie superalgebra homology group** of L with coefficient in V, denoted by $H_k(L,V)$, is defined to be the kth homology group of the following chain complex:

$$\cdots\longrightarrow C_n(L,V)\xrightarrow{d_n} C_{n-1}(L,V)\longrightarrow\cdots\xrightarrow{d_2} L\otimes V\xrightarrow{d_1} V\xrightarrow{d_0} 0,$$

that is, $H_k(L,V) = \mathrm{Ker}\, d_k/\mathrm{Im}\, d_{k+1}$, for $k\geq 0$.

Let us be more precise about the dual of a possibly infinite-dimensional space, as it is relevant to the Lie (super)algebra cohomology to be defined below.

Now suppose that L is a Lie superalgebra (or simply a vector superspace) that contains a sequence of finite-dimensional subalgebras (or subspaces) L_k, for $k\in\mathbb{N}$, such that $L_k\subseteq L_{k+1}$ for each k and $L=\cup_k L_k$. Furthermore, suppose that L contains a natural basis B such that $B\cap L_k$ is a basis for L_k, for all k. Then with respect to the basis B, we may extend by zero and regard $L_k^*\subseteq L_{k+1}^*$ for each k. In this way, the **restricted dual** of L, denoted by L^*, is given by $L^* := \cup_k L_k^*$. All the examples of Lie superalgebras in this book satisfy these assumptions, and there is such a basis for L consisting of root or weight vectors.

Fix $n\in\mathbb{Z}_+$, and let V be a vector superspace of countable dimension. Then, $\wedge^n L_k\subseteq\wedge^n L_{k+1}$, and the basis B of L induces a natural basis, denoted formally by $\wedge^n B$, of $\wedge^n L$ that is compatible with $\wedge^n L_k$ for all k. Regard $\mathrm{Hom}(\wedge^n L_k, V)\subseteq \mathrm{Hom}(\wedge^n L_{k+1}, V)$ by an extension by zero with respect to the basis $\wedge^n B$, for each k.

In this way, we define the (restricted) Hom-space

$$\mathrm{Hom}(\wedge^n L, V) := \bigcup_k \mathrm{Hom}(\wedge^n L_k, V). \tag{6.31}$$

In the same sense, a (restricted) $\mathbb{C}$-multilinear map $f : L \times \cdots \times L \to V$ means the extension by zero of a $\mathbb{C}$-multilinear map $f : L_k \times \cdots \times L_k \to V$ for some k.

The Lie superalgebra cohomology is defined as follows. Let L be a Lie superalgebra satisfying the assumptions above. Take an L-module V. For $n \in \mathbb{Z}_+$, let

$$C^n(L,V) = \mathrm{Hom}(\wedge^n L, V), \qquad C^\bullet(L,V) = \bigoplus_{n\geq 0} C^n(L,V), \tag{6.32}$$

(see (6.31)). An element $f \in C^n(L,V)$ is identified with a restricted $\mathbb{C}$-multilinear map $f : \overbrace{L \times \cdots \times L}^{n} \to V$, which is skew-supersymmetric in the following sense: for $1 \leq i \leq n-1$,

$$f(x_1, \ldots, x_i, x_{i+1}, \ldots, x_n) = -(-1)^{|x_i|\cdot|x_{i+1}|} f(x_1, \ldots, x_{i+1}, x_i, \ldots, x_n).$$

Define the **coboundary operator** $\partial = \bigoplus_{n\geq 0} \partial_n : C^\bullet(L,V) \to C^\bullet(L,V)$ by

(6.33)

$$\begin{aligned}
&(\partial_n f)(x_1, \ldots, x_{n+1}) \\
&:= \sum_{s=1}^{n+1} (-1)^{s+1+|x_s|(|f|+\sum_{i=1}^{s-1}|x_i|)} x_s f(x_1, \ldots, x_{s-1}, \widehat{x_s}, x_{s+1}, \ldots . x_{n+1}) \\
&+ \sum_{s<t} (-1)^{s+t+|x_s|\sum_{i=1}^{s-1}|x_i| + |x_t|\sum_{j=1}^{t-1}|x_j| + |x_s||x_t|} f([x_s, x_t], x_1, \ldots, \widehat{x_s}, \ldots, \widehat{x_t}, \ldots, x_{n+1}).
\end{aligned}$$

It is straightforward to verify that $\partial^2 = 0$. The kth (restricted) **Lie superalgebra cohomology group** of L with coefficient in the L-module V, denoted by $H^k(L,V)$, is by definition the kth cohomology group of the complex

$$0 \xrightarrow{\partial_{-1}} V \xrightarrow{\partial_0} C^1(L,V) \xrightarrow{\partial_1} \cdots \xrightarrow{\partial_{n-1}} C^n(L,V) \xrightarrow{\partial_n} C^{n+1}(L,V) \longrightarrow \cdots, \tag{6.34}$$

that is, $H^k(L,V) := \mathrm{Ker}\, \partial_k / \mathrm{Im}\, \partial_{k-1}$.

Remark 6.20. If we use the cochain complex (6.34) above by interpreting the Hom-space in the non-restricted sense to be the space of *all* linear maps from $\wedge^n L$ to V, the resulting nth Lie superalgebra cohomology group will be denoted by $\mathcal{H}^n(L,V)$. This version of cohomology will only appear once in this book, as it is used mildly in the proof of Theorem 6.34. Of course, the two versions of cohomology coincide when L is finite dimensional.

6.4.2. Kostant $\mathfrak{u}^-$-homology and $\mathfrak{u}$-cohomology. Recall the Lie algebra $\mathfrak{g}$ with parabolic subalgebra $\mathfrak{p}$, Levi subalgebra $\mathfrak{l}$, nilradical $\mathfrak{u}$, and opposite nilradical $\mathfrak{u}^-$ from Section 6.2.2. Let $M \in \mathcal{O}$. Then we have $M = \bigoplus_{\mu \in \mathfrak{h}^*} M_\mu$, with $\dim M_\mu < \infty$. In this case, the restricted dual M^* of M is given by $M^* = \bigoplus_{\mu \in \mathfrak{h}^*} M_\mu^*$. Let $\tau : \mathfrak{g} \to \mathfrak{g}$ be the Chevalley automorphism of $\mathfrak{g}$ from Section 1.1.2 in Chapter 1. We denote M^* with a τ-twisted $\mathfrak{g}$-action by $M^\vee$, i.e.,

$$M^\vee := \bigoplus_{\mu \in \mathfrak{h}^*} M_\mu^*,$$

where the action of $\mathfrak{g}$ is now given by $(x \cdot f)(v) = (-1)^{|x| \cdot |f| + 1} f(\tau(x) \cdot v)$. Then $M^\vee$ lies in $\mathcal{O}$, for $M \in \mathcal{O}$. We shall refer to $M^\vee$ as the **dual** of M in $\mathcal{O}$. Sending each $M \in \mathcal{O}$ to $M^\vee$ defines an exact functor, called the **duality functor**, on $\mathcal{O}$. The following is a summary of some standard properties of the duality functor on $\mathcal{O}$; cf. Humphreys [**54**, Theorem 3.2].

Proposition 6.21. *The duality functor is an exact contravariant functor, inducing a self-equivalence on the category $\mathcal{O}$. For each $M \in \mathcal{O}$, M and $M^\vee$ have the same character, and hence they have the same composition factor multiplicities. In particular, we have $L(\lambda)^\vee \cong L(\lambda)$, for $\lambda \in P^+$.*

Let $M \in \mathcal{O}$. We note that $\mathfrak{u}$ and $\mathfrak{u}^-$ are $\mathfrak{l}$-modules so that the chain and cochain complexes of the Lie superalgebra $\mathfrak{u}$ or $\mathfrak{u}^-$ with coefficients in M are semisimple $\mathfrak{l}$-modules. It is a standard fact that the boundary and coboundary operators are $\mathfrak{l}$-invariant, so the homology and cohomology groups of $\mathfrak{u}$ and $\mathfrak{u}^-$ are naturally semisimple $\mathfrak{l}$-modules.

Lemma 6.22. *Let $M \in \mathcal{O}$ and $n \in \mathbb{Z}_+$. As semisimple $\mathfrak{l}$-modules, we have*

$$H^n(\mathfrak{u}, M) \cong H_n(\mathfrak{u}, M^*)^*.$$

Proof. We have the following isomorphism of $\mathfrak{l}$-modules for each $n \geq 0$:

$$\mathrm{Hom}(\Lambda^n \mathfrak{u}, M) \cong \mathrm{Hom}(\Lambda^n \mathfrak{u} \otimes M^*, \mathbb{C}).$$

The Hom's here are understood as usual in the restricted sense as in (6.31). Furthermore, these isomorphisms are compatible with the coboundary operator ∂ of the complex $C^\bullet(\mathfrak{u}, M)$ and the coboundary operator $\mathrm{Hom}(d, \mathbb{C})$ of the complex $\mathrm{Hom}(C_\bullet(\mathfrak{u}, M^*), \mathbb{C})$. Since the operators are also $\mathfrak{l}$-module homomorphisms, the corresponding isomorphisms of the cohomology groups are also isomorphisms of $\mathfrak{l}$-modules. □

Lemma 6.23. *Let $M \in \mathcal{O}$ and $n \in \mathbb{Z}_+$. As semisimple $\mathfrak{l}$-modules, we have*

$$H_n(\mathfrak{u}, M^*)^* \cong H_n(\mathfrak{u}^-, M^\vee).$$

In particular, we have $H_n(\mathfrak{u}, L(\lambda)^*)^* \cong H_n(\mathfrak{u}^-, L(\lambda))$.

Proof. We use the superscript τ to indicate the $\mathfrak{l}$-action obtained by twisting the usual $\mathfrak{l}$-action by the Chevalley automorphism τ. Since $\mathfrak{u}^\tau \cong \mathfrak{u}^-$ and $(M^*)^\tau = M^\vee$, we see that

$$\left(\wedge^i \mathfrak{u} \otimes M^*\right)^\tau \cong \wedge^i \mathfrak{u}^- \otimes M^\vee.$$

Furthermore, these isomorphisms of chain complexes commute with the respective boundary operators as well. Thus, for each n we obtain an $\mathfrak{l}$-module isomorphism:

$$H_n(\mathfrak{u}, M^*)^\tau \cong H_n(\mathfrak{u}^-, M^\vee).$$

Now since the $\mathfrak{l}$-modules $H_n(\mathfrak{u}, M^*)^\tau$ and $H_n(\mathfrak{u}, M^*)^*$ are direct sums of simple $\mathfrak{l}$-modules with highest weights in P^+ and they have the same character, we have an $\mathfrak{l}$-module isomorphism $H_n(\mathfrak{u}, M^*)^\tau \cong H_n(\mathfrak{u}, M^*)^*$. The lemma follows. □

Clearly the counterparts for categories $\overline{\mathcal{O}}$ and $\widetilde{\mathcal{O}}$ of Proposition 6.21 and Lemmas 6.22 and 6.23 hold with verbatim arguments.

Theorem 6.24. *Let $M \in \mathcal{O}$, $\widetilde{M} \in \widetilde{\mathcal{O}}$, $\overline{M} \in \overline{\mathcal{O}}$, and $n \in \mathbb{Z}_+$. We have*

$$H_n\left(\mathfrak{u}^-, M^\vee\right) \cong H^n\left(\mathfrak{u}, M\right), \quad \text{as } \mathfrak{l}\text{-modules};$$

$$H_n\left(\bar{\mathfrak{u}}^-, \overline{M}^\vee\right) \cong H^n\left(\bar{\mathfrak{u}}, \overline{M}\right), \quad \text{as } \bar{\mathfrak{l}}\text{-modules};$$

$$H_n\left(\widetilde{\mathfrak{u}}^-, \widetilde{M}^\vee\right) \cong H^n\left(\widetilde{\mathfrak{u}}, \widetilde{M}\right), \quad \text{as } \widetilde{\mathfrak{l}}\text{-modules}.$$

In particular, we have $H_n\left(\mathfrak{u}^-, L(\lambda)\right) \cong H^n\left(\mathfrak{u}, L(\lambda)\right)$, $H_n\left(\bar{\mathfrak{u}}^-, \overline{L}(\lambda^\natural)\right) \cong H^n\left(\bar{\mathfrak{u}}, \overline{L}(\lambda^\natural)\right)$, and $H_n\left(\widetilde{\mathfrak{u}}^-, \widetilde{L}(\lambda^\theta)\right) \cong H^n\left(\widetilde{\mathfrak{u}}, \widetilde{L}(\lambda^\theta)\right)$, for $\lambda \in P^+$.

Proof. We only prove the statements for $\mathcal{O}$, while those for $\overline{\mathcal{O}}$ and $\widetilde{\mathcal{O}}$ are analogous.

Combining Lemma 6.22 and Lemma 6.23, we obtain an $\mathfrak{l}$-module isomorphism $H_n\left(\mathfrak{u}^-, M^\vee\right) \cong H^n\left(\mathfrak{u}, M\right)$. This implies that $H_n\left(\mathfrak{u}^-, L(\lambda)\right) \cong H^n\left(\mathfrak{u}, L(\lambda)\right)$, thanks to $L(\lambda)^\vee \cong L(\lambda)$ by Proposition 6.21. □

6.4.3. Comparison of Kostant homology groups. We start by setting up notation. Recall the functors $T : \widetilde{\mathcal{O}} \to \mathcal{O}$ and $\overline{T} : \widetilde{\mathcal{O}} \to \mathcal{O}$ from (6.25). For $\widetilde{M} \in \widetilde{\mathcal{O}}$, set $M = T(\widetilde{M}) \in \mathcal{O}$ and $\overline{M} = \overline{T}(\widetilde{M}) \in \overline{\mathcal{O}}$. Denote by

$$\widetilde{d} : \wedge(\widetilde{\mathfrak{u}}^-) \otimes \widetilde{M} \longrightarrow \wedge(\widetilde{\mathfrak{u}}^-) \otimes \widetilde{M},$$

$$d : \wedge(\mathfrak{u}^-) \otimes M \longrightarrow \wedge(\mathfrak{u}^-) \otimes M, \qquad \overline{d} : \wedge(\bar{\mathfrak{u}}^-) \otimes \overline{M} \longrightarrow \wedge(\bar{\mathfrak{u}}^-) \otimes \overline{M}$$

the boundary operators of the three chain complexes $C_\bullet(\widetilde{\mathfrak{u}}^-, \widetilde{M})$, $C_\bullet(\mathfrak{u}^-, M)$, and $C_\bullet(\bar{\mathfrak{u}}^-, \overline{M})$, respectively. Note that $\widetilde{d}$, d, and $\overline{d}$ are homomorphisms of $\widetilde{\mathfrak{l}}$-, $\mathfrak{l}$-, and $\bar{\mathfrak{l}}$-modules, respectively.

The $\mathfrak{l}$-module $\wedge(\mathfrak{u}^-)$ is a direct sum of $L(\mathfrak{l}, \mu)$, $\mu \in P^+$, each appearing with finite multiplicity. As an $\bar{\mathfrak{l}}$-module, $\wedge(\bar{\mathfrak{u}}^-)$ is a direct sum of $L(\bar{\mathfrak{l}}, \mu^\natural)$, $\mu \in P^+$, each appearing with finite multiplicity. By Theorem 6.4, the $\widetilde{\mathfrak{l}}$-module $\wedge(\widetilde{\mathfrak{u}}^-)$ is also a direct sum of $L(\widetilde{\mathfrak{l}},\mu^\theta)$, $\mu\in P^+$, each appearing with finite multiplicity. The $\mathfrak{l}$-module $\Lambda(\mathfrak{u}^-)\otimes M$, the $\overline{\mathfrak{l}}$-module $\Lambda(\bar{\mathfrak{u}}^-)\otimes\overline{M}$, and the $\widetilde{\mathfrak{l}}$-module $\Lambda(\widetilde{\mathfrak{u}}^-)\otimes\widetilde{M}$ are completely reducible by Theorem 6.4.

Lemma 6.25. *For $\widetilde{M}\in\widetilde{\mathcal{O}}$ and $\lambda\in P^+$, set $M=T(\widetilde{M})\in\mathcal{O}$ and $\overline{M}=\overline{T}(\widetilde{M})\in\overline{\mathcal{O}}$. Then,*

(1) $T\big(\Lambda(\widetilde{\mathfrak{u}}^-)\otimes\widetilde{M}\big)=\Lambda(\mathfrak{u}^-)\otimes M$, *and* $T\big(\Lambda(\widetilde{\mathfrak{u}}^-)\otimes\widetilde{L}(\lambda^\theta)\big)=\Lambda(\mathfrak{u}^-)\otimes L(\lambda)$. *Moreover,* $T\big[\widetilde{d}\big]=d$.

(2) $\overline{T}\big(\Lambda(\widetilde{\mathfrak{u}}^-)\otimes\widetilde{M}\big)=\Lambda(\bar{\mathfrak{u}}^-)\otimes\overline{M}$, *and* $\overline{T}\big(\Lambda(\widetilde{\mathfrak{u}}^-)\otimes\widetilde{L}(\lambda^\theta)\big)=\Lambda(\bar{\mathfrak{u}}^-)\otimes\overline{L}(\lambda^\natural)$. *Moreover,* $\overline{T}\big[\widetilde{d}\big]=\overline{d}$.

Proof. We will only prove (1), as the argument for (2) is parallel.

It follows from $\widetilde{\mathfrak{u}}^-\cap\mathfrak{g}=\mathfrak{u}^-$ and the definition of T that $T\big(\Lambda(\widetilde{\mathfrak{u}}^-)\big)=\Lambda(\mathfrak{u}^-)$. Now, since all modules involved have weights of the form $\sum_{i<0}a_i\varepsilon_i+\sum_{r>0}b_r\varepsilon_r$ (possibly modulo $d\Lambda_0$) with $b_r\in\mathbb{Z}_+$, we see that $T\big(\Lambda(\widetilde{\mathfrak{u}}^-)\otimes\widetilde{M}\big)$ and $\Lambda(\mathfrak{u}^-)\otimes M$ have the same character. Hence, it follows from $T\big(\Lambda(\widetilde{\mathfrak{u}}^-)\otimes\widetilde{M}\big)\supseteq\Lambda(\mathfrak{u}^-)\otimes M$ that $T\big(\Lambda(\widetilde{\mathfrak{u}}^-)\otimes\widetilde{M}\big)=\Lambda(\mathfrak{u}^-)\otimes M$. By Theorem 6.17, $T\big(\widetilde{L}(\lambda^\theta)\big)=L(\lambda)$, and so we have $T\big(\Lambda(\widetilde{\mathfrak{u}}^-)\otimes\widetilde{L}(\lambda^\theta)\big)=\Lambda(\mathfrak{u}^-)\otimes L(\lambda)$.

From the definitions of $\widetilde{d}$, d, and $\overline{d}$ in (6.30), we have $\widetilde{d}(v)=d(v)$ for all $v\in\Lambda(\mathfrak{u}^-)\otimes M$ and $\widetilde{d}(w)=\overline{d}(w)$ for all $w\in\Lambda(\bar{\mathfrak{u}}^-)\otimes\overline{M}$. Thus, $T\big[\widetilde{d}\big]=d$ and $\overline{T}\big[\widetilde{d}\big]=\overline{d}$. This completes the proof of (1). □

Proposition 6.26. *For $\widetilde{M}\in\widetilde{\mathcal{O}}$, set $M=T(\widetilde{M})\in\mathcal{O}$ and $\overline{M}=\overline{T}(\widetilde{M})\in\overline{\mathcal{O}}$. Suppose that $\Lambda(\widetilde{\mathfrak{u}}^-)\otimes\widetilde{M}\cong\bigoplus_{\mu\in P^+}L(\widetilde{\mathfrak{l}},\mu^\theta)^{m(\mu)}$, as $\widetilde{\mathfrak{l}}$-modules. Then*

(1) $\Lambda(\mathfrak{u}^-)\otimes M\cong\bigoplus_{\mu\in P^+}L(\mathfrak{l},\mu)^{m(\mu)}$, *as $\mathfrak{l}$-modules.*

(2) $\Lambda(\bar{\mathfrak{u}}^-)\otimes\overline{M}\cong\bigoplus_{\mu\in P^+}L(\overline{\mathfrak{l}},\mu^\natural)^{m(\mu)}$, *as $\overline{\mathfrak{l}}$-modules.*

Proof. We only prove (1), as (2) is similar. By Lemma 6.25, we have a surjective $\mathfrak{l}$-module homomorphism $T:\Lambda(\widetilde{\mathfrak{u}}^-)\otimes\widetilde{M}\to\Lambda(\mathfrak{u}^-)\otimes M$. By Lemma 6.14, $T\big(L(\widetilde{\mathfrak{l}},\mu^\theta)\big)=L(\mathfrak{l},\mu)$, for all $\mu\in P^+$. Now (1) follows. □

By Lemma 6.25 and (6.26), we have the following commutative diagram:

$$
\begin{array}{ccccccccc}
\cdots \longrightarrow & \Lambda^{n+1}(\widetilde{\mathfrak{u}}^-)\otimes\widetilde{M} & \xrightarrow{\widetilde{d}} & \Lambda^{n}(\widetilde{\mathfrak{u}}^-)\otimes\widetilde{M} & \xrightarrow{\widetilde{d}} & \Lambda^{n-1}(\widetilde{\mathfrak{u}}^-)\otimes\widetilde{M} & \xrightarrow{\widetilde{d}} & \cdots \\
 & \downarrow{\scriptstyle T_{\Lambda^{n+1}(\widetilde{\mathfrak{u}}^-)\otimes\widetilde{M}}} & & \downarrow{\scriptstyle T_{\Lambda^{n}(\widetilde{\mathfrak{u}}^-)\otimes\widetilde{M}}} & & \downarrow{\scriptstyle T_{\Lambda^{n-1}(\widetilde{\mathfrak{u}}^-)\otimes\widetilde{M}}} & & \\
\cdots \longrightarrow & \Lambda^{n+1}(\mathfrak{u}^-)\otimes M & \xrightarrow{d} & \Lambda^{n}(\mathfrak{u}^-)\otimes M & \xrightarrow{d} & \Lambda^{n-1}(\mathfrak{u}^-)\otimes M & \xrightarrow{d} & \cdots
\end{array}
\tag{6.35}
$$

Thus T induces an $\mathfrak{l}$-homomorphism from $H_n(\widetilde{\mathfrak{u}}^-,\widetilde{M})$ to $H_n(\mathfrak{u}^-,M)$. Similarly, $\overline{T}$ induces an $\overline{\mathfrak{l}}$-homomorphism from $H_n(\widetilde{\mathfrak{u}}^-,\widetilde{M})$ to $H_n(\bar{\mathfrak{u}}^-,\overline{M})$. We have the following more precise result.

Theorem 6.27. *For $\widetilde{M} \in \widetilde{\mathcal{O}}$, set $M = T(\widetilde{M}) \in \mathcal{O}$ and $\overline{M} = \overline{T}(\widetilde{M}) \in \overline{\mathcal{O}}$. Let $n \in \mathbb{Z}_+$. Then,*

(1) $T(H_n(\widetilde{\mathfrak{u}}^-, \widetilde{M})) \cong H_n(\mathfrak{u}^-, M)$, *as $\mathfrak{l}$-modules.*

(2) $\overline{T}(H_n(\widetilde{\mathfrak{u}}^-, \widetilde{M})) \cong H_n(\bar{\mathfrak{u}}^-, \overline{M})$, *as $\bar{\mathfrak{l}}$-modules.*

Proof. We shall only prove (1), as the argument for (2) is similar.

We regard $\wedge(\mathfrak{u}^-) \otimes M \subseteq \wedge(\widetilde{\mathfrak{u}}^-) \otimes \widetilde{M}$. By Lemma 6.25 and (6.35), we have

$$T(\mathrm{Ker}\, \widetilde{d}) = \mathrm{Ker}\, \widetilde{d} \cap \left(\wedge(\mathfrak{u}^-) \otimes M\right) = \mathrm{Ker}\, d,$$

and

$$T(\mathrm{Im}\, \widetilde{d}) = \mathrm{Im}\, \widetilde{d} \cap \left(\wedge(\mathfrak{u}^-) \otimes M\right) = \mathrm{Im}\, d.$$

Since T is an exact functor, we have

$$T\left(\bigoplus_{n\geq 0} H_n(\widetilde{\mathfrak{u}}^-, \widetilde{M})\right) = T(\mathrm{Ker}\, \widetilde{d})/T(\mathrm{Im}\, \widetilde{d}) = \mathrm{Ker}\, d/\mathrm{Im}\, d = \bigoplus_{n\geq 0} H_n(\mathfrak{u}^-, M).$$

This completes the proof. □

Corollary 6.28. *Let $\widetilde{M} \in \widetilde{\mathcal{O}}$, $\lambda \in P^+$, and $n \in \mathbb{Z}_+$. Furthermore, set $T(\widetilde{M}) = M$ and $\overline{T}(\widetilde{M}) = \overline{M}$. We have:*

(1) $T(H^n(\widetilde{\mathfrak{u}}, \widetilde{M})) \cong H^n(\mathfrak{u}, M)$, *and thus* $T(H^n(\widetilde{\mathfrak{u}}, \widetilde{L}(\lambda^\theta))) \cong H^n(\mathfrak{u}, L(\lambda))$, *as $\mathfrak{l}$-modules.*

(2) $\overline{T}(H^n(\widetilde{\mathfrak{u}}, \widetilde{M})) \cong H^n(\mathfrak{u}, \overline{M})$, *and thus* $\overline{T}(H^n(\widetilde{\mathfrak{u}}, \widetilde{L}(\lambda^\theta))) \cong H^n(\bar{\mathfrak{u}}, \overline{L}(\lambda^\natural))$, *as $\bar{\mathfrak{l}}$-modules.*

(3) *Moreover, we have*

$$\begin{aligned}\mathrm{Hom}_{\mathfrak{l}}\left(L(\mathfrak{l}, \mu), H^n(\mathfrak{u}, M)\right) &\cong \mathrm{Hom}_{\widetilde{\mathfrak{l}}}\left(L(\widetilde{\mathfrak{l}}, \mu^\theta), H^n(\widetilde{\mathfrak{u}}, \widetilde{M})\right)\\ &\cong \mathrm{Hom}_{\bar{\mathfrak{l}}}\left(L(\bar{\mathfrak{l}}, \mu^\natural), H^n(\bar{\mathfrak{u}}, \overline{M})\right),\end{aligned}$$

and so in particular

$$\begin{aligned}\mathrm{Hom}_{\mathfrak{l}}\left(L(\mathfrak{l}, \mu), H^n(\mathfrak{u}, L(\lambda))\right) &\cong \mathrm{Hom}_{\widetilde{\mathfrak{l}}}\left(L(\widetilde{\mathfrak{l}}, \mu^\theta), H^n(\widetilde{\mathfrak{u}}, \widetilde{L}(\lambda^\theta))\right)\\ &\cong \mathrm{Hom}_{\bar{\mathfrak{l}}}\left(L(\bar{\mathfrak{l}}, \mu^\natural), H^n(\bar{\mathfrak{u}}, \overline{L}(\lambda^\natural))\right).\end{aligned}$$

Proof. We shall prove (1). We have the following isomorphisms:

$$\begin{aligned}T\left(H^n(\widetilde{\mathfrak{u}}, \widetilde{M})\right) &\cong T\left(H_n(\widetilde{\mathfrak{u}}_-, \widetilde{M}^\vee)\right) \cong H_n\left(\mathfrak{u}_-, T(\widetilde{M}^\vee)\right)\\ &\cong H_n\left(\mathfrak{u}_-, T(\widetilde{M})^\vee\right) \cong H^n\left(\mathfrak{u}, T(\widetilde{M})\right).\end{aligned}$$

The first and the last isomorphisms follow from Theorem 6.24. The second isomorphism is due to Theorem 6.27(1), while the third isomorphism is due to the compatibility of the functors T and $\cdot^\vee$. This proves the first statement of (1). Setting $\widetilde{M} = \widetilde{L}(\lambda^\theta)$ and using Theorem 6.17, we get the second statement of (1). The first "$\cong$" in (3) follows from (1) and $T\big(L(\widetilde{\mathfrak{l}},\mu^\theta)\big) = L(\mathfrak{l},\mu)$ (see Lemma 6.14).

The argument for (2) is similar to (1), and the second "$\cong$" in (3) follows from (2) and Lemma 6.14. □

6.4.4. Kazhdan-Lusztig-Vogan (KLV) polynomials. For a basic Lie superalgebra, the Weyl group of its even subalgebra does not completely control the linkage, as we have seen in Section 2.2 in Chapter 2. So we cannot expect its associated Hecke algebra and corresponding Kazhdan-Lusztig polynomials to play fundamental roles for the category $\overline{\mathcal{O}}$. We instead adopt an approach via homological algebra.

Definition 6.29. The **(parabolic) Kazhdan-Lusztig-Vogan polynomials** (KLV polynomials, for short) in the categories $\mathcal{O}$, $\overline{\mathcal{O}}$, and $\widetilde{\mathcal{O}}$ are defined as follows: for $\mu,\lambda \in P^+$,

$$\ell_{\mu\lambda}(q) := \sum_{n=0}^{\infty}(-q)^{-n}\dim\mathrm{Hom}_{\mathfrak{l}}\left(L(\mathfrak{l},\mu),H_n(\mathfrak{u}^-,L(\lambda))\right),$$

$$\overline{\ell}_{\mu^\natural\lambda^\natural}(q) := \sum_{n=0}^{\infty}(-q)^{-n}\dim\mathrm{Hom}_{\overline{\mathfrak{l}}}\left(L(\overline{\mathfrak{l}},\mu^\natural),H_n(\overline{\mathfrak{u}}^-,L(\lambda^\natural))\right),$$

$$\widetilde{\ell}_{\mu^\theta\lambda^\theta}(q) := \sum_{n=0}^{\infty}(-q)^{-n}\dim\mathrm{Hom}_{\widetilde{\mathfrak{l}}}\left(L(\widetilde{\mathfrak{l}},\mu^\theta),H_n(\widetilde{\mathfrak{u}}^-,L(\lambda^\theta))\right).$$

Recall from (6.29) that $\mathrm{ch}\,L(\lambda) = \sum_{\mu\in P^+} a_{\mu\lambda}\mathrm{ch}\,\Delta(\mu)$, for $\lambda \in P^+$. By definition, $\ell_{\mu\lambda}(q), \overline{\ell}_{\mu^\natural\lambda^\natural}(q), \widetilde{\ell}_{\mu^\theta\lambda^\theta}(q)$ are power series in q^{-1}.

Proposition 6.30. *The $\ell_{\mu\lambda}(q), \overline{\ell}_{\mu^\natural\lambda^\natural}(q), \widetilde{\ell}_{\mu^\theta\lambda^\theta}(q)$ are polynomials in q^{-1}. Moreover, we have $\ell_{\mu\lambda}(1) = a_{\mu\lambda}$.*

Proof. We will only show that $\ell_{\mu\lambda}(q)$ is a polynomial in q^{-1}, and similar arguments can be applied to $\overline{\ell}_{\mu^\natural\lambda^\natural}(q), \widetilde{\ell}_{\mu^\theta\lambda^\theta}(q)$.

Take $\lambda \in P^+$, and recall that $C_n(\mathfrak{u}^-,L(\lambda)) = \wedge(\mathfrak{u}^-)\otimes L(\lambda)$. Since $C_n(\mathfrak{u}^-,L(\lambda))$ has finite-dimensional weight spaces, the module $L(\mathfrak{l},\mu)$ appears with finite multiplicity in $C_n(\mathfrak{u}^-,L(\lambda))$, for each n. Also observe that, for a given weight $\mu \in P^+$, μ cannot be a weight of $C_n(\mathfrak{u}^-,L(\lambda))$, for $n \gg 0$. This proves that $\ell_{\mu\lambda}(q)$ is a polynomial in q^{-1}.

We can now evaluate the polynomial $\ell_{\mu\lambda}(q)$ at $q=1$. Applying the Euler-Poincaré principle to the chain complex $C_\bullet(\mathfrak{u}^-,L(\lambda))$ of $\mathfrak{h}$-modules, we have

$$\begin{aligned}\mathrm{ch}\,L(\lambda)&\sum_{n\geq 0}(-1)^n\mathrm{ch}\wedge^n(\mathfrak{u}^-)\\ &=\sum_{n\geq 0}(-1)^n\mathrm{ch}\,H_n(\mathfrak{u}^-,L(\lambda))\end{aligned}$$

$$= \sum_{n\geq 0}(-1)^n \sum_{\mu\in P^+} \dim \mathrm{Hom}_{\mathfrak{l}}\big(L(\mathfrak{l},\mu), H_n(\mathfrak{u}^-, L(\lambda))\big)\,\mathrm{ch}\, L(\mathfrak{l},\mu)$$
$$= \sum_{\mu\in P^+} \ell_{\mu\lambda}(1)\mathrm{ch}\, L(\mathfrak{l},\mu).$$

The above identity can be rewritten as

$$\mathrm{ch}\, L(\lambda) = \sum_{\mu} \ell_{\mu\lambda}(1)\mathrm{ch}\,\Delta(\mu), \tag{6.36}$$

using the following character formula of $\Delta(\mu)$:

$$\mathrm{ch}\Delta(\mu) = \frac{\mathrm{ch}\, L(\mathfrak{l},\mu)}{\sum_{n\geq 0}(-1)^n \mathrm{ch}\, \wedge^n(\mathfrak{u}^-)}.$$

The equality $\ell_{\mu\lambda}(1) = a_{\mu\lambda}$ now follows by a comparison of (6.36) with (6.29) and the linear independence of ch $\Delta(\mu)$. □

The following reformulation of Theorem 6.27 is a generalization of the character formulas in Theorem 6.18.

Theorem 6.31. *For $\mu,\lambda \in P^+$ we have $\ell_{\mu\lambda}(q) = \widetilde{\ell}_{\mu^\theta\lambda^\theta}(q) = \overline{\ell}_{\mu^\natural\lambda^\natural}(q)$.*

6.4.5. Stability of KLV polynomials. For $n \in \mathbb{N}$ recall the category $\mathcal{O}_n$ of modules over the finite-dimensional Lie algebra $\mathfrak{g}_n$ from Section 6.2.3. We have denoted the corresponding finite-dimensional Levi subalgebra by $\mathfrak{l}_n$, and we shall denote the corresponding finite-dimensional nilradical and opposite nilradical by $\mathfrak{u}_n$ and $\mathfrak{u}_n^-$, respectively. Recall also that the set $\{\lambda \in P^+ | \ell(^+\lambda) \leq n\}$ is denoted by P_n^+, while $L_n(\lambda) \in \mathcal{O}_n$ denotes the irreducible $\mathfrak{g}_n$-module of highest weight $\lambda \in P_n^+$. We shall also write $L_\infty(\lambda) \equiv L(\lambda)$, $P_\infty^+ \equiv P^+$, etc. For $k \in \mathbb{N}\cup\infty$ with $k > n$, recall from Section 6.2.5 that the truncation functor $\mathfrak{tr}_n^k : \mathcal{O}_k \to \mathcal{O}_n$ is a $\mathfrak{g}_n$-module homomorphism.

Comparing the characters, we obtain, for $\lambda \in P_k^+$,

$$\mathfrak{tr}_n^k(L_k(\mathfrak{l}_k,\lambda)) = \begin{cases} L_n(\mathfrak{l}_n,\lambda), & \text{if } \lambda \in P_n^+, \\ 0, & \text{otherwise.} \end{cases} \tag{6.37}$$

The following lemma is easy.

Lemma 6.32. *For $n,k \in \mathbb{N}\cup\infty$ with $k > n$, we have the following $\mathfrak{l}_n$-module isomorphisms.*

(1) $\mathfrak{tr}_n^k\big(\wedge(\mathfrak{u}_k^-)\big) = \wedge(\mathfrak{u}_n^-)$.

(2) $\mathfrak{tr}_n^k\big(\wedge(\mathfrak{u}_k^-)\otimes L_k(\lambda)\big) = \wedge(\mathfrak{u}_n^-)\otimes L_n(\lambda)$.

We now employ the same method as in Section 6.4.3 to study the effect of the truncation functors $\mathfrak{tr}_n^k$ on the corresponding $\mathfrak{u}_k^-$-homology groups with coefficients in the module $L_k(\lambda)$, for $\lambda \in P_k^+$. Let $d^{(n)} : \wedge(\mathfrak{u}_n^-)\otimes L_n(\lambda) \to \wedge(\mathfrak{u}_n^-)\otimes L_n(\lambda)$ denote the corresponding boundary operator of the chains of the finite-dimensional Lie superalgebra $\mathfrak{u}_n^-$ with coefficients in $L_n(\lambda)$. It is evident that $d^{(k)}|_{\wedge(\mathfrak{u}_n^-)\otimes L_n(\lambda)} = d^{(n)}$, and furthermore we have $d^{(n)}\mathfrak{tr}_n^k = \mathfrak{tr}_n^k d^{(k)}$. The proof of Theorem 6.27 can be adapted to prove the following.

Theorem 6.33. *Let $\lambda \in P_k^+$ and $i \in \mathbb{Z}_+$. We have*

$$\mathfrak{tr}_n^k\left(H_i(\mathfrak{u}_k^-, L_k(\lambda))\right) = \begin{cases} H_i(\mathfrak{u}_n^-, L_n(\lambda)), & \text{if } \lambda \in P_n^+, \\ 0, & \text{otherwise.} \end{cases}$$

For $\lambda, \mu \in P_n^+$, the expression

$$\ell_{\mu\lambda}^{(n)}(q) := \sum_{i=0}^{\infty} (-q)^{-i} \dim \mathrm{Hom}_{\mathfrak{l}_n}\left(L(\mathfrak{l}_n,\mu), H_i(\mathfrak{u}_n^-, L_n(\lambda))\right)$$

equals the parabolic Kazhdan-Lusztig polynomial of the Lie algebra $\mathfrak{g}_n$ via Vogan's homological interpretation. Theorem 6.33 and (6.37) imply that the Kazhdan-Lusztig polynomials for finite-rank semisimple or reductive Lie algebras of classical type admit a remarkable stability, i.e., we have

$$\ell_{\mu\lambda}^{(n)}(q) = \ell_{\mu\lambda}(q), \quad \lambda, \mu \in P_n^+. \tag{6.38}$$

A similar argument leads to analogous stability of the KLV polynomials in categories $\overline{\mathcal{O}}_n$ and $\widetilde{\mathcal{O}}_n$, which coincide for $n \gg 0$ with $\widetilde{\ell}_{\mu^\theta\lambda^\theta}(q)$ and $\overline{\ell}_{\mu^\natural\lambda^\natural}(q)$, respectively. Theorem 6.31 then further allows us to suitably identify these KLV polynomials for finite-rank Lie (super)algebras $\mathfrak{g}_n$, $\overline{\mathfrak{g}}_n$, and $\widetilde{\mathfrak{g}}_n$.

6.5. Super duality as an equivalence of categories

In this section, we establish super duality, which is formulated as an equivalence of the two module categories $\mathcal{O}$ and $\overline{\mathcal{O}}$. More precisely, we shall show that the functors $T : \widetilde{\mathcal{O}} \to \mathcal{O}$ and $\overline{T} : \widetilde{\mathcal{O}} \to \overline{\mathcal{O}}$ are equivalences of categories. In the case when $\mathfrak{x} = \mathfrak{a}, \mathfrak{b}, \mathfrak{c}, \mathfrak{d}$, the category $\mathcal{O}$, as a variant of the BGG category for classical Lie algebras, is well understood, and super duality leads to new insight on the module category $\overline{\mathcal{O}}$ for Lie superalgebras.

Some basic facts on category theory and homological algebra needed in this section can be found in the book of Mitchell [**86**].

6.5.1. Extensions à la Baer-Yoneda. In this subsection, we shall review the Baer-Yoneda extension Ext^1 in a general abelian category $\mathcal{C}$. This is needed later on for the categories $\mathcal{O}$, $\overline{\mathcal{O}}$, $\widetilde{\mathcal{O}}$, since there may not be enough projective objects in these categories. A reader can also take for granted the exact sequence (6.42) and safely skip this subsection. We shall omit most of the proofs, and refer to Mitchell [**86**, Chapter VII] for details.

For two objects $A, C \in \mathcal{C}$, let E, E' be short exact sequences of the form

$$E: \quad 0 \longrightarrow A \xrightarrow{i} B \xrightarrow{j} C \longrightarrow 0, \tag{6.39}$$

$$E': \quad 0 \longrightarrow A \xrightarrow{i'} B' \xrightarrow{j'} C \longrightarrow 0.$$

We say that $E \sim E'$ if there exists $f : B \to B'$ such that $i' = f \circ i$ and $j' \circ f = j$. By the 5-lemma f is an isomorphism, and so $\sim$ is indeed an equivalence relation. The set of **extensions of degree** 1 of C by A is by definition the set of equivalence classes of exact sequences of the form (6.39) with respect to the relation $\sim$, and will be denoted by $\mathrm{Ext}^1_{\mathcal{C}}(C,A)$. We shall write $E \in \mathrm{Ext}^1_{\mathcal{C}}(C,A)$ for the class represented by the exact sequence E by abuse of notation.

Denote by E_0 the equivalence class corresponding to the short exact sequence

$$E_0: \quad 0 \longrightarrow A \longrightarrow A \oplus C \longrightarrow C \longrightarrow 0.$$

Given an object $C' \in \mathcal{C}$, a morphism $\gamma : C' \to C$, and an element $E \in \mathrm{Ext}^1_{\mathcal{C}}(C,A)$, we have the following diagram:

$$\begin{array}{ccccccccccc}
 & & & & & & & & C' & & \\
 & & & & & & & & \downarrow{\scriptstyle\gamma} & & \\
E: & 0 & \longrightarrow & A & \longrightarrow & B & \longrightarrow & & C & \longrightarrow & 0
\end{array}$$

Then, there exists a unique element $E' \in \mathrm{Ext}^1_{\mathcal{C}}(C',A)$ such that the following diagram is commutative:

$$\begin{array}{cccccccccc}
E': & 0 & \longrightarrow & A & \longrightarrow & B' & \longrightarrow & C' & \longrightarrow & 0 \\
 & & & \| & & \downarrow{\scriptstyle\beta} & & \downarrow{\scriptstyle\gamma} & & \\
E: & 0 & \longrightarrow & A & \longrightarrow & B & \longrightarrow & C & \longrightarrow & 0
\end{array}$$

Indeed, B' is simply the pullback of γ and the morphism $B \to C$. The morphism $A \to B'$ is then uniquely determined by the commutativity of the left square and the exactness of the first row of the diagram. We denote $E' \in \mathrm{Ext}^1(C',A)$ by $E\gamma$. Letting γ vary, we obtain a map $E\cdot : \mathrm{Hom}_{\mathcal{C}}(C',C) \to \mathrm{Ext}^1_{\mathcal{C}}(C',A)$ defined by $\gamma \xrightarrow{E\cdot} E\gamma$. On the other hand, letting E vary, one shows that a morphism $\gamma : C' \to C$ induces a well-defined map $\cdot\gamma_A : \mathrm{Ext}^1_{\mathcal{C}}(C,A) \to \mathrm{Ext}^1_{\mathcal{C}}(C',A)$. That is, for a fixed object A in $\mathcal{C}$, $\mathrm{Ext}^1_{\mathcal{C}}(\cdot,A)$ is a contravariant functor. With this notation $\cdot\gamma_A = \mathrm{Ext}^1_{\mathcal{C}}(\gamma,A)$ we shall write $\cdot\gamma_A = \cdot\gamma$ when A is clear from the context.

Dually, suppose we are given an object $A' \in \mathcal{C}$, a morphism $\alpha : A \to A'$, and an element $E \in \mathrm{Ext}^1_{\mathcal{C}}(C,A)$, so that we have the following diagram:

$$\begin{array}{cccccccccc}
E: & 0 & \longrightarrow & A & \longrightarrow & B & \longrightarrow & C & \longrightarrow & 0 \\
 & & & \downarrow{\scriptstyle\alpha} & & & & & & \\
 & & & A' & & & & & &
\end{array}$$

We obtain uniquely an element $E' \in \mathrm{Ext}^1_{\mathcal{C}}(C,A)$ such that the following diagram is commutative:

$$\begin{array}{ccccccccc} E: & 0 & \longrightarrow & A & \longrightarrow & B & \longrightarrow & C & \longrightarrow 0 \\ & & & \downarrow{\scriptstyle\alpha} & & \downarrow{\scriptstyle\beta} & & \| & \\ E': & 0 & \longrightarrow & A' & \longrightarrow & B' & \longrightarrow & C & \longrightarrow 0 \end{array}$$

The resulting element E' is denoted by αE. Fixing E we thus obtain a map $\cdot E$: $\mathrm{Hom}_{\mathcal{C}}(A,A') \to \mathrm{Ext}^1_{\mathcal{C}}(C,A')$ given by $\alpha \xrightarrow{\cdot E} \alpha E$. On the other hand, by letting E vary, we can show that a fixed morphism $\alpha : A \to A'$ induces a well-defined map ${}_C\alpha\cdot : \mathrm{Ext}^1_{\mathcal{C}}(C,A) \to \mathrm{Ext}^1_{\mathcal{C}}(C,A')$ so that $\mathrm{Ext}^1_{\mathcal{C}}(C,\cdot)$ is a covariant functor. With this notation we have ${}_C\alpha\cdot = \mathrm{Ext}^1_{\mathcal{C}}(C,\alpha)$. Thus, $\mathrm{Ext}^1_{\mathcal{C}}(\cdot,\cdot)$ gives a bi-functor on $\mathcal{C}$. We shall write ${}_C\alpha\cdot = \alpha\cdot$, when C is clear from the context.

For an extension E, with morphisms α and γ as above, we have $(\alpha E)\gamma = \alpha(E\gamma)$.

We define an abelian group structure on $\mathrm{Ext}^1_{\mathcal{C}}(C,A)$. For $A \in \mathcal{C}$, we define a morphism $\Delta = \Delta_A : A \to A \oplus A$, $\Delta(a) := (a,a)$, and a morphism $\nabla = \nabla_A : A \oplus A \to A$ by $\nabla(a,a') := a + a'$. Given two elements $E, E' \in \mathrm{Ext}^1_{\mathcal{C}}(C,A)$ we let $E \oplus E'$ be the obvious element in $\mathrm{Ext}^1_{\mathcal{C}}(C \oplus C, A \oplus A)$. To define the group operation we consider

$$\nabla_A(E \oplus E')\Delta_C, \tag{6.40}$$

which is an exact sequence of the form (6.39). Define $E + E'$ to be the element in $\mathrm{Ext}^1_{\mathcal{C}}(C,A)$ corresponding to (6.40). It can be shown that the operation $+ : \mathrm{Ext}^1_{\mathcal{C}}(C,A) \times \mathrm{Ext}^1_{\mathcal{C}}(C,A) \to \mathrm{Ext}^1_{\mathcal{C}}(C,A)$ is well-defined, abelian, and associative. Furthermore, it can be shown that the equivalence class of the split exact sequence E_0 is an additive identity and that $(-1_A)E$ is the additive inverse of E.

For $A, C \in \mathcal{C}$, it is convenient to set $\mathrm{Ext}^0_{\mathcal{C}}(C,A) := \mathrm{Hom}_{\mathcal{C}}(C,A)$.

Let X be an object in $\mathcal{C}$. Applying the functor $\mathrm{Hom}_{\mathcal{C}}(X,\cdot)$ to the exact sequence (6.39) gives us an exact sequence

$$0 \to \mathrm{Hom}_{\mathcal{C}}(X,A) \to \mathrm{Hom}_{\mathcal{C}}(X,B) \to \mathrm{Hom}_{\mathcal{C}}(X,C). \tag{6.41}$$

Now for each $\gamma \in \mathrm{Hom}_{\mathcal{C}}(X,C)$, we obtain $E\gamma \in \mathrm{Ext}^1_{\mathcal{C}}(X,A)$. It can be shown that the map $E\cdot$ extends (6.41) to the following exact sequence:

$$\begin{aligned} 0 \longrightarrow &\mathrm{Hom}_{\mathcal{C}}(X,A) \longrightarrow \mathrm{Hom}_{\mathcal{C}}(X,B) \longrightarrow \mathrm{Hom}_{\mathcal{C}}(X,C) \\ &\xrightarrow{E\cdot} \mathrm{Ext}^1_{\mathcal{C}}(X,A) \xrightarrow{i\cdot} \mathrm{Ext}^1_{\mathcal{C}}(X,B) \xrightarrow{j\cdot} \mathrm{Ext}^1_{\mathcal{C}}(X,C). \end{aligned} \tag{6.42}$$

6.5.2. Relating extensions in $\mathcal{O}$, $\overline{\mathcal{O}}$, and $\widetilde{\mathcal{O}}$. Now we return to the categories $\mathcal{O}$, $\overline{\mathcal{O}}$, and $\widetilde{\mathcal{O}}$, of $\mathfrak{g}$-, $\overline{\mathfrak{g}}$-, and $\widetilde{\mathfrak{g}}$-modules, respectively. Let $\widetilde{A}, \widetilde{C} \in \widetilde{\mathcal{O}}$ and let $\widetilde{E} \in \mathrm{Ext}^1_{\widetilde{\mathcal{O}}}(\widetilde{C},\widetilde{A})$. Then we have by Section 6.5.1 an exact sequence in $\widetilde{\mathcal{O}}$ of the form

$$\widetilde{E}: \quad 0 \longrightarrow \widetilde{A} \longrightarrow \widetilde{B} \longrightarrow \widetilde{C} \longrightarrow 0.$$

Applying the exact functors T and $\overline{T}$, we obtain exact sequences in $\mathcal{O}$ and $\overline{\mathcal{O}}$, respectively. It can be easily seen that, in this way, the functors T and $\overline{T}$ define natural homomorphisms of abelian groups from $\mathrm{Ext}^1_{\widetilde{\mathcal{O}}}(\widetilde{C},\widetilde{A})$ to $\mathrm{Ext}^1_{\mathcal{O}}\left(T(\widetilde{C}),T(\widetilde{A})\right)$ and $\mathrm{Ext}^1_{\overline{\mathcal{O}}}\left(\overline{T}(\widetilde{C}),\overline{T}(\widetilde{A})\right)$, respectively. The respective elements are denoted by $T[\widetilde{E}]$ and $\overline{T}[\widetilde{E}]$.

Recall the Lie superalgebra $\mathfrak{g}$ with parabolic subalgebra $\mathfrak{p}$, Levi subalgebra $\mathfrak{l}$ and nilradical $\mathfrak{u}$ from Section 6.2.2. We have the following relative Koszul resolution for the trivial $\mathfrak{p}$-module (see Knapp-Vogan [**72**, II.7]):

$$\cdots \longrightarrow C_k \xrightarrow{\delta_k} C_{k-1} \longrightarrow \cdots \xrightarrow{\delta_1} C_0 \xrightarrow{\varepsilon} \mathbb{C} \longrightarrow 0. \tag{6.43}$$

Here $C_k := U(\mathfrak{p}) \otimes_{U(\mathfrak{l})} \wedge^k(\mathfrak{p}/\mathfrak{l})$ is a $\mathfrak{p}$-module with $\mathfrak{p}$ acting on the left of the first factor, and ε is the augmentation map from $U(\mathfrak{p})$ to $\mathbb{C}$. The $\mathfrak{p}$-homomorphism δ_k is given by

$$\begin{aligned} \delta_k(a\otimes \overline{x}_1\overline{x}_2\cdots\overline{x}_k) := & \sum_{1\le s<t\le k} (-1)^{s+t} a\otimes \overline{[x_s,x_t]}\,\overline{x}_1\cdots\widehat{\overline{x}_s}\cdots\widehat{\overline{x}_t}\cdots\overline{x}_k \\ & + \sum_{s=1}^{k}(-1)^{s+1} a x_s \otimes \overline{x}_1\cdots\widehat{\overline{x}_s}\cdots\overline{x}_k. \end{aligned} \tag{6.44}$$

Here $a \in U(\mathfrak{p})$ and x_i are homogeneous elements in $\mathfrak{p}$, and $\overline{x}_i$ denotes $x_i + \mathfrak{l}$ in $\mathfrak{p}/\mathfrak{l}$.

For $\lambda \in P^+$, we extend the irreducible $\mathfrak{l}$-module $L(\mathfrak{l},\lambda)$ trivially to a $\mathfrak{p}$-module. For $k \ge 0$, $D_k := C_k \otimes L(\mathfrak{l},\lambda)$ is a $\mathfrak{p}$-module by the diagonal $\mathfrak{p}$-action. Tensoring (6.43) with $L(\mathfrak{l},\lambda)$ gives us the following exact sequence of $\mathfrak{p}$-modules, where $\delta_k \otimes 1$ is simply denoted by δ_k again for $k > 0$, and $\delta_0 := \varepsilon \otimes 1$:

$$\cdots \longrightarrow D_k \xrightarrow{\delta_k} D_{k-1} \longrightarrow \cdots \xrightarrow{\delta_1} D_0 \xrightarrow{\delta_0} L(\mathfrak{l},\lambda) \longrightarrow 0. \tag{6.45}$$

The above construction admits straightforward generalizations for the Lie superalgebras $\overline{\mathfrak{g}}$ and $\widetilde{\mathfrak{g}}$.

Theorem 6.34. *Let $\lambda \in P^+$, $V \in \mathcal{O}$, $\widetilde{V} \in \widetilde{\mathcal{O}}$, and $\overline{V} \in \overline{\mathcal{O}}$. We have, for $i = 0,1$,*

$$\begin{aligned} \mathrm{Hom}_{\mathfrak{l}}\left(L(\mathfrak{l},\lambda), H^i(\mathfrak{u},V)\right) &\cong \mathit{Ext}^i_{\mathcal{O}}(\Delta(\lambda),V), \\ \mathrm{Hom}_{\widetilde{\mathfrak{l}}}\left(L(\widetilde{\mathfrak{l}},\lambda^\theta), H^i(\widetilde{\mathfrak{u}},\widetilde{V})\right) &\cong \mathit{Ext}^i_{\widetilde{\mathcal{O}}}(\widetilde{\Delta}(\lambda^\theta),\widetilde{V}), \\ \mathrm{Hom}_{\overline{\mathfrak{l}}}\left(L(\overline{\mathfrak{l}},\lambda^\natural), H^i(\overline{\mathfrak{u}},\overline{V})\right) &\cong \mathit{Ext}^i_{\overline{\mathcal{O}}}(\overline{\Delta}(\lambda^\natural),\overline{V}). \end{aligned}$$

Proof. We shall only prove the first isomorphism for $\mathfrak{l}$ for $i = 0,1$, and the remaining two isomorphisms can be proved in an analogous fashion.

Let $V \in \mathcal{O}$. Recall the (unrestricted) cohomology group $\mathcal{H}^i(\mathfrak{u},V)$ from Remark 6.20 in contrast to the (restricted) cohomology group $H^i(\mathfrak{u},V)$ used mostly in this chapter. Let $\mathcal{H}^i(\mathfrak{u},V)^{\mathrm{ss}}$ be the maximal $\mathfrak{h}$-semisimple submodule of $\mathcal{H}^i(\mathfrak{u},V)$. We have the following

Claim. $\mathcal{H}^i(\mathfrak{u},V)^{\mathrm{ss}} \cong H^i(\mathfrak{u},V)$.

To see this, observe by definition that we have a natural injective map from $H^i(\mathfrak{u},V) \xrightarrow{\iota} \mathcal{H}^i(\mathfrak{u},V)^{\mathrm{ss}}$. Since the coboundary operator ∂_i is an $\mathfrak{h}$-homomorphism and $\mathrm{Hom}(\Lambda^i(\mathfrak{u}),V)$ is a direct product of finite-dimensional $\mathfrak{h}$-weight spaces, $\mathrm{Ker}\,\partial_i$ and $\mathrm{Im}\,\partial_i$ are also direct products of finite-dimensional $\mathfrak{h}$-weight spaces. Now given $x \in \mathrm{Ker}\,\partial_i$ representing a cohomology class in $\mathcal{H}^i(\mathfrak{u},V)^{\mathrm{ss}}$, there exists an element $y \in \mathrm{Im}\,\partial_{i-1}$ such that $x-y \in C^i(\mathfrak{u},V) \cap \mathrm{Ker}\,\partial_i$. We have then $\iota(x-y) \in x+\mathrm{Im}\,\partial_{i-1}$, and hence ι is also a surjection. This proves the claim.

Now since $L(\mathfrak{l},\lambda)$ is $\mathfrak{h}$-semisimple and $\mathcal{H}^i(\mathfrak{u},V)$ is a direct product of finite-dimensional $\mathfrak{h}$-weight spaces, it follows by the claim that

$$\mathrm{Hom}_{\mathfrak{l}}\left(L(\mathfrak{l},\lambda),\mathcal{H}^i(\mathfrak{u},V)\right) \cong \mathrm{Hom}_{\mathfrak{l}}\left(L(\mathfrak{l},\lambda),H^i(\mathfrak{u},V)\right). \tag{6.46}$$

This allows us to freely switch between H^i and $\mathcal{H}^i$ below.

Unraveling the definition, we have $H^0(\mathfrak{u},V)=V^{\mathfrak{u}}$, the $\mathfrak{u}$-invariants of V. Thus,

$$\begin{aligned}\mathrm{Hom}_{\mathfrak{l}}(L(\mathfrak{l},\lambda),H^0(\mathfrak{u},V)) &\cong \mathrm{Hom}_{\mathfrak{l}}(L(\mathfrak{l},\lambda),V^{\mathfrak{u}})\\ &\cong \mathrm{Hom}_{\mathfrak{p}}(L(\mathfrak{l},\lambda),V) \cong \mathrm{Hom}_{\mathcal{O}}(\Delta(\lambda),V),\end{aligned}$$

where the last "$\cong$" is due to Frobenius reciprocity. This proves the isomorphism for $\mathfrak{l}$ with $i=0$ in the theorem.

It remains to show that $\mathrm{Hom}_{\mathfrak{l}}\left(L(\mathfrak{l},\lambda),H^1(\mathfrak{u},V)\right) \cong \mathrm{Ext}^1_{\mathcal{O}}(\Delta(\lambda),V)$. We shall construct the isomorphism explicitly.

First, take $\bar{f} \in \mathrm{Hom}_{\mathfrak{l}}\left(L(\mathfrak{l},\lambda),H^1(\mathfrak{u},V)\right)$. Recall $H^1(\mathfrak{u},V)=\mathrm{Ker}\,d_1/\mathrm{Im}\,d_0$, with $\mathrm{Ker}\,d_1 \subseteq \mathrm{Hom}(\mathfrak{u},V)$. Let $f \in \mathrm{Hom}_{\mathfrak{l}}(L(\mathfrak{l},\lambda),\mathrm{Hom}(\mathfrak{u},V)) \cong \mathrm{Hom}_{\mathfrak{l}}(\mathfrak{u}\otimes L(\mathfrak{l},\lambda),V)$ be a representative that, by Frobenius reciprocity, is regarded as an element in $\mathrm{Hom}_{\mathfrak{p}}\left(\mathrm{Ind}_{\mathfrak{l}}^{\mathfrak{p}}(\mathfrak{u}\otimes L(\mathfrak{l},\lambda)),V\right)$.

Applying the exact functor $U(\mathfrak{g})\otimes_{U(\mathfrak{p})}$ - to (6.45), we obtain an exact sequence

$$\begin{aligned}U(\mathfrak{g})\otimes_{U(\mathfrak{l})}\left(\Lambda^2(\mathfrak{p}/\mathfrak{l})\otimes L(\mathfrak{l},\lambda)\right) &\xrightarrow{\partial_2'} U(\mathfrak{g})\otimes_{U(\mathfrak{l})}(\mathfrak{p}/\mathfrak{l}\otimes L(\mathfrak{l},\lambda)) \xrightarrow{\partial_1'}\\ &U(\mathfrak{g})\otimes_{U(\mathfrak{l})} L(\mathfrak{l},\lambda) \xrightarrow{\partial_0'} \Delta(\lambda) \longrightarrow 0,\end{aligned}$$

where ∂_i' denotes the differentials induced from the differentials δ_i. This gives rise to the following diagram

$$\begin{array}{ccccccc}U(\mathfrak{g})\otimes_{U(\mathfrak{l})}(\mathfrak{p}/\mathfrak{l}\otimes L(\mathfrak{l},\lambda)) & \xrightarrow{\partial_1'} & U(\mathfrak{g})\otimes_{U(\mathfrak{l})} L(\mathfrak{l},\lambda) & \longrightarrow & \Delta(\lambda) & \longrightarrow & 0\\ \downarrow f & & & & & & \\ V & & & & & & \end{array}$$

Taking the pushout E_f of f and ∂'_1, we obtain the following commutative diagram

$$\begin{array}{ccccccc}
\mathrm{Ind}_{U(\mathfrak{l})}^{U(\mathfrak{g})}(\mathfrak{p}/\mathfrak{l}\otimes L(\mathfrak{l},\lambda)) & \longrightarrow & \mathrm{Ind}_{U(\mathfrak{l})}^{U(\mathfrak{g})}L(\mathfrak{l},\lambda) & \longrightarrow & \Delta(\lambda) & \longrightarrow & 0 \\
\downarrow{\scriptstyle f} & & \downarrow & & \| & & \\
V & \xrightarrow{\alpha} & E_f & \longrightarrow & \Delta(\lambda) & \longrightarrow & 0
\end{array}$$

Since f represents a cocycle, we have $f\circ\partial'_2=0$. This implies that α is an injection. Now since V and $\Delta(\lambda)$ lie in $\mathcal{O}$, so does E_f, and hence we obtain an element in $\mathrm{Ext}^1_{\mathcal{O}}(\Delta(\lambda),V)$. One verifies in a straightforward manner that if f is a coboundary, then $E_f\cong V\oplus\Delta(\lambda)$, and hence the map $\bar{f}\mapsto E_{\bar{f}}$ is well-defined.

Conversely, suppose that we have an extension of the form

$$0\longrightarrow V\longrightarrow E\longrightarrow\Delta(\lambda)\longrightarrow 0.$$

Note that $E\in\mathcal{O}$, and hence there exists an $\mathfrak{l}$-submodule of E that is mapped isomorphically to $L(\mathfrak{l},\lambda)\subseteq\Delta(\lambda)$. We denote this $\mathfrak{l}$-module also by $L(\mathfrak{l},\lambda)$ and note that $\mathfrak{u}L(\mathfrak{l},\lambda)\subseteq V$. This allows us to define an element $f_E\in\mathrm{Hom}_{\mathfrak{l}}(\mathfrak{u}\otimes L(\mathfrak{l},\lambda),V)\cong\mathrm{Hom}_{\mathfrak{l}}(L(\mathfrak{l},\lambda),\mathrm{Hom}(\mathfrak{u},V))$ by $f_E(u\otimes w)=uw$. Now f_E corresponds to a cocycle in $\mathrm{Hom}_{\mathfrak{l}}(L(\mathfrak{l},\lambda),\mathrm{Hom}(\mathfrak{u},V))$, because V is a submodule of the $\mathfrak{u}$-module E. Also, one sees that if E is the trivial extension, then f_E must be a coboundary. Thus, sending E to $\bar{f}_E$ gives us a well-defined map $\mathrm{Ext}^1_{\mathcal{O}}(\Delta(\lambda),V)\to\mathrm{Hom}_{\mathfrak{l}}(L(\mathfrak{l},\lambda),\mathcal{H}^1(\mathfrak{u},V))\cong\mathrm{Hom}_{\mathfrak{l}}(L(\mathfrak{l},\lambda),H^1(\mathfrak{u},V))$; see (6.46).

It can be checked that these two maps are inverses to each other. Hence, we have $\mathrm{Hom}_{\mathfrak{l}}\big(L(\mathfrak{l},\lambda),H^1(\mathfrak{u},V)\big)\cong\mathrm{Ext}^1_{\mathcal{O}}(\Delta(\lambda),V)$. □

It makes sense to apply T and $\overline{T}$ to the short exact sequences (i.e., $\mathrm{Ext}^1_{\widetilde{\mathcal{O}}}$) in $\widetilde{\mathcal{O}}$.

Corollary 6.35. *Let $\lambda,\mu\in P^+$. We have, for $i=0,1$,*

$$T\Big(\mathrm{Ext}^i_{\widetilde{\mathcal{O}}}\big(\widetilde{\Delta}(\lambda^\theta),\widetilde{L}(\mu^\theta)\big)\Big)=\mathrm{Ext}^i_{\mathcal{O}}\big(\Delta(\lambda),L(\mu)\big),$$
$$\overline{T}\Big(\mathrm{Ext}^i_{\widetilde{\mathcal{O}}}\big(\widetilde{\Delta}(\lambda^\theta),\widetilde{L}(\mu^\theta)\big)\Big)=\mathrm{Ext}^i_{\overline{\mathcal{O}}}\big(\overline{\Delta}(\lambda^\natural),\overline{L}(\mu^\natural)\big).$$

Proof. We shall only prove the identity for T, as the proof for $\overline{T}$ is analogous.

By Theorem 6.34, for $i=0,1$, we have

$$\begin{aligned}
\mathrm{Ext}^i_{\mathcal{O}}\big(\Delta(\lambda),L(\mu)\big)&\cong\mathrm{Hom}_{\mathfrak{l}}\big(L(\mathfrak{l},\lambda),H^i(\mathfrak{u},L(\mu))\big),\\
\mathrm{Ext}^i_{\widetilde{\mathcal{O}}}\big(\widetilde{\Delta}(\lambda^\theta),\widetilde{L}(\mu^\theta)\big)&\cong\mathrm{Hom}_{\widetilde{\mathfrak{l}}}\big(L(\widetilde{\mathfrak{l}},\lambda^\theta),H^i(\widetilde{\mathfrak{u}},\widetilde{L}(\mu^\theta))\big).
\end{aligned}\tag{6.47}$$

Recall that $T\big(L(\widetilde{\mathfrak{l}},\lambda^\theta)\big)=L(\mathfrak{l},\lambda)$ by Lemma 6.14, and that $T(H^n(\widetilde{\mathfrak{u}},\widetilde{L}(\lambda^\theta)))\cong H^n(\mathfrak{u},L(\lambda))$ by Corollary 6.28. Now the identity for T in the corollary follows by applying the functor T to (6.47) and the naturality of T. □

6.5.3. Categories $\mathcal{O}^f$, $\overline{\mathcal{O}}^f$, and $\widetilde{\mathcal{O}}^f$. The following proposition is standard and well known. For a proof the reader may consult Kumar [**76**, Lemma 2.1.10]. Recall Π denotes the standard fundamental system of $\mathfrak{g}$.

Proposition 6.36. *Let $M \in \mathcal{O}$. Then there exists a (possibly infinite) increasing filtration of $\mathfrak{g}$-modules*

$$0 = M_0 \subseteq M_1 \subseteq M_2 \subseteq \cdots \subseteq M_{i-1} \subseteq M_i \subseteq \cdots \tag{6.48}$$

such that

(1) *$M = \bigcup_{i \geq 0} M_i$;*

(2) *M_i/M_{i-1} is a highest weight module of highest weight ν_i with $\nu_i \in P^+$, for $i \geq 1$;*

(3) *the condition $\nu_i - \nu_j \in \sum_{\alpha \in \Pi} \mathbb{Z}_+ \alpha$ implies that $i < j$;*

(4) *for any weight μ of M, there exists an $r \in \mathbb{N}$ such that $(M/M_r)_\mu = 0$.*

Similar statements hold for $\overline{M} \in \overline{\mathcal{O}}$ and $\widetilde{M} \in \widetilde{\mathcal{O}}$.

Let $\mathcal{O}^f$ denote the full subcategory of $\mathcal{O}$ consisting of finitely generated $U(\mathfrak{g})$-modules. The categories $\overline{\mathcal{O}}^f$ and $\widetilde{\mathcal{O}}^f$ are defined in a similar fashion.

Corollary 6.37. *Let $M \in \mathcal{O}$. Then $M \in \mathcal{O}^f$ if and only if there exists a finite increasing filtration $0 = M_0 \subseteq M_1 \subseteq M_2 \subseteq \cdots \subseteq M_k = M$ of $\mathfrak{g}$-modules such that M_i/M_{i-1} is a highest weight module of highest weight ν_i with $\nu_i \in P^+$, for $1 \leq i \leq k$. Similar statements hold for $\overline{M} \in \overline{\mathcal{O}}$ and $\widetilde{M} \in \widetilde{\mathcal{O}}$.*

Proof. If M is a $\mathfrak{g}$-module with a finite filtration of the form (6.48), then clearly M is finitely generated. Conversely, suppose that $u_1, u_2, \ldots, u_n$ are vectors in M of weights $\nu_1, \nu_2, \ldots, \nu_n$ generating M over $U(\mathfrak{g})$. By Proposition 6.36(4), M has a (possibly infinite) filtration of the form (6.48) such that there exists $r \in \mathbb{N}$ with M_r containing all vectors in M of weights $\nu_1, \nu_2, \ldots, \nu_n$. Thus $u_i \in M_r$, for $1 \leq i \leq n$, and hence $M_r = M$. The proofs for $\widetilde{\mathfrak{g}}$ and $\overline{\mathfrak{g}}$ are analogous and omitted. □

6.5.4. Lifting highest weight modules. The following proposition is the converse of Proposition 6.16.

Proposition 6.38. (1) *Suppose that $V(\lambda)$ is a highest weight $\mathfrak{g}$-module of highest weight $\lambda \in P^+$. Then there is a highest weight $\widetilde{\mathfrak{g}}$-module $\widetilde{V}(\lambda^\theta)$ of highest weight λ^θ such that $T(\widetilde{V}(\lambda^\theta)) = V(\lambda)$.*

(2) *Suppose that $\overline{U}(\lambda^\natural)$ is a highest weight $\overline{\mathfrak{g}}$-module of highest weight $\lambda^\natural$ with $\lambda \in P^+$. Then there is a highest weight $\widetilde{\mathfrak{g}}$-module $\widetilde{U}(\lambda^\theta)$ of highest weight λ^θ such that $\overline{T}(\widetilde{U}(\lambda^\theta)) = \overline{U}(\lambda^\natural)$.*

Proof. We shall only prove (1), as (2) is similar.

Let W be the kernel of the surjective $\mathfrak{g}$-homomorphism from $\Delta(\lambda)$ to $V(\lambda)$. Now Theorem 6.17 says that $T(\widetilde{\Delta}(\lambda^\theta)) = \Delta(\lambda)$. By the exactness of the functor T, it suffices to prove that W lifts to a submodule $\widetilde{W}$ of $\widetilde{\Delta}(\lambda^\theta)$ such that $T(\widetilde{W}) = W$.

We recall the standard Borel subalgebra $\mathfrak{b}$ and standard fundamental system Π for $\mathfrak{g}$. Recall the sequence of Borel subalgebras $\widetilde{\mathfrak{b}}^c(n)$ and corresponding fundamental systems $\widetilde{\Pi}^c(n)$ from Section 6.3.1, and that $\Pi_n = \Pi(\mathfrak{k}) \sqcup \Pi(\mathfrak{T}_n)$ from Section 6.1.1. We make the following.

Claim. Let v_μ be a $\mathfrak{b}$-singular vector of weight $\mu \in P^+$ in a $\widetilde{\mathfrak{g}}$-subquotient of $\widetilde{\Delta}(\lambda^\theta)$. Then v_μ is a $\widetilde{\mathfrak{b}}^c(n)$-singular vector, for $n \gg 0$.

This claim can be seen as follows. There exists $n \gg 0$ such that $\lambda - \mu$ is a non-negative sum of even simple roots lying in Π_n, and such that every weight in $\widetilde{\Delta}(\lambda^\theta)$ is of the form $\lambda - \nu$, where ν is a non-negative sum of simple roots in $\widetilde{\Pi}^c(n)$. Therefore, if α is a simple root in $\widetilde{\Pi}^c(n) \setminus \Pi_n$, then $\mu + \alpha$ cannot be a weight in $\widetilde{\Delta}(\lambda^\theta)$. This proves the claim.

There is an increasing filtration of $\mathfrak{g}$-modules for W, $0 = W_0 \subseteq W_1 \subseteq W_2 \subseteq \cdots$, satisfying the properties of Proposition 6.36. For each $i > 0$, let w_i be a weight vector of weight ν_i in W_i such that $w_i + W_{i-1}$ is a nonzero $\mathfrak{b}$-highest weight vector of W_i/W_{i-1}. Now by Theorem 6.17, $\widetilde{\Delta}(\lambda^\theta) = \bigoplus_{\mu \in P^+} L(\widetilde{\mathfrak{l}}, \mu^\theta)^{m(\mu)}$ if and only if $\Delta(\lambda) = \bigoplus_{\mu \in P^+} L(\mathfrak{l}, \mu)^{m(\mu)}$. Then, regarding $\Delta(\lambda) \subseteq \widetilde{\Delta}(\lambda^\theta)$, there is a highest weight vector $\widetilde{w}_i$ of the $\widetilde{\mathfrak{l}}$-module $U(\widetilde{\mathfrak{l}})w_i \cong L(\widetilde{\mathfrak{l}}, \nu_i^\theta)$ with respect to the Borel subalgebra $\widetilde{\mathfrak{b}} \cap \widetilde{\mathfrak{l}}$, for each i. Let $\widetilde{W}_i$ be the $\widetilde{\mathfrak{g}}$-submodule of $\widetilde{\Delta}(\lambda^\theta)$ generated by $\widetilde{w}_1, \widetilde{w}_2, \ldots, \widetilde{w}_i$ for $i \geq 1$, and set $\widetilde{W}_0 = 0$.

We will prove by induction on i that $T(\widetilde{W}_i) = W_i$, for all i. The case $i = 0$ is trivial. Assume that $i \geq 1$ and $T(\widetilde{W}_{i-1}) = W_{i-1}$. By construction, $w_i + \widetilde{W}_{i-1}$ is a $\mathfrak{b}$-singular vector in $\widetilde{W}_i/\widetilde{W}_{i-1}$. It follows by the claim above that $w_i + \widetilde{W}_{i-1}$ is a $\widetilde{\mathfrak{b}}^c(n)$-singular vector of weight ν_i for $n \gg 0$. By (6.22), we have $\widetilde{\mathfrak{b}} = \widetilde{\mathfrak{b}} \cap \widetilde{\mathfrak{l}} + \widetilde{\mathfrak{u}} \subseteq \widetilde{\mathfrak{b}} \cap \widetilde{\mathfrak{l}} + \widetilde{\mathfrak{b}}^c(n)$. Thus, by the construction of $\widetilde{w}_i$ above, $\widetilde{w}_i + \widetilde{W}_{i-1}$ is a $\widetilde{\mathfrak{b}}$-highest weight vector of $\widetilde{W}_i/\widetilde{W}_{i-1}$. Now $U(\mathfrak{u}^-)L(\mathfrak{l}, \nu_i) = W_i/W_{i-1}$ and $U(\widetilde{\mathfrak{u}}^-)L(\widetilde{\mathfrak{l}}, \nu_i^\theta) = \widetilde{W}_i/\widetilde{W}_{i-1}$, and thus $T(\widetilde{W}_i/\widetilde{W}_{i-1}) = W_i/W_{i-1}$. This together with the inductive assumption implies that $T(\widetilde{W}_i) = W_i$.

Finally, setting $\widetilde{W} = \bigcup_{i \geq 1} \widetilde{W}_i$, we obtain that $T(\widetilde{W}) = W$. □

6.5.5. Super duality and strategy of proof. Recall the functors T and $\overline{T}$ from Section 6.3.3. The following is a main result of this chapter.

Theorem 6.39. (1) *$T : \widetilde{\mathcal{O}} \to \mathcal{O}$ is an equivalence of categories.*

(2) *$\overline{T} : \widetilde{\mathcal{O}} \to \overline{\mathcal{O}}$ is an equivalence of categories.*

(3) *The categories $\mathcal{O}$ and $\overline{\mathcal{O}}$ are equivalent.*

The equivalence of categories in Theorem 6.39(3) is called **super duality**.

Define an equivalence relation $\sim$ on $\widetilde{\mathfrak{h}}^*$ by letting $\mu \sim \nu$ if and only if $\mu - \nu$ lies in the root lattice $\mathbb{Z}\widetilde{\Pi}$ of $\widetilde{\mathfrak{g}}$. For each such equivalence class $[\mu]$, fix a representative $[\mu]^o \in \widetilde{\mathfrak{h}}^*$ and declare $[\mu]^o$ to have $\mathbb{Z}_2$-grading $\bar{0}$. For $\varepsilon = \bar{0}, \bar{1}$, set

$$\widetilde{\mathfrak{h}}^*_\varepsilon = \Big\{\mu \in \widetilde{\mathfrak{h}}^* \mid \sum_{r \in 1/2+\mathbb{Z}_+} \langle \mu - [\mu]^o, \widehat{E}_r \rangle \equiv \varepsilon \ (\mathrm{mod}\ 2)\Big\}, \ \text{for } \mathfrak{x} = \mathfrak{a}, \mathfrak{b}, \mathfrak{c}, \mathfrak{d},$$

$$\widetilde{\mathfrak{h}}^*_\varepsilon = \Big\{\mu \in \widetilde{\mathfrak{h}}^* \mid \sum_{i=1}^{m} \langle \mu - [\mu]^o, \widehat{E}_{-i} \rangle + \sum_{r \in \mathbb{N}} \langle \mu - [\mu]^o, \widehat{E}_r \rangle \equiv \varepsilon \ (\mathrm{mod}\ 2)\Big\}, \ \text{for } \mathfrak{x} = \mathfrak{b}^\bullet.$$

Recall that $\widetilde{V} \in \widetilde{\mathcal{O}}$ is a semisimple $\widetilde{\mathfrak{h}}$-module with $\widetilde{V} = \bigoplus_{\gamma \in \widetilde{\mathfrak{h}}^*} \widetilde{V}_\gamma$. Then $\widetilde{V}$ gives rise to a $\mathbb{Z}_2$-graded $\widetilde{\mathfrak{g}}$-module $\widetilde{V}'$:

$$\widetilde{V}' := \widetilde{V}_{\bar{0}} \bigoplus \widetilde{V}_{\bar{1}}, \qquad \widetilde{V}_\varepsilon := \bigoplus_{\mu \in \widetilde{\mathfrak{h}}^*_\varepsilon} \widetilde{V}_\mu \quad (\varepsilon = \bar{0}, \bar{1}), \tag{6.49}$$

which is compatible with the $\mathbb{Z}_2$-grading on $\widetilde{\mathfrak{g}}$.

We define $\widetilde{\mathcal{O}}^{\bar{0}}$ and $\widetilde{\mathcal{O}}^{f,\bar{0}}$ to be the full subcategories of $\widetilde{\mathcal{O}}$ and $\widetilde{\mathcal{O}}^f$, respectively, consisting of objects equipped with $\mathbb{Z}_2$-gradation given by (6.49). Note that the morphisms in $\widetilde{\mathcal{O}}^{\bar{0}}$ and $\widetilde{\mathcal{O}}^{f,\bar{0}}$ are of degree $\bar{0}$. It is clear that for a given $\widetilde{V} \in \widetilde{\mathcal{O}}$, $\widetilde{V}'$ defined in (6.49) is isomorphic to $\widetilde{V}$ in $\widetilde{\mathcal{O}}$. It then follows by definition that the two categories $\widetilde{\mathcal{O}}$ and $\widetilde{\mathcal{O}}^{\bar{0}}$ are equivalent. Similarly, $\widetilde{\mathcal{O}}^f$ and $\widetilde{\mathcal{O}}^{f,\bar{0}}$ are equivalent categories.

We can now analogously define $\mathcal{O}^{\bar{0}}$, $\mathcal{O}^{f,\bar{0}}$, $\overline{\mathcal{O}}^{\bar{0}}$, and $\overline{\mathcal{O}}^{f,\bar{0}}$ to be the respective full subcategories of $\mathcal{O}$, $\mathcal{O}^f$, $\overline{\mathcal{O}}$, and $\overline{\mathcal{O}}^f$ consisting of objects equipped with $\mathbb{Z}_2$-gradation (6.49). Similarly, $\mathcal{O}^{\bar{0}} \cong \mathcal{O}$, $\overline{\mathcal{O}}^{\bar{0}} \cong \overline{\mathcal{O}}$, and also $\mathcal{O}^{f,\bar{0}} \cong \mathcal{O}^f$, $\overline{\mathcal{O}}^{f,\bar{0}} \cong \overline{\mathcal{O}}^f$.

Since $\mathcal{O}^{\bar{0}} \cong \mathcal{O}$, $\overline{\mathcal{O}}^{\bar{0}} \cong \overline{\mathcal{O}}$, and $\widetilde{\mathcal{O}}^{\bar{0}} \cong \widetilde{\mathcal{O}}$, it suffices to prove Theorem 6.39(1) for $T : \widetilde{\mathcal{O}}^{\bar{0}} \to \mathcal{O}^{\bar{0}}$ and $\overline{T} : \widetilde{\mathcal{O}}^{\bar{0}} \to \overline{\mathcal{O}}^{\bar{0}}$. In order to keep notation simple, we will from now on drop the superscript $\bar{0}$ and use $\overline{\mathcal{O}}$, $\widetilde{\mathcal{O}}$, $\overline{\mathcal{O}}^f$, and $\widetilde{\mathcal{O}}^f$ to denote the respective categories $\overline{\mathcal{O}}^{\bar{0}}$, $\widetilde{\mathcal{O}}^{\bar{0}}$, $\overline{\mathcal{O}}^{f,\bar{0}}$, and $\widetilde{\mathcal{O}}^{f,\bar{0}}$. Henceforth, when we write $\widetilde{\Delta}(\lambda^\theta), \widetilde{L}(\lambda^\theta) \in \widetilde{\mathcal{O}}^f$, $\lambda \in P^+$, we will mean the corresponding modules equipped with the $\mathbb{Z}_2$-gradation (6.49). Similar convention applies to $\overline{\Delta}(\lambda^\natural)$ and $\overline{L}(\lambda^\natural)$.

In the remainder of this subsection, we outline the strategy of the proof of Theorem 6.39. The detailed proof will be carried out in the next subsection. We shall restrict our proof for T entirely, since the arguments for the functors T and $\overline{T}$ are completely parallel. First, we study the relation between the morphisms in the categories $\widetilde{\mathcal{O}}$ and $\mathcal{O}$ under T. In particular, we show in Proposition 6.44 that T induces an isomorphism between morphisms of the two categories. In conjunction with Theorem 6.34 that relates extensions with homology groups, this isomorphism is then used to show that T induces an isomorphism between the first extension groups of the two categories. These, together with well-known facts from homological algebra, imply that T is an equivalence of categories.

6.5.6. The proof of super duality. We shall freely use the following short-hand notation below. Set $M = T(\widetilde{M})$ and $\overline{M} = \overline{T}(\widetilde{M})$, for $\widetilde{M} \in \widetilde{\mathcal{O}}$. Similar conventions apply to the images of T and $\overline{T}$ of $\widetilde{M}', \widetilde{M}'', \widetilde{N} \in \widetilde{\mathcal{O}}$. In addition, set $V(\lambda) = T(\widetilde{V}(\lambda^\theta))$ and $\overline{V}(\lambda^\natural) = \overline{T}(\widetilde{V}(\lambda^\theta))$, for a highest weight $\widetilde{\mathfrak{g}}$-module $\widetilde{V}(\lambda^\theta)$ of highest weight λ^θ with $\lambda \in P^+$ (see Proposition 6.16 for a justification of notation).

Lemma 6.40. *Let $\widetilde{N} \in \widetilde{\mathcal{O}}$, and let*

$$\widetilde{E}: \quad 0 \longrightarrow \widetilde{M}' \stackrel{\widetilde{i}}{\longrightarrow} \widetilde{M} \longrightarrow \widetilde{M}'' \longrightarrow 0 \tag{6.50}$$

be an exact sequence of $\widetilde{\mathfrak{g}}$-modules in $\widetilde{\mathcal{O}}$.

(1) *The exact sequence* (6.50) *induces the following commutative diagram with exact rows. (We will use subscripts to distinguish various maps induced by T.)*

$$\begin{array}{ccccccc}
0 \longrightarrow & \mathrm{Hom}_{\widetilde{\mathcal{O}}}(\widetilde{M}'',\widetilde{N}) & \longrightarrow & \mathrm{Hom}_{\widetilde{\mathcal{O}}}(\widetilde{M},\widetilde{N}) & \longrightarrow & \mathrm{Hom}_{\widetilde{\mathcal{O}}}(\widetilde{M}',\widetilde{N}) & \\
& \big\downarrow T_{\widetilde{M}'',\widetilde{N}} & & \big\downarrow T_{\widetilde{M},\widetilde{N}} & & \big\downarrow T_{\widetilde{M}',\widetilde{N}} & \\
0 \longrightarrow & \mathrm{Hom}_{\mathcal{O}}(M'',N) & \longrightarrow & \mathrm{Hom}_{\mathcal{O}}(M,N) & \longrightarrow & \mathrm{Hom}_{\mathcal{O}}(M',N) &
\end{array}$$

$$\begin{array}{ccccccc}
\stackrel{\cdot\widetilde{E}}{\longrightarrow} & \mathrm{Ext}^1_{\widetilde{\mathcal{O}}}(\widetilde{M}'',\widetilde{N}) & \longrightarrow & \mathrm{Ext}^1_{\widetilde{\mathcal{O}}}(\widetilde{M},\widetilde{N}) & \longrightarrow & \mathrm{Ext}^1_{\widetilde{\mathcal{O}}}(\widetilde{M}',\widetilde{N}) \\
& \big\downarrow T^1_{\widetilde{M}'',\widetilde{N}} & & \big\downarrow T^1_{\widetilde{M},\widetilde{N}} & & \big\downarrow T^1_{\widetilde{M}',\widetilde{N}} \\
\stackrel{\cdot T(\widetilde{E})}{\longrightarrow} & \mathrm{Ext}^1_{\mathcal{O}}(M'',N) & \longrightarrow & \mathrm{Ext}^1_{\mathcal{O}}(M,N) & \longrightarrow & \mathrm{Ext}^1_{\mathcal{O}}(M',N).
\end{array}$$

(2) *The analogous statement holds replacing T by $\overline{T}$ in (1), M by $\overline{M}$, etc.*

Proof. By (6.42), the rows are exact, and it remains to show that the following diagram is commutative:

$$\begin{array}{ccc}
\mathrm{Hom}_{\widetilde{\mathcal{O}}}(\widetilde{M}',\widetilde{N}) & \stackrel{\cdot\widetilde{E}}{\longrightarrow} & \mathrm{Ext}^1_{\widetilde{\mathcal{O}}}(\widetilde{M}'',\widetilde{N}) \\
\big\downarrow T_{\widetilde{M}',\widetilde{N}} & & \big\downarrow T^1_{\widetilde{M}'',\widetilde{N}} \\
\mathrm{Hom}_{\mathcal{O}}(M',N) & \stackrel{\cdot T(\widetilde{E})}{\longrightarrow} & \mathrm{Ext}^1_{\mathcal{O}}(M'',N).
\end{array} \tag{6.51}$$

Let $\widetilde{f} \in \mathrm{Hom}_{\widetilde{\mathcal{O}}}(\widetilde{M}',\widetilde{N})$. Then $\widetilde{f}\widetilde{E} \in \mathrm{Ext}^1_{\widetilde{\mathcal{O}}}(\widetilde{M}'',\widetilde{N})$ is the bottom exact row of the following commutative diagram:

$$\begin{array}{ccccccccc}
0 & \longrightarrow & \widetilde{M}' & \stackrel{\widetilde{i}}{\longrightarrow} & \widetilde{M} & \longrightarrow & \widetilde{M}'' & \longrightarrow & 0 \\
& & \big\downarrow \widetilde{f} & & \big\downarrow & & \big\| & & \\
0 & \longrightarrow & \widetilde{N} & \longrightarrow & \widetilde{F} & \longrightarrow & \widetilde{M}'' & \longrightarrow & 0.
\end{array} \tag{6.52}$$

Here we identify $\widetilde{f}\widetilde{E}$ with the module $\widetilde{F}$, which is the pushout of $\widetilde{f}$ and $\widetilde{i}$. Applying the functor T to (6.52) gives us a commutative diagram with exact rows:

$$\begin{array}{ccccccccc} 0 & \longrightarrow & M' & \xrightarrow{T_{\widetilde{M}',\widetilde{M}}[\widetilde{i}]} & M & \longrightarrow & M'' & \longrightarrow & 0 \\ & & \Big\downarrow T_{\widetilde{M}',\widetilde{N}}[\widetilde{f}] & & \Big\downarrow & & \Big\| & & \\ 0 & \longrightarrow & N & \longrightarrow & T(\widetilde{F}) & \longrightarrow & M'' & \longrightarrow & 0. \end{array}$$

We conclude that $T(\widetilde{F}) \equiv T^1_{\widetilde{M}'',\widetilde{N}}[\widetilde{f}\widetilde{E}]$ is the pushout of $T_{\widetilde{M}',\widetilde{N}}[\widetilde{f}]$ and $T_{\widetilde{M}',\widetilde{M}}[\widetilde{i}]$. Therefore, we obtain that

$$T^1_{\widetilde{M}'',\widetilde{N}}[\widetilde{f}\widetilde{E}] \equiv T(\widetilde{F}) \equiv T_{\widetilde{M}',\widetilde{N}}[\widetilde{f}]T[\widetilde{E}],$$

i.e., the diagram (6.51) is commutative. □

Lemma 6.41. *Let $\widetilde{M},\widetilde{N} \in \widetilde{\mathcal{O}}$. Then*

(1) $T : \mathrm{Hom}_{\widetilde{\mathcal{O}}}(\widetilde{M},\widetilde{N}) \to \mathrm{Hom}_{\mathcal{O}}(M,N)$ *is an injection.*

(2) $\overline{T} : \mathrm{Hom}_{\widetilde{\mathcal{O}}}(\widetilde{M},\widetilde{N}) \to \mathrm{Hom}_{\overline{\mathcal{O}}}(\overline{M},\overline{N})$ *is an injection.*

Proof. By Lemma 6.14, any $\widetilde{\mathfrak{l}}$-isomorphism $\widetilde{\varphi} : L(\widetilde{\mathfrak{l}},\mu^\theta) \to L(\widetilde{\mathfrak{l}},\mu^\theta)$, with $\mu \in P^+$, induces isomorphisms $T[\widetilde{\varphi}] : L(\mathfrak{l},\mu) \to L(\mathfrak{l},\mu)$ and $\overline{T}[\widetilde{\varphi}] : L(\overline{\mathfrak{l}},\mu^\natural) \to L(\overline{\mathfrak{l}},\mu^\natural)$. The lemma follows. □

Lemma 6.42. *Let $\widetilde{V}(\lambda^\theta)$ be a highest weight $\widetilde{\mathfrak{g}}$-module of highest weight λ^θ with $\lambda \in P^+$, and let $\widetilde{N} \in \widetilde{\mathcal{O}}$. Then*

(1) $T : \mathrm{Hom}_{\widetilde{\mathcal{O}}}(\widetilde{V}(\lambda^\theta),\widetilde{N}) \longrightarrow \mathrm{Hom}_{\mathcal{O}}(V(\lambda),N)$ *is an isomorphism.*

(2) $\overline{T} : \mathrm{Hom}_{\widetilde{\mathcal{O}}}(\widetilde{V}(\lambda^\theta),\widetilde{N}) \longrightarrow \mathrm{Hom}_{\overline{\mathcal{O}}}(\overline{V}(\lambda^\natural),\overline{N})$ *is an isomorphism.*

Proof. Consider the commutative diagram with exact rows

$$\begin{array}{ccccccccc} 0 & \longrightarrow & \widetilde{M} & \longrightarrow & \widetilde{\Delta}(\lambda^\theta) & \longrightarrow & \widetilde{V}(\lambda^\theta) & \longrightarrow & 0 \\ & & \Big\downarrow T_{\widetilde{M}} & & \Big\downarrow T_{\widetilde{\Delta}(\lambda^\theta)} & & \Big\downarrow T_{\widetilde{V}(\lambda^\theta)} & & \\ 0 & \longrightarrow & M & \longrightarrow & \Delta(\lambda) & \longrightarrow & V(\lambda) & \longrightarrow & 0. \end{array}$$

This gives rise to the following commutative diagram with exact rows:

$$\begin{array}{ccccccc} 0 & \longrightarrow & \mathrm{Hom}_{\widetilde{\mathcal{O}}}\big(\widetilde{V}(\lambda^\theta),\widetilde{N}\big) & \longrightarrow & \mathrm{Hom}_{\widetilde{\mathcal{O}}}\big(\widetilde{\Delta}(\lambda^\theta),\widetilde{N}\big) & \longrightarrow & \mathrm{Hom}_{\widetilde{\mathcal{O}}}\big(\widetilde{M},\widetilde{N}\big) \\ & & \Big\downarrow T_{\widetilde{V}(\lambda^\theta),\widetilde{N}} & & \Big\downarrow T_{\widetilde{\Delta}(\lambda^\theta),\widetilde{N}} & & \Big\downarrow T_{\widetilde{M},\widetilde{N}} \\ 0 & \longrightarrow & \mathrm{Hom}_{\mathcal{O}}\big(V(\lambda),N\big) & \longrightarrow & \mathrm{Hom}_{\mathcal{O}}\big(\Delta(\lambda),N\big) & \longrightarrow & \mathrm{Hom}_{\mathcal{O}}\big(M,N\big). \end{array}$$

Corollary 6.28(3) and Theorem 6.34 imply that $T_{\widetilde{\Delta}(\lambda^\theta),\widetilde{N}}$ is an isomorphism. By Lemma 6.41, $T_{\widetilde{M},\widetilde{N}}$ is an injection. Now a standard diagram chasing implies that $T_{\widetilde{V}(\lambda^\theta),\widetilde{N}}$ is an isomorphism. □

Lemma 6.43. *Let $\widetilde{V}(\lambda^\theta)$ be a highest weight $\widetilde{\mathfrak{g}}$-module of highest weight λ^θ with $\lambda \in P^+$, and let $\widetilde{N} \in \widetilde{\mathcal{O}}$. Then*

(1) $T : \mathrm{Ext}^1_{\widetilde{\mathcal{O}}}(\widetilde{V}(\lambda^\theta), \widetilde{N}) \to \mathrm{Ext}^1_{\mathcal{O}}(V(\lambda), N)$ *is an injection.*

(2) $\overline{T} : \mathrm{Ext}^1_{\widetilde{\mathcal{O}}}(\widetilde{V}(\lambda^\theta), \widetilde{N}) \to \mathrm{Ext}^1_{\overline{\mathcal{O}}}(\overline{V}(\lambda^\natural), \overline{N})$ *is an injection.*

Proof. Let

$$0 \longrightarrow \widetilde{N} \longrightarrow \widetilde{E} \xrightarrow{\widetilde{f}} \widetilde{V}(\lambda^\theta) \longrightarrow 0 \tag{6.53}$$

be an exact sequence of $\widetilde{\mathfrak{g}}$-modules. Suppose that (6.53) gives rise to a split exact sequence of $\mathfrak{g}$-modules $0 \to N \to E \xrightarrow{T[\widetilde{f}]} V(\lambda) \to 0$. Thus there exists $\psi \in \mathrm{Hom}_{\mathcal{O}}(V(\lambda), E)$ such that $T[\widetilde{f}] \circ \psi = 1_{V(\lambda)}$. By Lemma 6.42, there exists $\widetilde{\psi} \in \mathrm{Hom}_{\widetilde{\mathcal{O}}}(\widetilde{V}(\lambda^\theta), \widetilde{E})$ such that $T[\widetilde{\psi}] = \psi$. Thus $T[\widetilde{f} \circ \widetilde{\psi}] = T[\widetilde{f}] \circ T[\widetilde{\psi}] = 1_{V(\lambda)}$. By Lemma 6.42 again, we have $\widetilde{f} \circ \widetilde{\psi} = 1_{\widetilde{V}(\lambda^\theta)}$, and hence (6.53) splits. □

Proposition 6.44. *Let $\widetilde{M}, \widetilde{N} \in \widetilde{\mathcal{O}}$. Then*

(1) $T : \mathrm{Hom}_{\widetilde{\mathcal{O}}}(\widetilde{M}, \widetilde{N}) \longrightarrow \mathrm{Hom}_{\mathcal{O}}(M, N)$ *is an isomorphism.*

(2) $\overline{T} : \mathrm{Hom}_{\widetilde{\mathcal{O}}}(\widetilde{M}, \widetilde{N}) \longrightarrow \mathrm{Hom}_{\overline{\mathcal{O}}}(\overline{M}, \overline{N})$ *is an isomorphism.*

Proof. First we assume that $\widetilde{M} \in \widetilde{\mathcal{O}}^f$. We proceed by induction on the length of a filtration of $\widetilde{M}$. If $\widetilde{M}$ is a highest weight module, then it is true by Lemma 6.42. Let $0 = \widetilde{M}_0 \subset \widetilde{M}_1 \subset \widetilde{M}_2 \subset \cdots \subset \widetilde{M}_k = \widetilde{M}$ be an increasing filtration of $\widetilde{\mathfrak{g}}$-modules such that $\widetilde{M}_i/\widetilde{M}_{i-1}$ is a highest weight module of highest weight ν_i^θ with $\nu_i \in P^+$, for $1 \le i \le k$. Let $\widetilde{Z}_i := \widetilde{M}_i/\widetilde{M}_{i-1}$ and $Z_i := T(\widetilde{Z}_i)$.

Consider the following commutative diagram with an exact top row of $\widetilde{\mathfrak{g}}$-modules and an exact bottom row of $\mathfrak{g}$-modules, for $i \ge 1$.

$$\begin{array}{ccccccccc}
0 & \longrightarrow & \widetilde{M}_{i-1} & \xrightarrow{\widetilde{\iota}_i} & \widetilde{M}_i & \longrightarrow & \widetilde{Z}_i & \longrightarrow & 0 \\
 & & \big\downarrow T_{\widetilde{M}_{i-1}} & & \big\downarrow T_{\widetilde{M}_i} & & \big\downarrow T_{\widetilde{Z}_i} & & \\
0 & \longrightarrow & M_{i-1} & \xrightarrow{\iota_i} & M_i & \longrightarrow & Z_i & \longrightarrow & 0.
\end{array} \tag{6.54}$$

The sequence (6.54) induces the following commutative diagram with exact rows.

$$\begin{array}{ccccc}
0 & \longrightarrow & \mathrm{Hom}_{\widetilde{\mathcal{O}}}(\widetilde{Z}_i, \widetilde{N}) & \longrightarrow & \mathrm{Hom}_{\widetilde{\mathcal{O}}}(\widetilde{M}_i, \widetilde{N}) \\
 & & \big\downarrow T_{\widetilde{Z}_i, \widetilde{N}} & & \big\downarrow T_{\widetilde{M}, \widetilde{N}} \\
0 & \longrightarrow & \mathrm{Hom}_{\mathcal{O}}(Z_i, N) & \longrightarrow & \mathrm{Hom}_{\mathcal{O}}(M_i, N)
\end{array}$$

$$\begin{array}{ccc} \xrightarrow{\cdot\widetilde{\iota}_i} \mathrm{Hom}_{\widetilde{\mathcal{O}}}(\widetilde{M}_{i-1},\widetilde{N}) & \longrightarrow & \mathrm{Ext}^1_{\widetilde{\mathcal{O}}}(\widetilde{Z}_i,\widetilde{N}) \\ \Big\downarrow T_{\widetilde{M}_{i-1},\widetilde{N}} & & \Big\downarrow T^1_{\widetilde{Z}_i,\widetilde{N}} \\ \xrightarrow{\cdot\iota_i} \mathrm{Hom}_{\mathcal{O}}(M_{i-1},N) & \longrightarrow & \mathrm{Ext}^1_{\mathcal{O}}(Z_i,N). \end{array}$$

The map $T_{\widetilde{M}_{i-1},\widetilde{N}}$ is an isomorphism by induction. The map $T^1_{\widetilde{Z}_i,\widetilde{N}}$ is an injection by Lemma 6.43. Also $T_{\widetilde{Z}_i,\widetilde{N}}$ is an isomorphism by Lemma 6.42. Now a standard diagram chasing implies that $T_{\widetilde{M}_i,\widetilde{N}}$ is an isomorphism.

Now we consider the general case of $\widetilde{M} \in \widetilde{\mathcal{O}}$. By Proposition 6.36, we may choose an increasing filtration of $\widetilde{\mathfrak{g}}$-modules $0 = \widetilde{M}_0 \subset \widetilde{M}_1 \subset \widetilde{M}_2 \subset \cdots$ such that $\bigcup_{i\geq 0} \widetilde{M}_i = \widetilde{M}$ and $\widetilde{M}_i/\widetilde{M}_{i-1}$ is a highest weight module of highest weight ν_i^θ with $\nu_i \in P^+$, for $i \geq 1$. Then the direct limit of $\{\widetilde{M}_i\}_i$ is $\varinjlim \widetilde{M}_i \cong \widetilde{M}$ and the inverse limit is $\varprojlim \mathrm{Hom}_{\widetilde{\mathcal{O}}}(\widetilde{M}_i,\widetilde{N}) \cong \mathrm{Hom}_{\widetilde{\mathcal{O}}}(\widetilde{M},\widetilde{N})$. Similarly we have $\varinjlim M_i \cong M$ and $\varprojlim \mathrm{Hom}_{\mathcal{O}}(M_i,N) \cong \mathrm{Hom}_{\mathcal{O}}(M,N)$. Furthermore, we have the following commutative diagram (where $\varphi = \varprojlim T_{\widetilde{M}_i,\widetilde{N}}$):

$$\begin{array}{ccc} \mathrm{Hom}_{\widetilde{\mathcal{O}}}(\widetilde{M},\widetilde{N}) & \xrightarrow{\cong} & \varprojlim \mathrm{Hom}_{\widetilde{\mathcal{O}}}(\widetilde{M}_i,\widetilde{N}) \\ \Big\downarrow T_{\widetilde{M},\widetilde{N}} & & \Big\downarrow \varphi \\ \mathrm{Hom}_{\mathcal{O}}(M,N) & \xrightarrow{\cong} & \varprojlim \mathrm{Hom}_{\mathcal{O}}(M_i,N). \end{array}$$

Since φ is an isomorphism, so is $T_{\widetilde{M},\widetilde{N}}$. □

Lemma 6.45. *Let $\widetilde{M},\widetilde{N} \in \widetilde{\mathcal{O}}$. Then*

(1) $T : \mathrm{Ext}^1_{\widetilde{\mathcal{O}}}(\widetilde{M},\widetilde{N}) \to \mathrm{Ext}^1_{\mathcal{O}}(M,N)$ *is an injection.*

(2) $\overline{T} : \mathrm{Ext}^1_{\widetilde{\mathcal{O}}}(\widetilde{M},\widetilde{N}) \to \mathrm{Ext}^1_{\overline{\mathcal{O}}}(\overline{M},\overline{N})$ *is an injection.*

Proof. The proof is virtually identical to the proof of Lemma 6.43, where we use Proposition 6.44 in place of Lemma 6.42. □

Proposition 6.46. *Let $\widetilde{V}(\lambda^\theta)$ be a highest weight $\widetilde{\mathfrak{g}}$-module of highest weight λ^θ with $\lambda \in P^+$, and let $\widetilde{N} \in \widetilde{\mathcal{O}}$. Then*

(1) $T : \mathrm{Ext}^1_{\widetilde{\mathcal{O}}}(\widetilde{V}(\lambda^\theta),\widetilde{N}) \to \mathrm{Ext}^1_{\mathcal{O}}(V(\lambda),N)$ *is an isomorphism.*

(2) $\overline{T} : \mathrm{Ext}^1_{\widetilde{\mathcal{O}}}(\widetilde{V}(\lambda^\theta),\widetilde{N}) \to \mathrm{Ext}^1_{\overline{\mathcal{O}}}(\overline{V}(\lambda^\natural),\overline{N})$ *is an isomorphism.*

Proof. Consider the following commutative diagram with exact rows:

$$\begin{array}{ccccccccc} 0 & \longrightarrow & \widetilde{M} & \longrightarrow & \widetilde{\Delta}(\lambda^\theta) & \xrightarrow{\widetilde{\pi}} & \widetilde{V}(\lambda^\theta) & \longrightarrow & 0 \\ & & \Big\downarrow T_{\widetilde{M}} & & \Big\downarrow T_{\widetilde{\Delta}(\lambda^\theta)} & & \Big\downarrow T_{\widetilde{V}(\lambda^\theta)} & & \\ 0 & \longrightarrow & M & \longrightarrow & \Delta(\lambda) & \xrightarrow{\pi} & V(\lambda) & \longrightarrow & 0. \end{array} \tag{6.55}$$

The sequence (6.55) induces the following commutative diagram with exact rows:

$$\begin{array}{ccccc}
\mathrm{Hom}_{\widetilde{\mathcal{O}}}\big(\widetilde{\Delta}(\lambda^\theta),\widetilde{N}\big) & \longrightarrow & \mathrm{Hom}_{\widetilde{\mathcal{O}}}\big(\widetilde{M},\widetilde{N}\big) & \longrightarrow & \mathrm{Ext}^1_{\widetilde{\mathcal{O}}}\big(\widetilde{V}(\lambda^\theta),\widetilde{N}\big) \\
\big\downarrow T_{\widetilde{\Delta}(\lambda^\theta),\widetilde{N}} & & \big\downarrow T_{\widetilde{M},\widetilde{N}} & & \big\downarrow T^1_{\widetilde{V}(\lambda^\theta),\widetilde{N}} \\
\mathrm{Hom}_{\mathcal{O}}\big(\Delta(\lambda),N\big) & \longrightarrow & \mathrm{Hom}_{\mathcal{O}}\big(M,N\big) & \longrightarrow & \mathrm{Ext}^1_{\mathcal{O}}\big(V(\lambda),N\big)
\end{array}$$

$$\begin{array}{ccccc}
\xrightarrow{\cdot\widetilde{\pi}} & \mathrm{Ext}^1_{\widetilde{\mathcal{O}}}\big(\widetilde{\Delta}(\lambda^\theta),\widetilde{N}\big) & \longrightarrow & \mathrm{Ext}^1_{\widetilde{\mathcal{O}}}\big(\widetilde{M},\widetilde{N}\big) \\
& \big\downarrow T^1_{\widetilde{\Delta}(\lambda^\theta),\widetilde{N}} & & \big\downarrow T^1_{\widetilde{M},\widetilde{N}} \\
\xrightarrow{\cdot\pi} & \mathrm{Ext}^1_{\mathcal{O}}\big(\Delta(\lambda),N\big) & \longrightarrow & \mathrm{Ext}^1_{\mathcal{O}}\big(M,N\big).
\end{array}$$

The map $T^1_{\widetilde{M},\widetilde{N}}$ is an injection by Lemma 6.45. The map $T^1_{\widetilde{\Delta}(\lambda^\theta),\widetilde{N}}$ is an isomorphism by Corollary 6.35. Also $T_{\widetilde{\Delta}(\lambda^\theta),\widetilde{N}}$ and $T_{\widetilde{M},\widetilde{N}}$ are isomorphisms by Proposition 6.44. Now a standard diagram chasing implies that $T^1_{\widetilde{V}(\lambda^\theta),\widetilde{N}}$ is an isomorphism. □

Proof of Theorem 6.39. Let $\mathcal{C},\mathcal{C}'$ be abelian categories. Recall from Mitchell [**86**, II.4] that by definition a full and faithful functor $F:\mathcal{C}\mapsto\mathcal{C}'$ is an equivalence of categories if it satisfies the representative property that for every $M'\in\mathcal{C}'$ there exists $M\in\mathcal{C}$ with $F(M)\cong M'$.

Proposition 6.44 implies that the functor T is full and faithful. Now for every $M\in\mathcal{O}$, there is a filtration of $\mathfrak{g}$-modules for M, $0=M_0\subset M_1\subset M_2\subset\cdots$, with $M_i\in\mathcal{O}$ satisfying the properties of Proposition 6.36. The filtration $\{M_i\}_i$ of M lifts to a filtration $\{\widetilde{M}_i\}_i$ with $\widetilde{M}_i\in\widetilde{\mathcal{O}}^f$ such that $T(\widetilde{M}_i)\cong M_i$ by induction using Propositions 6.38 and 6.46. Set $\widetilde{M}:=\cup_{i\geq 0}\widetilde{M}_i$. It follows that $T(\widetilde{M})\cong M$.

We need to prove that $\widetilde{M}\in\widetilde{\mathcal{O}}$. To do that we proceed as in the proof of the claim in Proposition 6.15. Suppose that $\zeta_1,\ldots,\zeta_r\in P^+$ such that every weight of M is bounded by ζ_i, for some i. For $j=1,\ldots,r$, define P_j and $P(M)$ as in Proposition 6.15. Now since $T(\widetilde{M})=M$, it suffices to prove that if $\nu\in P^+$ and $\zeta_i-\nu\in\mathbb{Z}_+\Pi$ for some i, then there exists $\gamma\in P(M)$ such that $\gamma^\theta-\nu^\theta\in\mathbb{Z}_+\widetilde{\Pi}$. Let $|{}^+\zeta_i|=k_i$. We write

$$\zeta_i-\nu={}^-\kappa+p(\varepsilon_{-1}-\varepsilon_1)+{}^+\kappa,$$

where $p\in\mathbb{Z}_+$, ${}^-\kappa\in\sum_{\alpha\in\Pi(\mathfrak{k})}\mathbb{Z}_+\alpha$, and ${}^+\kappa\in\sum_{\beta\in\Pi(\mathfrak{T})\setminus\beta_\times}\mathbb{Z}_+\beta$. Then ${}^+\nu$ is a partition of size k_i+p. Hence, there exists $\gamma\in P(M)$ such that the partition ${}^+\nu$ is obtained from ${}^+\gamma$ by adding p boxes. We record the sub-indices of the boxes in $\theta({}^+\gamma)\setminus\theta({}^+\nu)$ in the multiset J^θ. Then we have $\gamma^\theta-\nu^\theta={}^-\kappa+\sum_{s\in J^\theta}(\varepsilon_{-1}-\varepsilon_s)$, and hence $\gamma^\theta-\nu^\theta\in\mathbb{Z}_+\widetilde{\Pi}$. This implies that all the weights of $\widetilde{M}$ are bounded above by the finite set $\{\gamma^\theta|\gamma\in P(M)\}$, and hence $\widetilde{M}\in\widetilde{\mathcal{O}}$.

We conclude that the functor T satisfies the representative property. Hence, $T:\widetilde{\mathcal{O}}\to\mathcal{O}$ is an equivalence of categories. This proves (1).

The proof of (2) is entirely parallel to (1), while (3) follows from (1) and (2) (see, e.g., [**86**, II.10]). □

Remark 6.47. Super duality helps to provide a categorical explanation and new proofs for results obtained in earlier chapters by other means: the irreducible polynomial character formula for $\mathfrak{gl}(m|n)$ in terms of super Schur functions via Schur-Sergeev duality in Chapter 3, and also the irreducible character formulas for the oscillator modules of Lie superalgebras via Howe duality in Chapter 5.

6.6. Exercises

Exercise 6.1. Prove Proposition 6.15 in the limiting case $m=0$.

Exercise 6.2. Prove the following identities of symmetric functions:

(1) $\omega\Big(\prod_{1\le i\le j}(1-x_ix_j)^{-1}\Big)=\prod_{1\le i<j}(1-x_ix_j)^{-1}$.

(2) $\omega\Big(\prod_{1\le i\le j}(1+x_ix_j)\Big)=\prod_{1\le i<j}(1+x_ix_j)$.

(Hint: Use $T(\widetilde{\mathfrak{u}}^-)=\mathfrak{u}^-$ and $\overline{T}(\widetilde{\mathfrak{u}}^-)=\overline{\mathfrak{u}}^-$.)

In Exercises 6.3 and 6.4 we shall use the following notation: Let Λ (respectively $\overline{\Lambda}$) denote the ring of symmetric functions in the variables $x_1,x_2,\ldots$ (respectively $x_{1/2},x_{3/2},\ldots$). Identifying Λ with $\overline{\Lambda}$, the involution of symmetric functions is identified with

$$\omega\left(\prod_{i=1}^{\infty}(1+x_it)\right)=\prod_{i=1}^{\infty}\frac{1}{(1-x_{i-\frac12}t)},$$

for an indeterminate t.

Exercise 6.3. Let $\Omega_{\mathfrak{k}}$ be the character ring of the BGG category of $(\mathfrak{k}+\mathbb{C}K)$-modules. For an indeterminate e we let $x_r=e^{\varepsilon_r}$, for $r\in\frac12\mathbb{Z}_+$, and let $\Omega_{\mathfrak{k}}\widehat{\otimes}\Lambda$ denote the topological completion of $\Omega_{\mathfrak{k}}\otimes\Lambda$ with respect to the degree filtration of Λ. Let $M\in\mathcal{O}$, $\overline{M}\in\overline{\mathcal{O}}$, and $\widetilde{M}\in\widetilde{\mathcal{O}}$. Prove:

(1) $\mathrm{ch}M\in\Omega_{\mathfrak{k}}\widehat{\otimes}\Lambda$ and $\mathrm{ch}\overline{M}\in\Omega_{\mathfrak{k}}\widehat{\otimes}\overline{\Lambda}$.

(2) $\omega\left(\mathrm{ch}\,T(\widetilde{M})\right)=\mathrm{ch}\,\overline{T}(\widetilde{M})$.

(3) $\omega\left(\mathrm{ch}H_i(\mathfrak{u}^-,T(\widetilde{M}))\right)=\mathrm{ch}H_i(\overline{\mathfrak{u}}^-,\overline{T}(\widetilde{M}))$, for all $i\in\mathbb{Z}_+$.

Exercise 6.4. Let $\widetilde{M}$ and $\widetilde{N}$ be two objects in $\widetilde{\mathcal{O}}$, and let $\lambda,\mu,\nu\in P^+$. Prove:

(1) The tensor product $\widetilde{M}\otimes\widetilde{N}$ is an object in $\widetilde{\mathcal{O}}$.

(2) $T(\widetilde{M}\otimes\widetilde{N})=T(\widetilde{M})\otimes T(\widetilde{N})$, and $\overline{T}(\widetilde{M}\otimes\widetilde{N})=\overline{T}(\widetilde{M})\otimes\overline{T}(\widetilde{N})$.

(3) $\omega\left(\mathrm{ch}\,T(\widetilde{M}\otimes\widetilde{N})\right)=\mathrm{ch}\,\overline{T}(\widetilde{M}\otimes\widetilde{N})$.

(4) $L(\lambda)$ appears as a composition factor in $L(\mu)\otimes L(\nu)$ with the same multiplicity as $\overline{L}(\lambda^\natural)$ appears as a composition factor in $\overline{L}(\mu^\natural)\otimes\overline{L}(\nu^\natural)$.

Exercises 6.5 and 6.6 show that a finite-dimensional simple module over the general linear Lie superalgebra may not have a BGG resolution in terms of Verma modules, but may have a resolution in terms of Kac modules. In the notation of this chapter we set $m=1$ and $\mathfrak{x}=\mathfrak{a}$. Let $\overline{\mathfrak{b}}_2$ be the standard Borel subalgebra of the Lie superalgebra $\overline{\mathfrak{g}}_2=\mathfrak{gl}(1|2)$ with Cartan subalgebra $\overline{\mathfrak{h}}_2$. By highest weight we always mean with respect to $\overline{\mathfrak{b}}_2$.

Exercise 6.5. Let $\overline{M}(\varepsilon_{-1})$ be the Verma module of highest weight $\varepsilon_{-1}\in\overline{\mathfrak{h}}_2^*$ with highest weight vector $v_{\varepsilon_{-1}}$. Let $\overline{L}_2(\varepsilon_{-1})$ be its unique irreducible quotient. Prove:

(1) If $N\subseteq\overline{M}(\varepsilon_{-1})$ is the $\overline{\mathfrak{g}}_2$-submodule generated by $E_{2,1}v_{\varepsilon_{-1}}$, then N contains all nontrivial singular weight vectors of $\overline{M}(\varepsilon_{-1})$.

(2) The $\overline{\mathfrak{g}}_2$-module $\overline{M}(\varepsilon_{-1})/N$ is reducible.

(3) There is no exact sequence of $\overline{\mathfrak{g}}_2$-modules of the form

$$\bigoplus_i \overline{M}(\mu_i)\longrightarrow\overline{M}(\varepsilon_{-1})\longrightarrow\overline{L}_2(\varepsilon_{-1})\longrightarrow 0,\quad \mu_i\in\overline{\mathfrak{h}}_2^*.$$

Exercise 6.6. Let w_λ denote a highest weight vector in the $\overline{\mathfrak{g}}_2$-Kac module $\overline{\Delta}_2(\lambda)$, for $\lambda\in\overline{P}_2^+$. Prove:

(1) The only nontrivial singular vectors of $\overline{\Delta}_2(\varepsilon_{-1})$ are scalar multiples of $E_{1,-1}E_{2,-1}w_{\varepsilon_{-1}}$.

(2) For $k\in\mathbb{N}$ the unique maximal submodule of $\overline{\Delta}_2(-k\varepsilon_{-1}+k\varepsilon_1+\varepsilon_2)$ is generated by the singular vector $E_{1,-1}w_{-k\varepsilon_{-1}+k\varepsilon_1+\varepsilon_2}$. Furthermore, it is isomorphic to $\overline{L}_2\left(-(k+1)\varepsilon_{-1}+(k+1)\varepsilon_1+\varepsilon_2\right)$.

(3) For every $k\in\mathbb{N}$ we have an exact sequence of $\overline{\mathfrak{g}}_2$-modules of the form

$$\overline{\Delta}_2\left(-(k+1)\varepsilon_{-1}+(k+1)\varepsilon_1+\varepsilon_2\right)\longrightarrow\overline{\Delta}_2(-k\varepsilon_{-1}+k\varepsilon_1+\varepsilon_2)\longrightarrow\cdots$$
$$\cdots\longrightarrow\overline{\Delta}_2(-\varepsilon_{-1}+\varepsilon_1+\varepsilon_2)\longrightarrow\overline{\Delta}_2(\varepsilon_{-1})\longrightarrow\overline{L}_2(\varepsilon_{-1})\longrightarrow 0.$$

Conclude that $\overline{L}_2(\varepsilon_{-1})$ can be resolved in terms of Kac modules.

(4) Recalling D from Section 2.2.3 and $W\cong\mathfrak{S}_2$ we have the following identity:

$$\mathrm{ch}\overline{L}_2(\varepsilon_{-1})=\frac{1}{D}\sum_{w\in W}(-1)^{\ell(w)}w\left(\frac{e^{\lambda+\rho}}{1+e^{-\varepsilon_{-1}+\varepsilon_2}}\right).$$

Exercise 6.7. Let $\mathfrak{g}=\mathfrak{g}^{\mathfrak{x}}$. Define

$$P^{++}:=\{\lambda\in P^+\,|\,\dim L_n(\lambda)<\infty,\ \text{for } n\gg 0\}. \tag{6.56}$$

(1) Let $\mathfrak{x}=\mathfrak{a}$. Prove that $\lambda=\sum_{i=-m}^{-1}\lambda_i\varepsilon_i+\sum_{j=1}^{\infty}{}^+\lambda_j\varepsilon_j\in P^{++}$ if and only if $(\lambda_{-m},\ldots,\lambda_{-1},{}^+\lambda_1,{}^+\lambda_2,\ldots)$ is a partition.

(2) Let $d \in \mathbb{Z}$ and $\mathfrak{x} = \mathfrak{c}$. Prove that $\lambda = d\Lambda_0 + \sum_{i=-m}^{-1} \lambda_i \varepsilon_i + \sum_{j=1}^{\infty} {}^+\lambda_j \varepsilon_j \in P^{++}$ if and only if $d \in -\mathbb{Z}_+$, $0 \geq \lambda_{-m}$, and $(\lambda_{-m} - d, \ldots, \lambda_{-1} - d, {}^+\lambda_1, {}^+\lambda_2, \ldots)$ is a partition (see (6.14)).

(3) For $\mathfrak{x} = \mathfrak{b}, \mathfrak{b}^\bullet, \mathfrak{d}$ and $d \in \frac{1}{2}\mathbb{Z}$, determine the set P^{++} as in (2).

Exercise 6.8. Let $\mathfrak{g} = \mathfrak{g}^{\mathfrak{x}}$, where $\mathfrak{x} = \mathfrak{a}, \mathfrak{b}, \mathfrak{c}, \mathfrak{d}$. Let $\mathfrak{l}$ and $\mathfrak{u}_-$ denote the Levi subalgebra and opposite nilradical, respectively. For λ in P^{++} as in (6.56) we have the following parabolic BGG resolution **[6, 79, 100]** (recall W^0 from Section 5.5):

(i) There exists an exact sequence of $\mathfrak{g}$-modules of the form

$$\cdots \longrightarrow \bigoplus_{w \in W_r^0} \Delta(w \circ \lambda) \longrightarrow \bigoplus_{w \in W_{r-1}^0} \Delta(w \circ \lambda) \longrightarrow \cdots$$
$$\cdots \longrightarrow \bigoplus_{w \in W_1^0} \Delta(w \circ \lambda) \longrightarrow \Delta(\lambda) \longrightarrow L(\lambda) \longrightarrow 0.$$

(ii) As $\mathfrak{l}$-modules we have $H_i(\mathfrak{u}^-, L(\lambda)) = \bigoplus_{w \in W_i^0} L(\mathfrak{l}, w \circ \lambda)$, for all $i \in \mathbb{Z}_+$.

Prove:

(1) The KLV polynomials $\ell_{\mu\lambda}(q)$ are monomials, for $\mu \in P^+$.

(2) Every $\overline{\mathfrak{g}}$-module $\overline{L}(\lambda^\natural)$ has a BGG resolution in terms of $\overline{\Delta}(\mu^\natural)$, $\mu \in P^+$.

(3) The irreducible $\mathfrak{gl}(m|n)$-tensor modules have resolutions in terms of Kac modules. Furthermore, if ν is such a highest weight and $\mu \in P_n^+$, then the KLV polynomial $\overline{\ell}_{\mu\nu}(q)$, if nonzero, is a monomial of q.

(4) Let $d \in \mathbb{Z}_+$ and assume that $\ell(\lambda) \leq \min\{n, d\}$. The irreducible highest weight $\mathfrak{osp}(1|2n)$-module of highest weight $\sum_{i=1}^n (\lambda_i + d)\varepsilon_i$ with respect to the fundamental system $\{-\varepsilon_1, \varepsilon_1 - \varepsilon_2, \ldots, \varepsilon_{n-1} - \varepsilon_n\}$ has a resolution in terms of parabolic Verma modules with Levi obtained by removing $-\varepsilon_1$. Also, similarly to (2), the corresponding nonzero KLV polynomials are monomials of q.

Exercise 6.9. Let $\mathfrak{g} = \mathfrak{g}^{\mathfrak{a}}$ with $Y_0 = \Pi(\mathfrak{k})$. Let $\lambda \in P^+$ with $\lambda_i \in \mathbb{Z}$. Suppose that for any $\lambda_j - j$ with $j < 0$ there exists an $i > 0$ such that $\lambda_j - j = \lambda_i - i$. Prove that $\Delta(\lambda)$ is irreducible. (Hint: Use Exercise 2.11 and super duality.)

Exercise 6.10. Let $\mathfrak{g} = \mathfrak{g}^{\mathfrak{a}}$ and let $\lambda, \mu \in P^+$. Prove:

(1) There exists $k \in \mathbb{N}$ such that for $n \geq k$ the multiplicity with which $\overline{L}_n(\mu^\natural)$ appears in a composition series of $\overline{\Delta}_n(\lambda^\natural)$ is independent of n.

(2) The parabolic Verma module $\overline{\Delta}(\lambda^\natural)$ has a finite composition series.

(3) The parabolic Verma module $\Delta(\lambda)$ has a finite composition series.

Exercise 6.11. Let L be a finite-dimensional irreducible $\mathfrak{osp}(k|2n)$-module, for $k = 2m$ or $k = 2m+1$. Prove that there exist $\mathfrak{g}^{\mathfrak{x}}$ with $\mathfrak{x} = \mathfrak{b}, \mathfrak{d}$ and $\lambda \in P^+$ such that the $\overline{\mathfrak{g}}_n$-module $\overline{L}_n(\lambda^\natural)$ is isomorphic to L.

Exercise 6.12. Recall $\mathfrak{a}_\infty$ from Section 5.4.1 and let $\mathfrak{l}$ be its Levi subalgebra obtained by removing the simple root $\varepsilon_0 - \varepsilon_1$ with opposite nilradical $\mathfrak{u}^-$. Then $\mathfrak{l} \cong \mathfrak{a}_{\leq 0} \oplus \mathfrak{a}_{>0} \oplus \mathbb{C}K$, where $\mathfrak{a}_{\leq 0}$ (respectively $\mathfrak{a}_{>0}$) consists of those matrices $(a_{ij}) \in \mathfrak{a}_\infty$ with $a_{ij} = 0$ unless $i, j \in -\mathbb{Z}_+$ (respectively $i, j \in \mathbb{N}$). For $d \in \mathbb{Z}$ let $\mathcal{C}_d$ be the parabolic BGG category of $\mathfrak{a}_\infty$-modules on which K acts as the scalar d and that, as $\mathfrak{l}$-modules, decompose into direct sums of irreducibles of the form $L(\mathfrak{a}_{\leq 0}, \nu) \otimes L(\mathfrak{a}_{>0}, \mu)$, where $\nu = \sum_{j \leq 0} \nu_j \varepsilon_j$ and $\mu = \sum_{j \geq 1} \mu_j \varepsilon_j$, with $(-\nu_0, -\nu_{-1}, \ldots) \in \mathcal{P}$ and $(\mu_1, \mu_2, \ldots) \in \mathcal{P}$. Denote the set of such weights $\nu + \mu$ by P^+. Prove:

(1) The categories $\mathcal{C}_d$ and $\mathcal{C}_{-d}$ are equivalent.

(2) For a generalized partition of length d of the form

$$\lambda = (\lambda_1 \geq \lambda_2 \geq \ldots \geq \lambda_i > 0 = \ldots = 0 > \lambda_{j+1} \geq \ldots \geq \lambda_d),$$

let $\Lambda(\lambda) := -d\Lambda_0 + \sum_{k=1}^{i} \lambda_i \varepsilon_i + \sum_{k=j+1}^{d} \lambda_k \varepsilon_{k-d} \in P^+$. Then the KLV polynomial $\ell_{\mu\Lambda(\lambda)}$ is a monomial for every $\mu \in P^+$.

Exercise 6.13. Let $\mathfrak{g} = \mathfrak{g}^{\mathfrak{a}}$ with $Y_0 = \Pi(\mathfrak{k})$. Let $d, m, n \in \mathbb{N}$ and $\nu, \mu \in \mathcal{P}$ with $\ell(\nu) \leq m$, $\ell(\mu) \leq n$, and $\ell(\nu) + \ell(\mu) \leq d$. Set $\xi := \sum_{i=1}^{m} (-d - \nu_i) \varepsilon_{-i} + \sum_{j=1}^{n} \mu_j \varepsilon_j$. Prove:

(1) For any $\eta \in P_n^+$ the KLV polynomial $\ell_{\eta\xi}$ for the finite-dimensional Lie algebra $\mathfrak{g}_n$ is a monomial.

(2) The $\mathfrak{g}_n$-module $L_n(\xi)$ has a resolution in terms of $\Delta_n(\eta)$ with $\eta \in P^+$.

(Hint: Use Exercise 6.12.)

Notes

This chapter is an exposition on the formulation and proof of super duality, which is an equivalence of the categories $\mathcal{O}$ and $\overline{\mathcal{O}}$, in Cheng-Lam-Wang [**24**], which in turn was built on Cheng-Lam [**23**] for type A. The super duality conjecture for type A in the maximal parabolic case was first formulated in Cheng-Wang-Zhang [**34**] and more generally in Cheng-Wang [**31**], motivated by and partially based on Brundan [**11**]. There is a different approach by Brundan-Stroppel [**15**] in the special case of the super duality conjecture formulated in [**34**].

Section 6.1. The presentation of the materials in this section follows closely Cheng-Lam-Wang [**24**].

Section 6.2. The truncation functors and their basic properties regarding Verma and irreducible modules (Proposition 6.9) have counterparts in the algebraic group setting, which was developed by Donkin [**40**]. In the super duality context, the truncation functors first appeared in Cheng-Wang-Zhang [**34**], and then were generalized in [**31, 23, 24**].

Section 6.3. The odd reflection approach leading to Theorems 6.17 and 6.18 was developed in [**23, 24**]. It is remarkable that, when combined with Proposition 6.9, this solves completely the irreducible character problem for modules in the category $\overline{\mathcal{O}}_n$, which include all finite-dimensional simple modules of type $\mathfrak{gl}$ and $\mathfrak{osp}$.

Totally different approaches and solutions for the *finite-dimensional* irreducible character problem have been developed in Serganova [**107**], Brundan [**11**], and Brundan-Stroppel [**15**] for $\mathfrak{gl}(m|n)$, and in Gruson-Serganova [**49**] for $\mathfrak{osp}$. For recent works, see also [**17, 82, 120, 121**].

Section 6.4. We follow Tanaka [**122**] for a formula of the (co)boundary operator for Lie superalgebras (see also Iohara-Koga [**55**]). The results on the identification of Kazhdan-Lusztig-Vogan polynomials in the categories $\mathcal{O}$, $\overline{\mathcal{O}}$, and $\widetilde{\mathcal{O}}$ are due to [**23, 24**]. The study of $\mathfrak{u}$-cohomology groups was initiated in the fundamental paper of Kostant [**73**]. We have followed Vogan's homological interpretation of the Kazhdan-Lusztig polynomials [**125**, Conjecture 3.4] (also see Serganova [**107**] in the setting of the finite-dimensional module category for $\mathfrak{gl}(m|n)$), as the usual approach via Hecke algebras is not applicable for Lie superalgebras. The Kazhdan-Lusztig conjectures [**70, 71**] for the BGG category of a semisimple Lie algebra were proved in Beilinson-Bernstein [**5**] and Brylinski-Kashiwara [**16**]. The polynomials $\ell_{\mu\lambda}(q)$ for the category $\mathcal{O}$ coincide with the usual parabolic Kazhdan-Lusztig polynomials [**39**]. Theorem 6.24, which describes an explicit connection between $\mathfrak{u}^-$-homology and $\mathfrak{u}$-cohomology groups, is based on [**81**, Section 4]. The stability of Kazhdan-Lusztig polynomials (6.38) even for classical Lie algebras has not been formulated in the literature until we needed it in the super duality framework.

Section 6.5. The formulation of super duality (Theorem 6.39) follows Cheng-Lam-Wang [**24**] (also see [**34, 31, 23**] for earlier formulation). The proof here using Ext^1 only is somewhat different from the original one, and it is based on Theorem 6.34, which is adapted from Rocha-Caridi and Wallach [**101**, §7, Theorem 2]. The definition of the module subcategory $\widetilde{\mathcal{O}}^{\bar{0}}$ and its variants for type A was first given in [**23**, Section 2.5], and in general in [**24**, Section 5.2].

The tilting modules, formulated earlier by Ringel and Donkin in different contexts, can be shown to exist in the categories $\mathcal{O}_n, \overline{\mathcal{O}}_n, \mathcal{O}_n$ (cf. Brundan [**11**], Cheng-Wang-Zhang [**34**] and Cheng-Wang [**28**] for type A). The functors $T : \widetilde{\mathcal{O}} \to \mathcal{O}$ and $\overline{T} : \widetilde{\mathcal{O}} \to \overline{\mathcal{O}}$ can be shown to respect tilting modules (see Cheng-Lam-Wang [**25**]). We do not treat tilting modules in the book, and refer the reader to the original papers for details. We also refer to [**26**] for a proof of the Brundan-Kazhdan-Lusztig conjecture for the full BGG category $\mathcal{O}$ of $\mathfrak{gl}(m|n)$-modules.

The results in Exercises 6.3(3) and 6.4(4) were taken from [**24**] (various special cases appeared earlier in [**34, 23**]), while Exercise 6.5 is taken from [**20**]. An analogous character formula as in Exercise 6.6(4) holds for finite-dimensional irreducible modules over the Lie superalgebras $\mathfrak{gl}(m|n)$ and $\mathfrak{osp}(2|2n)$ in the case when the highest weights have atypicality degree equal to 1, and it appeared for the first time in [**8**] (see also [**123, 124**]). The resolutions in Exercise 6.8(2) were first constructed in [**20**], while the KLV polynomials in this particular setting were first computed in [**35**], both without using super duality. The observation in Exercise 6.9 was first made in [**34**], while Exercise 6.10 is taken from [**23**] (a special case of (2) goes back to [**34**]). We are not aware of a direct proof for Exercise 6.10(3) without using super duality. The KLV polynomials in Exercise 6.12 were computed in [**19**] for the first time. Exercise 6.13(1) and (2) are classical results that appeared already in [**41**] and [**42**], respectively, stated in a different form (see [**53**]). They may now be regarded as direct applications of super duality.

Appendix A

Symmetric functions

A.1. The ring Λ and Schur functions

A.1.1. The ring Λ. A **partition** $\lambda = (\lambda_1, \lambda_2, \ldots, \lambda_\ell)$ is a sequence of decreasing nonnegative integers $\lambda_1 \geq \lambda_2 \geq \ldots \geq \lambda_\ell \geq 0$. Each λ_i is called a **part** of λ, and the number of nonzero parts of λ is called the **length** of λ, which is denoted by $\ell(\lambda)$. The sum of all its nonzero parts is called the **size** of λ, and is denoted by $|\lambda|$. We sometimes use the notation $\lambda = (1^{m_1} 2^{m_2} \ldots)$ to denote a partition λ, where m_i is the number of parts of λ that are equal to i for $i \geq 1$. We often identify a partition λ with its associated Young diagram, and denote by λ' the **conjugate partition** whose Young diagram is the transpose of the Young diagram of λ.

Let $\mathcal{P}$ denote the set of all partitions and let $\mathcal{P}_n$ denote the set of all partitions of n, for $n \geq 0$. For $\lambda, \mu \in \mathcal{P}$, we write $\lambda \supseteq \mu$ if $\lambda_i \geq \mu_i$ for all $i \geq 1$. The **dominance order** on $\mathcal{P}$, denoted by $\geq$, is the partial order defined by

$$\lambda \geq \mu \quad \Leftrightarrow \quad |\lambda| = |\mu| \text{ and } \lambda_1 + \ldots + \lambda_i \geq \mu_1 + \ldots + \mu_i, \text{ for all } i \geq 1.$$

Consider the $\mathbb{Z}$-module of formal power series $\mathbb{Z}[[x]]$ in the infinite set of indeterminates $x := \{x_1, x_2, x_3, \ldots\}$. For $\alpha = (\alpha_1, \alpha_2, \ldots, \alpha_k) \in \mathbb{Z}_+^k$, we write

$$x^\alpha := x_1^{\alpha_1} x_2^{\alpha_2} \cdots x_k^{\alpha_k} \in \mathbb{Z}[[x]].$$

The symmetric group $\mathfrak{S}_n$ acts on $\mathbb{Z}[[x]]$ by permuting the indeterminates, i.e., if $\sigma \in \mathfrak{S}_n$ and $f \in \mathbb{C}[[x]]$, then

$$\sigma f(x_1, x_2, \ldots) := f(x_{\sigma(1)}, x_{\sigma(2)}, \ldots),$$

where by definition $\sigma(i) = i$, for all $i > n$. Since the action of $\mathfrak{S}_{n+1}$ is compatible with that of $\mathfrak{S}_n$, we obtain an action of the direct limit, denoted by $\mathfrak{S}_\infty$, on $\mathbb{Z}[[x]]$.

The **monomial symmetric function** associated to a partition λ is

$$m_\lambda(x) = m_\lambda := \sum_{\sigma \in \mathfrak{S}_\infty/\mathfrak{S}_\lambda} \sigma x^\lambda \in \mathbb{Z}[[x]],$$

where $\mathfrak{S}_\lambda$ is the stabilizer subgroup of the monomial x^λ in $\mathfrak{S}_\infty$.

Definition A.1. The **ring of symmetric functions** Λ is the $\mathbb{Z}$-span of the elements $\{m_\lambda \mid \lambda \in \mathcal{P}\}$ in $\mathbb{Z}[[x]]$.

We remark that Λ is closed under multiplication and hence is indeed a ring. Also note that Λ is naturally graded by degree. We denote the $\mathbb{Z}$-submodule of degree n by Λ^n so that $\Lambda = \bigoplus_{n \geq 0} \Lambda^n$. Note that Λ^n is a free $\mathbb{Z}$-module of rank equal to the number of partitions of n. For a field $\mathbb{F}$ (which is often taken to be $\mathbb{Q}$ or $\mathbb{C}$), we denote the base change by $\Lambda_{\mathbb{F}} = \mathbb{F} \otimes_{\mathbb{Z}} \Lambda$ and $\Lambda_{\mathbb{F}}^n = \mathbb{F} \otimes_{\mathbb{Z}} \Lambda^n$.

There are several other bases for Λ that we use freely in this book.

Definition A.2. Let $n \in \mathbb{N}$. We define

$$\begin{aligned} p_n &:= m_{(n)} = \sum_{i \geq 1} x_i^n, \\ e_n &:= m_{(1^n)} = \sum_{i_1 < i_2 < \cdots < i_n} x_{i_1} x_{i_2} \cdots x_{i_n}, \\ h_n &:= \sum_{\lambda \in \mathcal{P}_n} m_\lambda = \sum_{i_1 \leq i_2 \leq \cdots \leq i_n} x_{i_1} x_{i_2} \cdots x_{i_n}, \end{aligned}$$

that are called the nth **power-sum**, **elementary**, and **complete symmetric functions**, respectively. Furthermore, we set $e_0 = h_0 = 1$ and $e_r = h_r = 0$ for $r < 0$.

Let t be a formal indeterminate. The generating series of the power, elementary, and complete symmetric functions are as follows:

$$\begin{aligned} P(t) &:= \sum_{n \geq 1} p_n t^{n-1} = \frac{d}{dt} \ln \prod_{i \geq 1} \frac{1}{1 - x_i t}, \\ E(t) &:= \sum_{n \geq 0} e_n t^n = \prod_{i \geq 1} (1 + x_i t), \\ H(t) &:= \sum_{n \geq 0} h_n t^n = \prod_{i \geq 1} \frac{1}{1 - x_i t}. \end{aligned}$$

It follows that

$$\begin{aligned} & E(-t)H(t) = 1, \\ & E(t) = \exp\left(\sum_{r \geq 1} (-1)^{r-1} \frac{p_r t^r}{r} \right), \\ & H(t) = \exp\left(\sum_{r \geq 1} \frac{p_r t^r}{r} \right). \end{aligned} \tag{A.1}$$

Definition A.3. For a partition $\lambda = (\lambda_1, \lambda_2, \dots, \lambda_\ell)$, we define

$$p_\lambda := \prod_{i=1}^{\ell} p_{\lambda_i}, \quad e_\lambda := \prod_{i=1}^{\ell} e_{\lambda_i}, \quad h_\lambda := \prod_{i=1}^{\ell} h_{\lambda_i}.$$

(A.2) *The set* $\{p_\lambda \mid \lambda \vdash n\}$ *forms a linear basis for* $\Lambda^n_{\mathbb{Q}}$, *for* $n \in \mathbb{N}$. *Moreover,*

$$\Lambda_{\mathbb{Q}} = \mathbb{Q}[p_1, p_2, \dots].$$

The sets $\{e_\lambda \mid \lambda \vdash n\}$ *and* $\{h_\lambda \mid \lambda \vdash n\}$ *are* $\mathbb{Z}$*-bases for* Λ^n, *for* $n \in \mathbb{N}$. *Moreover,*

$$\Lambda = \mathbb{Z}[e_1, e_2, \dots] = \mathbb{Z}[h_1, h_2, \dots].$$

Proof. For a partition λ, a monomial appearing in p_λ is of the form

$$x_{i_1}^{\lambda_1} x_{i_2}^{\lambda_2} \cdots x_{i_\ell}^{\lambda_\ell}, \tag{A.3}$$

for some $i_1, i_2, \dots, i_\ell \in \mathbb{N}$. Write $x^\alpha = x_{j_1}^{\alpha_1} \cdots x_{j_k}^{\alpha_k}$ for the expression (A.3) such that $\alpha_1 \geq \alpha_2 \geq \cdots \geq \alpha_k$ and all the j_i's are distinct. Then clearly $\alpha \geq \lambda$. Hence, we have

$$p_\lambda = \sum_{\mu \geq \lambda} c_{\lambda\mu} m_\mu,$$

with $c_{\lambda\mu} \in \mathbb{Z}_+$ and $c_{\lambda\lambda} > 0$. This implies that the matrix $(c_{\lambda\mu})$, with sub-indices ordered compatibly with the dominance order, is an invertible upper triangular matrix with coefficients in $\mathbb{Z}$. Since $\{m_\lambda \mid \lambda \vdash n\}$ is a basis for $\Lambda^n_{\mathbb{Q}}$, so is $\{p_\lambda \mid \lambda \vdash n\}$.

For a partition λ, every summand in $e_{\lambda'}$ is of the form

$$x_{i_1} \cdots x_{i_{\lambda'_1}} x_{j_1} \cdots x_{j_{\lambda'_2}} \cdots x_{k_\ell} \cdots x_{k_{\lambda'_\ell}}, \tag{A.4}$$

where $i_1 < i_2 < \cdots < i_{\lambda'_1}$, $j_1 < j_2 < \cdots < j_{\lambda'_2}$, etc. Let us write the expression in (A.4) as $x^\alpha = x_{j_1}^{\alpha_1} x_{j_2}^{\alpha_2} \cdots x_{j_s}^{\alpha_s}$, such that $\alpha_1 \geq \alpha_2 \geq \cdots \geq \alpha_s$ and all the j_i's are distinct. Note that $\lambda \geq \alpha$. Hence we conclude that

$$e_{\lambda'} = \sum_{\mu \leq \lambda} a_{\lambda\mu} m_\mu,$$

with $a_{\lambda\mu} \in \mathbb{Z}_+$ and $a_{\lambda\lambda} = 1$. This implies that the set $\{e_\lambda \mid \lambda \vdash n\}$ is a $\mathbb{Z}$-basis for Λ^n.

It follows by (A.1) that, for all $n \in \mathbb{N}$,

$$\sum_{r=0}^{n} (-1)^r e_r h_{n-r} = 0. \tag{A.5}$$

Thus we can express e_n as a polynomial of h_r's and e_{r-1}'s, for $r \leq n$. By induction on n it follows that every e_n is a polynomial of $\{h_1, \dots, h_n\}$ with integer coefficients. Therefore, the $\mathbb{Z}$-module $\mathbb{Z}[h_1, \dots, h_n] \cap \Lambda^n = \sum_{\lambda \vdash n} \mathbb{Z} h_\lambda$ contains the $\mathbb{Z}$-span of $\{e_\lambda \mid \lambda \vdash n\}$. Hence $\sum_{\lambda \vdash n} \mathbb{Z} h_\lambda = \Lambda^n$, and so $\{h_\lambda \mid \lambda \vdash n\}$ is a $\mathbb{Z}$-basis for Λ^n. □

We note the following immediate consequence of (A.2).

(A.6) *The sets* $\{p_1, p_2, \ldots\}$, $\{e_1, e_2, \ldots\}$, *and* $\{h_1, h_2, \ldots\}$ *are algebraically independent in* Λ.

By (A.6), we can define a ring homomorphism $\omega : \Lambda \to \Lambda$ by letting

$$\omega(e_n) = h_n, \quad \forall n \in \mathbb{N}.$$

(A.7) *We have* $\omega(h_n) = e_n$ *and* $\omega(p_n) = (-1)^{n-1} p_n$. *Thus,* ω *is an involution of the ring* Λ.

Proof. We prove the first statement by induction on n. From (A.5) we obtain

$$h_n = -\sum_{r=1}^{n} (-1)^r e_r h_{n-r}, \quad e_n = -\sum_{r=0}^{n-1} (-1)^{n-r} e_r h_{n-r}.$$

This implies by the induction hypothesis that

$$\omega(h_n) = -\sum_{r=1}^{n} (-1)^r h_r e_{n-r} = e_n.$$

Also, it follows by (A.1) that

$$\omega(P(t)) = \frac{d}{dt} \omega(\ln H(t)) = \frac{d}{dt} \ln E(t) = -P(-t),$$

from which we conclude that $\omega(p_n) = (-1)^{n-1} p_n$. □

For a partition $\lambda = (1^{m_1} 2^{m_2} \ldots)$ of n, we let

$$z_\lambda = \prod_{i \geq 1} i^{m_i} m_i!,$$

which is the order of the centralizer of a permutation in $\mathfrak{S}_n$ of cycle type λ. We compute

$$H(t) = \prod_{r \geq 1} \exp\left(\frac{p_r t^r}{r}\right) = \sum_{\lambda \in \mathcal{P}} z_\lambda^{-1} p_\lambda t^{|\lambda|}. \tag{A.8}$$

Equivalently, for each $n \geq 1$, we have

$$h_n = \sum_{\lambda \vdash n} z_\lambda^{-1} p_\lambda.$$

Applying ω to (A.8), we obtain

$$E(t) = \sum_{\lambda \in \mathcal{P}} (-1)^{|\lambda| - \ell(\lambda)} z_\lambda^{-1} p_\lambda t^{|\lambda|}. \tag{A.9}$$

Equivalently, for each $n \geq 1$, we have

$$e_n = \sum_{\lambda \vdash n} (-1)^{|\lambda| - \ell(\lambda)} z_\lambda^{-1} p_\lambda.$$

A.1.2. Schur functions. We first suppose that $x = \{x_1, x_2, \ldots, x_n\}$, and denote the ring of symmetric polynomials in n variables by Λ_n. There is a natural ring homomorphism $\pi_{m,n} : \Lambda_m \to \Lambda_n$, for $m > n$, given by sending each x_j for $j > n$ to 0, and Λ is the inverse limit of Λ_n.

For a partition λ of length no greater than n, we let

$$a_\lambda := \sum_{\sigma \in \mathfrak{S}_n} \operatorname{sgn}(\sigma) x_{\sigma(1)}^{\lambda_1} x_{\sigma(2)}^{\lambda_2} \cdots x_{\sigma(n)}^{\lambda_n},$$

where $\operatorname{sgn}(\sigma)$ denotes the sign of the permutation σ. Then a_λ is **skew-symmetric** in the sense that

$$\tau a_\lambda = \operatorname{sgn}(\tau) a_\lambda, \quad \forall \tau \in \mathfrak{S}_n.$$

Let $\rho = (n-1, n-2, \ldots, 1, 0)$, so that we have

$$a_{\lambda+\rho} = \sum_{\sigma \in \mathfrak{S}_n} \operatorname{sgn}(\sigma) x_{\sigma(1)}^{\lambda_1+n-1} x_{\sigma(2)}^{\lambda_2+n-2} \cdots x_{\sigma(n)}^{\lambda_n} = \det\left(x_i^{\lambda_j+n-j}\right)_{1 \le i,j \le n}.$$

In particular, a_ρ is the Vandermonde determinant. Since $a_{\lambda+\rho} = 0$ when $x_i = x_j$ for $i \neq j$, it follows that $a_{\lambda+\rho}$ is divisible by $x_i - x_j$, for all $i \neq j$. Therefore the expression

$$s_\lambda := \frac{a_{\lambda+\rho}}{a_\rho}$$

is a symmetric polynomial in $x_1, x_2, \ldots, x_n$, called the **Schur polynomial** associated to λ. Since $s_\lambda(x_1, \ldots, x_n, 0) = s_\lambda(x_1, \ldots, x_n)$, the inverse limit, denoted by $s_\lambda \in \Lambda$, exists and is called the **Schur function** associated to λ.

Let A_n denote the $\mathbb{Z}$-module of skew-symmetric polynomials in $x_1, x_2, \ldots, x_n$. Then the $a_{\lambda+\rho}$'s, as λ runs over partitions with $\ell(\lambda) \le n$, form a $\mathbb{Z}$-basis of A_n. Since every element in A_n is divisible by a_ρ, the map $\Lambda_n \to A_n$ given by multiplication by a_ρ is a $\mathbb{Z}$-module isomorphism. Thus, the s_λ's, as λ runs over all partitions with $\ell(\lambda) \le n$, form a $\mathbb{Z}$-basis for Λ_n. Therefore, we conclude that the set of Schur functions associated with partitions of size k is a basis for Λ^k, for each $k \ge 0$.

(A.10) *For a partition λ we have*

$$s_\lambda = \det\left(h_{\lambda_i - i + j}\right)_{1 \le i,j \le n}, \quad \ell(\lambda) \le n,$$

$$s_\lambda = \det\left(e_{\lambda'_i - i + j}\right)_{1 \le i,j \le n}, \quad \ell(\lambda') \le n.$$

In particular, we have $\omega(s_\lambda) = s_{\lambda'}$.

Proof. Let $e_r^{(k)}$ denote the rth elementary symmetric polynomial in the variables $x_1, x_2, \ldots, \widehat{x_k}, \ldots, x_n$, i.e., with x_k deleted. Set

$$M := \left((-1)^{n-i} e_{n-i}^{(k)}\right)_{1 \le i,k \le n}.$$

Let $\alpha = (\alpha_1, \dots, \alpha_n) \in \mathbb{Z}_+^n$. Let

$$A_\alpha = \left(x_j^{\alpha_i}\right)_{1\le i,j\le n}, \qquad H_\alpha := (h_{\alpha_i - n+j})_{1\le i,j\le n},$$

and let

$$E^{(k)}(t) = \sum_{r=0}^{n-1} e_r^{(k)} t^r = \prod_{i\neq k}(1+x_i t).$$

Then

$$H(t)E^{(k)}(-t) = \frac{1}{1-x_k t}.$$

This implies that

$$\sum_{j=1}^{n} h_{\alpha_i - n+j}(-1)^{n-j} e_{n-j}^{(k)} = x_k^{\alpha_i}.$$

This equation can be recast in a matrix form as $A_\alpha = H_\alpha M$. Taking the determinant of both sides with $\alpha = \lambda+\rho$, we obtain

$$a_{\lambda+\rho} = \det H_{\lambda+\rho} \det M.$$

Setting $\lambda = \emptyset$ and noting $\det H_\rho = 1$, we obtain that $a_\rho = \det M$, and so $a_{\lambda+\rho} = \det H_{\lambda+\rho} a_\rho$, which is an equivalent form of the first identity in (A.10) we wish to prove.

Now the matrices $H = (h_{i-j})_{0\le i,j\le N}$ and $E = \left((-1)^{i-j} e_{i-j}\right)_{0\le i,j\le N}$ are inverses of each other by (A.5), each with determinant equal to 1. This gives a relationship between the minors of H and the cofactors of E^t. Exploiting this relationship, one can prove the following identity:

$$\det\left(h_{\lambda_i - i+j}\right)_{1\le i,j\le n} = \det\left(e_{\lambda'_i - i+j}\right)_{1\le i,j\le n},$$

for $n \ge \max\{\ell(\lambda), \ell(\lambda')\}$. The second identity in (A.10) now follows by the first identity in (A.10) and the identity above.

The formula $\omega(s_\lambda) = s_{\lambda'}$ follows by applying ω to the first identity and using the second identity with λ and λ' switched. □

Take two sets of independent variables $x = \{x_1, x_2, \dots\}$ and $y = \{y_1, y_2, \dots\}$. Applying (A.8) to the set of variables $\{x_i y_j\}$ and setting $t = 1$, we obtain that

$$\prod_{i,j} \frac{1}{(1-x_i y_j)} = \sum_\lambda z_\lambda^{-1} p_\lambda(x) p_\lambda(y).$$

(A.11) *The following identities hold:*

$$\prod_{i,j} \frac{1}{1-x_i y_j} = \sum_{\lambda\in\mathcal{P}} h_\lambda(x) m_\lambda(y),$$

$$\prod_{i,j}(1+x_i y_j) = \sum_{\lambda\in\mathcal{P}} e_\lambda(x) m_\lambda(y),$$

$$\prod_{i,j}\frac{1}{1-x_iy_j}=\sum_{\lambda\in\mathcal{P}}s_\lambda(x)s_\lambda(y),$$
$$\prod_{i,j}(1+x_iy_j)=\sum_{\lambda\in\mathcal{P}}s_\lambda(x)s_{\lambda'}(y).$$

(The last two are known as **Cauchy identities.***)*

Proof. We compute

$$\prod_{i,j}\frac{1}{1-x_iy_j}=\prod_j H(y_j)=\prod_j\left(\sum_{r\geq 0}h_r(x)y_j^r\right)=\sum_\alpha h_\alpha(x)y_1^{\alpha_1}y_2^{\alpha_2}\cdots,$$

where the last sum is over all compositions $\alpha=(\alpha_1,\alpha_2,\ldots)$. Hence it equals $\sum_\lambda h_\lambda(x)m_\lambda(y)$, proving the first identity in (A.11). The second identity follows by applying ω in the x variables to the first identity and using (A.10).

Assume that the number of variables in x and y are both equal to n. The general case is proved by letting n go to infinity, as usual. Consider the $n\times n$ matrix $\left((1-x_iy_j)^{-1}\right)_{1\leq i,j\leq n}$. If we multiply the ith row of this matrix by $\prod_{j=1}^n(1-x_iy_j)$, then we obtain a matrix whose (i,k)th entry is equal to

$$\prod_{r\neq k}(1-x_iy_r)=\sum_{j=1}^n(-1)^{n-j}e_{n-j}^{(k)}(y)x_i^{n-j}.$$

That is, it is the (i,k)th entry of the matrix $A_\rho(x)^tM(y)$, where A_ρ and M are as in the proof of (A.10). This implies that

(A.12) $$\det\left((1-x_iy_j)^{-1}\right)_{1\leq i,j\leq n}=a_\rho(x)a_\rho(y)\prod_{i,j}\frac{1}{1-x_iy_j}.$$

Next, we compute that

$$\det\left((1-x_iy_j)^{-1}\right)_{1\leq i,j\leq n}=\det\Big(\sum_{k=0}^\infty x_i^ky_j^k\Big)_{1\leq i,j\leq n}=\sum_\alpha\sum_{\sigma\in\mathfrak{S}_n}\operatorname{sgn}(\sigma)\prod_{i=1}^n x_i^{\alpha_i}y_{\sigma(i)}^{\alpha_i},$$

where the first summation is over compositions α. Thus, we can write

$$\det\left((1-x_iy_j)^{-1}\right)_{1\leq i,j\leq n}=\sum_\alpha x^\alpha a_\alpha(y)=\sum_\lambda\sum_{\tau\in\mathfrak{S}_n}(-1)^{\ell(\tau)}x^{\tau(\lambda+\rho)}a_{\lambda+\rho}(y)$$

(A.13) $$=\sum_\lambda a_{\lambda+\rho}(x)a_{\lambda+\rho}(y),$$

where the summation over λ is over partitions of length no greater than n.

The first Cauchy identity follows now by combining (A.12) and (A.13). The second Cauchy identity follows by applying ω in the y variables to the first Cauchy identity and using (A.10). □

We define a $\mathbb{Z}$-valued bilinear form $(\cdot,\cdot)$ on Λ by declaring

(A.14) $$(h_\lambda,m_\mu)=\delta_{\lambda\mu}.$$

This extends to a $\mathbb{Q}$-valued bilinear form on $\Lambda_{\mathbb{Q}}$.

(A.15) *Let* $\{u_\lambda \mid \lambda \in \mathcal{P}\}$ *and* $\{v_\lambda \mid \lambda \in \mathcal{P}\}$ *be two bases for* $\Lambda_{\mathbb{Q}}$. *Then the following conditions are equivalent.*

(1) $(u_\lambda, v_\mu) = \delta_{\lambda\mu}$.

(2) $\sum_\lambda u_\lambda(x)v_\lambda(y) = \prod_{i,j} \frac{1}{(1-x_iy_j)}$.

Proof. We write

$$u_\lambda = \sum_\nu a_{\lambda\nu}h_\nu, \quad v_\mu = \sum_\eta b_{\mu\eta}m_\eta, \quad a_{\lambda\nu}, b_{\mu\eta} \in \mathbb{Q}.$$

Then (1) is equivalent to $\sum_\nu a_{\lambda\nu}b_{\mu\nu} = \delta_{\lambda\mu}$, which is equivalent to $\sum_\lambda a_{\lambda\nu}b_{\lambda\eta} = \delta_{\nu\eta}$.

Now (2) is equivalent to $\sum_\lambda \sum_{\nu,\eta} a_{\lambda\nu}b_{\lambda,\eta}h_\nu(x)m_\eta(y) = \sum_\lambda h_\lambda(x)m_\lambda(y)$ by using (A.11). This is equivalent to $\sum_\lambda a_{\lambda\nu}b_{\lambda\eta} = \delta_{\nu\eta}$. □

It follows from (A.15) and (A.11) that

(A.16) $$(s_\lambda, s_\mu) = \delta_{\lambda\mu}.$$

A.1.3. Skew Schur functions. For partitions μ and ν, we write the product $s_\mu s_\nu$ as

(A.17) $$s_\mu s_\nu = \sum_\lambda c^\lambda_{\mu\nu}s_\lambda, \quad c^\lambda_{\mu\nu} \in \mathbb{Z}.$$

The $c^\lambda_{\mu\nu}$ associated with a triple of partitions are the **Littlewood-Richardson coefficients**, and they allow us to define the **skew Schur function** associated to partitions λ, μ by the formula

$$s_{\lambda/\mu} := \sum_\nu c^\lambda_{\mu\nu}s_\nu.$$

This definition is equivalent to the following identity:

(A.18) $$(s_\mu f, s_\lambda) = (f, s_{\lambda/\mu}), \quad \forall f \in \Lambda.$$

This can be seen by checking for $f = s_\nu$. It follows from the definitions that $s_{\lambda/\emptyset} = s_\lambda$, and by (A.10) that

$$\omega(s_\lambda) = s_{\lambda'}.$$

Since ω is a ring isomorphism, we obtain by the definition of the Littlewood-Richardson coefficients above that $c^{\lambda'}_{\mu'\nu'} = c^\lambda_{\mu\nu}$. It now follows from the definition of the skew Schur functions that

$$\omega(s_{\lambda/\mu}) = s_{\lambda'/\mu'}.$$

Let $\mu \subseteq \lambda$ be partitions which we may regard as Young diagrams. The **skew diagram** λ/μ is obtained from λ by removing the sub-diagram μ. A **tableau** of shape λ/μ is a filling of the boxes of λ/μ with elements from the set $\mathbb{N}$. A tableau T of λ/μ is called **semistandard** if the entries in T are strictly increasing from top to bottom along each column and weakly increasing from left to right along each row. Given such a semistandard tableau T with entries $t \in T$, we define $x^T := \prod_{t \in T} x_t$.

The skew Schur function $s_{\lambda/\mu}$ admits the following combinatorial description, a proof of which can be found in [**83**, (5.12)].

(A.19) *For partitions $\mu \subseteq \lambda$, we have*

$$s_{\lambda/\mu} = \sum_T x^T,$$

where the summation is over all semistandard tableaux of shape λ/μ.

The following is an easy consequence of (A.19).

(A.20) *Let $x = \{x_1, x_2, \ldots\}$ and $y = \{y_1, y_2, \ldots\}$ be two independent sets of variables. We have*

$$s_\lambda(x,y) = \sum_{\mu \subseteq \lambda} s_\mu(x) s_{\lambda/\mu}(y).$$

Let $\mu \subseteq \lambda$ be partitions. Recall that the weight of a semistandard tableau T of shape λ/μ is $(1^{m_1}2^{m_2}\cdots)$, where m_i is the number of times i appears in T. Denote by $K_{\lambda-\mu,\nu}$ the number of semistandard tableaux of shape λ/μ and weight ν. Then (A.19) implies that

$$s_{\lambda/\mu} = \sum_{\nu \in \mathcal{P}} K_{\lambda-\mu,\nu} m_\nu. \tag{A.21}$$

It follows from (A.18) and (A.21) that

$$s_\mu h_\nu = \sum_\lambda K_{\lambda-\mu,\nu} s_\lambda.$$

Taking $\nu = (r)$ to be the one-part partition, we obtain **Pieri's formula**:

$$s_\mu h_r = \sum_\lambda s_\lambda, \tag{A.22}$$

where the summation is over partitions λ such that λ/μ is a skew diagram of size r whose columns all have length at most one.

Recall that a partition is called **even** if all of its parts are even. We have the following two symmetric function identities ([**83**, I.5, Example 5]):

$$\begin{aligned} \sum_{\mu \text{ even}} s_\mu &= \prod_{1 \le i \le j} (1 - x_i x_j)^{-1}, \\ \sum_{\nu \text{ even}} s_{\nu'} &= \prod_{1 \le i < j} (1 - x_i x_j)^{-1}. \end{aligned} \tag{A.23}$$

Since the involution ω interchanges s_λ with $s_{\lambda'}$, we conclude from these identities that (see also Exercise 6.2)

$$\omega\Big(\prod_{1 \le i \le j} (1 - x_i x_j)^{-1} \Big) = \prod_{1 \le i < j} (1 - x_i x_j)^{-1}. \tag{A.24}$$

A.1.4. The Frobenius characteristic map. We shall freely use the notations from Section 3.2.2. Let R_n denote the Grothendieck group of the category of $\mathfrak{S}_n$-modules, which has $[S^\lambda]$ for $\lambda \in \mathcal{P}_n$ as a $\mathbb{Z}$-basis. Let

$$R = \bigoplus_{n=0}^{\infty} R_n.$$

Then R is a graded algebra with multiplication given by

$$fg = \mathrm{Ind}_{\mathfrak{S}_m \times \mathfrak{S}_n}^{\mathfrak{S}_{m+n}} (f \otimes g), \qquad \text{for } f \in R_m, g \in R_n.$$

The **Frobenius characteristic map** $\mathrm{ch}^{\mathrm{F}} : R \to \Lambda$ is the $\mathbb{Z}$-linear map such that, for $n \geq 0$,

$$\mathrm{ch}^{\mathrm{F}}(\chi) = \sum_{\mu \in \mathcal{P}_n} z_\mu^{-1} \chi_\mu p_\mu, \quad \chi \in R_n, \tag{A.25}$$

where χ_μ denotes the character value of χ at a permutation of cycle type μ. Denote by $\mathbf{1}_\lambda$ and sgn_λ the trivial and sign representations or characters of the Young subgroup $\mathfrak{S}_\lambda$, respectively. The following basic properties of ch^{F} are well known.

(A.26) *The characteristic map* ch^{F} *is an isomorphism of graded rings. Moreover, for* $n \geq 0$ *and* $\lambda \in \mathcal{P}_n$, *we have that*

$$\mathrm{ch}^{\mathrm{F}}\left(\left[\mathrm{Ind}_{\mathfrak{S}_\lambda}^{\mathfrak{S}_n} \mathbf{1}_\lambda\right]\right) = h_\lambda,$$
$$\mathrm{ch}^{\mathrm{F}}\left(\left[\mathrm{Ind}_{\mathfrak{S}_\lambda}^{\mathfrak{S}_n} \mathrm{sgn}_\lambda\right]\right) = e_\lambda,$$
$$\mathrm{ch}^{\mathrm{F}}([S^\lambda]) = s_\lambda.$$

Proof. One can use the induced character formula to prove that ch^{F} is a ring homomorphism. By (A.8) and (A.9) we have, respectively, $\mathrm{ch}^{\mathrm{F}}(\mathbf{1}_n) = h_n$ and $\mathrm{ch}^{\mathrm{F}}(\mathrm{sgn}_n) = e_n$. This immediately implies the first two identities since ch^{F} is a homomorphism.

It is known (e.g., as a special case of Lemma 3.12) that

$$\mathrm{Ind}_{\mathfrak{S}_\mu}^{\mathfrak{S}_n} \mathbf{1}_n \cong \bigoplus_{\lambda \geq \mu} K_{\lambda\mu} S^\lambda, \tag{A.27}$$

where the Kostka number $K_{\lambda\mu}$ is equal to the number of semistandard λ-tableaux of content μ. Note that $K_{\mu\mu} = 1$.

Also from (A.21) we conclude that

$$s_\lambda = \sum_{\mu \leq \lambda} K_{\lambda\mu} m_\mu.$$

Using the relations of dual bases (A.14) and (A.16), this identity admits the following dual version:

$$h_\mu = \sum_{\lambda \geq \mu} K_{\lambda\mu} s_\lambda. \tag{A.28}$$

It follows by comparing (A.27) and (A.28) and using the first identity in (A.26) that $\mathrm{ch}^{\mathrm{F}}([S^\lambda]) = s_\lambda$. Since $[S^\lambda]$ and s_λ for $\lambda \in \mathcal{P}$ form $\mathbb{Z}$-bases for R and Λ respectively, we conclude that ch^{F} is an isomorphism. □

(A.29) *For $\mu \in \mathcal{P}_n$, we have an $\mathfrak{S}_n$-module isomorphism: $S^\mu \otimes \mathrm{sgn}_n \cong S^{\mu'}$.*

Proof. Let ψ denote the character of S^μ. By (A.7) and (A.25), we have that

$$\begin{aligned}\mathrm{ch}^{\mathrm{F}}([S^\mu \otimes \mathrm{sgn}_n]) &= \sum_{\nu\in\mathcal{P}_n} (-1)^{n-\ell(\nu)} z_\nu^{-1} \psi_\nu p_\nu \\ &= \omega\Big(\sum_{\nu\in\mathcal{P}_n} z_\nu^{-1} \psi_\nu p_\nu \Big) \\ &= \omega(s_\mu) = s_{\mu'} = \mathrm{ch}^{\mathrm{F}}([S^{\mu'}]).\end{aligned}$$

Now the isomorphism follows from (A.26) that ch^{F} is an isomorphism. □

A.2. Supersymmetric polynomials

A.2.1. The ring of supersymmetric polynomials. Let $x = \{x_1, x_2, \ldots, x_m\}$ and $y = \{y_1, y_2, \ldots, y_n\}$ be independent indeterminates. A polynomial f in $\mathbb{Z}[x,y]$ is called **supersymmetric** if the following conditions are satisfied:

(1) f is symmetric in $x_1, \ldots, x_m$.
(2) f is symmetric in $y_1, \ldots, y_n$.
(3) The polynomial obtained from f by setting $x_m = y_n = t$ is independent of t.

Denote by $\Lambda_{(m|n)}$ the ring of supersymmetric polynomials in $\mathbb{Z}[x,y]$. For a field $\mathbb{F}$, we denote the base change by $\Lambda_{(m|n),\mathbb{F}} = \mathbb{F} \otimes_{\mathbb{Z}} \Lambda_{(m|n)}$. We call an element $f \in \mathbb{Z}[x,y]$ satisfying (1) and (2) above **doubly symmetric**.

For $r \geq 1$ set

$$\sigma_{m,n}^r := \sum_{i=1}^m x_i^r - \sum_{j=1}^n y_j^r.$$

Then $\sigma_{m,n}^r$ is supersymmetric. Let

$$\square_{m,n} := \prod_{i=1}^m \prod_{j=1}^n (x_i - y_j).$$

Note that $\square_{m,n}$ is supersymmetric. Also, if g is doubly symmetric, then $g\square_{m,n}$ is supersymmetric.

(A.30) *Let q be doubly symmetric. Then $q\square_{m,n}$ is a polynomial in $\{\sigma_{m,n}^r \mid r \geq 1\}$.*

Proof. We proceed by induction on n (for arbitrary m), with the case $n = 0$ being clear.

Let $n \geq 1$. We may assume, without loss of generality, that q is homogeneous of degree $k \geq 0$, and we proceed by induction on k. Consider

$$q\square_{m,n}|_{y_n=0} = q|_{y_n=0}x_1 \cdots x_m \square_{m,n-1}.$$

By the induction hypothesis on n, there exists a polynomial g in $\{\sigma^1_{m,n-1}, \dots, \sigma^r_{m,n-1}\}$ such that

$$q\square_{m,n}|_{y_n=0} = g(\sigma^1_{m,n-1}, \dots, \sigma^r_{m,n-1}).$$

Note that

$$g(\sigma^1_{m,n}, \dots, \sigma^r_{m,n})|_{x_m=y_n=t} = q\square_{m,n}|_{x_m=y_n=0} = 0.$$

Thus $x_m - y_n$ divides $g(\sigma^1_{m,n}, \dots, \sigma^r_{m,n})$. Since $g(\sigma^1_{m,n}, \dots, \sigma^r_{m,n})$ is doubly symmetric, it follows that

$$g(\sigma^1_{m,n}, \dots, \sigma^r_{m,n}) = q^*\square_{m,n}, \tag{A.31}$$

for some doubly symmetric polynomial q^*. Now

$$q^*\square_{m,n}|_{y_n=0} = g(\sigma^1_{m,n-1}, \dots, \sigma^r_{m,n-1}) = q\square_{m,n}|_{y_n=0}.$$

Hence y_n divides $q^* - q$. Since $q^* - q$ is symmetric in $y_1, \dots, y_n$, we conclude that there exists a doubly symmetric polynomial f such that

$$q\square_{m,n} = q^*\square_{m,n} + y_1 \cdots y_n f \square_{m,n}. \tag{A.32}$$

Note that f has degree $k-n$ if $k \geq n$ and $f = 0$ if $k < n$.

In the case when $k < n$ we have $f = 0$. Hence we are done by (A.31).

Suppose that $k \geq n$. Write f as a polynomial in terms of $\sigma^u_{m,0}$ and $\sigma^v_{0,n}$, for $u, v \geq 1$. Replace every $\sigma^u_{m,0}$ in this very polynomial by $\sigma^u_{m+1,0}$ and denote by f^* the resulting homogeneous doubly symmetric polynomial in $\mathbb{Z}[x_1, \dots, x_{m+1}, y_1, \dots, y_n]$. Since f^* has degree $k - n < k$, we have by the induction hypothesis on k that there exists some polynomial h such that

$$f^*\square_{m+1,n} = h(\sigma^1_{m+1,n}, \dots, \sigma^s_{m+1,n}).$$

This implies that

$$y_1 \cdots y_n f \square_{m,n} = (-1)^n f^* \square_{m+1,n}|_{x_{m+1}=0} = (-1)^n h(\sigma^1_{m,n}, \dots, \sigma^s_{m,n}),$$

and hence (A.30) follows from (A.32). $\square$

(A.33) *The algebra $\Lambda_{(m|n),\mathbb{Q}}$ of supersymmetric polynomials is generated by $\sigma^r_{m,n}$ for all $r \geq 1$.*

Proof. We proceed by induction on n to show that every supersymmetric polynomial $p \in \mathbb{Q}[x, y]$ is a polynomial in $\{\sigma^r_{m,n} \mid r \geq 1\}$. The case when $n = 0$ is clear.

Let $n \geq 1$ and $p \in \mathbb{Q}[x,y]$ be a supersymmetric polynomial. Then, the polynomial $p|_{x_m=y_n=0}$ in $x_1,\ldots,x_{m-1}$ and $y_1,\ldots,y_{n-1}$ is supersymmetric. By the induction hypothesis on n, there exists a polynomial f such that

$$p|_{x_m=y_n=0} = f(\sigma^1_{m-1,n-1},\ldots,\sigma^r_{m-1,n-1}).$$

Set $f^* := f(\sigma^1_{m,n},\ldots,\sigma^r_{m,n})$. Then $(p-f^*)|_{x_m=y_n=t} = 0$, and hence $x_m - y_n$ divides $p - f^*$. By double symmetry $\square_{m,n}$ divides $p - f^*$, and hence there exists a doubly symmetric polynomial q such that

$$p = f^* + q\square_{m,n}.$$

By (A.30), $q\square_{m,n}$ is generated by $\{\sigma^r_{m,n} \mid r \geq 1\}$, and hence so is p. $\square$

We similarly define the ring of supersymmetric functions $\Lambda_{(m|\infty)}$ and prove the following $n = \infty$ version of (A.33).

(A.34) *The algebra* $\Lambda_{(m|\infty),\mathbb{Q}}$ *of supersymmetric functions is generated by* $\sigma^r_{m,\infty}$ *for all* $r \geq 1$.

Proof. Let $f \in \Lambda_{(m|\infty),\mathbb{Q}}$. Assume that f is homogeneous of degree k. Then $f_{(n)} := f|_{y_{n+1}=y_{n+2}=\cdots=0}$ is supersymmetric and hence, by (A.33),

$$f_{(n)} = g(\sigma^1_{m,n},\ldots,\sigma^r_{m,n}).$$

Consider the element

$$f - g(\sigma^1_{m,\infty},\ldots,\sigma^r_{m,\infty}) \in \Lambda_{(m|\infty),\mathbb{Q}}.$$

The expression $f - g(\sigma^1_{m,\infty},\ldots,\sigma^r_{m,\infty})$ is homogeneous of degree k and contains no monomials made up of only $x_1,\ldots,x_m,y_1,\ldots,y_n$. It follows by the symmetry of the y variables that $f - g(\sigma^1_{m,\infty},\ldots,\sigma^r_{m,\infty}) = 0$. $\square$

A.2.2. Super Schur functions. Define $\widetilde{\Lambda}_{(m|n)}$ to be the subring of the ring of doubly symmetric polynomials over $\mathbb{Z}$ in the two sets of variables $\{x_1,\ldots,x_m\}$ and $\{y_1,\ldots,y_n\}$ consisting of polynomials f such that $f|_{x_m=t=-y_n}$ is independent of t. Evidently $f(x,y) \in \widetilde{\Lambda}_{m|n}$ if and only if $f(x,-y)$ is supersymmetric; hence, we have an isomorphism of rings

(A.35) $$\rho_y : \Lambda_{(m|n)} \xrightarrow{\cong} \widetilde{\Lambda}_{(m|n)},$$

given by $\rho_y(f(x,y)) = f(x,-y)$. We define the ring $\widetilde{\Lambda}_{(m|\infty)}$ accordingly, and it is isomorphic to $\Lambda_{(m|\infty)}$.

Regard Λ as the ring of symmetric functions in $x_1,x_2,\ldots$, and identify the variables $x_{m+i} = y_i$, for $i \geq 1$. By applying the involution ω on the y variables, denoted by ω_y, to the set of infinite variables $\{y_1,y_2,\ldots\}$, we obtain a ring homomorphism $\omega_y : \Lambda \to \widetilde{\Lambda}_{(m|\infty)}$. Indeed, by (A.7), we have

(A.36) $$\omega_y(p_r) = x_1^r + \ldots + x_m^r + (-1)^{r-1}(y_1^r + y_2^r + \ldots) \in \widetilde{\Lambda}_{(m|\infty)}.$$

Since the p_r's, for $r \geq 1$, are algebraically independent generators of $\Lambda_{\mathbb{Q}}$ by (A.2) and (A.6), we conclude that $\omega_y(\Lambda_{\mathbb{Q}}) \subseteq \widetilde{\Lambda}_{(m|\infty),\mathbb{Q}}$, and so $\omega_y(\Lambda) \subseteq \widetilde{\Lambda}_{(m|\infty),\mathbb{Q}} \cap \mathbb{Z}[x,y] = \widetilde{\Lambda}_{(m|\infty)}$. Now (A.34) and the fact that $\rho_y^{-1}(\omega_y(p_r)) = \sigma^r_{m,\infty}$, $r \geq 1$, imply that $\rho_y^{-1} \circ \omega_y : \Lambda_{\mathbb{Q}} \to \Lambda_{(m|\infty),\mathbb{Q}}$ is a ring isomorphism. This implies that $\omega_y : \Lambda_{\mathbb{Q}} \to \widetilde{\Lambda}_{(m|\infty),\mathbb{Q}}$ is an isomorphism. The inverse isomorphism $\omega_y^{-1} : \widetilde{\Lambda}_{(m|\infty),\mathbb{Q}} \to \Lambda_{\mathbb{Q}}$, which is again given by the involution ω on the y variables, clearly satisfies that $\omega_y^{-1}(\widetilde{\Lambda}_{(m|\infty)}) \subseteq \Lambda$. Hence, $\omega_y : \Lambda \to \widetilde{\Lambda}_{(m|\infty)}$ is a ring isomorphism.

Define the **super Schur function** hs_λ associated to a partition λ by

(A.37)
$$\mathrm{hs}_\lambda(x;y) = \sum_{\mu \subseteq \lambda} s_\mu(x) s_{\lambda'/\mu'}(y).$$

The definition of $\mathrm{hs}_\lambda(x;y)$ makes sense for x and y being finite and infinite. By (A.20), we have for y infinite

(A.38)
$$\omega_y\big(s_\lambda(x,y)\big) = \mathrm{hs}_\lambda(x;y).$$

(A.39) *The map* $\omega_y : \Lambda \to \widetilde{\Lambda}_{(m|\infty)}$ *is an isomorphism of rings. In addition, the set* $\{\mathrm{hs}_\lambda(x_1,\dots,x_m;y_1,y_2,\dots) \mid \lambda \in \mathcal{P}\}$ *is a* $\mathbb{Z}$*-basis for* $\widetilde{\Lambda}_{(m|\infty)}$.

The map $\widetilde{\Lambda}_{(m|\infty)} \to \widetilde{\Lambda}_{(m|n)}$ obtained by setting the variables $y_i = 0$, for $i > n$, is a ring epimorphism. Furthermore, it sends the super Schur function corresponding to an $(m|n)$-hook partition to the respective super Schur polynomial corresponding to the same $(m|n)$-hook partition. The other super Schur functions are sent to zero. Evidently, the super Schur polynomials corresponding to $(m|n)$-hook partitions are linearly independent. Thus we have the following.

(A.40) *The set* $\{\mathrm{hs}_\lambda\}$, *as* λ *runs over all* $(m|n)$*-hook partitions, forms a* $\mathbb{Z}$*-basis for* $\widetilde{\Lambda}_{(m|n)}$.

The Cauchy identities in (A.11) admit the following super generalization.

(A.41) *Let* $x = \{x_1,x_2,\dots\}, y = \{y_1,y_2,\dots\}, z = \{z_1,z_2,\dots\}$ *be three sets of (possibly infinite) indeterminates. The following identity holds:*

$$\frac{\prod_{j,k}(1+y_j z_k)}{\prod_{i,k}(1-x_i z_k)} = \sum_{\lambda \in \mathcal{P}} \mathrm{hs}_\lambda(x;y) s_\lambda(z).$$

Proof. It suffices to prove the claim in the case when all three sets of indeterminates are infinite. A variant of the Cauchy identity in (A.11) can be written as

$$\prod_{i,k} \frac{1}{(1-x_i z_k)} \prod_{j,k} \frac{1}{(1-y_j z_k)} = \sum_{\lambda \in \mathcal{P}} s_\lambda(x,y) s_\lambda(z).$$

Now the claim follows by applying the involution ω_y on the y variables to both sides of the above equation and using (A.38). □

A.3. The ring Γ and Schur Q-functions

A.3.1. The ring Γ. Let $x = \{x_1, x_2, \ldots\}$. Define a family of symmetric functions $q_r = q_r(x)$, $r \geq 0$, via the generating function

$$Q(t) := \sum_{r\geq 0} q_r(x) t^r = \prod_i \frac{1+tx_i}{1-tx_i}. \tag{A.42}$$

Note that $q_0(x) = 1$, and that $Q(t)$ satisfies the relation

$$Q(t)Q(-t) = 1, \tag{A.43}$$

which is equivalent to the identities:

$$\sum_{r+s=n} (-1)^r q_r q_s = 0, \quad n \geq 1.$$

These identities are vacuous for n odd. When $n = 2m$ is even, we have

$$q_{2m} = \sum_{r=1}^{m-1} (-1)^{r-1} q_r q_{2m-r} - \frac{1}{2}(-1)^m q_m^2. \tag{A.44}$$

Let Γ be the $\mathbb{Z}$-subring of Λ generated by the q_r's:

$$\Gamma = \mathbb{Z}[q_1, q_2, q_3, \ldots].$$

The ring Γ is graded by the degree of functions: $\Gamma = \bigoplus_{n\geq 0} \Gamma^n$, where $\Gamma^n = \Gamma \cap \Lambda^n$. We set

$$\Gamma_{\mathbb{Q}} = \mathbb{Q} \otimes_{\mathbb{Z}} \Gamma, \quad \Gamma_{\mathbb{C}} = \mathbb{C} \otimes_{\mathbb{Z}} \Gamma.$$

For a partition $\mu = (\mu_1, \ldots, \mu_\ell)$, we denote

$$q_\mu = \prod_{i=1}^{\ell} q_{\mu_i}.$$

We shall denote by $\mathcal{OP}$ and $\mathcal{SP}$ the sets of odd and strict partitions, respectively.

(A.45) *The ring $\Gamma_{\mathbb{Q}}$ enjoys the following remarkable properties:*

(1) *$\Gamma_{\mathbb{Q}}$ is a polynomial algebra with polynomial generators p_{2r-1} for $r \geq 1$.*

(2) *$\Gamma_{\mathbb{Q}}$ is a polynomial algebra with polynomial generators q_{2r-1} for $r \geq 1$.*

(1′) *$\{p_\mu \mid \mu \in \mathcal{OP}\}$ forms a linear basis for $\Gamma_{\mathbb{Q}}$.*

(2′) *$\{q_\mu \mid \mu \in \mathcal{OP}\}$ forms a linear basis for $\Gamma_{\mathbb{Q}}$.*

Proof. Recall the generating function for the power-sums: $P(t) = \sum_{r\geq 1} p_r t^{r-1}$. We have

$$\frac{Q'(t)}{Q(t)} = \frac{d}{dt} \ln Q(t) = P(t) + P(-t) = 2\sum_{r\geq 0} p_{2r+1} t^{2r}.$$

It follows that $Q'(t) = 2Q(t)\sum_{r\geq 0} p_{2r+1} t^{2r}$, from which we deduce that

$$r q_r = 2(p_1 q_{r-1} + p_3 q_{r-3} + \ldots),$$

with the last term on the right-hand side being $p_{r-1}q_1$ for r even and p_r for r odd.

By using induction on r, we conclude that (i) each q_r is expressible as a polynomial in terms of p_s's with odd s; (ii) each p_r with odd r is expressible as a polynomial in terms of q_s's, which can be further restricted to odd s (note that each q_r can be written as a polynomial of q_s with odd s by applying (A.44) and induction on r). So

$$\Gamma_{\mathbb{Q}} = \mathbb{Q}[p_1, p_3, \ldots] = \mathbb{Q}[q_1, q_3, \ldots].$$

Since the p_r's (for r odd) are algebraically independent by (A.6), we have proved (1). (2) follows from (1) by the above equation and dimension counting.

Clearly, (1′) is equivalent to (1), while (2′) is equivalent to (2). □

(A.46) $\{q_\mu \mid \mu \in \mathcal{SP}\}$ *forms a* $\mathbb{Z}$*-basis for* Γ. *Moreover, for any partition* λ, *we have*

$$q_\lambda = \sum_{\mu \in \mathcal{SP}, \mu \geq \lambda} a_{\mu\lambda} q_\mu,$$

for some $a_{\mu\lambda} \in \mathbb{Z}$.

Proof. We claim that q_λ lies in the $\mathbb{Z}$-span of $\{q_\mu \mid \mu \in \mathcal{SP}_n, \mu \geq \lambda\}$, for each partition λ of a given n. This can be seen by induction downward on the dominance order on λ as follows. The initial step for $\lambda = (n)$ is clear, as (n) is strict. For nonstrict λ, we have $\lambda_i = \lambda_j = m$ for some $i < j$ and some $m > 0$. Applying (A.44) to rewrite q_λ easily provides the inductive step.

According to a formula of Euler, $|\mathcal{SP}_n| = |\mathcal{OP}_n|$. Now (A.46) follows from Part (2) of (A.45) and the above claim. We record here a short proof of Euler's formula:

$$\begin{aligned}\sum_{n\geq 0} |\mathcal{SP}_n| q^n &= \prod_{r\geq 1}(1+q^r) = \frac{\prod_{r\geq 1}(1+q^r)(1-q^r)}{\prod_{r\geq 1}(1-q^r)} \\ &= \frac{\prod_{r\geq 1}(1-q^{2r})}{\prod_{r\geq 1}(1-q^r)} = \frac{1}{\prod_{r\geq 1,\text{odd}}(1-q^r)} = \sum_{n\geq 0} |\mathcal{OP}_n| q^n.\end{aligned}$$

□

(A.47) *We have* $q_n = \sum_{\alpha \in \mathcal{OP}_n} 2^{\ell(\alpha)} z_\alpha^{-1} p_\alpha$.

Proof. As seen in the proof of (A.45) above, we have

$$Q(t) = \exp\left(2\sum_{r\geq 1} \frac{1}{2r-1} p_{2r-1} t^{2r-1}\right).$$

Now (A.47) follows by comparing the coefficients of t^n on both sides. □

A.3.2. Schur Q-functions. We shall define the Schur Q-functions Q_λ, for $\lambda \in \mathcal{SP}$. Let

$$Q_{(n)} = q_n, \quad n \geq 1.$$

Consider the generating function

$$Q(t_1,t_2) := (Q(t_1)Q(t_2) - 1)\frac{t_1 - t_2}{t_1 + t_2}.$$

By (A.43), $Q(t_1,t_2)$ is a power series in t_1 and t_2, and we write

$$Q(t_1,t_2) = \sum_{r,s\geq 0} Q_{(r,s)} t_1^r t_2^s.$$

The following can be checked easily from the definition by noting $Q(t_1,t_2) = -Q(t_2,t_1)$.

(A.48) *We have* $Q_{(r,s)} = -Q_{(s,r)}, Q_{(r,0)} = q_r$. *In addition,*

$$Q_{(r,s)} = q_r q_s + 2\sum_{i=1}^{s}(-1)^i q_{r+i} q_{s-i}, \quad r > s.$$

We recall the following classical facts (see Wikipedia). Any $2n \times 2n$ skew-symmetric matrix $A = (a_{ij})$ satisfies $\det(A) = \mathrm{Pf}(A)^2$, where $\mathrm{Pf}(A)$ is the **Pfaffian** of A given by

$$\mathrm{Pf}(A) = \sum_{\sigma} \mathrm{sgn}(\sigma) a_{\sigma(1)\sigma(2)} \cdots a_{\sigma(2n-1)\sigma(2n)}, \tag{A.49}$$

summed over $\sigma \in \mathfrak{S}_{2n}$ such that $\sigma(2i-1) < \sigma(2i)$ and $\sigma(2i-1) < \sigma(2i+1)$ for all admissible i. Equivalently, $\mathrm{Pf}(A)$ can be defined as the above sum over the whole group $\mathfrak{S}_{2n}$ divided by $2^n n!$.

Definition A.4. The Schur Q-function Q_λ, for $\lambda \in \mathcal{SP}$ with $\ell(\lambda) \leq 2n$, is defined to be the Pfaffian of the $2n \times 2n$ skew-symmetric matrix $(Q_{(\lambda_i,\lambda_j)})_{1\leq i,j\leq 2n}$.

Example A.5. For indeterminates $t_1,\ldots,t_{2n}$, let $A = (a_{ij})$ be the $2n \times 2n$ skew-symmetric matrix with $a_{ij} = \frac{t_i - t_j}{t_i + t_j}$. Then,

$$\mathrm{Pf}(A) = \prod_{1\leq i<j\leq 2n} \frac{t_i - t_j}{t_i + t_j}.$$

Remark A.6. Equivalently, Q_λ is the coefficient of $t_1^{\lambda_1} \ldots t_{2n}^{\lambda_{2n}}$ in

$$Q(t_1,\ldots,t_{2n}) := \mathrm{Pf}\,(Q(t_i,t_j)).$$

It can be shown that $Q(t_1,\ldots,t_{2n})$ is alternating in the sense that

$$Q(t_{\sigma(1)},\ldots,t_{\sigma(2n)}) = \mathrm{sgn}(\sigma) Q(t_1,\ldots,t_{2n}), \quad \forall \sigma \in \mathfrak{S}_{2n}.$$

It follows that $Q_\lambda = 0$ unless all parts of λ are distinct.

(A.50) *For* $\lambda \in \mathcal{SP}$ *with* $\ell(\lambda) \leq n$, Q_λ *is equal to the coefficient of* $t_1^{\lambda_1} t_2^{\lambda_2} \ldots$ *in*

$$Q(t_1)\ldots Q(t_n) \prod_{1\leq i<j\leq n} \frac{t_i - t_j}{t_i + t_j},$$

where it is understood that $\frac{t_i - t_j}{t_i + t_j} = 1 + 2\sum_{r\geq 1}(-1)^r (t_i^{-1} t_j)^r$.

The statement of (A.50) can be found in [**83**, (8.8), pp. 253]. The following recursive relations for Q_λ are obtained directly by the Pfaffian definition of Q_λ and the Laplacian expansion of Pfaffians (see Wikipedia).

(A.51) *For a strict partition* $\lambda = (\lambda_1, \ldots, \lambda_m)$,

$$Q_\lambda = \sum_{j=2}^{m} (-1)^j Q_{(\lambda_1,\lambda_j)} Q_{(\lambda_2,\ldots,\hat{\lambda}_j,\ldots,\lambda_m)}, \quad \textit{for } m \textit{ even},$$

$$Q_\lambda = \sum_{j=1}^{m} (-1)^{j-1} Q_{\lambda_j} Q_{(\lambda_1,\ldots,\hat{\lambda}_j,\ldots,\lambda_m)}, \quad \textit{for } m \textit{ odd}.$$

(A.52) *For* $\lambda \in \mathcal{SP}_n$, *we have*

$$Q_\lambda = q_\lambda + \sum_{\mu\in\mathcal{SP}_n, \mu>\lambda} a_{\lambda\mu} q_\mu.$$

It follows from the recursive relation (A.51) that Q_λ is equal to q_λ plus a linear combination of q_ν for $\nu > \lambda$ (not necessarily strict). Now (A.52) follows since each q_ν is expressible as a linear combination of q_μ with strict partitions $\mu \geq \nu$ by (A.44) (this fact has been used in the induction step in the proof of (A.46)).

As an immediately corollary of (A.46) and (A.52), we have

(A.53) $\{Q_\lambda \mid \lambda \in \mathcal{SP}\}$ *is a* $\mathbb{Z}$*-basis for* Γ. *Moreover, for any partition* μ, *we have*

$$q_\mu = \sum_{\lambda\in\mathcal{SP}, \lambda\geq\mu} \widehat{K}_{\lambda\mu} Q_\lambda,$$

where $\widehat{K}_{\lambda\mu} \in \mathbb{Z}$ *and* $\widehat{K}_{\mu\mu} = 1$. *(Clearly, one can further assume that* μ *is a composition instead of a partition in the statement.)*

Indeed from representation-theoretical consideration it can be seen that $\widehat{K}_{\lambda\mu} \geq 0$ (cf. Chapter 3, Lemma 3.50) .

A.3.3. Inner product on Γ. Let $x = \{x_1, x_2, \ldots\}$ and $y = \{y_1, y_2, \ldots\}$ be two independent sets of variables.

(A.54) *We have*

$$\prod_{i,j} \frac{1 + x_i y_j}{1 - x_i y_j} = \sum_{\alpha\in\mathcal{OP}} 2^{\ell(\alpha)} z_\alpha^{-1} p_\alpha(x) p_\alpha(y)$$
$$= \sum_{\mu\in\mathcal{P}} m_\mu(x) q_\mu(y).$$

Proof. Let $xy = \{x_i y_j \mid i, j = 1, 2, \ldots\}$. It follows by (A.42) that

$$\prod_{i,j} \frac{1 + x_i y_j}{1 - x_i y_j} = \sum_{n \geq 0} q_n(xy) = \sum_{\alpha \in \mathcal{OP}} 2^{\ell(\alpha)} z_\alpha^{-1} p_\alpha(xy).$$

Now the first identity follows since $p_\alpha(xy) = p_\alpha(x) p_\alpha(y)$. We further compute that

$$\begin{aligned} \prod_{i,j} \frac{1 + x_i y_j}{1 - x_i y_j} &= \prod_i \sum_{r_i = 0}^{\infty} q_{r_i}(y) x_i^{r_i} \\ &= \sum_{\mu \in \mathcal{P}} m_\mu(x) q_\mu(y). \end{aligned}$$

The second identity is proved. □

We define an inner product $\langle \cdot, \cdot \rangle$ on $\Gamma_{\mathbb{Q}}$ by letting

(A.55) $$\langle p_\alpha, p_\beta \rangle = 2^{-\ell(\alpha)} z_\alpha \delta_{\alpha\beta}.$$

(A.56) *Let $\{u_\lambda\}, \{v_\lambda\}$ be dual bases for $\Gamma_{\mathbb{Q}}$ with respect to $\langle \cdot, \cdot \rangle$. The following are equivalent:*

$$(i) \qquad \langle u_\lambda, v_\mu \rangle = \delta_{\lambda\mu} \quad \forall \lambda, \mu;$$

$$(ii) \qquad \sum_\lambda u_\lambda(x) v_\lambda(y) = \prod_{i,j} \frac{1 + x_i y_j}{1 - x_i y_j}.$$

Such an equivalence can be established by the same standard argument as (A.15), now based on (A.54).

(A.57) *We have*

$$\langle Q_\lambda, Q_\mu \rangle = 2^{\ell(\lambda)} \delta_{\lambda\mu}, \quad \lambda, \mu \in \mathcal{SP}.$$

Equivalently, we have

$$\prod_{i,j} \frac{1 + x_i y_j}{1 - x_i y_j} = \sum_{\lambda \in \mathcal{SP}} 2^{-\ell(\lambda)} Q_\lambda(x) Q_\lambda(y).$$

A rather nontrivial direct proof of (A.57) can be found in Józefiak [**57**]. Alternatively, one first works on the generality of Hall-Littlewood symmetric functions $Q_\lambda(x;t)$ for all partitions λ, and then (A.57) can be obtained as a specialization at $t = -1$. For details we refer to [**83**].

From the Hall-Littlewood approach, the Schur Q-function, for a strict partition λ with $\ell(\lambda) = \ell$ and $m \geq \ell$, is equal to

(A.58) $$Q_\lambda(x_1, \ldots, x_m) = 2^\ell \sum_{w \in \mathfrak{S}_m / \mathfrak{S}_{m-\ell}} w\Big(x_1^{\lambda_1} \cdots x_\ell^{\lambda_\ell} \prod_{1 \leq i \leq \ell} \prod_{i < j \leq m} \frac{x_i + x_j}{x_i - x_j} \Big),$$

where the symmetric group $\mathfrak{S}_m$ acts by permuting the variables $x_1, \ldots, x_m$ and $\mathfrak{S}_{m-\ell}$ is the subgroup acting on $x_{\ell+1}, \ldots, x_m$.

A.3.4. A characterization of Γ. Denote by Γ_m the counterpart of Γ in m variables, which consists of symmetric polynomials obtained by setting all but the first m variables to be zero in the functions in Γ. We define

$$\square := \prod_{1\le i<j\le m} (x_i + x_j).$$

(A.59) *Let f be any symmetric polynomial in m variables. Then $f\square$ lies in Γ_m.*

Proof. Let λ be a partition with $\ell(\lambda) \le m$ and let $\rho = (m-1, m-2, \dots, 1, 0)$ so that $\lambda+\rho$ is a strict partition. Then $Q_{\lambda+\rho}(x_1,\dots,x_m) \in \Gamma_m$. Now by (A.58) we see that the Schur Q-polynomial $Q_{\lambda+\rho}(x_1,\dots,x_m)$, up to a 2-power, is equal to

$$\begin{aligned}
\sum_{w\in\mathfrak{S}_m} w\Big(x_1^{\lambda_1+m-1}\cdots x_m^{\lambda_m} \prod_{1\le i<j\le m} \frac{x_i+x_j}{x_i-x_j}\Big) \\
&= \prod_{1\le i<j\le m} \frac{(x_i+x_j)}{(x_i-x_j)} \sum_{w\in\mathfrak{S}_m} \operatorname{sgn}(w) w(x_1^{\lambda_1+m-1}\cdots x_m^{\lambda_m}) \\
&= \frac{a_{\lambda+\rho}}{a_\rho} \prod_{1\le i<j\le m} (x_i+x_j) = s_\lambda(x_1,\dots,x_m)\square.
\end{aligned}$$

Since the Schur polynomials $s_\lambda(x_1,\dots,x_m)$ form a basis for the space Λ_m, this proves the lemma. □

We have the following characterization of Γ.

(A.60) *Let $g \in \Lambda$. Then $g \in \Gamma$ if and only if it satisfies the cancelation property that $g|_{x_i=-x_j=t}$ is independent of t for some $i \ne j$.*

Proof. Clearly, the power sums p_r for r odd satisfy the above cancelation property, and so does any element in Γ.

To prove the converse, it suffices to do so in the setting Γ_m of m variables. We shall show that if $g \in \Lambda_m$ is a symmetric polynomial in $x_1,\dots,x_m$ such that $g|_{x_i=-x_j=t}$ is independent of t, then g can be written as a polynomial in the odd power sums $\{p_{2k+1} \mid k \in \mathbb{Z}_+\}$. Observe that $g|_{x_i=-x_j=t}$ is independent of t for some particular choice of i,j with $i \ne j$ if and only if it is independent of t for all (nonidentical) pairs i,j.

We first note by (A.59) that $q\square \in \Gamma_m$ for any symmetric polynomial q. We proceed by induction on m. If $m = 0, 1$, then the cancelation property is vacuous.

Assume that $m \ge 2$. Let p'_{2k+1} denote the $(2k+1)$st power-sum in the variables $x_1,\dots,x_{m-2}$. Then the polynomial $g|_{x_{m-1}=-x_m=0}$ is symmetric in $x_1,\dots,x_{m-2}$, and it satisfies the cancelation property. By the induction hypothesis on m, there exists a polynomial f such that

$$g|_{x_{m-1}=-x_m=0} = f(p'_1,\dots,p'_{2r+1}).$$

Set $f^* := f(p_1, \ldots, p_{2r+1})$. Then $(g - f^*)|_{x_{m-1}=-x_m=t} = 0$, and hence $x_{m-1} + x_m$ divides $g - f^*$. Since $g - f^*$ is symmetric in $x_1, \ldots, x_m$, $\square$ divides $g - f^*$, and thus there exists a symmetric polynomial q such that

$$g = f^* + q\square.$$

By (A.59), $q\square$ is generated by the odd power sums in the variables $x_1, \ldots, x_m$, and hence so is g. $\square$

A.3.5. Relating Λ and Γ. There is an intimate connection between the rings Λ and Γ. Recall a homomorphism φ (cf. [**83**, III, §8, Example 10]) defined by

$$\varphi : \Lambda \longrightarrow \Gamma,$$

(A.61)
$$\varphi(p_r) = \begin{cases} 2p_r, & \text{for } r \text{ odd}, \\ 0, & \text{otherwise}, \end{cases}$$

where p_r denotes the rth power sum. One checks by definition that

$$\varphi\big(H(t)\big) = Q(t).$$

This can be reformulated as follows.

(A.62) *We have*

$$\varphi(h_n) = q_n \ (\forall n \geq 0), \qquad \varphi(h_\mu) = q_\mu, \ (\forall \mu \in \mathcal{P}).$$

The homomorphism φ admits a categorification involving representations of the symmetric group; see Exercise 3.13.

Suppose that λ is a strict partition of n. Let λ^* be the associated shifted diagram that is obtained from the ordinary Young diagram by shifting the kth row to the right by $k-1$ squares, for each k. Denoting $\ell(\lambda) = \ell$, we define the *double partition* $\widetilde{\lambda}$ to be $\widetilde{\lambda} = (\lambda_1, \ldots, \lambda_\ell | \lambda_1 - 1, \lambda_2 - 1, \ldots, \lambda_\ell - 1)$ in Frobenius notation.

Example A.7. Let $\lambda = (4,3,1)$. The corresponding shifted diagram λ^* and double diagram $\widetilde{\lambda}$ are

$\lambda^* =$ 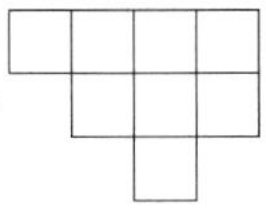 $\widetilde{\lambda} =$ 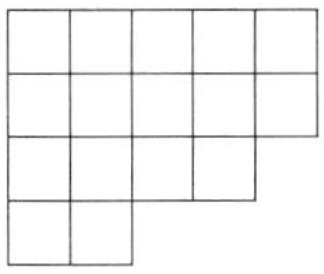

We have the following basic property that relates Schur and Q-Schur functions. (see [**83**, III, §8, 10]).

(A.63)
$$\varphi(s_{\widetilde{\lambda}}) = 2^{-\ell(\lambda)} Q_\lambda^2, \quad \forall \lambda \in \mathcal{SP}_n.$$

A.4. The Boson-Fermion correspondence

A.4.1. The Maya diagrams. The **Maya diagrams** are by definition the functions $f : \frac{1}{2} + \mathbb{Z} \to \{\pm\}$ such that $f(n) = +$ and $f(-n) = -$, for $n \gg 0$. In other words, a Maya diagram is an assignment of the signs $\pm$ to the unit intervals on the real line associated to the lattice $\mathbb{Z}$ with fixed asymptotic values $\pm$ at infinity. These intervals are naturally labeled by $1/2 + \mathbb{Z}$, and we shall identify them.

A Maya diagram f can be reconstructed by knowing the subset $\mathbf{m}$ of $\frac{1}{2} + \mathbb{Z}$ which is the preimage of $+$ for f, which is simply an increasing sequence $\mathbf{m} = \{m_j\}_{j \geq 1}$ in $\frac{1}{2} + \mathbb{Z}$, such that

$$m_1 < m_2 < m_3 < \ldots, \text{ and } m_j = m_{j-1} + 1 \text{ for } j \gg 0.$$

In this appendix, we shall identify Maya diagrams with such sequences $\mathbf{m}$.

Example A.8. Consider the following Maya diagrams:

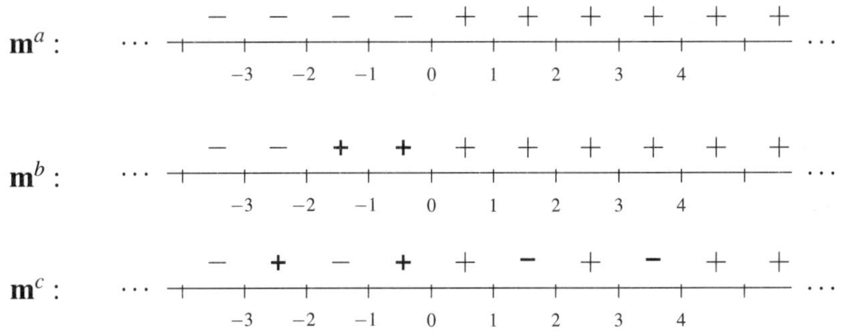

The Maya diagram $\mathbf{m}^b$ is obtained from $\mathbf{m}^a$ by changing signs from $-$ to $+$ at intervals $-\frac{3}{2}$ and $-\frac{1}{2}$. The diagram $\mathbf{m}^c$ is obtained from $\mathbf{m}^a$ by changing signs from $-$ to $+$ at intervals $-\frac{5}{2}, -\frac{1}{2}$ and from $+$ to $-$ at $\frac{3}{2}, \frac{7}{2}$. In terms of infinite sequences, we have $\mathbf{m}^a = \frac{1}{2} + \mathbb{Z}_+$, $\mathbf{m}^b = (-\frac{3}{2}, -\frac{1}{2}, \frac{1}{2}, \frac{3}{2}, \frac{5}{2}, \ldots)$, and $\mathbf{m}^c = (-\frac{5}{2}, -\frac{1}{2}, \frac{1}{2}, \frac{5}{2}, \frac{9}{2}, \frac{11}{2}, \ldots)$.

A.4.2. Partitions. To each Maya diagram $\mathbf{m} = \{m_j\}_{j \geq 1}$, we assign an integer, called the **charge** of $\mathbf{m}$,

$$\ell_{\mathbf{m}} = \lim_{j \to \infty} \left(j - m_j - \frac{1}{2}\right)$$

and a partition

$$\lambda_{\mathbf{m}} = (1/2 - m_1 - \ell_{\mathbf{m}}, 3/2 - m_2 - \ell_{\mathbf{m}}, 5/2 - m_3 - \ell_{\mathbf{m}}, \ldots). \tag{A.64}$$

Recall that $\mathcal{P}$ denotes the set of all partitions. The following can be viewed as the **Boson-Fermion correspondence at the combinatorial level**.

(A.65) *The map* $\mathbf{m} \mapsto (\ell_{\mathbf{m}}, \lambda_{\mathbf{m}})$ *is a bijection between the set of Maya diagrams and the set* $\mathbb{Z} \times \mathcal{P}$ *of "charged partitions".*

Proof. We will use primarily the example of $\mathbf{m}^c$ in Example A.8 to illustrate how to visualize the Young diagram associated to $\lambda_{\mathbf{m}^c}$. Let us draw a new diagram following the signs on a Maya diagram from far left to far right as follows. We move to the southeast direction by one unit for each unit interval with a minus sign on the Maya diagram of $\mathbf{m}^c$, and move to the northeast by one unit for each unit interval with a plus sign on. As a result, we obtain the zigzag path in the first diagram below. The path is labeled by the unit intervals on the Maya diagram. Extending the two line segments coming from minus infinity and going to plus infinity, respectively, a finite bounded region emerges. This region can be further partitioned by solid lines as in the second diagram, and from which we read off the partition $\lambda_{\mathbf{m}^c} = (3,2,2,1)$ starting from the southwest row. It is easy to identify the partition obtained in this way for a Maya diagram $\mathbf{m}$ with the partition $\lambda_{\mathbf{m}}$ defined in (A.64). The vertical dashed line on the first diagram indicates that the intersection of the two infinite line segments corresponds to the coordinate 0 separating the two unit intervals $-\frac{1}{2}$ and $\frac{1}{2}$. From this we read off the charge $\ell_{\mathbf{m}^c} = 0$.

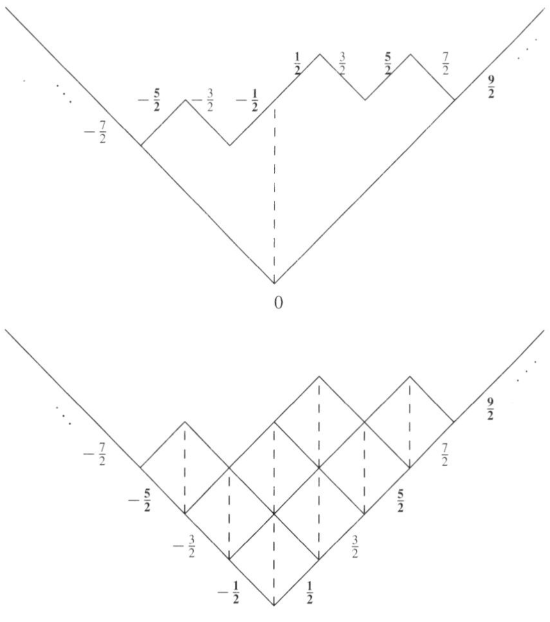

Observe that uniformly shifting every sign on a Maya diagram $\mathbf{m}$ one unit interval to the left gives rise to a Maya diagram whose charge is $\ell_{\mathbf{m}}+1$. For example, the $\mathbf{m}^b$ can be obtained from $\mathbf{m}^a$ by such a uniform shift to the left twice, and so $\ell_{\mathbf{m}^b} = \ell_{\mathbf{m}^a} + 2$. Such a shift clearly corresponds to a uniform shift of the labels on the corresponding zigzag path in the first diagram. Hence, it suffices to show that the set of Maya diagrams with charge 0 is in bijection with the set of partitions via $\mathbf{m} \mapsto \lambda_{\mathbf{m}}$. This bijection follows since we can easily reconstruct the labeled zigzag path (and hence the Maya diagram) from a given partition λ by first using a dashed diagonal line on the upside-down Young diagram for λ (whose first southwest row has λ_1 boxes) to locate the coordinate 0 point. □

A.4.3. Fermions and fermionic Fock space. Let $\mathcal{F}$ be the vector space with a linear basis given by the symbols $|\mathbf{m}\rangle$ parameterized by the Maya diagrams $\mathbf{m}$. We will call $\mathcal{F}$ the **fermionic Fock space**.

Note that different Maya diagrams are related by changing signs at finitely many intervals, and let us consider changing signs to a Maya diagram one interval at a time. Accordingly, we define the linear operators ψ_n^+, ψ_n^- on $\mathcal{F}$, for $n \in \frac{1}{2}+\mathbb{Z}$, as follows:

$$\psi_n^-|\mathbf{m}\rangle = \begin{cases} (-1)^{i-1}|\ldots, m_{i-1}, \widehat{m}_i, m_{i+1}, \ldots\rangle, & \text{if } m_i = -n \text{ for some } i, \\ 0, & \text{otherwise,} \end{cases}$$

$$\psi_n^+|\mathbf{m}\rangle = \begin{cases} (-1)^{i}|\ldots, m_i, n, m_{i+1}, \ldots\rangle, & \text{if } m_i < n < m_{i+1} \text{ for some } i, \\ 0, & \text{otherwise,} \end{cases}$$

where, as usual, $\widehat{m}_i$ denotes omission of the term m_i. That is, ψ_n^- corresponds to the sign change at $-n$ if the initial sign at $-n$ is $+$, and ψ_n^+ corresponds to the sign change at n if the initial sign at n is $-$.

Remark A.9. Let V be a vector space with a standard basis $v_i, i \in \frac{1}{2}+\mathbb{Z}$, and let $\{v_i^*\}$ denote its dual basis. It is instructive to regard each vector $|\mathbf{m}\rangle$ as a semi-infinite wedge $v_{\mathbf{m}} := v_{m_1} \wedge v_{m_2} \wedge v_{m_3} \wedge \cdots$. The signs in the above definition of $\psi_n^{\pm}$ can be explained naturally by identifying ψ_n^- as the contraction operator v_{-n}^* and ψ_n^+ as taking the exterior product with v_n. In this way, the Maya diagram model is canonically identified with a semi-infinite wedge model of the fermionic Fock space $\mathcal{F}$.

It is possible to use the subset $\overline{\mathbf{m}} := f^{-1}(-)$ of $\frac{1}{2}+\mathbb{Z}$, which is the complementary subset of $\mathbf{m} = f^{-1}(+)$, to identify with a Maya diagram f. Then $\overline{\mathbf{m}}$ would be naturally identified with a semi-infinite wedge that goes to $-\infty$ (instead of ∞). We will not adopt this convention in this book.

We shall refer to $|\frac{1}{2}, \frac{3}{2}, \frac{5}{2}, \ldots\rangle$ as the **vacuum vector** of $\mathcal{F}$ and denote it by $|0\rangle$. Let us quote Dirac from his book "The Theory of Quantum Mechanics": *The perfect vacuum is a region where all the states of positive energy are unoccupied and all those of negative energy are occupied.* We took the liberty in switching "positive" with "negative" in this exposition.

Our convention is to consider the operators $\psi_n^{\pm}$ as odd operators and the commutators among them to be anti-commutators: $[A,B]_+ = AB + BA$.

(A.66) *The following anti-commutation relations hold for all $m, n \in \frac{1}{2}+\mathbb{Z}$:*

$$[\psi_n^+, \psi_m^-]_+ = \delta_{n,-m} I, \quad [\psi_n^+, \psi_m^+]_+ = 0, \quad [\psi_n^-, \psi_m^-]_+ = 0.$$

In particular, $(\psi_n^+)^2 = (\psi_n^-)^2 = 0$.

Proof. This follows by a direct verification using the definitions of $\psi_n^\pm$. It corresponds to the fact that the processes of changing signs at two (possibly identical) intervals are interchangeable. □

One also verifies directly the following.

(A.67) *We have* $\psi_n^+|0\rangle = \psi_n^-|0\rangle = 0$, *for* $n > 0$.

Denote by $\widehat{\mathcal{C}}$ the Clifford (super)algebra generated by $\psi_n^\pm$, for $n \in \frac{1}{2}+\mathbb{Z}$, subject to the relations in (A.66).

(A.68) *A linear basis for* $\mathcal{F}$ *can be given by*

(A.69) $$\psi^+_{-p_1}\psi^+_{-p_2}\cdots\psi^+_{-p_r}\psi^-_{-q_1}\psi^-_{-q_2}\cdots\psi^-_{-q_s}|0\rangle,$$

for $p_1 > p_2 > \ldots > p_r > 0$ *and* $q_1 > q_2 > \ldots > q_s > 0$ *with* $r,s \geq 0$. *The basis element* (A.69) *is equal to* $\mathbf{m}$ *(up to a sign), where* $\mathbf{m}$ *is obtained from the set* $\frac{1}{2}+\mathbb{Z}_+$ *by deleting* $q_1,\ldots,q_s$ *and adding* $-p_1,\ldots,-p_r$.

The Fock space $\mathcal{F}$ *is an irreducible module of the Clifford algebra* $\widehat{\mathcal{C}}$ *generated by the vacuum vector* $|0\rangle$.

Proof. The identification of the vector (A.69) with $\mathbf{m}$ up to a sign follows from the definitions of $\psi_n^\pm$, which correspond to the additions and removals of the $+$ signs. The irreducibility follows from the fact that one Maya diagram can be transformed into another by changing the signs one interval at a time. □

Example A.10. We have $|\mathbf{m}^c\rangle = -\psi^+_{-5/2}\psi^+_{-1/2}\psi^-_{-7/2}\psi^-_{-3/2}|0\rangle$, for $\mathbf{m}^c$ from Example A.8.

We define the **half-integral Frobenius coordinates** for a partition λ as follows. Draw the Young diagram for λ in the English convention, and draw a diagonal line from the vertex $(0,0)$ that divides the Young diagram into two halves. We then record the lengths of rows of the upper half and the columns of the lower half as $(p_1,p_2,\ldots,p_r|q_1,q_2,\ldots,q_r)$, where each p_i,q_i lie in $\frac{1}{2}+\mathbb{Z}_+$, and

$$p_1 > p_2 > \ldots > p_r > 0, \qquad q_1 > q_2 > \ldots > q_r > 0.$$

The sequence $(p_1-\frac{1}{2},p_2-\frac{1}{2},\ldots,p_r-\frac{1}{2}|q_1-\frac{1}{2},\ldots,q_r-\frac{1}{2})$ is Frobenius' original notation for the partition λ (see Macdonald [**83**, p. 3]). For example, the modified Frobenius coordinates for the partition $(3,2,2,1)$ are $(\frac{5}{2},\frac{1}{2}|\frac{7}{2},\frac{3}{2})$.

It turns out that the indices p_i,q_j in (A.69) (at least when $r=s$) have natural interpretations in terms of partitions.

(A.70) *Let* $\mathbf{m}$ *be a Maya diagram of charge* 0. *Write* $(p_1,p_2,\ldots,p_r|q_1,q_2,\ldots,q_r)$ *for the modified Frobenius coordinates for the partition* $\lambda_{\mathbf{m}}$. *Then, up to a sign,* $|\mathbf{m}\rangle$ *is equal to* $\psi^+_{-p_1}\psi^+_{-p_2}\cdots\psi^+_{-p_r}\psi^-_{-q_1}\psi^-_{-q_2}\cdots\psi^-_{-q_r}|0\rangle$.

The sign in (A.70) can be computed to be $(-1)^{q_1+\ldots+q_r-\frac{r^2}{2}}$.

Proof. The proof follows from two observations that work for any general Maya diagram of charge 0, and we will illustrate below using our running example $\mathbf{m}^c$ from Example A.8: (i) In the diagrams appearing in the proof of (A.65), the two 2 corresponding to the $+$ sign to the left of the dashed vertical line are $-\frac{5}{2}, -\frac{1}{2}$, and the opposites of the 2 labels corresponding to the $-$ sign to the right of the dashed vertical line are $-\frac{3}{2}, -\frac{7}{2}$; they are exactly the indices of the $\psi^{\pm}$s in the formula $|\mathbf{m}^c\rangle = -\psi^+_{-5/2}\psi^+_{-1/2}\psi^-_{-7/2}\psi^-_{-3/2}|0\rangle$ by (A.68). (ii) The modified Frobenius coordinates for the partition $(3,2,2,1)$ are $(\frac{5}{2},\frac{1}{2}|\frac{7}{2},\frac{3}{2})$, and they precisely correspond to the $2+2$ labels specified in (i). $\square$

We form generating functions (called fermionic vertex operators) $\psi^{\pm}(z)$ in a formal variable z by letting

(A.71) $$\psi^+(z) = \sum_{n\in\frac{1}{2}+\mathbb{Z}} \psi^+_n z^{-n-\frac{1}{2}}, \qquad \psi^-(z) = \sum_{n\in\frac{1}{2}+\mathbb{Z}} \psi^-_n z^{-n-\frac{1}{2}}.$$

A.4.4. Charge and energy. We define the **energy operator** L_0 as a linear operator on $\mathcal{F}$ that diagonalizes the basis elements (A.69) with eigenvalues $p_1+\ldots+p_r+q_1+\ldots+q_s$. In addition, we define a **charge operator** α_0 as a linear operator on $\mathcal{F}$ that diagonalizes the basis elements (A.69) with eigenvalues $r-s$. The element (A.69) will be said to have **energy** $p_1+\ldots+p_r+q_1+\ldots+q_s$ and **charge** $r-s$. We have the following charge decomposition:

$$\mathcal{F} = \bigoplus_{\ell\in\mathbb{Z}} \mathcal{F}^{(\ell)},$$

where $\mathcal{F}^{(\ell)}$ is spanned by the basis vectors (A.69) of charge ℓ. We shall see that the notion of charge here matches the notion of charge $\ell_{\mathbf{m}}$ for a Maya diagram $\mathbf{m}$.

(A.72) *For each $\ell\in\mathbb{Z}$, $\mathcal{F}^{(\ell)}$ has a basis $|\mathbf{m}\rangle$ parameterized by the Maya diagrams $\mathbf{m}=\{m_j\}_{j\geq 1}$ that satisfy $\ell_{\mathbf{m}}=\ell$.*

Proof. By definition, ψ^-_n of charge -1 corresponds to the removal of a term from a Maya diagram $\mathbf{m}=(m_1,m_2,\ldots)$, which decreases the integer $\ell_{\mathbf{m}}$ by 1 . On the other hand, the operator ψ^+_n of charge 1 corresponds to the creation of a new sequence (i.e. Maya diagram) by the insertion of a term into a sequence $\mathbf{m}$, which increases the integer $\ell_{\mathbf{m}}$ by 1. This means that the two notions of charge are identical up to a possible constant shift. This constant must be zero once we observe that $|0\rangle = |\mathbf{m}^a\rangle$ is clearly in $\mathcal{F}^{(0)}$ and also $\ell_{\mathbf{m}^a}=0$ by definition. $\square$

For $\ell\in\mathbb{Z}$, we let

(A.73) $$|\ell\rangle = \begin{cases} \psi^+_{\frac{1}{2}-\ell}\cdots\psi^+_{-\frac{3}{2}}\psi^+_{-\frac{1}{2}}|0\rangle, & \text{if } \ell>0 \\ |0\rangle, & \text{if } \ell=0 \\ \psi^-_{\frac{1}{2}+\ell}\cdots\psi^-_{-\frac{3}{2}}\psi^-_{-\frac{1}{2}}|0\rangle, & \text{if } \ell<0. \end{cases}$$

In particular, $|1\rangle = \psi^+_{-\frac{1}{2}}|0\rangle$ and $|\text{-}1\rangle = \psi^-_{-\frac{1}{2}}|0\rangle$. The following is easily verified.

(A.74) *For each $\ell \in \mathbb{Z}$, the element $|\ell\rangle \in \mathcal{F}^{(\ell)}$ has the minimal energy, which is $\ell^2/2$, among all vectors* (A.69) *of charge ℓ.*

A.4.5. From Bosons to Fermions. The Heisenberg algebra $\mathfrak{Heis}$ is the Lie algebra generated by a_k $(k \in \mathbb{Z})$ and a cental element c with the commutation relation:

$$[a_m, a_n] = m\delta_{m,-n}c, \qquad m,n \in \mathbb{Z}. \tag{A.75}$$

In particular, a_0 is central. The Heisenberg algebra $\mathfrak{Heis}$ acts irreducibly on a polynomial algebra $\mathcal{B}^{(\ell)} := \mathbb{C}[p_1, p_2, \ldots]$ in infinitely many variables $p_1, p_2, \ldots$, by letting a_{-k} $(k > 0)$ act as the multiplication operator by kp_k, a_k $(k > 0)$ act as the differential operator $\frac{\partial}{\partial p_k}$, c act as the identity I, and a_0 act as ℓI, for $\ell \in \mathbb{C}$. The $\mathfrak{Heis}$-module $\mathcal{B}^{(\ell)}$ is irreducible, since any nonzero polynomial in $\mathcal{B}^{(\ell)}$ can be transformed to a nonzero constant polynomial by a suitable sequence of differential operators and then produce arbitrary monomials by the multiplication operators. The $\mathfrak{Heis}$-module $\mathcal{B}^{(\ell)}$ is a highest weight module in the sense that $a_n.1 = 0$ for $n > 0$, and any highest weight $\mathfrak{Heis}$-module generated by a highest weight vector v such that $c.v = v$ and $a_0.v = \ell v$ is isomorphic to $\mathcal{B}^{(\ell)}$.

Similarly to (5.45), we introduce the normal ordered product

$$:\psi^+_m\psi^-_n: = \begin{cases} -\psi^-_n\psi^+_m, & \text{if } n = -m < 0, \\ \psi^+_m\psi^-_n, & \text{otherwise.} \end{cases}$$

Define the bosonic vertex operator $\alpha(z) = \sum_{k\in\mathbb{Z}} \alpha_k z^{-k-1}$ by letting

$$\alpha(z) = \sum_{k\in\mathbb{Z}} \alpha_k z^{-k-1} = :\psi^+(z)\psi^-(z):, \tag{A.76}$$

which is equivalent to defining componentwise

$$\alpha_k = \sum_{n\in\frac{1}{2}+\mathbb{Z}} :\psi^+_n\psi^-_{k-n}:, \qquad k \in \mathbb{Z}. \tag{A.77}$$

Note that

$$\alpha_0 = \sum_{n>0} \psi^+_{-n}\psi^-_n - \sum_{n<0} \psi^-_n\psi^+_{-n}$$

is a well-defined linear operator on $\mathcal{F}$.

The following is (one half of) the Boson-Fermion correspondence.

(A.78) *Let $\ell \in \mathbb{Z}$. The Heisenberg algebra $\mathfrak{Heis}$ acts on $\mathcal{F}^{(\ell)}$ by letting $c \mapsto I, \alpha_0 \mapsto \ell \cdot I$, and $a_k \mapsto \alpha_k$, for $k \in \mathbb{Z}$. That is, the following commutation relation holds:*

$$[\alpha_m, \alpha_n] = m\delta_{m,-n}I, \qquad m,n \in \mathbb{Z}. \tag{A.79}$$

Moreover, the $\mathfrak{Heis}$-module $\mathcal{F}^{(\ell)}$ is irreducible and it is isomorphic to $\mathcal{B}^{(\ell)}$.

Proof. We shall be free to use the following (anti-)commutator identities:

$$[AB,C] = A[B,C]_+ - [A,C]_+B = A[B,C] + [A,C]B.$$

By a direct computation using the anti-commutator identity, we have

$$[\alpha_k, \psi_n^{\pm}] = \pm\psi_{k+n}^{\pm}, \qquad k \in \mathbb{Z},\ n \in \frac{1}{2} + \mathbb{Z}.$$

The identity (A.79) follows by another computation using the commutator identity.

By definition, the operators α_k have charge 0. Hence, $\alpha_k(\mathcal{F}^{(\ell)}) \subseteq \mathcal{F}^{(\ell)}$; that is, $\mathcal{F}^{(\ell)}$ is a $\mathfrak{Heis}$-module.

For $p, q > 0$ and $\mathbf{m}$ of charge ℓ, $\psi^+_{-p}\psi^-_{-q}|\mathbf{m}\rangle = \pm|\mathbf{m}'\rangle$ if it is nonzero, where $\mathbf{m}'$ is obtained from $\mathbf{m}$ by replacing a term q appearing in $\mathbf{m}$ by a new term $-p$, and the charge of $\mathbf{m}'$ remains ℓ. Recalling the definition (A.64) of the partition $\lambda_{\mathbf{m}}$, we see that $|\lambda_{\mathbf{m}'}| = |\lambda_{\mathbf{m}}| + p + q$. This implies that the energy of $\mathbf{m}$ is equal to $|\lambda_{\mathbf{m}}|$ up to a constant shift that depends on the charge ℓ. The constant is determined to be $\ell^2/2$ by recalling (A.74) and noting $\lambda_{|\ell\rangle} = \emptyset$. Hence, the energy of the element $|\mathbf{m}\rangle$ is $|\lambda_{\mathbf{m}}| + \ell_{\mathbf{m}}^2/2$. From this it follows that the q-dimension (or graded dimension) of $\mathcal{F}^{(\ell)}$ is given by

$$\mathrm{tr}|_{\mathcal{F}^{(\ell)}} q^{L_0} = \frac{q^{\frac{\ell^2}{2}}}{\prod_{k=1}^{\infty}(1-q^k)}.$$

On the other hand, note that the energy operator L_0 satisfies (and is indeed characterized by) the following properties: (i) $L_0|0\rangle = 0$; (ii) $[L_0, \psi^{\pm}_{-n}] = n\psi^{\pm}_{-n}$, for all $n \in \frac{1}{2} + \mathbb{Z}$. It follows from (A.77) that

(A.80) $$[L_0, \alpha_{-k}] = k\alpha_{-k}, \qquad k \in \mathbb{Z}.$$

Note that $\alpha_k|\ell\rangle = 0$, for $k > 0$, and $\alpha_0|\ell\rangle = \ell|\ell\rangle$. Hence, the irreducible $\mathfrak{Heis}$-submodule of $\mathcal{F}^{(\ell)}$ generated by $|\ell\rangle$ is isomorphic to $\mathcal{B}^{(\ell)}$ and has q-dimension equal to $\frac{q^{\frac{\ell^2}{2}}}{\prod_{k=1}^{\infty}(1-q^k)}$, the same as the graded dimension of $\mathcal{F}^{(\ell)}$. It follows that the $\mathfrak{Heis}$-module $\mathcal{F}^{(\ell)}$ is irreducible. □

Remark A.11. From the proof, we have $[\alpha_0, \psi_n^{\pm}] = \pm\psi_n^{\pm}$ for $n \in \frac{1}{2} + \mathbb{Z}$. This together with $\alpha_0|0\rangle = 0$ implies that α_0 here can be identified with the charge operator defined earlier.

The other half of the Boson-Fermion correspondence allows one to reconstruct the fermions $\psi^{\pm}(z)$ in terms of the boson $\alpha(z)$. We will formulate the statement but skip its proof. Denote by $S : \mathcal{F}^{(\ell)} \to \mathcal{F}^{(\ell+1)}$ the shift operator that sends $|\ell\rangle$ to $|\ell+1\rangle$ and commutes with the action of α_k for all $k \neq 0$. Then

$$\begin{aligned} \psi^{\pm}(z) &= S^{\pm 1}{:}\exp(\pm \int \alpha(z)dz){:} \\ &= S^{\pm 1} z^{\pm\alpha_0} e^{\mp\sum_{j<0}\frac{z^{-j}}{j}\alpha_j} e^{\mp\sum_{j>0}\frac{z^{-j}}{j}\alpha_j}. \end{aligned}$$

A.4.6. Fermions and Schur functions. By the Boson-Fermion correspondence (A.78), we have an isomorphism $\mathcal{F}^{(0)} \cong \mathcal{B}^{(0)}$. On the other hand, we can naturally identify $\mathcal{B}^{(0)}$ with the ring of symmetric functions Λ by identifying $\alpha_{-\lambda_1} \dots \alpha_{-\lambda_\ell}|0\rangle$ with the power-sum symmetric functions p_λ, for all partitions $\lambda = (\lambda_1, \dots, \lambda_\ell)$. Let us denote by $\Theta : \mathcal{F}^{(0)} \to \Lambda$ the composition of the above two isomorphisms. Recall that the Schur functions s_λ form a linear basis for Λ, and recall from (A.64) that a partition $\lambda_{\mathbf{m}}$ is associated to a Maya diagram $\mathbf{m}$.

(A.81) *The linear isomorphism* $\Theta : \mathcal{F}^{(0)} \to \Lambda$ *sends a Maya diagram* $\mathbf{m}$ *of charge* 0 *to the Schur function* $s_{\lambda_{\mathbf{m}}}$.

A proof of this can be found in [**65**]. The following supporting examples can be verified via a direct computation by repeatedly using (A.77); the elements in terms of $\psi^{\pm}$'s can be converted to the elements in terms of Maya diagrams via (A.70).

Example A.12. We have

$$\frac{1}{2}(\alpha_{-1}^2 + \alpha_{-2})|0\rangle = \psi^+_{-\frac{3}{2}} \psi^-_{-\frac{1}{2}} |0\rangle$$
$$\frac{1}{2}(\alpha_{-1}^2 - \alpha_{-2})|0\rangle = -\psi^+_{-\frac{1}{2}} \psi^-_{-\frac{3}{2}} |0\rangle.$$

In addition, we have

$$\left(\frac{1}{6}\alpha_{-1}^3 + \frac{1}{2}\alpha_{-2}\alpha_{-1} + \frac{1}{3}\alpha_{-3}\right)|0\rangle = \psi^+_{-\frac{5}{2}} \psi^-_{-\frac{1}{2}} |0\rangle$$
$$\left(\frac{1}{6}\alpha_{-1}^3 - \frac{1}{2}\alpha_{-2}\alpha_{-1} + \frac{1}{3}\alpha_{-3}\right)|0\rangle = \psi^+_{-\frac{1}{2}} \psi^-_{-\frac{5}{2}} |0\rangle$$
$$\frac{1}{3}(\alpha_{-1}^3 - \alpha_{-3})|0\rangle = -\psi^+_{-\frac{3}{2}} \psi^-_{-\frac{3}{2}} |0\rangle.$$

A.4.7. Jacobi triple product identity.

(A.82) *Let* q, y *be formal variables. The Jacobi triple product identity holds:*

$$\prod_{k=1}^{\infty} (1 - q^k)(1 + q^{k-\frac{1}{2}} y)(1 + q^{k-\frac{1}{2}} y^{-1}) = \sum_{\ell \in \mathbb{Z}} q^{\frac{\ell^2}{2}} y^\ell.$$

Proof. We compute the trace of the operator $q^{L_0} y^{\alpha_0}$ on $\mathcal{F}$ in two ways.

First, by the linear basis (A.68) of $\mathcal{F}$ and knowing each fermionic operator $\psi^{\pm}_{-\ell}$ contributes ℓ to the energy and ± 1 to the charge, we have

$$\mathrm{tr}|_{\mathcal{F}} q^{L_0} y^{\alpha_0} = \prod_{n \in \frac{1}{2} + \mathbb{Z}_+} (1 + q^n y)(1 + q^n y^{-1}).$$

On the other hand, $\mathcal{F}^{(\ell)}$ for each $\ell \in \mathbb{Z}$ is identified with the bosonic Fock space $\mathcal{B}^{(\ell)}$ by (A.78). Using (A.74), we compute that

$$\mathrm{tr}|_{\mathcal{F}} q^{L_0} y^{\alpha_0} = \sum_{\ell \in \mathbb{Z}} \mathrm{tr}|_{\mathcal{B}^{(\ell)}} q^{L_0} y^{\alpha_0} = \sum_{\ell \in \mathbb{Z}} \frac{q^{\frac{\ell^2}{2}} y^{\ell}}{\prod_{k=1}^{\infty}(1-q^k)}.$$

The Jacobi triple product identity follows now by equating the two formulas for the trace and clearing the denominator. □

Notes

Section A.1. The materials on symmetric functions and the Frobenius characteristic map are fairly standard, and they can be found in Macdonald [**83**] in possibly different order.

Section A.2. The characterization of supersymmetric polynomials in A.2.1 appeared in Stembridge [**117**]. The results on super Schur functions and super Cauchy identity in A.2.2 can be found in [**7, 110**].

Section A.3. The materials can be found in Macdonald [**83**] and in Józefiak [**57**]. The characterization of the ring Γ in A.3.4 appeared in Pragacz [**98**, Theorem 2.11], and is used in Section 2.3.2.

Section A.4. The materials on Boson-Fermion correspondence are standard. The use of Maya diagrams follows Miwa-Jimbo-Date [**87**]. Additional applications of Boson-Fermion correspondence, most notably to soliton equations, can be found in Kac-Raina [**65**] and Miwa-Jimbo-Date [**87**].

Bibliography

[1] A. Alldridge, *The Harish-Chandra isomorphism for reductive symmetric superpairs*, Transform. Groups, DOI: 10.1007/S00031-012-9200-y.

[2] F. Aribaud, *Une nouvelle démonstration d'un théorème de R. Bott et B. Kostant*, Bull. Soc. Math. France **95** (1967), 205–242.

[3] M. Atiyah, R. Bott, and V.K. Patodi, *On the heat equation and the index theorem*, Invent. Math. **19** (1973), 279–330.

[4] S. Azam, H. Yamane, and M. Yousofzadeh, *Classification of Finite Dimensional Irreducible Representations of Generalized Quantum Groups via Weyl Groupoids*, preprint, arXiv:1105.0160.

[5] A. Beilinson and J. Bernstein, *Localisation de $\mathfrak{g}$-modules*, C.R. Acad. Sci. Paris Ser. I Math. **292** (1981), 15–18.

[6] I. Bernstein, I. Gelfand and S. Gelfand: *Differential operators on the base affine space and a study of $\mathfrak{g}$-modules*. Lie groups and their representations (Proc. Summer School, Bolyai Janos Math. Soc., Budapest, 1971), pp. 21–64. Halsted, New York, 1975.

[7] A. Berele and A. Regev, *Hook Young Diagrams with Applications to Combinatorics and to Representations of Lie Superalgebras*, Adv. Math. **64** (1987), 118–175.

[8] I.N. Bernstein and D.A. Leites, *A formula for the characters of the irreducible finite-dimensional representations of Lie superalgebras of series $\mathfrak{gl}$ and $\mathfrak{sl}$* (Russian), C. R. Acad. Bulgare Sci. **33** (1980), 1049–1051.

[9] B. Boe, J. Kujawa, and D. Nakano, *Cohomology and support varieties for Lie superalgebras*, Trans. Amer. Math. Soc. **362** (2010), 6551–6590.

[10] A. Brini, A. Palareti, and A. Teolis, *Gordan-Capelli series in superalgebras*, Proc. Natl. Acad. Sci. USA **85** (1988), 1330–1333.

[11] J. Brundan, *Kazhdan-Lusztig polynomials and character formulae for the Lie superalgebra* $\mathfrak{gl}(m|n)$, J. Amer. Math. Soc. **16** (2003), 185–231.

[12] J. Brundan, *Kazhdan-Lusztig polynomials and character formulae for the Lie superalgebra* $\mathfrak{q}(n)$, Adv. Math. **182** (2004), 28–77.

[13] J. Brundan and A. Kleshchev, *Projective representations of symmetric groups via Sergeev duality*, Math. Z. **239** (2002), 27–68.

[14] J. Brundan and J. Kujawa, *A new proof of the Mullineux conjecture*, J. Algebraic Combin. **18** (2003), 13–39.

[15] J. Brundan and C. Stroppel, *Highest weight categories arising from Khovanov's diagram algebras IV: the general linear supergroup*, J. Eur. Math. Soc. **14** (2012), 373–419.

[16] J.L. Brylinski and M. Kashiwara, *Kazhdan-Lusztig conjecture and holonomic systems*, Invent. Math. **64** (1981), 387–410.

[17] B. Cao and L. Luo, *Generalized Verma Modules and Character Formulae for* $\mathfrak{osp}(3|2m)$, preprint, arXiv:1001.3986.

[18] R. Carter, *Lie Algebras of Finite and Affine Type*. Cambridge Studies in Advanced Mathematics **96**. Cambridge University Press, Cambridge, 2005. xviii+632 pp.

[19] S.-J. Cheng and J.-H. Kwon, *Howe duality and Kostant homology formula for infinite-dimensional Lie superalgebras*, Int. Math. Res. Not. **2008**, Art. ID rnn 085, 52 pp.

[20] S.-J. Cheng, J.-H. Kwon, and N. Lam, *A BGG-type resolution for tensor modules over general linear superalgebra*, Lett. Math. Phys. **84** (2008), 75–87.

[21] S.-J. Cheng, J.-H. Kwon, and W. Wang, *Kostant homology formulas for oscillator modules of Lie superalgebras*, Adv. Math. **224** (2010), 1548–1588.

[22] S.-J. Cheng and N. Lam, *Infinite-dimensional Lie superalgebras and hook Schur functions*, Commun. Math. Phys. **238** (2003), 95–118.

[23] S.-J. Cheng and N. Lam, *Irreducible characters of general linear superalgebra and super duality*, Commun. Math. Phys. **280** (2010), 645–672.

[24] S.-J. Cheng, N. Lam and W. Wang, *Super duality and irreducible characters of ortho-symplectic Lie superalgebras*, Invent. Math. **183** (2011), 189–224.

[25] S.-J. Cheng, N. Lam and W. Wang, *Super duality for general linear Lie superalgebras and applications*, arXiv:1109.0667, Proc. Symp. Pure Math. (to appear).

[26] S.-J. Cheng, N. Lam and W. Wang, *Brundan-Kazhdan-Lusztig conjecture for general linear Lie superalgebras*, arXiv:1203.0092.

[27] S.-J. Cheng, N. Lam, and R.B. Zhang, *Character formula for infinite dimensional unitarizable modules of the general linear superalgebra*, J. Algebra **273** (2004), 780–805.

[28] S.-J. Cheng and W. Wang, *Remarks on Schur-Howe-Sergeev duality*, Lett. Math. Phys. **52** (2000), 143–153.

[29] S.-J. Cheng and W. Wang, *Howe duality for Lie superalgebras*, Compositio Math. **128** (2001), 55–94.

[30] S.-J. Cheng and W. Wang, *Lie subalgebras of differential operators on the super circle*, Publ. Res. Inst. Math. Sci. **39** (2003), 545–600.

[31] S.-J. Cheng and W. Wang, *Brundan-Kazhdan-Lusztig and Super Duality Conjectures*, Publ. Res. Inst. Math. Sci. **44** (2008), 1219–1272.

[32] S.-J. Cheng and W. Wang, *Dualities for Lie superalgebras*, Lie Theory and Representation Theory, 1–46, Surveys of Modern Mathematics **2**, International Press and Higher Education Press, 2012.

[33] S.-J. Cheng, W. Wang, and R.B. Zhang, *A Fock space approach to representation theory of* $\mathfrak{osp}(2|2n)$, Transform. Groups **12** (2007), 209–225.

[34] S.-J. Cheng, W. Wang, and R.B. Zhang, *Super duality and Kazhdan-Lusztig polynomials*, Trans. Amer. Math. Soc. **360** (2008), 5883–5924.

[35] S.-J. Cheng and R.B. Zhang, *Analogue of Kostant's* $\mathfrak{u}$*-homology formula for general linear superalgebras*, Int. Math. Res. Not. **2004**, 31–53.

[36] S.-J. Cheng and R.B. Zhang, *Howe duality and combinatorial character formula for orthosymplectic Lie superalgebras*, Adv. Math. **182** (2004), 124–172.

[37] E. Date, M. Jimbo, M. Kashiwara, and T. Miwa: *Transformation groups for soliton equations. III. Operator approach to the Kadomtsev-Petviashvili equation*, J. Phys. Soc. Japan 50 (1981), 3806–3812. *Transformation groups for soliton equations. IV. A new hierarchy of soliton equations of KP-type*, Phys. D 4 (1981/82), 343–365.

[38] M. Davidson, E. Enright, and R. Stanke, *Differential Operators and Highest Weight Representations*, Mem. Amer. Math. Soc. **94** (1991), no. 455.

[39] V. Deodhar, *On some geometric aspects of Bruhat orderings II: the parabolic analogue of Kazhdan-Lusztig polynomials*, J. Algebra **111** (1987), 483–506.

[40] S. Donkin, *On tilting modules for algebraic groups*, Math. Z. **212** (1993), 39–60.

[41] T. Enright, *Analogues of Kostant's* $\mathfrak{u}$*-cohomology formulas for unitary highest weight modules*, J. Reine Angew. Math. **392** (1988), 27–36.

[42] T. Enright and J. Willenbring, *Hilbert series, Howe duality, and branching rules for classical groups*, Ann. of Math. **159** (2004), 337–375.

[43] L. Frappat, A. Sciarrino, and P. Sorba, *Dictionary on Lie algebras and superalgebras*. Academic Press, Inc., San Diego, CA, 2000.

[44] I. Frenkel, *Representations of affine Lie algebras, Hecke modular forms and Kortweg-de Vries type equations*, Lect. Notes Math. **933** (1982), 71–110.

[45] J. Germoni, *Indecomposable representations of* $\mathrm{osp}(3,2)$, $D(2,1;\alpha)$ *and* $G(3)$. Colloquium on Homology and Representation Theory (Spanish) (Vaquerias, 1998). Bol. Acad. Nac. Cienc. (Cordoba) **65** (2000), 147–163.

[46] R. Goodman and N. Wallach, *Representations and invariants of the classical groups*. Encyclopedia of Mathematics and its Applications, **68**. Cambridge University Press, Cambridge, 1998.

[47] M. Gorelik, *Strongly typical representations of the basic classical Lie superalgebras*, J. Amer. Math. Soc. **15** (2002), 167–184.

[48] M. Gorelik, *The Kac construction of the centre of* $U(\mathfrak{g})$ *for Lie superalgebras*, J. Nonlinear Math. Phys. **11** (2004), 325–349.

[49] C. Gruson and V. Serganova, *Cohomology of generalized supergrassmannians and character formulae for basic classical Lie superalgebras*, Proc. Lond. Math. Soc. **101** (2010), 852–892.

[50] D. Hill, J. Kujawa, and J. Sussan, *Degenerate affine Hecke-Clifford algebras and type Q Lie superalgebras*, Math. Z. **268** (2011), 1091–1158.

[51] R. Howe, *Remarks on classical invariant theory*, Trans. Amer. Math. Soc. **313** (1989), 539–570.

[52] R. Howe, *Perspectives on invariant theory: Schur duality, multiplicity-free actions and beyond*, The Schur Lectures, Israel Math. Conf. Proc. **8**, Tel Aviv (1992), 1–182.

[53] P.-Y. Huang, N. Lam, and T.-M. To, *Super duality and homology of unitarizable modules of Lie algebras*, Publ. Res. Inst. Math. Sci. **48** (2012), 45–63.

[54] J. E. Humphreys, *Representations of Semisimple Lie Algebras in the BGG Category* $\mathcal{O}$, Graduate Studies in Mathematics, **94**. American Mathematical Society, Providence, RI, 2008.

[55] K. Iohara and Y. Koga, *Second homology of Lie superalgebras*, Math. Nachr. **278** (2005), 1041–1053.

[56] T. Józefiak, *Semisimple superalgebras*, In: Algebra–Some Current Trends (Varna, 1986), pp. 96–113, Lect. Notes in Math. **1352**, Springer-Verlag, Berlin-New York, 1988.

[57] T. Józefiak, *Characters of projective representations of symmetric groups*, Expo. Math. **7** (1989), 193–247.

[58] T. Józefiak, *A class of projective representations of hyperoctahedral groups and Schur Q-functions*, Topics in Algebra, Banach Center Publ., **26**, Part 2, PWN-Polish Scientific Publishers, Warsaw (1990), 317–326.

[59] V. Kac, *Classification of simple Lie superalgebras* (in Russian), Funkcional. Anal. i Priložen **9** (1975), 91–92.

[60] V. Kac, *Lie superalgebras*, Adv. Math. **26** (1977), 8–96.

[61] V. Kac, *Characters of typical representations of classical Lie superalgebras*, Comm. Algebra **5** (1977), 889–897.

[62] V. Kac, *Representations of classical Lie superalgebras*. Differential geometrical methods in mathematical physics, II (Proc. Conf., Univ. Bonn, Bonn, 1977), pp. 597–626, Lecture Notes in Math. **676**, Springer, Berlin, 1978.

[63] V. Kac, *Laplace operators of infinite-dimensional Lie algebras and theta functions*, Proc. Nat. Acad. Sci. USA **81** (1984), 645–647.

[64] V. Kac, *Infinite dimensional Lie algebras*. Third edition. Cambridge University Press, Cambridge, 1990.

[65] V. Kac and A. Raina, *Bombay Lectures on Highest Weight Representations of Infinite Dimensional Lie Algebras*, Advanced Series in Mathematical Physics **2**. World Scientific Publishing Co., Inc., Teaneck, NJ, 1987.

[66] V. Kac and M. Wakimoto, *Integrable highest weight modules over affine superalgebras and Appell's function*, Commun. Math. Phys. **215** (2001), 631–682.

[67] V. Kac, W. Wang, and C. Yan, *Quasifinite representations of classical Lie subalgebras of* $W_{1+\infty}$, Adv. Math. **139** (1998), 56–140.

[68] S.-J. Kang and J.-H. Kwon, *Graded Lie superalgebras, supertrace formula, and orbit Lie superalgebras*, Proc. London Math. Soc. **81** (2000), 675–724.

[69] M. Kashiwara and M. Vergne, *On the Segal-Shale-Weil representations and harmonic polynomials*, Invent. Math. **44** (1978), 1–47.

[70] D. Kazhdan and G. Lusztig, *Representations of Coxeter groups and Hecke algebras*, Invent. Math. **53** (1979), 165–184.

[71] D. Kazhdan and G. Lusztig, *Schubert varieties and Poincaré duality*. Geometry of the Laplace operator (Proc. Sympos. Pure Math., Univ. Hawaii, Honolulu, Hawaii, 1979), pp. 185–203, Proc. Sympos. Pure Math., XXXVI, Amer. Math. Soc., Providence, R.I., 1980.

[72] A. Knapp and D. Vogan, *Cohomological Induction and Unitary Representations*, Princeton Mathematical Series, **45**. Princeton University Press, Princeton, NJ, 1995.

[73] B. Kostant, *Lie Algebra Cohomology and the Generalized Borel-Weil Theorem*, Ann. Math. **74** (1961), 329–387.

[74] H. Kraft and C. Procesi, *Classical Invariant Theory, A Primer*. 1996, 128 pp. Available at http://www.math.unibas.ch/~kraft/Papers/KP-Primer.pdf.

[75] S. Kudla, *Seesaw reductive pairs*, in Automorphic forms in several variables, Taniguchi Symposium, Katata 1983, Birkhäuser, Boston, 244–268.

[76] S. Kumar, *Kac-Moody groups, their flag varieties and representation theory*. Progress in Mathematics, **204**. Birkhauser Boston, Inc., Boston, MA, 2002.

[77] N. Lam and R.B. Zhang, *Quasi-finite modules for Lie superalgebras of infinite rank*, Trans. Amer. Math. Soc. **358** (2006), 403–439.

[78] D. Leites, M. Saveliev, and V. Serganova, *Embedding of* $\mathfrak{osp}(N/2)$ *and the associated nonlinear supersymmetric equations*. Group theoretical methods in physics, Vol. I (Yurmala, 1985), 255–297, VNU Sci. Press, Utrecht, 1986.

[79] J. Lepowsky: *A generalization of the Bernstein-Gelfand-Gelfand resolution*, J. Algebra **49** (1977), 496–511.

[80] J.-A. Lin, *Categories of* $\mathfrak{gl}(m|n)$ *with typical central characters*, Master's degree thesis, National Taiwan University, 2009.

[81] L. Liu, *Kostant's Formula for Kac-Moody Lie Algebras*, J. Algebra **149** (1992), 155–178.

[82] L. Luo, *Character Formulae for Ortho-symplectic Lie Superalgebras* $\mathfrak{osp}(n|2)$, J. Algebra **353** (2012), 31–61.

[83] I. G. Macdonald, *Symmetric functions and Hall polynomials*, Second Edition, Oxford Mathematical Monographs. Oxford Science Publications. The Clarendon Press, Oxford University Press, New York, 1995.

[84] Yu. Manin, *Gauge field theory and complex geometry*, Grundlehren der mathematischen Wissenschaften **289**, Second Edition, Springer-Verlag, Berlin, 1997.

[85] J. Milnor and J. Moore, *On the Structure of Hopf Algebras*, Ann. Math. **81** (1965), 211–264.

[86] B. Mitchell, *Theory of categories*, Pure and Applied Mathematics XVII, Academic Press, New York-London, 1965.

[87] T. Miwa, M. Jimbo, and E. Date, *Solitons. Differential equations, symmetries and infinite-dimensional algebras*, Cambridge Tracts in Mathematics **135**. Cambridge University Press, Cambridge, 2000.

[88] E. Moens and J. van der Jeugt, *A determinantal formula for supersymmetric Schur polynomials*, J. Algebraic Combin. **17** (2003), 283–307.

[89] I. Musson, *Lie superalgebras, Clifford algebras, induced modules and nilpotent orbits*, Adv. Math. **207** (2006), 39–72.

[90] I. Musson, *Lie superalgebras and enveloping algebras*, Graduate Studies in Mathematics, **131**. American Mathematical Society, Providence, RI, 2012.

[91] M. Nazarov, *Capelli identities for Lie superalgebras*, Ann. Sci. École Norm. Sup. **30** (1997), 847–872.

[92] M. Nazarov, *Young's symmetrizers for projective representations of the symmetric group*, Adv. Math. **127** (1997), 190–257.

[93] I. Penkov, *Characters of typical irreducible finite-dimensional* $\mathfrak{q}(n)$*-modules*, Funct. Anal. App. **20** (1986), 30–37.

[94] I. Penkov and V. Serganova, *Cohomology of* G/P *for classical complex Lie supergroups* G *and characters of some atypical* G*-modules*, Ann. Inst. Fourier **39** (1989), 845–873.

[95] I. Penkov and V. Serganova, *Generic irreducible representations of finite-dimensional Lie superalgebras*, Internat. J. Math. **5** (1994), 389–419.

[96] I. Penkov and V. Serganova, *Characters of finite-dimensional irreducible* $\mathfrak{q}(n)$*-modules*, Lett. Math. Phys. **40** (1997), 147–158.

[97] N. Popescu, *Abelian categories with applications to rings and modules*. London Mathematical Society Monographs **3**, Academic Press, London-New York, 1973.

[98] P. Pragacz, *Algebro-geometric applications of Schur S- and Q-polynomials*. Topics in invariant theory (Paris, 1989/1990), 130–191, Lecture Notes in Math. **1478**, Springer, Berlin, 1991.

[99] E. W. Read, *The* α*-regular classes of the generalized symmetric groups*, Glasgow Math. J. **17** (1976), 144–150.

[100] A. Rocha-Caridi, *Splitting Criteria for* $\mathfrak{g}$*-Modules Induced from a Parabolic and the Bernstein-Gelfand-Gelfand Resolution of a Finite Dimensional Irreducible* $\mathfrak{g}$*-Module*, Trans. Amer. Math. Soc. **262** (1980) 335–366.

[101] A. Rocha-Caridi and N. Wallach, *Projective modules over graded Lie algebras I*, Math. Z. **180** (1982), 151–177.

[102] L. Ross, *Representations of graded Lie algebras*, Trans. Amer. Math. Soc. **120** (1965), 17–23.

[103] B. Sagan, *Shifted tableaux, Schur Q-functions, and a conjecture of R. Stanley*, J. Combin. Theory Ser. A **45** (1981), 62–103.

[104] J. Santos, *Foncteurs de Zuckermann pour les superalgébres de Lie*, J. Lie Theory **9** (1999), 69–112.

[105] M. Scheunert, *The Theory of Lie Superalgebras*, Lect. Notes in Math. **716**, Springer, Berlin, 1979.

[106] M. Scheunert, W. Nahm, and V. Rittenberg, *Classification of all simple graded Lie algebras whose Lie algebra is reductive I, II. Construction of the exceptional algebras*, J. Math. Phys. **17** (1976), 1626–1639, 1640–1644.

[107] V. Serganova, *Kazhdan-Lusztig polynomials and character formula for the Lie superalgebra* $\mathfrak{gl}(m|n)$, Selecta Math. (N.S.) **2** (1996), 607–651.

[108] V. Serganova, *Kac-Moody superalgebras and integrability*, In: Developments and trends in infinite-dimensional Lie theory, 169–218, Progr. Math. **288**, Birkhäuser, 2011.

[109] A. Sergeev, *The centre of enveloping algebra for Lie superalgebra* $Q(n,\mathbb{C})$, Lett. Math. Phys. **7** (1983), 177–179.

[110] A. Sergeev, *The tensor algebra of the identity representation as a module over the Lie superalgebras* $gl(n,m)$ *and* $Q(n)$, Math. USSR Sbornik **51** (1985), 419–427.

[111] A. Sergeev, *The invariant polynomials of simple Lie superalgebras*, Represent. Theory **3** (1999), 250–280 (electronic).

[112] A. Sergeev, *The Howe duality and the projective representation of symmetric groups*, Represent. Theory **3** (1999), 416–434.

[113] A. Sergeev, *An Analog of the Classical Invariant Theory, I, II*, Michigan J. Math. **49** (2001), 113–146, 147–168.

[114] A. Sergeev and A. Veselov, *Grothendieck rings of basic classical Lie superalgebras*, Ann. of Math. **173** (2011), 663-703.

[115] B. Shu and W. Wang, *Modular representations of the ortho-symplectic supergroups*, Proc. London Math. Soc. **96** (2008), 251–271.

[116] W. Soergel, *Character formulas for tilting modules over Kac-Moody algebras*, Represent. Theory (electronic) **2** (1998), 432–448.

[117] J. Stembridge, *A characterization of supersymmetric polynomials*, J. Algebra **95** (1985), 439–444.

[118] J. Stembridge, *Shifted tableaux and the projective representations of symmetric groups*, Adv. Math. **74** (1989), 87–134.

[119] Y. Su, *Composition factors of Kac modules for the general linear Lie superalgebras*, Math. Z. **252** (2006), 731–754.

[120] Y. Su and R.B. Zhang, *Character and dimension formulae for general linear superalgebra*, Adv. Math. **211** (2007), 1–33.

[121] Y. Su and R.B. Zhang, *Generalised Verma modules for the orthosymplectic Lie superalgebra* $\mathfrak{osp}(k|2)$, J. Algebra **357** (2012), 94–115.

[122] J. Tanaka, *On homology and cohomology of Lie superalgebras with coefficients in their finite-dimensional representations*, Proc. Japan Acad. Ser. A Math. Sci. **71** (1995), 51–53.

[123] J. Van der Jeugt, *Character formulae for Lie superalgebra* $C(n)$, Comm. Algebra **19** (1991), 199–222.

[124] J. Van der Jeugt, J.W.B. Hughes, R. C. King, and J. Thierry-Mieg, *Character formulas for irreducible modules of the Lie superalgebras* $\mathfrak{sl}(m/n)$, J. Math. Phys. **31** (1990), 2278–2304.

[125] D. Vogan, *Irreducible characters of semisimple Lie Groups II: The Kazhdan-Lusztig Conjectures*, Duke Math. J. **46** (1979), 805–859.

[126] C.T.C. Wall, *Graded Brauer groups*, J. Reine Angew. Math. **213** (1964), 187–199.

[127] J. Wan and W. Wang, *Lectures on spin representation theory of symmetric groups*, Bull. Inst. Math. Acad. Sin. (N.S.) **7** (2012), 91–164.

[128] J. Wan and W. Wang, *Spin Kostka polynomials*, J. Algebraic Combin., DOI: 10.1007/s10801-012-0362-4.

[129] W. Wang, *Duality in infinite dimensional Fock representations*, Commun. Contem. Math. **1** (1999), 155–199.

[130] W. Wang and L. Zhao, *Representations of Lie superalgebras in prime characteristic I*, Proc. London Math. Soc. **99** (2009), 145–167.

[131] H. Weyl, *The classical groups. Their invariants and representations.* Fifteenth printing. Princeton Landmarks in Mathematics. Princeton University Press, Princeton, NJ, 1997.

[132] M. Yamaguchi, *A duality of the twisted group algebra of the symmetric group and a Lie superalgebra*, J. Alg. **222** (1999), 301–327.

[133] Y. Zou, *Categories of finite-dimensional weight modules over type I classical Lie superalgebras*, J. Algebra **180** (1996), 459–482.

Index